Limit Analysis
and Soil Plasticity

TITLES IN THE SERIES

Architectural Acoustics
M. David Egan
ISBN 13: 978-1-932159-78-3, ISBN 10: 1-932159-78-9, 448 pages

Earth Anchors
By Braja M. Das
ISBN 13: 978-1-932159-72-1, ISBN 10: 1-932159-72-X, 242 pages

Limit Analysis and Soil Plasticity
By Wai-Fah Chen
ISBN 13: 978-1-932159-73-8, ISBN 10: 1-932159-73-8, 638 pages

Plasticity in Reinforced Concrete
By Wai-Fah Chen
ISBN 13: 978-1-932159-74-5, ISBN 10: 1-932159-74-6, 474 pages

Plasticity for Structural Engineers
By Wai-Fah Chen & Da-Jian Han
ISBN 13: 978-1-932159-75-2, ISBN 10: 1-932159-75-4, 606 pages

Theoretical Foundation Engineering
By Braja M. Das
ISBN 13: 978-1-932159-71-4, ISBN 10: 1-932159-71-1, 440 pages

Theory of Beam-Columns, Volume 1: In-Plane Behavior and Design
By Wai-Fah Chen & Toshio Atsuta
ISBN 13: 978-1-932159-76-2, ISBN 10: 1-932159-76-9, 513 pages

Theory of Beam-Columns, Volume 2: Space Behavior and Design
By Wai-Fah Chen & Toshio Atsuta
ISBN 13: 978-1-932159-77-6, ISBN 10: 1-932159-77-0, 732 pages

Limit Analysis and Soil Plasticity

by Wai-Fah Chen

J.ROSS PUBLISHING

Copyright ©2008 by Wai-Fah Chen

ISBN-10: 1-932159-73-8
ISBN-13: 978-1-932159-73-8

Printed and bound in the U.S.A. Printed on acid-free paper
10 9 8 7 6 5 4 3 2 1

This J. Ross Publishing edition, first published in 2008, is an unabridged republication of the work originally published by Elsevier Scientific Publishing Co., Amsterdam, in 1975.

Library of Congress Cataloging-in-Publication Data

Chen, Wai-Fah, 1936–
    Limit analysis and soil plasticity / by Wai-Fah Chen.
       p. cm.
    Reprint. Originally published: Amsterdam : Elsevier, 1975.
    Includes bibliographical references and index.
    ISBN 978-1-932159-73-8 (paperback : alk. paper)
    1. Soil mechanics. 2. Plastic analysis (Engineering) I. Title.
    TA710.C533 2007
    624.1'5136—dc22   2007044138

Phone: (954) 727-9333
Fax: (561) 892-0700
Web: www.jrosspub.com

To my wife, Lily

FOREWORD

It is a great pleasure to commend to you, the reader, this remarkably comprehensive book combining fundamental theory and practical applications. You will find an informative historical review along with a careful development of the basic concepts of limit analysis, perfect plasticity, and more general plasticity theory in the context of soil mechanics and concrete. An engineering practitioner concerned directly with soil and foundation problems will find many tables of useful numbers and the rationale for their validity in the standard terminology of soil mechanics.

Perhaps you have only marginal interest in soils but wish to learn about and use conventional plastic limit analysis for structures and machines. You will find the material presented to be excellent for your purpose. The author has brought together in a unified manner so much of what until now was known only to a few in the field, who like Professor W.F. Chen contributed fruitful ideas and techniques. Student, researcher, or practitioner, novice and expert alike, will profit much from reading this book and having it for reference in the years to come.

D.C. DRUCKER

PREFACE

The present book is devoted to the theory and applications of limit analysis as applied to soil mechanics. It contains also an introduction to the modern development of theory of soil plasticity. In fact, some of the material on soil plasticity and rock-like material such as concrete in Chapters 5 and 12 is here published for the first time.

The book divides roughly into four parts. The first part, from Chapter 1 to Chapter 4, describes the technique of limit analysis in detail, beginning with the historical review of the subject and the assumptions on which it is based, and covering various aspects of modern techniques of limit analysis. These techniques are illustrated by many examples.

The second part, from Chapter 6 to Chapter 9, deals with the applications of limit analysis to what may be termed "classical soil mechanics problems" which include: (1) bearing capacity of footings; (2) lateral earth pressure problems; and (3) stability of slopes. Many results of recent development are summarized in graphical or tabular form suitable for direct use. In many cases, comparisons of limit analysis solution and conventional limit equilibrium and slip-line solutions are presented and discussed.

The third part, Chapters 10 and 11, deals with new advances on bearing-capacity problems of concrete blocks or rock which have never previously been discussed in a book form. Both theoretical and experimental results of various concrete bearing problems are presented.

The fourth part, Chapter 5 and Chapter 12, deals with modern development of theory of soil plasticity. Two elastic—plastic soil models and one elastic—plastic—fracture model for concrete are treated in Chapter 12. Numerical results obtained by the finite element method along with limit analysis solutions are presented in Chapter 5. Some of the finite element solutions for concrete blocks are presented in Chapter 11.

Particular care has been taken to present the subject in easy stages from the most elementary to the most advanced. The reader is assumed to have some knowledge of soil mechanics and the theory of elasticity for reading of Chapter 1 to Chapter 11. Chapter 12 requires some knowledge of the theory of plasticity and finite element method. I have endeavored to give reasonably complete literature reference to the topics covered in the book. For convenience, the references are given in a separate list at the end of the book and, as usual, reference to any work is indicated by the publication year in brackets.

This book owes its existence to the fact that the National Science Foundation sponsored research work in soil plasticity at Fritz Engineering Laboratory, Lehigh University. A survey report entitled *Limit Analysis and Limit Equilibrium Solutions in Soil Mechanics* prepared under this project constituted the first draft to this book. Moreover, I benefited greatly from many of my students on this research project, in particular Drs. H.L. Davidson, A.C.T. Chen, T. Atsuta, N. Snitbhan, J.L. Rosenfarb and Messrs. M.W. Giger, S. Covarrubias, C.R. Scawthorn, B.E. Trumbauer, T.A. Colgrove, J.L. Carson and M.W. Hyland. This book contains many results which were first presented in the form of Technical Reports prepared under this project. Chapters 5 and 12 are primarily based on Drs. Davidson and A.C.T. Chen's dissertations. Mr. Snitbhan read the entire manuscript and gave me many useful suggestions.

The development of this book was strongly influenced by my colleagues, students, and friends as well as the experience I had in teaching a plasticity course offered to graduate students at the Department of Civil Engineering, Lehigh University. The privilege of studying under D.C. Drucker at Brown University remains memorable. The warm encouragement and suggestions by Professor H.Y. Fang were much appreciated.

I express my sincere thanks to Miss Shirley Matlock for her rapid and careful typing of the manuscript.

February 1974   W.F. CHEN
Bethlehem, Pennsylvania

CONTENTS

Chapter 1

INTRODUCTION

1.1. INTRODUCTION

Partly for simplicity in practice and partly because of the historical development of mechanics of deformable solids, the problems of soil mechanics are often divided into two distinct groups — the *stability problems* and the *elasticity problems.* They are then treated in two separate and unrelated ways. The stability problems deal with the conditions of *ultimate failure* of a mass of soil. Problems of earth pressure, bearing capacity, and stability of slopes most often are considered in this category. The most important feature of such problems is the determination of the loads which will cause failure of the soil mass. Solutions to these problems are often obtained using the theory of *perfect plasticity.* The elasticity problems on the other hand deal with stress or deformation of the soil when no failure of the soil is involved. Stresses at points in a soil mass under a footing, or behind a retaining wall, deformations around tunnels or excavations, and all settlement problems belong in this category. Solutions to these problems are often obtained by using the theory of *linear elasticity.*

The theory of linear elasticity is based on Hooke's law which establishes a linear relation between stress and strain. The theory of perfect plasticity takes account of the fact that real soil exhibits the mechanical behavior stipulated by Hooke's law only as long as the *stress intensity* remains sufficiently small. When the stress intensity first reaches a certain critical value, which is called the *yield value,* the soil leaves the *elastic range* and enters the *plastic flow range,* which indicates continuous deformation at a constant state of stress. In order to maintain plastic flow, the stress intensity must be at the yield value but can never exceed this value. As soon as the stress intensity drops below this value, any change of strain is of a purely elastic nature.

Intermediate between the elasticity problems and stability problems mentioned above are the problems known as *progressive failure.* Progressive failure problems deal with the elastic—plastic transition from the initial linear elastic state to the ultimate failure state of the soil by plastic flow.

An illustrative example

As an illustration, Fig. 1.1 shows a deep soil stratum loaded by a strip footing.

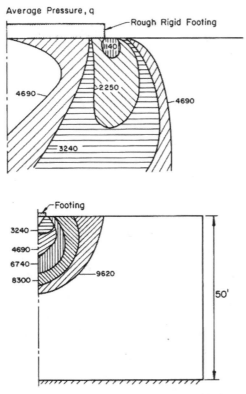

Fig. 1.1. Spread of yield zone. Internal friction angle $\varphi = 20°$ and cohesion $c = 500$ psf (23.8 kN/m²). All numbers in the figure are in psf = lb/ft².

The base of the soil stratum is rigid and perfectly rough, and the footing is also rigid and perfectly rough. The dimension perpendicular to the plane of the paper will be taken as unity, but all motion is supposed in the plane. As the applied load or the average pressure q over the footing is gradually increased starting from zero, the soil stratum is first stressed in a purely elastic manner. Eventually, the stress intensity reaches the yield value mentioned above at the footing corners, and plastic regions begin to form there. Theoretically, some plastic yielding should occur at the footing corners for any load level since the true elastic solution contains a *singularity* at the corner. Numerically, this implies that yielding at the footing corners should occur in the first increment of loading.

According to the numerical work of Davidson (1974), at a load of 1140 psf (54.6 kN/m²) some yielding has occurred near the footing corner and a small yield zone near the footing corner can be seen in Fig. 1.1. As the load increases yielding spreads downward and toward the footing centerline. The yield zone reaches the footing centerline at a load just below 3240 psf (155 kN/m²).

While finite amounts of soil are stressed plastically at this stage, plastic flow in the yielded zones is still contained by the surrounding elastic regions. This type of elastic–plastic behavior characterizes the range of *contained plastic deformation*. Throughout this range, the footing undergoes a definite overall displacement for each value of the applied pressure just as in the elastic range. Whereas, in the elastic range, the displacement is proportional to the applied pressure, this is no longer the case in the range of contained plastic deformation.

As we keep increasing the average pressure q, the yield zone increases in size until at an average pressure of 6740 psf (322 kN/m^2) all of the soil immediately below the footing is yielded, and eventually, at the numerical maximum bearing pressure 9620 psf (460 kN/m^2), a significant portion of the soil stratum around the footing has yielded (Fig. 1.1). At this instant of *impending plastic flow*, the footing is first able to move downward under constant pressure. The maximum applied pressure 9620 psf for impending plastic flow characterizes the maximum load carrying capacity of the footing. When this occurs, the progressive failure of the footing reaches its ultimate limit. Any further increase in load, no matter how small, will cause the footing to collapse. The problem now becomes the stability problem mentioned above. At this stability stage, the yielding of soil has spread to such an extent that the remaining elastic soil plays a relatively insignificant role in sustaining the load. It has been termed *uncontained* or *unrestricted plastic flow*, to distinguish it from *contained plastic flow* mentioned earlier, in which elastic action still plays a major role. Here we will use the term *plastic collapse* or simply *collapse*. This plastic collapse load can be used as a realistic basis for design. Once the footing is designed with a suitable factor of safety against collapse, the designer can then check the settlement of the footing under working load using linear elastic analysis (elasticity problems).

Limit analysis

In view of the facts that: (1) a complete analysis including the range of contained plastic flow is far too complicated and impractical for application; and (2) failure by plastic collapse is the governing condition in so many problems in soil mechanics, the development of efficient methods for computing the collapse load in a more *direct* manner is, therefore, of intense practical interest to engineers.

Limit analysis is concerned with the development and applications of such methods. Although the limit analysis methods were established firmly less than twenty years ago, there has been an enormous number of applications in a wide variety of fields from metal deformation processing to the design of reinforced concrete structures. Applications to beams and frames are the most highly developed aspect of limit analysis and design so that the basic techniques are given in several texts. The ASCE Manual 41 (ASCE-WRC, 1971) on plastic design in steel gives a good account of this application. Reference to the work of many investigators is given also in this manual. Applications of limit analysis to plates and shells for both metal and reinforced concrete materials are given in the recent book by Save and Massonnet (1972). A great deal of attention has been paid recently to soil mechanics in addition to concrete and rock. An appreciable amount of practical information is now available as the result of this and allied work. Perhaps the most striking feature of the limit analysis method is that no matter how complex the geometry of a problem or loading condition, it is always possible to obtain a realistic value of the collapse load. When this is coupled with its other merits, namely, that it is relatively simple to apply, that it provides engineers with a clear physical picture of the mode of failure, and that many of the solutions obtained by the method have been substantiated numerically by comparing with the existing results for which satisfactory solutions already exist, it can be appreciated that it is a working tool with which every engineer should be conversant.

The main objective of this book, therefore, is to describe the technique of limit analysis in soil mechanics in detail, beginning with the basic assumptions on which it is based, and covering almost every aspect of typical stability problems in soil mechanics that are at present known. The text not only includes what may be termed "classical stability problems" which include: (1) bearing capacity of footings; (2) lateral earth pressure problems; and (3) stability of slopes, but also includes new advances on bearing capacity problems of concrete blocks or rock which have never previously been discussed.

Before the technique is described in detail, it should be realized that limit analysis is not the only method of assessing the collapse load of a stability problem in soil mechanics. The other standard and widely known techniques used in the solutions of soil mechanics problems may be divided into two principal groups — the *slip-line method* and the *limit equilibrium method*. The slip-line method attempts first to derive the basic differential equations which then make it possible to obtain the solutions of various problems by the determination of the so-called slip-line network. The limit equilibrium method attempts first to create a simplified mode of failure which then makes it possible to solve various problems by simple statics. Since all the methods utilize the concept of perfect plasticity, the relation between these solutions, corresponding to these three different methods, involves terminology and special concepts that are not commonly used in the field of soil mechanics. A brief description of the salient features of these methods and their historical developments follows.

1.2. SLIP-LINE METHOD AND LIMIT EQUILIBRIUM METHOD

Slip-line method

The progressive failure of a footing under increasing load has been described in the preceding section. Impending plastic flow of the soil occurs when a sufficiently large region of the soil beneath the footing is stressed to its limiting or yield condition, resulting in an unrestricted plastic flow of the soil beneath the footing. At the instant of impending plastic flow, both equilibrium and yield conditions are satisfied in the region near the footing. For soils, the Coulomb criterion is widely used for this yield condition. Combining the Coulomb criterion with the equations of equilibrium gives a set of differential equations of *plastic equilibrium* in this region. Together with the stress boundary conditions, this set of differential equations can be used to investigate the stresses in the soil beneath the footing or behind a retaining wall at the instant of impending plastic flow. In order to solve specific problems, it is convenient to transform this set of equations to curvilinear coordinates whose directions at every point in this yielded region coincide with the directions of failure or slip plane. These slip directions are known as *slip lines* and the network is called the *slip-line field*.

Kötter (1903) was the first to derive these slip-line equations for the case of plane deformations. Prandtl (1920) was the first to obtain an analytical closed form solution to these equations for a footing on a *weightless* soil. In the analysis, he developed the solution with a singular point with a pencil of straight slip-lines passing through it. These results were subsequently applied by Reissner (1924) and Novotortsev (1938) to certain particular problems on the bearing capacity of footings on a weightless soil, when the slip-lines of at least one family are straight and the solutions have closed form.

However, the important inclusion of soil weight considerably complicates the mathematical solution. Consequently, many approximate methods have been developed. Sokolovskii (1965) adopted a numerical procedure based on a *finite difference* approximation of the slip-line equations. He obtained a number of interesting problems on the bearing capacity of footings or slopes as well as the pressure of a fill on retaining walls, for which it is impossible to find closed form solutions. De Jong (1957) on the other hand adopted a different approach and developed a graphical procedure for solutions. Other forms of approximate solution include the applications of perturbation methods (Spencer, 1962) and series expansion methods (Dembicki et al., 1964).

Comments on the slip-line method

One weakness of the slip-line method is the neglect of the stress—strain relationship of the soil. According to the mechanics of deformable solids, this condition must be satisfied for a *valid* solution. Here, only the equilibrium and yield conditions are utilized. For the case of plane deformations, the two equations of equilibrium and the yield condition produce what appears to be and sometimes is *static determinancy* in the sense that there are the same number of equations as unknown stress components. In most practical problems however, the boundary conditions involve stresses and rates of displacement and the static determinancy is then misleading. Stress–strain relationship of the soil must be considered in order to obtain a solution.

In general, in a slip-line solution, only a part of the soil mass near a footing or behind a retaining wall is assumed to be in the state of plastic equilibrium. The solution consists of constructing a slip-line field in the region, which satisfies all the stress boundary conditions that directly concern the region, as well as the equilibrium and yield conditions at every point inside the region. The stress field so obtained has been termed *partial stress field.* The stress distribution outside this partial stress field region is not defined. For the solution to be *valid*, it is obvious that it must at least be able to show that there *exists* an associated stress distribution in the non-yielded region, which is in equilibrium with the partial stress field and nowhere violates the yield criterion. The "partial" stress field so extended will be called *extended stress field.* The extended stress field defines the stress distribution over the whole body concerned. Even if such an extended stress field can be found, there is no guarantee that this extended stress solution will give the correct answer. Fortunately, this important question of the correctness of a solution can now be answered fully in the light of the theorems of limit analysis. When the rigorous rules for limit analysis are discussed in the following section, it will be seen that the extended stress field is precisely that required by the lower-bound rules of limit analysis. Hence the above-mentioned, extended stress solution gives only a *lower* bound to the collapse load.

Once the partial stress distribution in the slip-line solution is acceptable as a lower-bound solution, use has to be made of the stress–strain relations to determine whether given stress and displacement states correspond. If the solution is the correct one, the associated displacement mode will be compatible with a continuous distortion satisfying the displacement boundary conditions and everywhere the rate of plastic work will be positive. Provided that this is the case, then, the solutions so obtained is also an *upper* bound to the collapse value according to the upper-bound rules of limit analysis, and hence is identical with the correct value. Further discussion on this point will be given later when the rules of limit analysis are presented.

Here, we must keep in mind that a partial stress field obtained from the slip-line equations is not necessarily the correct solution nor is it known when it is an upper-bound or a lower-bound solution. If a compatible displacement or velocity field can be associated with the partial stress field through a given stress–strain relation, the slip-line solution is an upper-bound solution. If, in addition, the partial stress distribution in the plastic zone can be extended to the entire body, satisfying the equilibrium equations, the yield criterion and the stress boundary conditions, the slip-line solution is also a lower bound and is hence the correct solution.

Limit equilibrium method

The so-called limit equilibrium method has traditionally been used to obtain approximate solutions for the stability problems in soil mechanics. Examples of this approach are the solutions presented in the book by Terzaghi (1943). The method can probably best be described as an approximate approach to the construction of a slip-line field and generally entails an assumed failure surface of various simple shapes – plane, circular or logspiral. With this assumption, each of the stability problems is now reduced to one of finding the most dangerous position for the failure or slip surface of the shape chosen which may not be particularly well-founded, but quite often gives acceptable results. In this method, it is also necessary to make sufficient assumptions regarding the stress distribution along the failure surface such that an overall equation of equilibrium, in terms of stress resultants, may be written for a given problem. Therefore, this simplified approach makes it possible to solve various problems by simple statics. Various solutions obtained by this method are summarized in graphical or tabular form in the texts by Terzaghi (1943) and by Taylor (1948) and are now quite widely used in practice.

It is worth mentioning here that none of the equations of solid mechanics is explicitly satisfied everywhere inside or outside of the failure surface. Since the stress distribution is not precisely defined anywhere inside and outside of the assumed failure surface, one cannot say definitely that an acceptable stress distribution which satisfies equilibrium, stress boundary conditions and the yield criterion, exists such that the solution meets the requirements of the lower-bound rules of limit analysis. Although the limit equilibrium technique utilizes the basic philosophy of the upper-bound rules of limit analysis, that is, a failure surface is assumed and a least answer is sought, it does not meet the precise requirements of the upper-bound rules so that it is not an upper bound. The method basically gives no consideration to soil kinematics, and equilibrium conditions are satisfied only in a limited sense.

It is clear then that a solution obtained using the limit equilibrium method is not necessarily an upper or a lower bound. However, any upper-bound limit analysis solution will obviously be *a* limit equilibrium solution.

1.3. LIMIT ANALYSIS METHOD

Before proceeding to a general discussion of the limit analysis method, let us review the conditions required for a valid solution in the mechanics of deformable solids. Three basic conditions are needed in the solution, namely, the stress equilibrium equations, the stress–strain relations and the compatibility equations relating strain and displacement. In general an infinity of stress states will satisfy the stress boundary conditions, the equilibrium equations and the yield criterion *alone,* and an infinite number of displacement modes will be compatible with a continuous distortion satisfying the displacement boundary conditions. As in the theory of elasticity, use has to be made of the stress–strain relations to determine whether given stress and displacement states correspond and a *unique* solution results. In an elastic–plastic material, however, there is as a rule a three-stage development in a solution (when the applied loads are gradually increased in magnitude from zero), namely the initial elastic response, the intermediate contained plastic flow and finally the unrestricted plastic flow. The complete solution by this approach is likely to be cumbersome for all but the simplest problems, and methods are needed to furnish the load-carrying capacity in a more direct manner. Limit analysis is the method which enables definite statement to be made about the collapse load without carrying out the step-by-step elastic–plastic analysis.

In contrast to slip-line and limit equilibrium methods, the limit analysis method considers the stress–strain relationship of a soil in an idealized manner. This idealization, termed *normality* (or the *flow rule*), establishes the *limit theorems* on which limit analysis is based. Within the framework of this assumption, the approach is rigorous and the techniques are competitive with those of limit equilibrium, in some instances being much simpler. The plastic limit theorems of Drucker et al. (1952) may conveniently be employed to obtain upper and lower bounds of the collapse load for stability problems, such as the critical heights of unsupported vertical cuts, or the bearing capacity of nonhomogeneous soils.

The conditions required to establish an upper- or lower-bound solution are essentially as follows:

Lower-bound theorem

The loads, determined from a distribution of stress *alone*, that satisfies: (a) the equilibrium equations; (b) stress boundary conditions; and (c) nowhere violates the yield criterion, are *not greater* than the actual collapse load. The distribution of stress satisfying items (a), (b) and (c) has been termed a *statically admissible* stress field for the problem under consideration. Hence the lower-bound theorem may be restated as follows: If a statically admissible stress distribution can be found, uncontained plastic flow will *not* occur at a lower load. From these rules it can be seen that the lower-bound technique considers only equilibrium and yield. It gives no consideration to soil kinematics.

Upper-bound theorem

The loads, determined by equating the external rate of work to the internal rate of dissipation in an assumed deformation mode (or velocity field) that satisfies: (a) velocity boundary conditions; and (b) strain and velocity compatibility conditions, are *not less* than the actual collapse load. The dissipation of energy in plastic flow associated with such a field can be computed from the idealized stress/strain rate relation (or the so-called flow rule). A velocity field satisfying the above conditions has been termed a *kinematically admissible* velocity field. Hence, the upper-bound theorem states that if a kinematically admissible velocity field can be found, uncontained plastic flow must impend or have taken place previously. The upper-bound technique considers only velocity or failure modes and energy dissipations. The stress distribution need *not* be in equilibrium, and is only defined in the deforming regions of the mode.

By suitable choice of stress and velocity fields, the above two theorems thus enable the required collapse load to be bracketed as closely as seems necessary for the problem under consideration.

The meaning of slip-line solutions

The partial stress field obtained by the slip-line method does not meet the requirements of the lower-bound theorem since it is not demonstrated that an equilibrium stress distribution satisfying the boundary conditions and not exceeding the yield criterion exists, *outside* the slip-line network regions. Suppose, however, that the partial stress distribution in the slip-line network regions can be extended to the entire body in a statically admissible manner, the slip-line solution is then a lower-bound solution. This statically admissible extension of the partial stress field is in general the most difficult to deal with, since no general methods of developing such statically admissible fields are available, though Bishop (1953) has suggested a particular approach involving the construction of a fully plastic stress field in parts of the region outside the slip-line network regions. Such stress fields, and those for lower bounds, frequently involve curves across which the stress is discontinuous.

From the flow rule or stress—strain relations, a velocity field associated with the partial stress field in the plastic or slip-line network regions may be constructed. If the velocity field so constructed satisfies the velocity boundary conditions, then, this velocity field is kinematically admissible. The slip-line solution is also an upper-bound solution. Hence the actual collapse load will have been found.

The stress and velocity are now compatible in the slip-line network region in which deformation is occurring. A unique theorem due to Hill (1951) states that where deformation is actually occurring in any mode, the stress found is the actual state. The particular deformation mode obtained is not necessarily the actual one, however.

The meaning of the slip-line solutions in the light of the upper- and lower-bound theorems of limit analysis can now be summarized. The acceptability of a slip-line field solution depends on whether it is possible to find a kinematically admissible velocity field and a statically admissible stress field for the problem under consideration. A *complete* solution must satisfy both conditions mentioned above. Most slip-line field solutions presented consist of a stress solution for the region containing the slip-line fields and a compatible deformation mode associated with this partial stress field. Hence, these slip-line solutions are strictly only upper-bound solutions and will be called *incomplete solutions.* If these partial stress solutions can be extended to the whole body in a statically admissible manner, the slip-line solutions are also lower-bound solutions and hence *complete solutions.*

An illustrative example

As an example of the above discussion, let us consider the column of material shown in Fig. 1.2 of width $(2A + B)$, subjected to a band of pressure of width B.

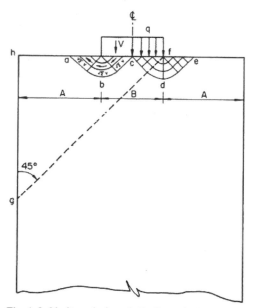

Fig. 1.2. Limit analysis and slip-line solutions.

With Tresca's criterion of yielding (a special case of Coulomb's criterion when the friction angle $\varphi = 0$), which is also identical with Von Mises' criterion for plane strain deformation, the slip-line field can be constructed from the slip-line equations mentioned earlier and a *partial* stress distribution in equilibrium with the applied pressure q and satisfying the yield criterion is then known in the regions abc and cde. The corresponding pressure q is equal to $5.14c$ where c is the cohesion of the soil (Hencky, 1923). Outside these partial stress regions, the stress distribution is not defined. Using the flow rule, the slip-line field shown also determines a velocity field (Hill, 1950). The incipient velocity field associated with this stress solution is shown by small arrows in Fig. 1.2. Plastic flow is confined to the region above the line of discontinuity $abcde$. The streamlines are parallel to this line and if the initial downward velocity of the pressure is v, the velocity has the constant value $\sqrt{2}\,v$ along each streamline. The value $5.14c$ is therefore an upper bound to the collapse pressure. Since the velocity field is kinematically admissible, it can be used directly to obtain an upper bound for the collapse load of the elastic–plastic problem. It is useful to observe that in this case the collapse pressure is the same as that of $5.14c$, and the direct upper-bound solution involves a much simpler computation. The slip-line solution shown in Fig. 1.2 gives an *incomplete* solution and will represent a *complete* solution if the partial stress fields can be extended throughout the body in a satisfactory manner.

The partial stress field in the plastic regions $abcdefa$ has been extended by Bishop (1953) into the rigid regions to satisfy boundary conditions without violating the yield criterion provided that the minimum width of the column of material is not less than $8.67B$ or $A/B \geqslant 3.84$. When this is the case ($A/B \geqslant 3.84$), the slip-line solution gives a *complete* solution and the value $5.14c$ is the correct collapse pressure.

When the width of the material is small ($A/B < 3.84$), some doubt exists that the stress field is in fact statically admissible in the rigid regions. For this case in which it is not possible with presently available direct methods to check the stress field in the assumed rigid regions for violating of the yield criterion, this can sometimes be accomplished indirectly by application of the upper-bound technique of limit analysis.

An upper bound for the collapse pressure q can be obtained from the velocity field in which the triangle hfg slides as a rigid body along gf. Application of the upper-bound theorem to be described in details in Chapter 3 determines the upper bound:

$$q^u = 2c\left(1 + \frac{A}{B}\right) \tag{1.1}$$

Let us assume that the slip-line configuration of Fig. 1.2 represents a complete solution. Then the collapse pressure $5.14c$ must be less than the upper bound q^u

given in [1.1]. This is true only when the ratio A/B is greater than 1.57. Thus if $A/B < 1.57$, the value 5.14c cannot be the actual collapse pressure, and so the slip-line field shown in Fig. 1.2 must imply violation of the yield criterion below the boundary *abcde*. Thus we have deduced indirectly that a statically admissible stress distribution does not exist outside the partial stress field, even though the *average* stress in the column of the material is only just over 0.6 of the yield stress in simple compression (2c) when $A/B = 1.57$. The dimensions of Fig. 1.2 represent such a case where the plastic stress distribution across *abcde* cannot be spread to a condition of uniform compression without violating yield.

General remarks

The knowledge of the collapse load for the elastic–plastic problem is also of interest in assessing the stage of development of elastic–plastic solutions which can be obtained numerically by the *finite element* method (see for example, Zienkiewicz, 1967, 1970, 1971) or by the *finite difference* method. In these numerical analyses, the solution becomes more and more difficult to obtain as the plastic regions spread, and the continuation of the techniques used in the elastic region may have to be revised considerably as new difficulties arise. When the plastic regions first meet and merge together, the plastic–elastic boundaries start to spread rapidly and further development of the solution becomes extremely difficult. Furthermore, unavoidable errors must be introduced by replacing differential equations by difference equations or continuous media by a finite number of elements, especially when one or more displacement components may change very rapidly across a thin layer in the plastic regions in the solution. In many cases the numerical method has to be abandoned before uncontained plastic flow occurs. A study of the elastic–plastic solutions for small strains, combined with the determination of the collapse load by the limit analysis methods of this book, is of value in obtaining a better understanding of the development of uncontained plastic flow through contained plastic flow.

1.4. A BRIEF HISTORICAL ACCOUNT OF SOIL PLASTICITY

The theory of soil plasticity is concerned with the analysis of stresses and strains in the plastic range of soil media. Applied to the design of foundations and retaining structures, it represents a necessary extension of the theory of elasticity in that it furnishes more realistic estimates of load-carrying capacities against failure, and in addition, it provides better estimates of settlements or displacements when subjected to its working load.

The earliest contribution to a theory of soil plasticity was made in 1773 by

Coulomb who proposed the Coulomb yield criterion for soils. He also established the important concept of *limiting plastic equilibrium* to a continuum and applied it to determine the pressure of a fill on a retaining wall. Later, in 1857, Rankine investigated the limiting plastic equilibrium of an infinite body and introduced the concept of slip surfaces. In 1899, Massau used the method of characteristics for the approximate determination of stress fields in soils. In particular, for problems of plane strain in cohesionless soils, Massau established the basic geometric property of the net of slip-line field, recognized the possibility of limiting lines and gave an exhaustive discussion of stress discontinuities.

In the development of the theory of earth pressure mentioned above, the introduction of stress—strain relations was obviated by the restriction to the consideration of limiting plastic equilibrium and the appeal to a heuristic extremum principle implied in Coulomb's work and more clearly formulated by Moseley (1833). It was Massau who first clearly recognized that some proofs in the theory of earth pressure were unsatisfactory because they openly or tacitly implied the validity of an extremum principle of Moseley.

Subsequently the development of the theory of soil plasticity proceeded slowly. Some progress was made at the beginning of this century, when the works of Kötter (1903) and Fellenius (1926) were published. The former was an attempt to obtain a set of differential equations of plastic equilibrium and then transform them to curvilinear co-ordinates (slip-line method). In Fellenius's work, a simplified theory of plastic equilibrium was clearly described (limit equilibrium method). He was attempting to solve problems by making the assumption of slip-surfaces of various simple shapes — plane or circular cylindrical. The Kötter's equations were subsequently applied by Sokolovskii to various stability problems in soil mechanics. His works have recently been summarized in book form (1965). Fellenius's works were developed further by many investigators and summarized in the well-known Terzaghi's book on soil mechanics (1943). This development altogether neglects the important fact that stress—strain relations are the essential constituent of a complete theory of any branch of the continuum mechanics of deformable solids.

During the last twenty years the theory of *metal plasticity* has been intensively developed. The development of metal plasticity has been strongly influenced by the much older theory of earth pressure. For instance, Tresca's yield condition (1868) is a special case of Coulomb's yield condition (1773) which is nearly a century before Tresca. Similarly, Rankine's (1857) investigation of plastic states of equilibrium in loose earth preceded De Saint Venant's (1870) investigation of such states of equilibrium in plastic solids. As remarked by Prager (1955a), it is fortunate that the modern development of metal plasticity did not copy the unsatisfactory feature of the theory of earth pressure but introduced instead a flow rule relating the stress to the velocity strain. Research in metal plasticity was thereby forced to pursue an independent course and, as a result of this, the situation has been reversed. The theory of metal plasticity is now able to pay some of the debt of gratitude it owes to the theory of soil plasticity: the general theory of *limit analysis*, developed in 1950s as a subject in metal plasticity, has shed much needed light on the foundations of the theory of limiting plastic equilibrium.

Between 1950 and 1965 the concept of perfect plasticity (i.e., no work-hardening) and the theorems of limit analysis form the central and most extensively developed part of the theory of metal plasticity. However, the corresponding extension to problems in soil mechanics is more recent. Nevertheless, the general theory of metal plasticity is now appreciated in the development of a modern theory of soil plasticity. It is therefore appropriate to mention here the recent works of Roscoe and his students (1958–1963) on a simple isotropic work-hardening theory of soil plasticity and also that of subsequent developments and applications. These developments have recently been summarized in the book entitled *Stress–Strain Behavior of Soils* (Cambridge University Press, 1971). This extension marks the beginning of the modern development of a consistent theory of soil plasticity.

Much of this book is devoted to the theory and applications of limit analysis as applied to soil mechanics. It contains the general theorems of limit analysis, describes their techniques in detail, and solves various problems. Many results of recent developments are summarized in graphical or tabular form suitable for direct use. It also deals with recent achievements in the theory of soil plasticity with emphasis on its use in finite elements. It does not, however, aim to cover the whole field of soil plasticity since much of the work on the theory of the slip-line field and the classical theory of earth pressure have been published in several texts as well as review papers.

Chapter 2

THE ASSUMPTIONS AND THEOREMS USED IN LIMIT ANALYSIS

2.1. INTRODUCTION

As has been indicated in Chapter 1, a complete progressive failure analysis of stress and strain in a soil mechanics problem as the load is increased from zero to failure is almost always impracticable. It is often satisfactory, however, to know the load at which the soil mass will collapse or deform excessively. For this purpose, the slip-line field analysis and limit equilibrium method have been the methods of solving various soil stability problems. However, the slip-line field analysis described in the preceding chapter is often considered to be too esoteric and complex to be a useful tool for engineers. In fact, so much depends on the skill and intuition of the engineer that solving a mixed (displacement and stress) boundary-value problem appears to be an art rather than an engineering method. On the other hand, the success of the relatively crude approach of the limit equilibrium method requires a blend of theoretical deduction, engineering intuition and experimental confirmation. Under these circumstances, it is hardly surprising that often the engineer employs well-known texts, such as Terzaghi (1943) or Sokolovskii (1965), or the more recent technical literature as a magic handbook and tries to fit his problem to the particular solutions he finds. Intuition and innovation seem discouraged by unfamiliarity and apparent complexity.

The basic theorems of limit analysis which have been stated in Chapter 1, give upper and lower bounds on the collapse load. Close approximation to the collapse load with known limits of error, as well as exact answers in some cases, have been found for a wide variety of practical problems in soil mechanics and the method can always produce a realistic value of the collapse load for a problem no matter how complex it may be. The method is rational and completely self-consistent, being based on a few well-defined assumptions to be discussed here. The basic theorems of limit analysis are conceptually simple, and often enable useful results to be achieved with remarkably simple and quick calculations. It usually admits a closed form expression for a given problem in terms of the governing parameters and geometry of the problem. It gives engineers a clear physical picture and thus can be easily utilized by the engineer as a working tool to obtain particular solutions he needs for his problem. Further, the theorems of limit analysis shed quite a bit of light on the meaning and validity of slip-line solutions.

In view of the uncertainties inherent in all engineering problems, and the essential role of judgement in their solution, it is clear that the approximated nature of the method is no basic handicap. The real difficulty is the possible discrepancy between the plastic deformation properties of the ideal and the real material, which often exhibits some degree of work softening, and may not be a perfectly plastic material. Since the assumptions regarding the mechanical properties of the material under investigation determine the range of validity of theory of limit analysis, a complete and concise statement of the assumptions used in this theory will be presented and illustrated before the techniques of limit analysis are described.

2.2. PERFECTLY PLASTIC ASSUMPTION AND COULOMB YIELD CRITERION

Elastic–perfectly plastic assumption

Fig. 2.1 shows a typical stress–strain diagram for soils. The stress–strain behavior of most real soils is characterized by an initial linear portion and a *peak* or *failure stress* followed by softening to a *residual stress*. In limit analysis, it is necessary to ignore the *strain softening* (or *work softening*) feature of the stress–strain diagram and to take the stress–strain diagram to consist of two straight lines as shown by the dashed lines in Fig. 2.1. A hypothetical material exhibiting this property of continuing plastic flow at constant stress is called an *ideally plastic* or *perfectly plastic* material.

It should be noted that the yield stress level used in limit analysis applications where *perfect plasticity* assumption is made may be chosen to represent the average stress in an appropriate range of strain. Thus the validity of the assumption of perfect plasticity may be wider than might appear possible at first glance. The choice of the level of the yield stress is not an absolute one, but is determined by the most significant features of the problem to be solved. For stability problems, this assumption may be more justifiable than for other problems in soil mechanics. As

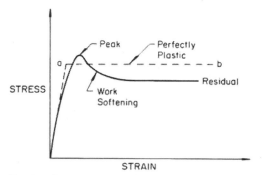

Fig. 2.1. Stress–strain relationship for ideal and real soils.

an illustration, the shear failure in soil under the footing shown in Fig. 1.2 is a phenomenon of progressive failure at quite variable stress levels. Consequently when the potential slip surface *abcde* reaches the points *a* and *e*, just mobilizing the peak shearing stress at these two points, the soil strength at the beginning of the slip surface, point *c*, must be well below the peak and near the residual shearing stress. At the instant of collapse, the maximum resistance to shear available above this surface must have a value somewhere between the peak stress and the residual stress. Further displacement of the footing would cause this average shear stress to drop towards the residual value.

In all stability problems, the maximum average shear stress mobilized over the whole of the failure surface in a real soil will be less than the peak value and more than the residual value, its relative position between these two limits being determined both by the properties of the soil and by the geometry and boundary stresses in the problem to be analyzed.

Coulomb yield criterion

The stress–strain diagram given above is associated with a simple shear test or a triaxial compression test. It is important to know the behavior of the soil for a *complex stress state*. In particular it is necessary to have an idea of what conditions characterize the change of the material from an elastic state to a yield or flow state (the horizontal line *ab*, Fig. 2.1). Here the question arises of a possible form of the condition which characterizes the transition of a soil from an elastic state to a plastic flow state with a complex stress state. This condition, satisfied in the yield state, is called the *yield criterion* (*perfect plasticity condition*).

It is generally assumed that plastic flow occurs when, on any plane at any point in a mass of soil, the shear stress τ reaches an amount that depends linearly upon the cohesion stress c, and the normal stress, σ:

$$\tau = c + \sigma \tan \varphi \qquad [2.1]$$

provided σ is a compressive stress. This equation was first suggested by Coulomb (1773). The angle φ is known as the angle of internal friction of a soil. The constants c and φ can be looked upon simply as parameters which characterize the total resistance of the soil media to shear.

It is usual to call a soil medium in which cohesion is absent ($c = 0$) a *cohesionless soil*, and one in which internal friction is absent ($\varphi = 0$) a *purely* (or *ideally*) *cohesive soil*. Now, Tresca's yield criterion, which applies widely to ductile metals, corresponds to the particular case of Coulomb's yield criterion when there is no internal friction. In other words, the Tresca yield criterion in metals can be alternatively represented by [2.1] with $c = k$ and $\varphi = 0$ where k is the shear yield stress for metals. In this book the terms "purely cohesive soil" and "Tresca material" will be used interchangeably.

Coulomb yield surface

In order to express the Coulomb yield condition [2.1] in terms of the principal stress components σ_1, σ_2 and σ_3, appropriate for general treatment of three-dimensional problems, we use the graphical representation of stress due to Mohr (1882). In the Mohr diagram (Fig. 2.2) the normal stress σ and the shearing stress τ are used as coordinates. The stresses (σ,τ) acting at a point in the soil on any plane parallel to the second principal stress direction lie on the stress circle which passes through the points $(\sigma_1,0)$, $(\sigma_3,0)$, and which has its center on the σ-axis. Similarly the stress circle through the stress points $(\sigma_1,0)$, $(\sigma_2,0)$ and the circle through $(\sigma_2,0)$, $(\sigma_3,0)$ represent states of stress acting on planes parallel to the third and first principal stress directions, respectively. In Fig. 2.2, we have taken $\sigma_1 > \sigma_2 > \sigma_3$ for definiteness. At a point of the soil where the principal stresses have the values σ_1, σ_2, σ_3, the stress point (σ,τ) representing the normal and shear stresses across any section through the point lies on or within the shaded areas of the largest stress circle in the Mohr diagram.

Values of σ,τ satisfying the Coulomb yield criterion [2.1] are represented in Fig. 2.2 by the two straight lines which start from the point $(c \cot \varphi,0)$ and inclined at angles of amount φ to the positive σ-axis. If a state of stress σ_1, σ_2, σ_3 is such that the Mohr circles lie within the wedge-shaped region, the soil remains in the linear elastic range. Plastic flow of the soil can occur when the largest of the circles touches the two straight lines. More generally, when the principal stress components σ_1, σ_2 and σ_3 are used as coordinates, the Coulomb yield *curve* in two-dimensional pictures becomes the Coulomb yield *surface* in the three-dimensional representation.

In the following we outline a geometrical method of constructing such a surface in principal stress space showing that this yield surface is a right hexagonal pyramid

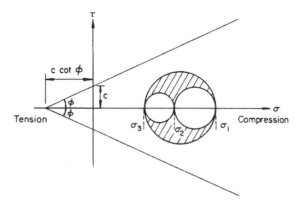

Fig. 2.2. Mohr's representation of a stress and the Coulomb yield criterion.

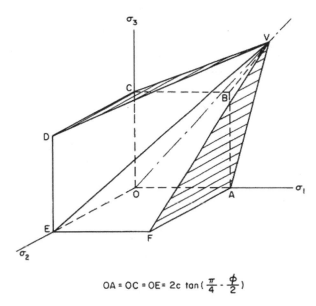

$$OA = OC = OE = 2c \tan\left(\frac{\pi}{4} - \frac{\phi}{2}\right)$$

Fig. 2.3. The Coulomb yield surface in principal stress space.

equally inclined to the σ_1, σ_2, σ_3-axes (see Fig. 2.3). Its intersection with the plane of Fig. 2.4, perpendicular to the σ_2-axis and at a distance σ_2 from the origin, is a hexagon. In the following the method of constructing the yield curve in the plane of Fig. 2.4 will be chosen. The value of σ_2 is fixed for the curve in the (σ_1, σ_3) coordinate plane.

Since plastic flow of the soil can occur only when the largest of the circles touches the two straight lines shown in Fig. 2.2 while the intermediate principal stress can have any value between the largest and smallest principal stresses, it follows that the determination of the critical circle requires the consideration of the relative magnitudes of the three principal stresses. There are six possible orderings of the relative magnitudes of the stresses σ_1, σ_2, σ_3 which determine the six yield lines $A'B'$, $B'C'$, $C'D'$, $D'E'$, $E'F$ and $F'A'$ as shown in Fig. 2.4. For example, for the line $C'D'$ in the figure, the stress points on the line having the ordering $\sigma_1 \geqslant \sigma_2 \geqslant \sigma_3$ and the critical circle is the one which passes through the points σ_1, σ_3 of Fig. 2.4(a). The geometrical relations shown in the Mohr diagram in this case give:

$$\sigma_1 - \sigma_3 = 2c \cos \varphi + (\sigma_1 + \sigma_3) \sin \varphi \qquad [2.2]$$

Equation [2.2] can be written:

$$\sigma_1 = \sigma_3 \tan^2 \left(\tfrac{1}{4}\pi + \tfrac{1}{2}\varphi\right) + 2c \tan \left(\tfrac{1}{4}\pi + \tfrac{1}{2}\varphi\right) \qquad [2.3]$$

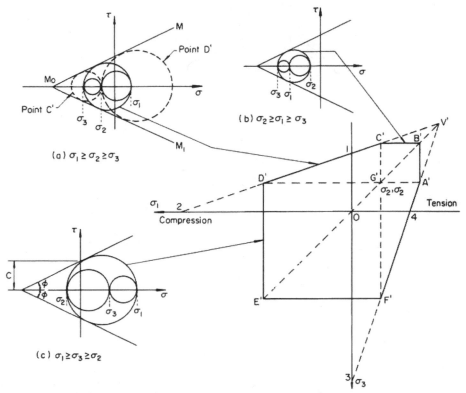

Fig. 2.4. Mohr's representation of a stress and the Coulomb yield curve in the plane $\sigma_2 =$ constant.

$$0-1 = 0-4 = 2c \tan(\tfrac{1}{4}\pi - \tfrac{1}{2}\phi)$$
$$0-2 = 0-3 = 2c \tan(\tfrac{1}{4}\pi + \tfrac{1}{2}\phi)$$

This equation is represented by the straight line $V'-2$ in the figure. Geometrically, Coulomb's criterion of yielding states that the circles which represent critical states of stress can move to the right or to the left in the wedge-shaped region and have the lines M_oM, M_oM_1 as common envelope as shown in Fig. 2.4(a). Since the principal stress σ_2 has a fixed value, the critical circles can not move arbitrarily in the wedge-shaped region but are restricted to the right as well as to the left. The two right and left extreme circles are shown in the figure by dotted lines which correspond to the stress points D' and C' respectively on the straight line $V'-2$. By proceeding similarily for the lines $B'C'$, $D'E'$, $E'F'$, $F'A'$, $A'B'$ it can be found that the yield surface intersects the (σ_1, σ_3) plane along the hexagon $A'-B'-C'-D'-E'-F'-A'$; and from the interchangeability of the $\sigma_1, \sigma_2, \sigma_3$, for example σ_1, σ_3 in (a), σ_2, σ_3 in (b) and σ_1, σ_2 in (c), it can be concluded that the yield hexagon in the figure is symmetrical in shape about the line $\sigma_1 = \sigma_3$. It can be seen now that the yield surface is a right hexagonal pyramid equally inclined to the $\sigma_1, \sigma_2, \sigma_3$-axes and with

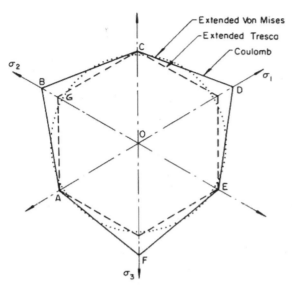

Fig. 2.5. Section of the yield surface by the π-plane ($\sigma_1 + \sigma_2 + \sigma_3 = 0$).

its vertex V at the point $\sigma_1 = \sigma_2 = \sigma_3 = -c \cot \varphi$ (see Fig. 2.3). Clearly, such a pyramid is fully defined by the hexagon A'-B'-C'-D'-E'-F'-A' in the (σ_1, σ_3) coordinate plane. Fig. 2.3 shows the hexagonal pyramid with the line VO as its center line and every two faces of the pyramid opposite to each other are parallel to a corresponding axis. The stress point V in the figure corresponds to a state of triaxial tensile stresses $\sigma_1 = \sigma_2 = \sigma_3 = -c \cot \varphi$, that is the point M_o of Fig. 2.4(a)

The hexagon drawn with a full line in Fig. 2.5 is the section of the pyramid by the plane whose equation is:

$$\sigma_1 + \sigma_2 + \sigma_3 = 0 \tag{2.4}$$

This plane is called π-plane (or deviatoric plane) which passes through the origin and perpendicular to the straight line $\sigma_1 = \sigma_2 = \sigma_3$. The hexagon is irregular since the yield stress in tension differs from the yield stress in compression.

Tresca's yield criterion, which applies to ductile metals, may be considered as a soil for which $\varphi = 0$ and $c = k$. If we take advantage of the geometrical interpretation of the Coulomb yield surface developed above, then Tresca yield surface in principal stress coordinates will be a right hexagonal cylinder instead of a pyramid. The axis of the cylinder is the straight line $\sigma_1 = \sigma_2 = \sigma_3$. Figure 2.6 shows the section of the Tresca yield surface by the plane $\sigma_2 = $ constant. The cross section is a regular hexagon.

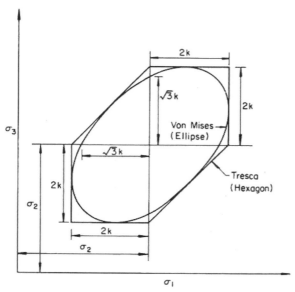

Fig. 2.6. Cross-sections of Von Mises and Tresca yield surfaces in principal stress space with σ_2 = constant.

Extended Tresca yield criterion

Fundamentally, Coulomb's yield criterion for soils contrasts with Tresca's for metals in dependence upon the *hydrostatic pressure* or the mean normal stress:

$$p = \tfrac{1}{3}(\sigma_1 + \sigma_2 + \sigma_3) = \tfrac{1}{3}(\sigma_x + \sigma_y + \sigma_z) \tag{2.5}$$

On the basis of the Tresca criterion for metals, Drucker (1953) proposed a modified Tresca criterion appropriate for the general treatment of three-dimensional problems of soils. The proposed yield surface in principal stress space is a *right hexagonal pyramid*. In contrast to the *irregular* hexagonal pyramid of Coulomb shown in Fig. 2.3, the yield surface proposed by Drucker is a right *regular* hexagonal pyramid. The intersection of the π-plane [2.4] with this yield surface is a *regular* hexagon shown by the broken line in Fig. 2.5. This yield surface lies within the Coulomb yield surface and is called *extended Tresca yield criterion*.

If the dimensions of the extended Tresca yield surface are properly chosen as shown in Fig. 2.5, it can be shown that this extended yield criterion reduces to the Coulomb rule [2.1] in the case of plane strain.

Extended Von Mises yield criterion

The use of the Tresca type of yield criterion, expressed by the hexagonal type of surface, is associated with certain mathematical difficulties in three-dimensional problems. In metals, this situation gave Von Mises the idea of replacing the hexahedral prism with the inscribed circular cylinder. The intersection of this cylinder with the plane σ_2 = constant is the *ellipse* inscribing the hexagon (Fig. 2.6).

Von Mises' criterion may be written in the usual notation:

$$J_2 = \tfrac{1}{6}[(\sigma_1 - \sigma_2)^2 + (\sigma_2 - \sigma_3)^2 + (\sigma_3 - \sigma_1)^2] = k^2$$

or:

$$J_2 = \tfrac{1}{6}[(\sigma_x - \sigma_y)^2 + (\sigma_y - \sigma_z)^2 + (\sigma_z - \sigma_x)^2] + \tau_{xy}^2 + \tau_{yz}^2 + \tau_{zx}^2 = k^2 \qquad [2.6a,b]$$

A generalization of [2.6] as a modified Von Mises criterion for soils has been proposed by Drucker and Prager (1952). The yield surface of the modified Von Mises criterion in principal stress space is a right *circular cone* equally inclined to the principal axes. The intersection of the π-plane [2.4] with this yield surface is a circle shown by the dotted line in Fig. 2.5. The yield surface lies between the two right circular cones inscribing and circumscribing the extended Tresca hexagonal pyramid, respectively. The yield function used by Drucker and Prager to describe this cone in applying the limit theorems to perfectly plastic soils has the form:

$$\alpha p + J_2^{1/2} = k \qquad [2.7]$$

where α and k are material constants. If α is zero, [2.7] reduces to the Von Mises yield condition for metal [2.6]. This yield criterion is called the *extended Von Mises criterion*. The yield function [2.7] reduces to the Coulomb rule [2.1] in the case of plane strain if:

$$\alpha = \frac{3 \tan \varphi}{\sqrt{(9 + 12 \tan^2 \varphi)}} \quad k = \frac{3c}{\sqrt{(9 + 12 \tan^2 \varphi)}} \qquad [2.8]$$

General remarks

There is, of course, no limit to the number of yield surfaces or functions which may be devised as a proper generalization of the Coulomb rule to three dimensions, and which reduce to the Coulomb rule in two-dimensional plane strain problems. The extended Tresca as well as the extended Von Mises yield criterion can both be interpreted as a proper generalization of the Coulomb rule to three dimensions. However, the preceding interpretation of the Coulomb rule by the Mohr diagram (Fig. 2.2) leads to only *one* yield surface for three-dimensional stress fields (Fig. 2.3). This unique yield surface is first obtained by Shield (1955a) and generally considered to be the correct formulation of the Coulomb yield criterion to three dimensions.

Bishop (1966) has attempted to correlate all three criteria with experimental data and has concluded that the Coulomb criterion best predicts soil failure or yield. However, for the plane strain case it can be shown that in the limit or collapse state (where elastic strains are identically zero) both the extended Von Mises and the extended Tresca criteria reduce to the Coulomb rule in two-dimensional problems. This implies that we can adjust the constants of the extended Von Mises ([2.8]) and extended Tresca criterion (Fig. 2.5) such that all three criteria will give identical collapse loads. We note however that the three yield criteria for soils will give different predictions for soil response below collapse load.

It is worth mentioning here that Palmer (1966) has concluded that the Coulomb yield criterion represents a lower yield condition for real soils.

Yield function

With the above discussions concerning the three specific yield surfaces in mind, we give now a summary of the basic properties of a yield function required in limit analysis of general bodies.

Each element of the body is assumed to be governed by a yield function f. For a perfectly plastic material, f depends only on the set of stress components $\sigma_{ij} = (\sigma_x, \sigma_y, \sigma_z, \tau_{xy}, \tau_{yz}, \tau_{zx})$ but not on the strain components $\epsilon_{ij} = (\epsilon_x, \epsilon_y, \epsilon_z, \epsilon_{xy}, \epsilon_{yz}, \epsilon_{zx})$. Plastic flow can occur only when the yield function is satisfied:

$$f(\sigma_{ij}) = 0 \tag{2.9}$$

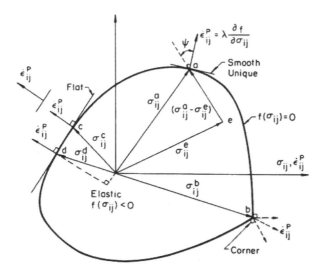

Fig. 2.7. Pictorial representation of yield surface and flow rules.

Stress states for which $f(\sigma_{ij}) > 0$ are excluded, and $f(\sigma_{ij}) < 0$ corresponds to elastic behavior.

The term yield surface is used to emphasize the fact that three or more components of stress σ_{ij} may be taken as coordinate axes; a two-dimensional picture only is drawn, however, as shown in Fig. 2.7. The yield surface thus is represented by a yield curve or actually becomes a yield curve when two independent components of stress are studied. It is helpful to visualize a state of stress in a nine-dimensional stress space as a point in the two-dimensional picture, shown in Fig. 2.7, as a vector whose components are the nine σ_{ij} (quantities such as τ_{xy}, τ_{yx} are treated as formally independent variables). Here we will assume that the yield surface is *convex* This is discussed later.

2.3. THE KINEMATIC ASSUMPTION ON SOIL DEFORMATIONS AND THE CONCEPT OF FLOW RULE

The theorems of limit analysis are also concerned with the plastic deformations of a soil. Plastic flow occurs when a stress-point in stress space, represented by a vector from the origin, reaches the perfectly plastic yield surface. What is the kinematics of the plastic flow? It is immediately clear that we cannot say anything about the *total* plastic strain ϵ_{ij}^p because the magnitude of the plastic flow is unlimited. We must always therefore think in terms of the *strain rates* $\dot\epsilon_{ij}$. The total strain rate $\dot\epsilon_{ij}$ is composed of elastic and plastic parts, $\dot\epsilon_{ij} = \dot\epsilon_{ij}^e + \dot\epsilon_{ij}^p$. The $\dot\epsilon_{ij}^e$ are related to the $\dot\sigma_{ij}$ through Hooke's law. The $\dot\epsilon_{ij}^p$ depend on the state of stress through an appropriate kinematic assumption on the deformations.

In discussing plastic strain rates we need to know the directions of the axes of principal strain rates. For isotropic materials we expect these to *coincide* with the axes of principal stress. In other words, a rectangular element of isotropic material under simple compression would be expected during any plastic flow to deform in such a way that its faces remained mutually perpendicular. Clearly, the assumption of coincidence of the principal axes of strain rate and stress determines only the *directions* of the strain rate components, not their relative magnitude, i.e., the ratios of the strain rate components.

Tresca material and flow rule

For most metals, the condition of incompressibility is very closely obeyed. This condition may be used to derive the ratios of the plastic strain rate components. For example, for problems of plane plastic flow, this incompressibility condition is:

$$\dot\epsilon_{max}^p + \dot\epsilon_{min}^p = 0 \qquad\qquad [2.10]$$

where $\dot{\epsilon}^p_{max}$ and $\dot{\epsilon}^p_{min}$ denote the principal strain rates. This condition [2.10] thus determines the relative magnitude of $\dot{\epsilon}^p_{ij}$ for problems of plane plastic flow.

As Von Mises (1928) pointed out, the assumption of coincidence of the principal axes of strain rate and stress along with [2.10] implies that the ratio of the principal plastic strain rate components $\dot{\epsilon}^p_{max}$ and $\dot{\epsilon}^p_{min}$ can be written in the form:

$$\frac{\dot{\epsilon}^p_{max}}{\dot{\epsilon}^p_{min}} = \frac{\partial f / \partial \sigma_{max}}{\partial f / \partial \sigma_{min}} \qquad [2.11]$$

where $f = \sigma_{max} - \sigma_{min} - 2k$ denotes the Tresca yield criterion (Fig. 2.6) and σ_{max}, σ_{min} denote the principal stresses. The yield condition thus furnishes the so-called *plastic potential f* from which the stress and plastic strain rate relation can be established. This sort of stress flow relation established through the yield condition is known as the *flow rule*.

Coulomb material and flow rule

In discussing the yield conditions and flow rules for crystals, Von Mises introduced the concept of the plastic potential primarily for reasons of mathematical convenience. Only in recent years has the fundamental importance of this concept been fully realized. In particular, the basic theorems of limit analysis require this particular connection between yield condition and flow rule. In the following, the purely kinematic implications of the concept of the plastic potential or the flow rule will be discussed for the case of plane flow of Coulomb material. This will be followed then by a discussion in terms of a general yield function.

With the notation introduced above, Coulomb's yield condition [2.1] may be written in the form:

$$f = \sigma_{max}(1 - \sin \varphi) - \sigma_{min}(1 + \sin \varphi) - 2c \cos \varphi = 0 \qquad [2.12]$$

The concept of plastic potential or flow rule of [2.11] requires:

$$\frac{\dot{\epsilon}^p_{max}}{\dot{\epsilon}^p_{min}} = -\frac{1 - \sin \varphi}{1 + \sin \varphi} \qquad [2.13]$$

or:

$$\dot{\epsilon}^p_{max} = -\dot{\epsilon}^p_{min} \frac{1 - \sin \varphi}{1 + \sin \varphi} = -\dot{\epsilon}^p_{min} \tan^2(45 - \tfrac{1}{2}\varphi) \qquad [2.14]$$

Equation [2.14] implies that *any plastic deformation of Coulomb material must be accompanied by an increase in volume if $\varphi \neq 0$.* This property is known as *dilatancy.*

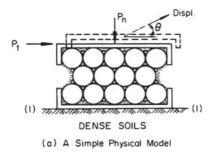

DENSE SOILS

(a) A Simple Physical Model

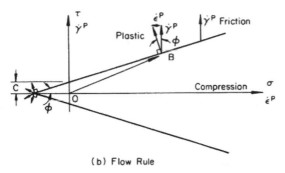

(b) Flow Rule

Fig. 2.8. Plastic strain rate is normal to yield curve for perfectly plastic theory, but parallel to τ-axis for frictional theory.

For further illustration on this point, a simple physical model, shown in Fig. 2.8(a), may be helpful. In the figure, a layer of dense granular material is subjected to the action of two forces. The vertical or normal force P_n acts at right angles to the plane $l-l$ whereas the other, the horizontal force P_t, acts tangentially to that plane. Let us further assume that the normal force P_n remains constant during the experiment, whereas, the horizontal force P_t gradually increases from zero to the value which will produce sliding. At the instant of sliding, the value P_t must not only overcome cohesion, but also must exceed the resistance furnished by two types of friction. The first of these arises on the contact surfaces of adjoining particles and is termed *surface friction*. The second, offered by the interference of the particles themselves to changes of their relative position, is termed *interlocking friction*. It is this interlocking friction that requires the displacement upward, as well as the usual displacement to the side. The displacement vector must, therefore, make an angle θ to the slip plane.

If soil were idealized as perfectly plastic with Coulomb's rule of yielding, then [2.1] defines the yield curve in the stress space σ, τ (Fig. 2.8b). If now a stress state, represented by a vector from the origin, is increased from zero, yield will be

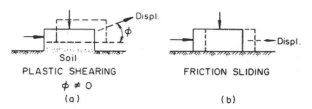

Fig. 2.9. Difference between Coulomb sliding and Coulomb shear.

incipient when the vector reaches the curve (two straight lines). For a perfectly plastic material, the vector representing the stress state at any given point can never protrude beyond the curve, since it is an unattainable stress state in granular media.

We shall now obtain in a convenient fashion a geometrical interpretation of the flow rule. In the (σ, τ) stress coordinates, we set out along the axes of another set of coordinates, the values of the corresponding plastic deformations, namely, the plastic normal strain rate $\dot{\epsilon}^p$ and plastic shear strain rate $\dot{\gamma}^p$. The corresponding stresses and plastic strain rates are set out parallel and in one and the same direction. In order to associate each strain rate vector with the corresponding stress vector, we plot it with the corresponding stress point as a floating origin. Fig. 2.8(b) shows this combined stress and strain rate plot. The direct consequence of the flow rule is that the plastic strain rate vector must be *normal* to the yield curve when their corresponding axes are superimposed. It can be seen from the figure that this is equivalent to assuming $\theta = \varphi$ in Fig. 2.8(a). The perfectly plastic idealization with associated flow rule (normality) is illustrated by a block shearing on a horizontal plane, Fig. 2.9(a). Volume expansion is seen to be a necessary accompaniment to shearing deformation according to the idealizations. This theory was proposed by Drucker and Prager (1952) and generalized later by Drucker (1953), and Shield (1955a).

Frictional material

In contrast to the above-mentioned effort, one may idealize soil as a frictional material for which the interlocking friction is ignored ($\theta = 0$). Deformation occurs by the smooth sliding of adjacent surfaces of material points (see Fig. 2.9b). If $\tan \varphi$ in [2.1] denotes the coefficient of friction between adjacent surfaces of material points along the plane, [2.1] becomes the well-known *Coulomb friction limit condition* for the shear strength of soil. The important difference between Coulomb friction and perfectly plastic Coulomb action is seen in Fig. 2.9, where frictional sliding is horizontal while perfectly plastic shearing involves large upward vertical motion. If the plastic strain rate vector is superimposed to the Coulomb limit curves (assumed as yield curve), the normality rule does *not* hold (see Fig. 2.8b). The extent of this endeavor is described in a recent work by Dais (1970a).

General remarks

Real soils are quite complex and are still imperfectly understood. They are neither truly frictional in behavior, nor are they plastic. Hence, any such idealized treatment, as discussed above, will either result in some differences between predictions and experimental facts, or will entertain certain mathematical difficulties. For example, the dilatation which is predicted by perfect plastic theory to accompany the shearing action will usually be larger than that found in practice (Brinch Hansen, 1953). The inadequacy of a perfectly plastic idealization has been discussed by Drucker (1966a) and De Jong (1964) among others. The lack of uniqueness for solutions to problems using friction theory has been exhibited and explored by Dais (1970b).

In order to improve upon the perfectly plastic theory, Drucker et al. (1957) introduced the strain-hardening theories of soil plasticity, which were later extended by Jenike and Shield (1959), and more recently by Roscoe and his students (1958–63). The work-hardening plastic action may involve upward or downward vertical motion, or neither, of the sliding block as illustrated in Fig. 2.9, which qualitatively agrees with experimental data. Recently, more sophisticated theories have been proposed by Weidler and Paslay (1969, 1970), A.J.M. Spencer (1964), and Sobotka (1958, 1960) on non-homogeneous soils in an attempt to overcome some of the known deficiencies in previous theories. Clearly the development of a more sophisticated theory will almost always bring a more elaborate stress–strain relation. Solutions to practically important problems, on the contrary, become exceedingly difficult to obtain, if the stress–strain relation is too involved. A compromise must, therefore, be made between convenience and physical reality.

In certain circumstances, such as in the stability problems of soil mechanics, the deformation conditions of the problem are often insufficiently restrictive for the soil deformation properties to affect the collapse load to a great extent. The adoption of a limit analysis approach based upon Coulomb's yield criterion and its associated flow rule in soils appears to be reasonably justified.

Index notation

For the present, we have denoted the three mutually perpendicular coordinate axes by the familiar notation x, y, and z; for future convenience, these three mutually perpendicular axes will be denoted by x_1, x_2 and x_3 as a dual notation. Accordingly, each of the components of a force vector (or displacement vector) and the vector itself is represented by the symbol T_i (or u_i) where i takes on the values 1, 2, 3:

$$T_i = \begin{bmatrix} T_1 \\ T_2 \\ T_3 \end{bmatrix} = \begin{bmatrix} T_x \\ T_y \\ T_z \end{bmatrix} \qquad [2.15]$$

Similarly, each of the components of a stress tensor (or strain tensor) and the tensor itself will be represented by the symbol σ_{ij} (or ϵ_{ij}) where i and j take on the values 1, 2, 3:

$$\sigma_{ij} = \begin{bmatrix} \sigma_{11} & \sigma_{12} & \sigma_{13} \\ \sigma_{21} & \sigma_{22} & \sigma_{23} \\ \sigma_{31} & \sigma_{32} & \sigma_{33} \end{bmatrix} = \begin{bmatrix} \sigma_x & \tau_{xy} & \tau_{xz} \\ \tau_{yx} & \sigma_y & \tau_{yz} \\ \tau_{zx} & \tau_{zy} & \sigma_z \end{bmatrix} \qquad [2.16]$$

The symmetry of the stress tensor, $\sigma_{12} = \sigma_{21}$ or $\tau_{xy} = \tau_{yx}$, etc., is symbolized by $\sigma_{ij} = \sigma_{ji}$.

In the following, we will often encounter sums in which a certain subscript pair is "*summed*" from 1 to 3. It will be inconvenient to write summation signs and so we here introduce a summation convention which consists essentially in merely dropping the summation sign.

Summation convention

Whenever a subscript occurs twice in the same term, it will be understood that the subscript is to be summed from 1 to 3. Thus:

$$\sum_{i=1}^{3} T_i u_i = T_i u_i = T_1 u_1 + T_2 u_2 + T_3 u_3$$

$$\sum_{i=1}^{3} \sum_{j=1}^{3} \sigma_{ij}\epsilon_{ij} = \sigma_{ij}\epsilon_{ij} = \sigma_{11}\epsilon_{11} + \sigma_{12}\epsilon_{12} + \sigma_{13}\epsilon_{13}$$
$$\sigma_{21}\epsilon_{21} + \sigma_{22}\epsilon_{22} + \sigma_{23}\epsilon_{23}$$
$$\sigma_{31}\epsilon_{31} + \sigma_{32}\epsilon_{32} + \sigma_{33}\epsilon_{33} \qquad [2.17a,b]$$

with $T_1 = T_x$, $\sigma_{11} = \sigma_x$, $\sigma_{12} = \tau_{xy}$, $\epsilon_{11} = \epsilon_x$, $\epsilon_{12} \equiv \epsilon_{xy}$, $u_1 = u_x$, etc. Such repeated subscripts are often called *dummy* subscripts because of the fact that the particular letter used in the subscript is not important; thus $T_i u_i = T_j u_j$ or $\sigma_{ij} \epsilon_{ij} = \sigma_{mn} \epsilon_{mn}$. The subscript index which occurs only "*once*" in a term is called *free subscript*. A free subscript also takes the values 1, 2, 3. Thus the equation $T_i = \sigma_{ji} n_j$ (or $T_i = \sigma_{mi} n_m$) implies the following three simultaneous equations:

$$T_1 = \sigma_{11}n_1 + \sigma_{21}n_2 + \sigma_{31}n_3$$

$$T_2 = \sigma_{12}n_1 + \sigma_{22}n_2 + \sigma_{32}n_3$$

$$T_3 = \sigma_{13}n_1 + \sigma_{23}n_2 + \sigma_{33}n_3 \hspace{3cm} [2.18a,b,c]$$

with $T_1 = T_x$, $\sigma_{11} = \sigma_x$, $\sigma_{12} = \tau_{xy}$, $n_1 = n_x$, etc.

Yield function and flow rule

The geometrical significance of the flow rule in terms of general stress coordinates will now be examined. Suppose the coordinate axes of the stress space already referred to represent simultaneously plastic strain rates as well as stresses, each axis σ_{ij} being an axis of the corresponding strain rate component $\dot{\epsilon}_{ij}^p$. Thus, a point specifies a plastic strain rate state. Fig. 2.7 shows this combined stress and strain rate plot. Now the significance of the flow rule is clear: the vector representing the plastic strain rates has the direction of the outward normal to the yield surface.

This *normality* of the associated plastic strain rate vector, $\dot{\epsilon}_{ij}^p$, to the yield surface $f(\sigma_{ij}) = 0$ for a plastic material is the direct consequence of the flow rule which applies not only to plastic *material* but also, in the appropriate load-space, to *structures* made of plastic *material*. In many cases the yield surface may have corners or vertices where there is not a *unique* normal direction. (The Coulomb yield criterion shown in Fig. 2.4 has such corners.) For such cases, the flow rule only demands that the plastic strain rate vector may have any direction within the fan defined by the normals of the contiguous surfaces. The lack of a unique plastic strain rate is found to be no handicap in limit analysis. This is discussed later. Fig. 2.7 shows the yield and flow relations pictorially. At point a, the yield surface is *smooth* with a *uniquely* defined normal and the plastic strain rate vector is normal to the yield surface associated with the plastic stress state σ_{ij}^a. At point b, with stress state σ_{ij}^b, there is a corner, and the plastic strain rate vector can have any direction between the two normals defined by the adjacent surfaces. The stress/strain rate relation derived from the normality condition corresponding to a yield function can be written in the general form:

$$\dot{\epsilon}_{ij}^p = \lambda \frac{\partial f}{\partial \sigma_{ij}} \hspace{3cm} [2.19]$$

where $\lambda > 0$ is a scalar proportionality factor.

We give now a summary of the results which are required in limit analysis of general bodies. These results are the direct consequences of the assumptions mentioned above.

Since the definitions of *convexity* and *normality* in analytical geometry are essentially the same in a space with any number of dimensions, a two-dimensional schematic representation of Fig. 2.7 is perfectly adequate for general discussion. This schematic convex yield surface is drawn deliberately nonsymmetrical, because neither symmetry nor the yield stress in tension equal or different from the yield stress in compression has any direct relevance to the proof of the theorems of limit analysis: the only critical requirement of the *shape* of the yield surface is that the yield surface must be *convex* in the sense that the entire yield surface lies on one side of any plane which is tangent to the surface. Fig. 2.7 also illustrates the convexity property.

Suppose that the stress state σ_{ij}^a is a point on the yield surface, while σ_{ij}^e is any other point on or within the yield surface. The vectors representing these stress states are shown in Fig. 2.7, together with the vector $(\sigma_{ij}^a - \sigma_{ij}^e)$. It is seen that, for a convex yield surface, the scalar product of the vector $(\sigma_{ij}^a - \sigma_{ij}^e)$ with the normal vector $\dot{\epsilon}_{ij}^p = \lambda \partial f / \partial \sigma_{ij}$ cannot be negative; because the angle ψ between these two vectors cannot be greater than $\frac{1}{2}\pi$; hence, the *convexity* property of the yield surface together with the *flow rule* for the plastic deformation [2.19] requires:

$$(\sigma_{ij}^a - \sigma_{ij}^e)\,\dot{\epsilon}_{ij}^p \geqslant 0 \qquad\qquad\qquad [2.20]$$

If the state σ_{ij}^e is inside the yield surface, $f(\sigma_{ij}^e) < 0$, then the inequality sign holds in [2.20].

Dissipation function

The stress state σ_{ij}^a is the *actual* stress state which is associated with the given plastic strain rate $\dot{\epsilon}_{ij}^p$. The relations given above [2.20] express a **principle of maximum local energy dissipation** due to Von Mises: **the rate at which plastic work is done on a given plastic strain rate system has a maximum value for the actual stress state.** This statement applies equally for yield surface with flat faces or corners. In the case of a flat yield face, a given plastic strain rate vector determines the *rate of dissipation* $D(\dot{\epsilon}_{ij}^p)$ uniquely even when the stress itself is not uniquely determined. Referring to Fig. 2.7, since the scalar product of two vectors is found by multiplying the length of the given $\dot{\epsilon}_{ij}^p$ vector by the projection of the σ_{ij}^c or σ_{ij}^d onto it, which is represented graphically in the figure, it follows that the rate of dissipation of energy per unit volume associated with the given strain rate vector $\dot{\epsilon}_{ij}^p$ has the unique value:

$$D(\dot{\epsilon}_{ij}^p) = \sigma_{ij}^c \dot{\epsilon}_{ij}^p = \sigma_{ij}^d \dot{\epsilon}_{ij}^p = \sigma_{ij}^p \dot{\epsilon}_{ij}^p \qquad\qquad\qquad [2.21]$$

where σ_{ij}^p denotes in general terms the actual plastic stress state associated with a given strain rate $\dot{\epsilon}_{ij}^p$. For a point on the yield surface which has a uniquely defined normal (point a, Fig. 2.7), the associated actual stress state σ_{ij}^p is σ_{ij}^a. If the stress point coincides with the vertex or the corner b, then $\sigma_{ij}^p = \sigma_{ij}^b$. For points on the flat yield surface, however, the values of σ_{ij}^p can be either σ_{ij}^c or σ_{ij}^d.

2.4. THE ASSUMPTION OF SMALL CHANGE IN GEOMETRY AND THE EQUATION OF VIRTUAL WORK

The proofs of the theorems of limit analysis require the assumption that changes in geometry of the body that occur at the instant of collapse are small, in the sense that, in all calculations, original undeformed dimensions will be used in the equilibrium equations. That is, if equilibrium equations are established for the original state of the problem, it will be assumed that the overall dimensions at the incipient of collapse will alter by negligible amounts, so that the same equations can be used to describe the deformed state of the problem.

The key to prove the limit theorems and to apply them is the use of the *equation of virtual work*. The assumption of no appreciable change in geometry implies that the virtual work equation is applicable. For future reference, we give here two suitable forms of this equation.

Virtual work equations

The equation of virtual work deals with two separate and unrelated sets: *equilibrium set and compatible set*. Equilibrium set and compatible (or geometry) set are brought together, side by side but independently, in the equation of virtual work:

Equilibrium Set

$$\int_A T_i \dot{u}_i^* \, dA + \int_V F_i \dot{u}_i^* \, dV = \int_V \sigma_{ij} \dot{\epsilon}_{ij}^* \, dV \qquad [2.22]$$

Compatible set

Here integration is over the whole area, A or volume V, of the body. The quantities T_i, F_i are external forces on the surface and body forces in a body, respectively. σ_{ij} are any set of stresses, *real or otherwise*, in equilibrium with F_i in the body and with the external forces T_i on the surface. Referring to Fig. 2.10(a), a valid equilibrium set must therefore satisfy the following equilibrium equations:

At surface points $\qquad T_i = \sigma_{ji} n_j \qquad\qquad\qquad\qquad\qquad$ [2.23a]

At interior points $\qquad \dfrac{\partial \sigma_{ji}}{\partial x_j} + F_i = 0 \qquad\qquad\qquad\qquad$ [2.23b]

$\qquad\qquad\qquad\qquad \sigma_{ji} = \sigma_{ij} \qquad\qquad\qquad\qquad\qquad$ [2.23c]

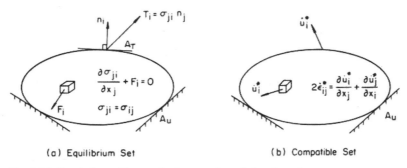

(a) Equilibrium Set   (b) Compatible Set

Fig. 2.10. Two independent sets in the equation of virtual work.

where n_i is the outward-drawn unit normal vector to a surface element.

Similarly, the strain rate $\dot{\epsilon}_{ij}^*$ represents any set of strains or deformations compatible with the *real* or *imagined* (virtual) displacement rate $\dot{u}_i^*$ of the points of applications of the external forces T_i or the points of displacements corresponding to the body forces F_i. Referring to Fig. 2.10(b), a continuous distortion of a body compatible with an assumed displacement field must satisfy the following strain and displacement rate compatibility relation:

$$2\,\dot{\epsilon}_{ij}^* = \frac{\partial \dot{u}_i^*}{\partial x_j} + \frac{\partial \dot{u}_j^*}{\partial x_i} \qquad\qquad [2.24]$$

The important point to keep in mind is that neither the equilibrium set T_i, F_i, σ_{ij} (Fig. 2.10a) nor the compatible set $\dot{u}_i^*, \dot{\epsilon}_{ij}^*$ (Fig. 2.10b) need be the actual state, nor need the equilibrium and compatible sets be related in any way to each other. Here, we use the asterisk for the compatible set to emphasize the point that these two sets are completely independent. When the actual or real states (which satisfy both equilibrium and compatibility) are substituted in the equation of virtual work, the asterisk will be omitted.

Any equilibrium set may be substituted in [2.22]. In particular, an increment or rate of change of forces and interior stresses $\dot{T}_i, \dot{F}_i, \dot{\sigma}_{ij}$ may be used as an equilibrium set.

$$\int_A \dot{T}_i \dot{u}_i^* dA + \int_V \dot{F}_i \dot{u}_i^* dV = \int_V \dot{\sigma}_{ij} \dot{\epsilon}_{ij}^* \, dV \qquad\qquad [2.25]$$

is an equation of virtual work in *rate form*. The two forms of virtual work [2.22] and [2.25] will be used later in proving the theorems of limit analysis. It should be remembered that the virtual work equation carries the implication, which will hold throughout our work (except in some parts of Chapters 5 and 12), that all displacements are sufficiently small for the original undeformed configuration of the problem to be used in setting up the equations of the system.

2.5. THEOREMS OF LIMIT ANALYSIS

Definition of plastic limit load

Fig. 2.11 shows a typical load-displacement curve as it might be measured for a surface footing test. The curve consists of an elastic portion, a region of transition from mainly elastic to mainly plastic behavior; a plastic region, in which the load increases very little while the deflection increases manifold; and, finally, a region in which either work hardening of the soil or the changes in geometry from a surface footing to a subsurface footing (or both) effectively stiffen the bearing strength of the footing. In a case such as this, there exists no physical collapse load. However, to know the load at which the footing will deform excessively has obvious practical importance. For this purpose, idealizing the soil as a perfectly plastic medium and neglecting the changes in geometry lead to the condition in which displacements can increase without limit while the load is held constant as shown in Fig. 2.11. A load computed on the basis of this ideal situation is called *plastic limit load* or *collapse load*. This hypothetical limit load usually gives a good approximation to the physical plastic collapse load or the load at which deformations become excessive. The methods of limit analysis furnish bounding estimates of this *hypothetical limit load*.

Ideal plastic body

The theorems of limit analysis can be established directly for a general body if the body possesses the following properties:

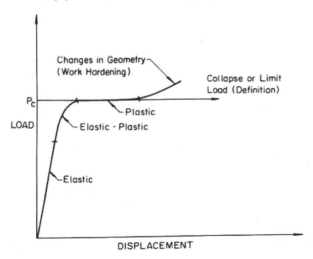

Fig. 2.11. A typical plastic collapse phenomenon and definition of limit load.

(1) The material exhibits *perfect* or *ideal* plasticity; i.e., work hardening or work softening does not occur. This implies that stress point can not move outside the yield surface, so the vector $\dot{\sigma}_{ij}$ must be tangential to the yield surface whenever plastic strain rates are occurring.

(2) The yield surface is *convex* and the plastic strain rates are derivable from the yield function through the *flow rule* (or the *normality condition*). It follows from the flow rule and the property mentioned above that $\dot{\sigma}_{ij}\,\dot{\epsilon}_{ij}^{p} = 0$.

(3) *Changes in geometry* of the body that occur at the limit load are insignificant; hence the equations of virtual work [2.22] and [2.25] can be applied.

In summary, the limit load is defined as the plastic collapse load of an *ideal* body replacing the actual one, and having the ideal properties listed above.

The limit theorems

The foundations of limit analysis are the two limit theorems stated previously in Chapter 1. Before we proceed to prove the theorems, we will first prove the following statement: **When the limit load is reached and the deformation proceeds under constant load, all stresses remain constant; only plastic (not elastic) increments of strain occur.** Thus the application of the elastic—perfectly plastic stress—strain rate relation becomes formally the same as the use of the rigid—perfectly plastic stress—strain rate relation. However, in this case the elastic strain increments are not neglected because they are assumed to be small; they are proved to be zero in the above statement.

A direct proof of this statement starts with the equation of virtual work [2.25] in rate form for the stress rates $\dot{\sigma}_{ij}^{c}$ and strain rates $\dot{\epsilon}_{ij}^{c}$ in the body at the collapse or limit load and continuous displacement rates $\dot{u}_{i}^{c}$:

$$\int_{A_T} \dot{T}_i^c \dot{u}_i^c \, dA + \int_{A_u} \dot{T}_i^c \dot{u}_i^c \, dA + \int_V \dot{F}_i^c \dot{u}_i^c \, dV = \int_V \dot{\sigma}_{ij}^c \dot{\epsilon}_{ij}^c \, dV \qquad [2.26]$$

In this equation the load system at collapse consists of body force rates $\dot{F}_i^c$ (per unit volume) and surface traction rates $\dot{T}_i^c$ (per unit area). Each component $\dot{T}_i^c$ is specified on the surface area A_T and each component of displacement rate is prescribed to be zero on area A_u. The superscript c emphasizes the fact that all the quantities used in [2.26] are the actual state at collapse.

Now, at the limit load, the left-hand side of [2.26] vanishes, by our definition; $\dot{F}_i^c = 0$ everywhere; and $\dot{T}_i^c = 0$ on A_T and $\dot{u}_i^c = 0$ on A_u. Since total strain rate $\dot{\epsilon}_{ij}^c$ consists of elastic and plastic parts, $\dot{\epsilon}_{ij}^c = \dot{\epsilon}_{ij}^{ec} + \dot{\epsilon}_{ij}^{pc}$, it follows from [2.26] that:

$$\int_V \dot{\sigma}_{ij}^c (\dot{\epsilon}_{ij}^{ec} + \dot{\epsilon}_{ij}^{pc}) \, dV = 0 \qquad [2.27]$$

But it follows from the properties (1) and (2) that $\dot{\sigma}_{ij}^{C} \dot{\epsilon}_{ij}^{pc} = 0$. Therefore:

$$\int_V \dot{\sigma}_{ij}^{C} \dot{\epsilon}_{ij}^{ec} \, dV = 0 \qquad [2.28]$$

Since $\dot{\sigma}_{ij}^{C} \dot{\epsilon}_{ij}^{ec}$ is a positive quantity when $\dot{\sigma}_{ij}^{C} \neq 0$ for any elastic materials, the vanishing of the integral in [2.28] requires that $\dot{\sigma}_{ij}^{C} = 0$ throughout the body. Therefore, there is no change in stress, and correspondingly no elastic change in strain during deformation at the limit load. All deformation is plastic. This statement states that *elastic characteristic plays no part in the collapse at the limit load.*

The lower- and upper-bound limit theorems will now be restated here and proved (Drucker et. al., 1952).

Theorem I (lower bound) – If an equilibrium distribution of stress σ_{ij}^{E} covering the whole body can be found which balances the applied loads T_i on the stress boundary A_T and is everywhere below yield $f(\sigma_{ij}^{E}) < 0$; then the body at the loads T_i, F_i will not collapse.

Proof. To prove the theorem, assume it false. We show that this leads to a contradiction. If the body at the loads T_i, F_i collapses, a collapse pattern associated with the actual stresses, strain rates and displacement rates, σ_{ij}^{C}, $\dot{\epsilon}_{ij}^{C}$ and $\dot{u}_i^{C}$ exists. This collapse pattern corresponds to the collapse loads T_i on A_T and F_i in V, with $\dot{u}_i^{C} = 0$ on A_u. Two equilibrium systems would exist, T_i, F_i, σ_{ij}^{C} and T_i, F_i, σ_{ij}^{E}. From virtual work equation [2.22]:

$$\int_{A_T} T_i^C \dot{u}_i^C \, dA + \int_V F_i^C \dot{u}_i^C \, dV = \int_V \sigma_{ij}^C \dot{\epsilon}_{ij}^C \, dV$$

$$\int_{A_T} T_i^C \dot{u}_i^C \, dA + \int_V F_i^C \dot{u}_i^C \, dV = \int_V \sigma_{ij}^E \dot{\epsilon}_{ij}^C \, dV \qquad [2.29a,b]$$

Hence:

$$\int_V (\sigma_{ij}^C - \sigma_{ij}^E) \, \dot{\epsilon}_{ij}^C \, dV = 0 \qquad [2.30]$$

Since at collapse, all deformation is plastic, it follows that:

$$\int_V (\sigma_{ij}^C - \sigma_{ij}^E) \, \dot{\epsilon}_{ij}^{pc} \, dV = 0 \qquad [2.31]$$

In view of the fact that convexity and normality properties require $(\sigma_{ij}^C - \sigma_{ij}^E)$ $\dot{\epsilon}_{ij}^{pc} > 0$ for σ_{ij}^E below yield ([2.20]; see Fig. 2.7). A sum of positive terms cannot vanish. Therefore [2.31] cannot be true and the lower bound theorem is proved. If $f(\sigma_{ij}^E) = 0$ is permitted the body may be at the point of collapse.

The lower-bound theorem expresses *the ability of the ideal body to adjust itself to carry the applied loads if at all possible.*

Theorem II (upper bound) – If a compatible mechanism of plastic deformation $\dot{\epsilon}_{ij}^{p*}$, $\dot{u}_i^{p*}$ is assumed, which satisfies the condition $\dot{u}_i^{p*} = 0$ on the displacement boundary A_u; then, the loads T_i, F_i determined by equating the rate at which the external forces do work:

$$\int_{A_T} T_i \dot{u}_i^{p*} \, dA + \int_V F_i \dot{u}_i^{p*} \, dV \qquad\qquad [2.32]$$

To the rate of internal dissipation

$$\int_V D(\dot{\epsilon}_{ij}^{p*}) \, dV = \int_V \sigma_{ij}^{p*} \dot{\epsilon}_{ij}^{p*} \, dV \qquad\qquad [2.33]$$

will be either higher or equal to the actual limit load.

Proof. Again, assume the theorem false, then we show that this leads to a contradiction. If the loads so computed are less than the actual limit load, then the body will not collapse at this load. An equilibrium distribution of stress σ_{ij}^E everywhere below yield $f(\sigma_{ij}^E) < 0$ must therefore exist (converse of lower-bound theorem mentioned above). From virtual work equation [2.22]:

$$\int_{A_T} T_i \dot{u}_i^{p*} \, dA + \int_V F_i \dot{u}_i^{p*} \, dV = \int_V \sigma_{ij}^E \dot{\epsilon}_{ij}^{p*} \, dV \qquad\qquad [2.34]$$

Since T_i and F_i are computed by equating [2.32] to [2.33], it follows that:

$$\int_V (\sigma_{ij}^{p*} - \sigma_{ij}^E) \dot{\epsilon}_{ij}^{p*} \, dV = 0 \qquad\qquad [2.35]$$

The convexity and normality properties require, however, $(\sigma_{ij}^{p*} - \sigma_{ij}^E) \dot{\epsilon}_{ij}^{p*} > 0$ for σ_{ij}^E below yield. This leads to a contradiction and thus proves theorem II.

The upper-bound theorem states that *if a path of failure exists the ideal body will not stand up.*

Some corollaries follow immediately from the lower-bound theorem because the original stress distribution is admissible in the modified situation.

Theorem III – Initial stresses or deformations have no effect on the plastic limit or collapse load provided the geometry is essentially unaltered.

Theorem IV – Addition of (weightless) material to a body without any change in the position of the applied loads cannot result in a lower collapse load.

Theorem V − Increasing (decreasing) the yield strength of the material in any region cannot weaken (strengthen) the body.

The negative approach in the last two corollaries is required. It is not necessarily true that increasing a local yield strength or adding material will strengthen the elastic perfectly plastic body. The last corollary can be put in another form which is of great practical use in calculations.

Theorem VI − A limit load computed from a convex yield surface which circumscribes the actual surface will be an upper bound on the actual limit load. A limit load computed from an inscribed surface will be a lower bound on the actual collapse load.

For example, Fig. 2.5 shows an extended Tresca surface inscribed in a Coulomb yield surface. Thus, an inscribed extended Tresca material would give a lower bound on the actual collapse load for a body made of a Coulomb material.

An illustrative example

In the application of the upper- and lower-bound theorems; discontinuous stress and velocity fields prove very useful. Fig. 2.12 shows a simple example of discontinuous stress and velocity fields for the problem of a slab material of thickness h pressed between perfectly smooth flat rigid punches of width b. A discontinuous stress field satisfying the Tresca yield condition is shown in Fig. 2.12(a) which gives $2kb$ as a lower bound to the actual collapse load P_c. Fig. 2.12(b) shows an assumed mechanism of plastic deformation consisting of rigid body translations separated by velocity discontinuities. The number m of intersection M of velocity discontinuities between the rigid regions is chosen to give the least upper bound. The rate of work at a velocity discontinuity Δv is $k\Delta v$ per unit length for unit width of the slab. By equating the external rate of work due to punch loadings to the total internal dissipation along the discontinuous surfaces we get $k(mh + b^2/mh)$ as an upper bound to the limit load. The resultant bounds:

$$1 \leqslant \frac{P_c}{2kb} \leqslant \frac{1}{2}\left(\frac{mh}{b} + \frac{b}{mh}\right) \tag{2.36}$$

where $m(m-1) \leqslant b/h \leqslant m(m+1)$, are shown as dashed lines in Fig. 2.12(c) and compared with the exact solution due to Green (1951). Basic techniques on applications of the upper- and lower-bound theorems are given in the following two chapters.

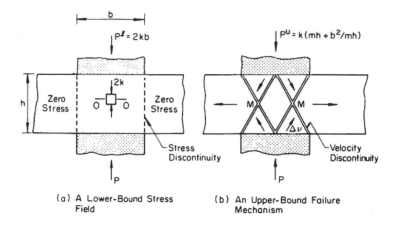

(a) A Lower-Bound Stress   (b) An Upper-Bound Failure
    Field   Mechanism

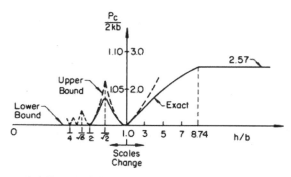

(c) Exact and Upper- and Lower-Bound Solutions

Fig. 2.12. Compression between smooth rigid punches.

2.6. LIMIT THEOREMS FOR MATERIALS WITH NON-ASSOCIATED FLOW RULES

An essential point in the proofs of the limit theorems given earlier is the inequality [2.20]. This inequality is a direct consequence of the *normality* relationship between a yield surface and its associated plastic strain rate vector. Without
this inequality the theorems cannot be proven in general. This normality condition
or the so-called *associated flow rule* is known to be a property of several wide
classes of materials satisfying certain thermodynamic conditions (Drucker, 1951;
Il'Yushin, 1961). As discussed previously, *friction* is not analogous to *plastic* material. The inequality [2.20] does not hold for frictional materials and systems, and
hence the limit theorems of plastic bodies do not apply to bodies with frictional
interfaces. We shall examine first the frictional material, and then go on to see how
we can, nevertheless, obtain a limited amount of useful information about friction

effects by use of the theorems derived for materials with associated flow rule. Finally, limit theorems for a class of materials with non-associated flow rule are derived.

Frictional material

A simple frictional system to which normality does not apply is illustrated in Fig. 2.13. Fig. 2.13(a) shows a block resting on a rough horizontal surface and subjected to two forces, a horizontal force Q and a vertical force P. The coefficient of Coulomb friction between the block and the surface is μ. Then the "yield surface" in P,Q-load space for the system is $Q = \mu P$, which is a straight line. If the forces on the block are represented by a point below the straight line the block will not move; if they are represented by a point on the line an infinitesimal force increment directed upward will cause the block to slide. This is analogous to a *yield* surface for a perfectly plastic material. Any sliding of the block along the horizontal plane gives a corresponding increment of irreversible displacement. Since the displacement of the block is in the direction of the horizontal force Q and there is no displacement in the direction corresponding to the vertical force P, as indicated in Fig. 2.13(b), the displacement increment vector in P,Q-space has the direction parallel to the Q-axis, and is thus *not* normal to the "yield surface", except in the special case $\mu = 0$, i.e., frictionless sliding.

Referring now to Fig. 2.8(a), a frictional material corresponds to $\theta = 0$ and the rate of volume change is zero during frictional sliding. For all $\theta > 0$, the rate of plastic volume change is negative, i.e., the material expands or dilates during plastic deformation. We see from Fig. 2.8(b) that only when $\theta = \varphi$ is the plastic strain rate vector normal to the Columb yield line. A Coulomb material with $\theta = \varphi$ is referred to as having an *"associated flow rule"*. A frictional material for which $\theta = 0$ belongs to the class of materials which have *non-associated flow rules* in the sense that the plastic strain rate vector at a point on the yield surface is not in general normal to the surface. Such materials have been discussed by Mroz (1963, 1964), Palmer

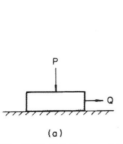
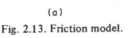
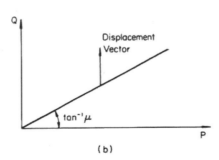

Fig. 2.13. Friction model.

(1966), De Jong (1964) and Dais (1970a, 1970b). In the following somewhat restricted limit theorems for perfectly plastic materials of this kind will be developed.

The frictional theorems

There are special conditions for frictional interfaces for which the limit theorems of plastic bodies fully apply. These conditions are: (1) the coefficient of friction is zero; (2) there is no relative motion or separation at the interface. As we have seen, normality does apply for a smooth or frictionless material and no relative movement at the interface implies that any "sliding" must be of the plastic kind. With this background it is intuitively clear that

Theorem VII — **Any set of loads which produces collapse of an assemblage of elastic–plastic bodies with frictional interfaces for the condition of no relative motion at the interfaces will produce collapse for the case of finite friction.**

No relative motion is a more inclusive term than infinite friction because separation is not permitted.

Theorem VIII — **Any set of loads which will not cause collapse of an assemblage of elastic–plastic bodies with frictional interfaces when all coefficients of friction at the interfaces are zero will not produce collapse with any values of the coefficients.**

According to the frictional theorems, the limit load is bounded below by the limit load for the same bodies with zero friction on the interfaces. It is bounded above by the limit load for no relative motion at the interfaces. Hence, in a lower-bound calculation, if we take the footing base or retaining wall interface to be a plane of principal stress; then, our calculation is *"safe"* for a *"smooth"* footing or wall and hence also for a footing or wall with finite friction. Further, in an upper-bound calculation, if we assume a mechanism having no relative motion at the footing base or wall interface; then, our calculation is *"unsafe"* for a *"rough"* footing or wall and hence also for a footing or wall with finite friction.

In some of the stability problems we have considered in this book, the range between upper and lower bounds corresponding to these two "extreme" conditions is not affected much by the question of whether the footing or retaining wall is rough or smooth. This indicates that friction is of only secondary importance in the determination of the limit load. In some other cases, however, we find that there is a large difference between the bounds corresponding to rough and smooth footings or walls; this then indicates that friction plays an important role, and it may be necessary to use approximate intuitive methods for assessing the effect of any particular coefficient of friction.

If there is *finite* friction with friction angle δ it is advantageous to include the concept of an *interface cement* composed of a soil-like *cohesionless* material with friction angle $\Phi = \delta$ obeying the Coulomb yield criterion and its associated flow rule so that the limit theorems now apply to the modified body with *"cemented"* interfaces. With this background, the following upper-bound theorem due to Drucker (1954a) may be of further help.

Theorem IX – Any set of loads which will not cause collapse of an assemblage of elastic–plastic bodies with frictional interfaces will not produce collapse when the interfaces are "cemented" together with a cohesionless soil of friction angle $\Phi = \delta = \tan^{-1}\mu$.

The proof of this theorem follows essentially from the observation that the state of stress in the friction case satisfies the condition $f(\sigma_{ij}) \leq 0$ and $|\tau| \leq \mu|\sigma|$ at impending collapse. A safe state of stress exists, therefore, for the soil case when collapse does not occur in the friction problem. Clearly, according to this theorem, the limit load for an assemblage of bodies with frictional interfaces is bounded above *also* by the limit load for the same assemblage cemented at the interfaces by *a cohesionless "soil"* obeying the Coulomb yield criterion and its associated flow rule.

Frictional dissipation

The upper-bound theorem II for an ideal plastic body strictly is not applicable in general for a body with frictional interfaces in which energy is dissipated by friction. However, there is a strong temptation to ignore this consideration and to include the frictional dissipation in the total rate of internal dissipation of energy in an upper-bound calculation. This calculation will provide useful information, if not a full answer.

The difficulty in this approach, however, comes from the fact that at the frictional interfaces, the normal stress on the plane of sliding governs the frictional dissipation and an assumed failure mechanism does not determine this normal stress at the interfaces. Fortunately, in all of the stability problems we have considered in this book, the frictional interface is either at the footing base or the rear face of a retaining wall. In such a case, the normal force at the interface is either the applied force itself or the component of the applied force. The rate of dissipation of energy due to friction at the interface can therefore be computed by multiplying the discontinuous *"sliding"* velocity across the interface by μ (friction coefficient at the interface) times the normal force acting on this interface. The total rate of dissipation of energy is then obtained by adding this frictional dissipation to the plastic dissipation as given previously by [2.33].

Materials with non-associated flow rules

The frictional materials mentioned above belong to the class of materials which have non-associated flow rules. The limit theorems developed earlier for materials with associated flow rule are inapplicable for materials with non-associated flow rules. However, we can prove the following theorem which may have practical relevance to the material having the same yield criterion but with a non-associated flow rule.

Theorem X (upper bound) – Any set of loads which produces collapse for the material with associated flow rule will produce collapse for the same material with non-associated flow rules.

This follows readily from the fact that statically admissible stress solutions are independent of the flow rule (see [2.23]) so that the stress field corresponding to the actual collapse load for the material with non-associated flow rules must also be statically admissible for the same material with associated flow rule. It follows immediately from the lower-bound theorem I that the actual collapse load for the material with non-associated flow rules must therefore be less than or equal to the actual collapse load for the same material with associated flow rule. The result has been discussed and applied by a number of investigators, e.g., Radenkovic (1961); Sacchi and Save (1968); Mroz and Drescher (1969); and Collins (1969, 1973).

In what follows a lower-bound theorem for perfectly plastic materials with non-associated flow rules will be developed (Palmer, 1966; and De Jong, 1964). In Fig. 2.14, a single yield surface, not necessarily smooth or convex, is shown. It is assumed that at each point on this yield surface the directions of the plastic strain rate vector, not necessarily normal to the yield surface, are known. Through each point on the yield surface, $f(\sigma_{ij}) = 0$ we construct the hyperplane perpendicular to the direction of the plastic strain rate vector at that point (Fig. 2.14). If the direction at a point is non-unique, construct hyperplanes perpendicular to each of the admissible plastic strain rate directions. Either these hyperplanes have an *envelope* which is a surface completely within the yield surface, or they do not. If they do not, the limit theorem which follows cannot be applied. If they do, the surface is necessarily convex, by a well-known theorem in convex set theory (Eggleston, 1958). Denoting stress by σ_{ij} this envelope or new surface can be represented by $g(\sigma_{ij}) = 0$, in such a way that at points within it $g(\sigma_{ij}) < 0$ and outside it $g(\sigma_{ij}) > 0$; it will be called the *g-surface*. We now state and prove the following lower-bound theorem:

Theorem XI (lower bound) – If an equilibrium stress distribution σ_{ij}^E covering the whole body can be found which balances the applied loads on the stress boundary surface and is everywhere below yield $g(\sigma_{ij}^E) < 0$, then the body will not collapse.

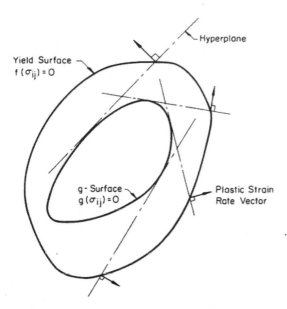

Fig. 2.14. Construction of g-surface.

Proof. The proof follows exactly the proof of the lower-bound theorem I given earlier. From the definition of the g-surface, it follows that the inequality [2.20] is applicable for the new yield surface $g(\sigma_{ij}) = 0$. If the normality condition does hold, then the g-surface is identical with the yield surface $f(\sigma_{ij}) = 0$ and this theorem reduces to the lower-bound theorem I.

As discussed previously, the behavior of a real soil is partly plastic and partly frictional in character, and where normality is sometimes known not to relate the observed plastic strain rate directions to yield conditions of the Coulomb type. However, as pointed out by Palmer (1966), if we construct a g-surface from the measured yield points and the observed deformation increment vectors, as reported by Haythornthwaite (1960), this g-surface is found to lie quite close to the yield locus predicted from the Coulomb yield criterion. This suggests that the Coulomb yield criterion represents a *lower-bound* yield condition for real soils.

Chapter 3

LIMIT ANALYSIS BY THE UPPER-BOUND METHOD

3.1. INTRODUCTION

The preceding chapter was concerned with basic assumptions and theorems used in limit analysis. In the following two chapters we shall discuss in more detail some of the basic techniques of applying these upper- and lower-bound theorems. Herein we shall illustrate the applications of these techniques by means of relatively simple examples; more complex applications will be taken up in later chapters.

The upper-bound theorem

We shall begin by re-examining the rules of the upper-bound theorem. As stated in the upper-bound theorem, the imposed loads cannot be carried by the soil mass if for any assumed failure mechanism the rate of work done by the external forces exceeds the internal rate of dissipation. Equating of external to internal rate of work for any such valid mechanism thus gives an unsafe upper bound on the collapse or limit load. The equation formed in this way is called the work equation for a particular assumed mechanism. The conditions required to establish such an upper-bound solution are essentially as follows:

(1) A valid mechanism of collapse must be assumed which satisfies the mechanical boundary conditions.

(2) The expenditure of energy by the external loads (including soil weights) due to the small displacement defined by the assumed mechanism must be calculated.

(3) The internal dissipation of energy by the plastically deformed regions which is associated with the mechanism must be calculated.

(4) The most critical or least upper-bound solution corresponding to a particular layout of the assumed mechanism must be obtained via the work equation.

Any mechanism is said to be *"valid"* if the small change in displacement within the body (or velocity field) due to the mechanism is *"compatible"* or *"kinematically admissible"*. In other words, the mechanism must be continuous in the sense that no gaps or overlaps develop within the body and the direction of the strains which is defined by the mechanism must in turn define the yield stresses required to calculate the dissipation. (This is known as the yield criterion and associated flow rule.)

It should be mentioned that discontinuous fields of stress and velocity may be used in applying lower- and upper-bound theorems. Discontinuous stress fields are actually very useful in deriving lower bounds. Surfaces of stress discontinuity are clearly possible provided the equilibrium equations are satisfied at all points of these surfaces. Details of constructing such discontinuous stress fields will be discussed in the following chapter. Surfaces of velocity discontinuity can also be admitted, provided the energy dissipation is properly computed. Rigid body sliding of one part of the body against the other part is a well-known example. This discontinuous surface should be regarded as the limiting case of continuous velocity fields, in which one or more velocity components change very rapidly across a narrow transition layer, which is replaced by a discontinuity surface as a matter of convenience. Discontinuous velocity fields not only prove convenient but often are contained in actual collapse mode or mechanism. This is in marked contrast to the stress situation where discontinuity is useful and permissible but rarely resembles the actual state.

Before the solution of a particular mechanism can be found, however, the work equation must be formed by equating the external rate of work due to the external applied loads and soil weight to the internal dissipation of energy in the plastically deformed region. Since these two quantities of work or energy have to be calculated separately before they are equated, the way in which these quantities are calculated shall be presented separately in what follows. A number of familiar deformation zones, which are useful as the basic building blocks of various mechanisms for later applications will be presented first as illustrated examples, and this will be followed by a separate presentation of the techniques for calculating the external rate of work. These results are adequate in connection with later applications.

Least upper bound

Once the work equation is formed, the collapse or limit load may be solved in terms of the variables that define the assumed mechanism. The final step in the analysis is to seek the particular layout or the value of the variables which is the least or the most critical. By the use of differential calculus, the magnitudes of the variables which give the most critical solution can generally be found. The *algebraic technique,* when it can be applied, gives a general solution applicable to all sizes of body of the particular mechanism assumed. This method can only be used, however, when the plot of load vs. variable parameters has a stationary minimum value. Sometimes, as will be demonstrated later, because of the physical conditions imposed on the parameters, there will not be a stationary minimum value within the valid range of a particular parameter. In such a case, the value of the least upper bound is not governed by the stationary minimum condition.

An alternative to the differential calculus technique is to try certain values of distances or angles which are treated as variables and several values of upper-bound solution can be obtained directly via the work equation. Visual inspection of the magnitude of the various solutions enables the most critical answer to be selected. Since many solutions are not very sensitive to a particular layout of a mechanism and further the valid ranges of variable parameters are already considered in the trial values, the method may be used conveniently in all circumstances.

As an alternative to the algebraic technique, an *arithmetic process* can be used in which several particular layouts corresponding to a particular assumed mechanism are each examined in turn, each solution being obtained directly and arithmetically via the work equation. The most critical answer can then be selected. Since this technique can be combined with graphical constructions of various layouts and mechanisms, it can be used conveniently in problems involving complex geometry.

Both the algebraic and arithmetic methods, and indeed a combination of both methods, are each most suitable for certain types of problems. Here we attempt not only to demonstrate these techniques, but also to give an indication where each is most suitable. More elaborate techniques such as the method of steepest descent for several variable parameters cases will be used in later chapters where more complex problems are treated.

Since Tresca's yield criterion is the particular case of Coulomb's yield criterion with $c = k$ and $\varphi = 0$, all techniques of analysis and mathematical derivations corresponding to Tresca material are much simpler than that of Coulomb material. In order to have a deeper appreciation of the upper-bound method, the techniques and derivations of certain equations used in this chapter will be presented first in terms of Tresca material. This will then be followed by the more general case of Coulomb material.

3.2. RIGID BLOCK SLIDING SEPARATED BY NARROW TRANSITION LAYER

Here, and throughout the next three chapters unless otherwise noted, we shall restrict our discussion to the plane strain case. The dimension perpendicular to the plane of the book will be taken as unity, but all motion is supposed in the plane.

3.2.1. Narrow transition layer of Tresca material

Dissipation of energy (Fig. 3.1)

The simplest discontinuity surface for a Tresca material is one across which the tangential velocity changes, say by δu, as shown in Fig. 3.1. The rigid top part of this block is moving to the right with velocity δu relative to the rigid bottom part, and the two parts are separated by a narrow transition layer of plastic deformation of unit area and thickness t, in which the shearing strain rate is uniform. Since the

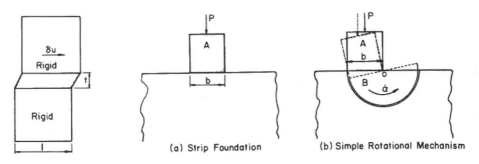

Fig. 3.1. Narrow transition layer of Tresca material.

Fig. 3.2. A strip footing on a purely cohesive soil.

mode of deformation is simple or pure shear with the large shear strain rate $\dot{\gamma}$ in the layer being equal to $\delta u/t$, the rate of energy dissipation in the narrow transition layer is computed from:

$$D = k\dot{\gamma}t = k\frac{\delta u}{t}t = k\delta u \qquad\qquad [3.1]$$

where k is the yield stress in pure shear. The rate of energy dissipation in the narrow layer is seen to be independent of the thickness t, so t may be as small as we please, including zero thickness as a matter of convenience. The zero-thickness idealization is certainly simple but we should always keep in mind that it is the limiting case of a rapidly varying velocity layer. The rate of energy dissipation per unit area of this velocity discontinuity is simply the product of the yield stress in pure shear and the relative velocity of the two rigid blocks.

Example 3.1: A strip footing on a purely cohesive soil

(a) Rotational mechanisms (Figs. 3.2, 3.3). As a first example, consider the bearing capacity of a purely cohesive soil or the Tresca material in which the angle of internal friction, φ, is zero. This is illustrated, schematically, in Fig. 3.2(a), where a long "rigid" strip foundation A is pressed by a force P per unit length into a large block of perfectly plastic Tresca material.

The first step in our analysis must be to postulate a valid mechanism which satisfies the geometrical boundary conditions. Since the foundation is assumed to be rigid, the movements of the contact plane between the foundation and soil mass must always remain plane.

In Fig. 3.2(b) is shown a simple postulated rigid body rotational mechanism about O. This mechanism is geometrically permissible if there are no external constraints to hold the foundation vertical. The block of material B rotates with angular velocity $\dot{\alpha}$ as a rigid body, and there is a semi-circular narrow transition layer between it and the remainder of the body. Since the angular velocity is $\dot{\alpha}$, the rate of work done by the external force P is a downward velocity at the center of the foundation, $\dot{\alpha}b/2$, multiplied by the applied force P, while the total rate of energy dissipation along the semi-circular discontinuity surface is found by multiplying the length of this discontinuity, πb, by the yield stress in pure shear, k, times the discontinuity in velocity across the surface $b\dot{\alpha}$. Equating the rate of external work to the rate of total internal energy dissipation gives:

$$P^{\text{u}} \left(\tfrac{1}{2}b\dot{\alpha}\right) = k(b\dot{\alpha})(\pi b) \quad \cdot \tag{3.2}$$

Note that the net rate of work done by the soil weight of the body B is zero since its motion is symmetric with respect to the vertical line passing through the corner O and thus the positive work done by the left part of body B cancels the negative work done by the right part of it. The superscript reminds us that we are finding an upper bound on P. From [3.2] we get:

$$P^{\text{u}} = 2\pi \, kb = 6.28 \, kb \tag{3.3}$$

As we can see, the upper-bound load calculation is independent of the magnitude of the angular velocity $\dot{\alpha}$, so we can regard $\dot{\alpha}$ as sufficiently small not to disturb the overall geometry. In other words, the proofs of the limit theorems as described in Chapter 2 can carry through using the initial geometry of the problem. The introduction of velocity discontinuity does not affect the upper-bound theorem.

Now the rotational mechanism of Fig. 3.2(b) may be generalized by regarding the radius and the position of the center of the circle as two independent variables; and we hope to find a lower, and therefore better, upper-bound solution by such a shifting of the center of rotation. If the center is shifted to O', as illustrated in Fig. 3.3(a), the rate of external work is:

$$P(r \cos \theta - \tfrac{1}{2}b) \, \dot{\alpha} \tag{3.4}$$

where r is the radius of the discontinuity surface and θ the angle between the face of the foundation and the line AO'. The rate of internal work is given by:

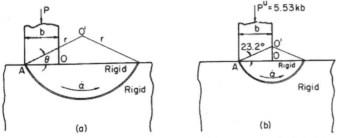

Fig. 3.3. Rotating soil mass with center at O' for purpose of minimizing upper bound.

$$kr(\pi - 2\theta)\, r\dot{\alpha} \qquad\qquad\qquad\qquad\qquad\qquad\qquad\qquad [3.5]$$

and the resulting upper-bound solution is:

$$P^u = \frac{k(\pi - 2\theta) r^2}{r \cos \theta - \frac{1}{2} b} \qquad\qquad\qquad\qquad\qquad\qquad [3.6]$$

The upper bound may be reduced by minimizing the solution with respect to the variables r and θ:

$$\frac{\partial P^u}{\partial r} = 0 \quad \text{and} \quad \frac{\partial P^u}{\partial \theta} = 0 \qquad\qquad\qquad\qquad\qquad [3.7]$$

Hence: $b = r \cos \theta_{cr}$ \qquad\qquad\qquad\qquad\qquad\qquad\qquad [3.8a]

and: $b = r[2 \cos \theta_{cr} - (\pi - 2\theta_{cr}) \sin \theta_{cr}]$ \qquad\qquad\qquad [3.8b]

Equation [3.8a] states that the location of the center of rotation O' is directly above O as shown in Fig. 3.3(b). Equating [3.8b] to [3.8a] yields:

$$\cos \theta_{cr} = (\pi - 2\theta_{cr}) \sin \theta_{cr} \qquad\qquad\qquad\qquad\qquad [3.9]$$

The critical value of the angle θ_{cr} can be determined from [3.9] by trial and error. This is found to be $\theta_{cr} = 23.2°$ The most critical layout of the mechanism is shown in Fig. 3.3(b), and the corresponding critical upper-bound value is:

$$P^u = \frac{4\,kb}{\sin 2\theta_{cr}} = 5.53\ kb \qquad\qquad\qquad\qquad\qquad [3.10]$$

which agrees with the well-known Fellenius (1927) solution that had been obtained by using the conventional method of limit equilibrium. This implies that the Fellenius solution is an upper bound, unless the solution is exact.

(b) Translational mechanisms (Figs. 3.4, 3.5). Now we move to consider the mechanisms which involve only rigid body translation. Figs. 3.4(a) and 3.5(a) show

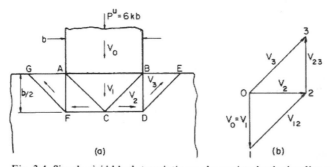

Fig. 3.4. Simple rigid block translation and associated velocity diagram.

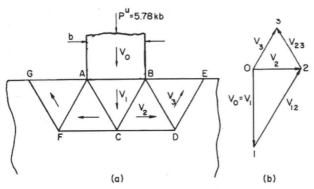

Fig. 3.5. Improved simple rigid block translation and associated velocity diagram.

two such examples of rigid block sliding separated by plane velocity discontinuities. In Fig. 3.4(a) the depth of the mechanism is taken to be one-half of the width of the foundation and in Fig. 3.5(a) all of the blocks are shown equilateral, both for convenience of calculation, but they could obviously have been drawn in many different ways in accord with the intuitive feeling of the designer or analyst for the appropriate layout of the mechanism.

Since plastic flow is confined to the lines of velocity discontinuity, we need to know all the relative sliding velocities between adjacent blocks in order to evaluate the dissipation of energy. The most direct method of determining these is by means of *velocity diagrams* (or *hodograph*). Before we move to construct these diagrams we shall briefly describe first the kinematics of the mechanisms.

Since these mechanisms are symmetrical about the center line, it is only necessary to consider the movement on the right-hand side of Figs. 3.4(a) and 3.5(a). The triangular region $A–B–C$ moves downward with the foundation as a rigid body. If the initial downward velocity of the foundation is taken to be V_0, the triangle in contact with the foundation must move with the same velocity, $V_1 = V_0$. The two triangular regions of material $B–C–D$ and $B–D–E$ move as rigid bodies in the directions parallel to CD and DE, respectively. The velocity of the triangle BCD is determined by the condition that the relative velocity, V_{12}, between this triangle and the triangle in contact with the foundation must have the direction BC. The velocity of the third triangle is determined in a similar manner. The information regarding velocities is represented in the velocity diagrams as shown in Figs. 3.4(b) and 3.5(b). The velocity of the foundation is first drawn as V_0 to a convenient arbitrary scale, and in the correct downward direction with respect to the layout of Figs. 3.4(a) and 3.5(a). Since the triangle in contact with the foundation has the same velocity as the foundation, the velocities V_1 and V_0 coincide. The directions of the velocity of triangle BCD relative to triangle ABC and to the zone which remains stationary are the same as the directions of the corresponding interfaces, i.e. line BC and line CD. The magnitude of the velocity V_2 and the relative velocity V_{12} is therefore determined uniquely from points labelled 0 and 1, as shown. Similarly, the point labelled 3 is located from 0 and 2. It can be seen that the velocity of each block is represented by a vector from the origin O and the vector joining two end points represents the relative velocity of the corresponding blocks.

Using the notation l_{bc} for the length of the interface BC, etc., we write down the work equation for these two mechanisms, making use of symmetry:

$$P^u V_0 = 2k(l_{bc} V_{12} + l_{cd} V_2 + l_{bd} V_{23} + l_{de} V_3) \tag{3.11}$$

Since energy dissipation must always be positive, each of the terms on the right-hand side of the above equation presents no difficulty over signs when the expression comes to be evaluated.

Considering first the mechanism shown in Fig. 3.4(a), the trigonometry of the velocity relation is simple, and we write down by inspection:

$$V_1 = V_2 = V_{23} = \frac{V_{12}}{\sqrt{2}} = \frac{V_3}{\sqrt{2}} = V_0 \tag{3.12}$$

also: $l_{cd} = l_{bd} = \dfrac{l_{bc}}{\sqrt{2}} = \dfrac{l_{de}}{\sqrt{2}} = \tfrac{1}{2}b$ $\tag{3.13}$

Expressing all the velocities and lengths in terms of V_0 and $\tfrac{1}{2}b$, respectively, and then substituting into [3.11], we have:

$$P^u V_0 = 2k \tfrac{1}{2}b(2 + 1 + 1 + 2) V_0 \tag{3.14a}$$

and: $P^u = 6\,kb$ $\tag{3.14b}$

Here for the convenience of discussion we ignore the self-weight of the soil.

Referring now to the mechanism and velocity diagram shown in Fig. 3.5, we have, by inspection:

$$V_2 = V_3 = V_{23} = \frac{V_{12}}{2} = \frac{V_0}{\sqrt{3}} \tag{3.15}$$

also: $l_{bc} = l_{cd} = l_{bd} = l_{de} = b$ $\tag{3.16}$

Thus: $P^u V_0 = 2kb\left(\dfrac{2}{\sqrt{3}} + \dfrac{1}{\sqrt{3}} + \dfrac{1}{\sqrt{3}} + \dfrac{1}{\sqrt{3}} \right) V_0$ $\tag{3.17a}$

or: $P^u = \dfrac{10\,kb}{\sqrt{3}} = 5.78\,kb$ $\tag{3.17b}$

The improvement of the previous solutions is only about 4%.

Different layouts of this five-triangle mechanism may be employed to reduce the upper bound. A partial minimization may be accomplished analytically by keeping all the triangles as equal isosceles triangles with $< ABC = < EBD = \theta$, then $P^u = 6\,kb$ when $\theta = \frac{1}{4}\pi$ and the best value of $P^u = 4\sqrt{2}\,kb$, when $\theta = \tan^{-1}\sqrt{2} \approx 55°$. The value of P is seen further decreased by only about 2% in this way.

It is clear from these calculations that the collapse or limit load corresponding to this five-triangle mechanism is not sensitive to the particular layout of the mechanism. In other words the minimum is rather a flat curve, and a wide variety of geometric layouts within the same mechanism would be expected to give only slightly different upper bounds.

(c) Remarks on the solutions. Although the upper bounds, $P^u = 6.28\,kb$, $6\,kb$, $5.78\,kb$ and $5.53\,kb$, are not extremely close to the actual answer $(2 + \pi)kb$ which will be discussed in the later part of this Chapter, they are obtained quickly and easily. Were it not for the fact that this foundation problem is so well known, any one of these upper bounds would be considered as providing useful information. Each can prove even more valuable when the soil being loaded is inhomogeneous and not so easily amenable to exact solution.

Before we move to consider Coulomb material which involves c and φ, we note first that in the upper-bound calculation the absolute value of the velocity diagram, and indeed the magnitude of the foundation velocity are immaterial. Further, the question may arise whether the five-triangle mechanism is really valid or kinematically admissible because the triangle in contact with the foundation must move into the soil in order to produce small displacement, since its lower edge labelled C is already in contact with the zone which remains stationary and rigid. The answer to this is that the energy dissipated at this edge tip is insignificant because if the edge tip is allowed to deform plastically, this plastically deformed area is of second order in comparison with the areas of interfaces and can therefore be neglected in the upper-bound calculations. Since the initial stresses and geometric imperfections such as a very small hole existing in the soil mass will not change the value of the collapse or limit load of this problem (Theorem III, Chapter 2), this difficulty can then be removed physically by inserting such a small hole at the edge C. We know that the surface of velocity discontinuity should always be regarded as a small but finite transition layer; the contradiction again disappears from this viewpoint. The magnitude of the incremental motion of the mechanism is in any case irrelevant to the calculation.

3.2.2. Narrow transition layer of Coulomb material

In Chapter 2 we have discussed the importance of the concept of normality or flow rule in limit analysis. Here, the kinematic implications of this concept as applied to Coulomb material will be re-examined in somewhat greater detail for the case of plane flow.

The perfectly plastic idealization of Coulomb yield criterion with associated flow rule implies that any plastic deformation must be accompanied by an increase in volume. This prediction of volume expansion applies only to the initial phase of the flow under a limiting state of equilibrium; it must not be interpreted as predicting that an indefinite increase of volume can be obtained by subjecting a mass of soil to repeated shear in opposite directions. This restriction of volume expansion to the initial phase of flow does not reduce its practical importance, however, because stability problems in soil mechanics are concerned only with this initial flow or its prevention.

Flow patterns where large masses of soil move as rigid bodies while deformation occurs only in narrow transition zones play an important role in soil mechanics as we have already demonstrated for the special case of Tresca material. If one mass of soil performs a translation with respect to another, as shown in Fig. 3.6, the transition zone between these masses is bounded by two planes which are parallel to each other. Plastic volume expansion during simple shearing demands that the tangential velocity change, δu, of the rigid top part of this block relative to the rigid bottom part must always be accompanied by a separation velocity, $\delta v = \delta u \tan \varphi$, for $\varphi \neq 0$. This separation behavior is extremely important since it makes the ideal plastic soil fundamentally different from that of Coulomb friction sliding for which the limit theorems, proved previously for assemblage of perfectly plastic bodies, do not always apply.

As was discussed in Chapter 2, the flow rule or normality concept requires that the tangential velocity change, δu, must be accompanied by the separation velocity, $\delta v = \delta u \tan \varphi$. This kinematic slip condition simply states that the relative velocity

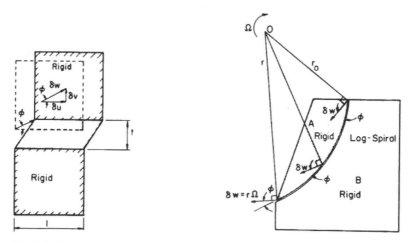

Fig. 3.6. Simple slip accompanied by a separation for $\varphi \neq 0$.

Fig. 3.7. Rigid body rotation of body A rotating with respect to body B.

change, δw, in a narrow transition zone bounded by two parallel planes must form the angle φ with the slip planes. More generally, if one mass of soil performs a rigid-body rotation with respect to another, the trace of the transition zone in a plane normal to the axis of rotation is bounded *not* by two concentric circles but rather by two logarithmic spirals which intersect the radii from the center of rotation under the constant angle $\frac{1}{2}\pi + \varphi$. This is illustrated in Fig. 3.7. The rigid top part of A is rotating with respect to the center O with an angular velocity Ω relative to the bottom part B, and the two parts are separated by a thin layer of logarithmic spiral. The relative velocity vector, δw, is seen making a constant angle φ with the narrow transition layer along the entire length.

In short, we should keep in mind that in upper-bound limit analysis, the familiar circular surface of discontinuity is *not* a permissible surface any more for a rigid body sliding because of the separation requirement for $c - \varphi$ soils. The *plane* surface and the *logarithmic-spiral* surface of angle φ, however, are the only two surfaces of velocity discontinuity which are permitted in limit analysis for rigid body motions relative to a fixed surface.

As was already remarked for the case of Tresca material, these kinematic relations lead to a curious result when we attempt to let the thickness of the transition layer approach zero. In the limit, the two masses of soil cannot remain in contact but must separate as they move with respect to each other. Of course, this merely means that the mathematical abstraction of an indefinitely thin transition layer is not physically acceptable. The replacement of a thin transition layer by a discontinuity surface in limit analysis is therefore purely a matter of convenience for practical use. It enables us to consider simple mechanisms consisting of rigid blocks sliding over each other.

Dissipation of energy (Fig. 3.6).

Let us now evaluate the rate of dissipation of energy, D, in this layer of plastic shearing. The mode of deformation in the transition layer shown in Fig. 3.6 is a combination of shear flow parallel to the layer with extension normal to it. The shear strain rate $\dot{\gamma}$ which is assumed to be uniform in the layer is equal to $\delta u/t$ and the normal strain rate $\dot{\epsilon}$ is equal to $\delta v/t$, so the rate of dissipation of energy is equal to $\tau\dot{\gamma} - \sigma\dot{\epsilon}$ per unit volume, τ and σ (here taken to be positive in compression) being the shear and normal stresses, respectively. The volume of the layer is numerically equal to t, so:

$$D = (\tau\dot{\gamma} - \sigma\dot{\epsilon})\, t = (\tau\delta u - \sigma\delta v)$$

or: $D = \delta u(\tau - \sigma \tan \varphi)$ [3.18]

Since the Coulomb yield criterion must be satisfied in the plastic layer it follows from [2.1] that:

$$D = c \, \delta u \qquad\qquad\qquad\qquad\qquad\qquad\qquad\qquad [3.19]$$

This equation states that the rate of dissipation of energy per unit area of discontinuous surfaces of the narrow transition layer of $c - \varphi$ soils is simply the product of the cohesion stress, c, and the tangential velocity change, δu, across the layer. Again, here, as for Tresca material, the expression is independent of the layer thickness, t, so t may be as small as we please, including zero as a matter of convenience.

Example 3.2: Critical height of a vertical cut

(a) Translational mechanism (Fig. 3.8). As an example of the application of this narrow transition layer of $c - \varphi$ soils, we obtain the critical height, H_{cr} of a vertical cut in a cohesive soil (Fig. 3.8). The unit weight of the soil is γ. The critical height is defined here as the height at which the unsupported vertical cut, as illustrated in the figure, will collapse due to its own weight.

We assume first that the failure occurs by sliding along a plane making an angle β with the vertical. A limiting condition is reached when the rate at which the gravity forces are doing work is equal to the rate of energy dissipation along the surface of sliding. The rate of work done by the gravity forces is the vertical component of the velocity multiplied by the weight of the soil wedge:

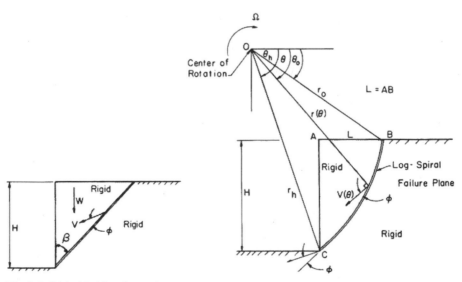

Fig. 3.8. Critical height of a vertical cut.

Fig. 3.9. Rotational failure mechanism for the critical height of a vertical cut.

$$\tfrac{1}{2}\gamma H^2 \tan \beta \, V \cos (\varphi + \beta) \qquad\qquad\qquad [3.20]$$

while the rate of energy dissipated along the discontinuity surface in [3.19] is:

$$c \frac{H}{\cos \beta} V \cos \varphi \qquad\qquad\qquad [3.21]$$

Equating the rate of external work to the rate of internal energy dissipation gives:

$$H = \frac{2c}{\gamma} \frac{\cos \varphi}{\sin \beta \cos (\varphi + \beta)} \qquad\qquad\qquad [3.22]$$

If β is minimized then:

$$\beta_{cr} = \tfrac{1}{4}\pi - \tfrac{1}{2}\varphi \qquad\qquad\qquad [3.23]$$

and: $\quad H_{cr} = \dfrac{4c}{\gamma} \tan (\tfrac{1}{4}\pi + \tfrac{1}{2}\varphi) \qquad\qquad\qquad [3.24]$

This is the same value obtained by the conventional Rankine analysis (limit equilibrium method). See for example Terzaghi (1943). This implies that the well-known Rankine solution is an upper bound, unless the solution is exact.

(b) Rotational Mechanism (Fig. 3.9). An improved upper-bound solution may be obtained by considering a rotational discontinuity (logarithmic spiral) instead of the translation discontinuity (plane surface) used above. Such an analysis requires an expression for the rate of dissipation of energy along a logarithmic spiral surface as well as an expression for the external rate of work done by the weight of the rotating soil mass. Since these expressions are useful for many applications in soil mechanics, details of the calculations are treated here as an illustrative example. The results will be taken up again in Chapter 9 when more complex slope stability problems are attempted.

The rotational discontinuity mechanism used here is shown in Fig. 3.9. The triangular-shaped region $A-B-C$ rotates as a rigid body about the center of rotation O (as yet undefined) with the materials below the logarithmic surface BC remaining at rest. Thus, the surface BC is a thin layer surface of velocity discontinuity. The assumed mechanism can be specified completely by three variables. For the sake of convenience, we shall select the slope angles θ_o and θ_h of the chords OB and OC, respectively and the height H of the vertical cut. Since the equation for the logarithmic spiral surface is given by:

$$r(\theta) = r_o \exp [(\theta - \theta_o) \tan \varphi] \qquad\qquad\qquad [3.25]$$

the length of the chord OC is:

$$r_h = r(\theta_h) = r_o \exp [(\theta_h - \theta_o) \tan \varphi] \qquad\qquad\qquad [3.26]$$

From the geometrical relations it can easily be shown that the ratios, H/r_o and L/r_o, may be expressed in terms of the angles θ_o and θ_h in the forms:

$$H = r_h \sin \theta_h - r_o \sin \theta_o \tag{3.27a}$$

$$\text{or:} \quad \frac{H}{r_o} = \sin \theta_h \exp [(\theta_h - \theta_o) \tan \varphi] - \sin \theta_o \tag{3.27b}$$

$$\text{and:} \quad L = r_o \cos \theta_o - r_h \cos \theta_h \tag{3.28a}$$

$$\text{or:} \quad \frac{L}{r_o} = \cos \theta_o - \cos \theta_h \exp [(\theta_h - \theta_o) \tan \varphi] \tag{3.28b}$$

A direct integration of the rate of external work due to the soil weight in the region $A-B-C$ is very complicated. An easier alternative is to use the method of superposition by first finding the rates of work, $\dot{W}_1$, $\dot{W}_2$ and $\dot{W}_3$ due to the soil weight in the regions $O-B-C$, $O-A-B$, and $O-A-C$, respectively. The rate of external work for the required region $A-B-C$ is then found by the simple algebraic summation, $\dot{W}_1 - \dot{W}_2 - \dot{W}_3$. We now proceed to compute the respective expressions for each of the three regions.

Considering first the logarithmic spiral region $O-B-C$, a differential element of the region is shown in Fig. 3.10(a). The rate of external work done by this differential element is:

$$d\dot{W}_1 = (\Omega \tfrac{2}{3} r \cos \theta) (\gamma \tfrac{1}{2} r^2 \, d\theta) \tag{3.29a}$$

integrating over the entire area, we obtain:

$$\dot{W}_1 = \tfrac{1}{3} \gamma \Omega \int_{\theta_o}^{\theta_h} r^3 \cos \theta \, d\theta = \gamma r_o^3 \Omega \int_{\theta_o}^{\theta_h} \tfrac{1}{3} \exp [3(\theta - \theta_o) \tan \varphi] \cos \theta \, d\theta \tag{3.29b}$$

$$\text{or:} \quad \dot{W}_1 = \gamma r_o^3 \Omega f_1(\theta_h, \theta_o) \tag{3.30}$$

where the function $f_1(\theta_h, \theta_o)$ is defined as:

$$f_1(\theta_h, \theta_o) =$$

$$\frac{\{(3 \tan \varphi \cos \theta_h + \sin \theta_h) \exp [3(\theta_h - \theta_o) \tan \varphi] - 3 \tan \varphi \cos \theta_o - \sin \theta_o \}}{3 (1 + 9 \tan^2 \varphi)} \tag{3.31}$$

Consider next the triangular region $O-A-B$ shown separately in Fig. 3.10(b). The rate of work done by the weight of the region is:

$$\dot{W}_2 = (\tfrac{1}{2} \gamma L r_o \sin \theta_o) [\tfrac{1}{3} (2r_o \cos \theta_o - L)] \Omega \tag{3.32}$$

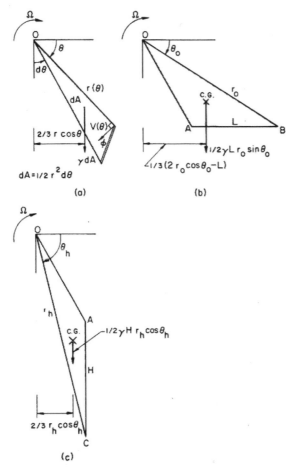

Fig. 3.10. Detail calculations for Fig. 3.9.

where the first bracket represents the total weight of the region and the other represents the vertical component of the velocity at the center of gravity of the region. The horizontal distance from the center of gravity to the vertical line passing through the point O is obtained by taking the mean horizontal distances of the points O, A and B. This is represented in the second bracket above. Rearranging the terms in [3.32], we obtain:

$$\dot{W}_2 = \gamma r_o^3 \Omega f_2(\theta_h, \theta_o) \qquad [3.33]$$

where the function $f_2(\theta_h, \theta_o)$ is defined as:

$$f_2(\theta_h, \theta_o) = \frac{1}{6} \frac{L}{r_o} \left(2 \cos \theta_o - \frac{L}{r_o} \right) \sin \theta_o \qquad [3.34]$$

and L/r_o is a function of θ_h and θ_o (see [3.28b]).

A similar technique can be used for the triangular area $O-A-C$ as shown separately in Fig. 3.10(c). It is found that:

$$\dot{W}_3 = \gamma r_o^3 \Omega f_3(\theta_h, \theta_o) \tag{3.35}$$

where the function $f_3(\theta_h, \theta_o)$ is defined as:

$$f_3(\theta_h, \theta_o) = \frac{1}{3}\frac{H}{r_o} \cos^2 \theta_h \exp\left[2(\theta_h - \theta_o)\tan\varphi\right] \tag{3.36}$$

and H/L_o is a function of θ_h and θ_o (see [3.27b]).

The magnitude of the rate of work done by the soil weight in the required region $O-B-C$ is now obtained by the simple algebraic summation:

$$\dot{W}_1 - \dot{W}_2 - \dot{W}_3 = \gamma r_o^3 \Omega(f_1 - f_2 - f_3) \tag{3.37}$$

The internal dissipation of energy occurs along the discontinuity surface BC (Fig. 3.9). The differential rate of dissipation of energy along the surface may be found by multiplying the differential area, $rd\theta/\cos\varphi$, of this surface by the cohesion c times the tangential discontinuity in velocity, $V\cos\varphi$, across the surface of discontinuity, [3.19]. The total internal dissipation of energy is then found by integration over the whole surface:

$$\int_{\theta_o}^{\theta_h} c(V\cos\varphi)\frac{rd\theta}{\cos\varphi} = \frac{cr_0^2\,\Omega}{2\tan\varphi}\left\{\exp\left[2(\theta_h - \theta_o)\tan\varphi\right] - 1\right\} \tag{3.38}$$

Equating the external rate of work, [3.37], to the rate of internal energy dissipation, [3.38], gives:

$$H = \frac{c}{\gamma} f(\theta_h, \theta_o) \tag{3.39}$$

where $f(\theta_h, \theta_o)$ is defined as:

$$f(\theta_h, \theta_o) = \frac{\left[\exp\{2(\theta_h - \theta_o)\tan\varphi\} - 1\right]\left[\sin\theta_h \exp\{(\theta_h - \theta_o)\tan\varphi\} - \sin\theta_o\right]}{2\tan\varphi(f_1 - f_2 - f_3)} \tag{3.40}$$

By the upper-bound theorem of limit analysis, [3.39] gives an upper bound for the critical value of the height. The function $f(\theta_h, \theta_o)$ has a minimum value when θ_h and θ_o satisfy the conditions:

$$\frac{\partial f}{\partial \theta_h} = 0 \quad \text{and} \quad \frac{\partial f}{\partial \theta_o} = 0 \tag{3.41}$$

Solving these equations and substituting the values of θ_h and θ_o thus obtained into

[3.39], we obtain a least upper bound for the critical height, H_{cr}, of the vertical cut.

To avoid lengthy computations, these simultaneous equations may be solved by a semi-graphic method. Details of the method will be presented in the later part of this chapter where various techniques for obtaining a critical answer are described. These equations will be taken up again there as an illustrative example.

The function $f(\theta_h,\theta_o)$ is found to have a minimum value near the point $\theta_o = 40°$, $\theta_h = 65°$ for the case $\varphi = 20°$, where it has the value $3.83 \tan(\frac{1}{4}\pi + \frac{1}{2}\varphi)$ for all values of φ, so that:

$$H_{cr} = \frac{3.83\,c}{\gamma} \tan\left(\tfrac{1}{4}\pi + \tfrac{1}{2}\varphi\right) \qquad\qquad [3.42]$$

is an upper bound for the critical height of the vertical cut. The value 3.83 in [3.42] is an improvement of the previous solution 4.0 as given in [3.24].

This is the same value obtained by Fellenius (1927) using the conventional limit equilibrium method. This implies that the Fellenius solution is an upper bound, unless the solution is exact.

3.3. INTERMIXING OF HOMOGENEOUS DEFORMING REGIONS AND RIGID BLOCK SLIDING

In real problems with complicated shapes and loading, the essential step in upper-bound limit analysis is to get any reasonable velocity pattern of mechanism for a solution. In the preceding section, rigid block sliding is the first choice. However, rigid block sliding is not always appealing nor is it always the best or most convenient choice. Intermixing of simple deforming regions and rigid block sliding offers far more scope. Homogeneous deformation in addition to rigid block sliding is presented here as an aid in bringing into play the feeling of the analyst for the likely deformation of the body he is studying. This will be followed in the next section by a separated presentation of some inhomogeneous deformation fields which are frequently used in soil mechanics.

3.3.1. Homogeneous deforming regions of Tresca material

The simplest homogeneous deforming fields are simple compression and simple shear flow. Shown in Fig. 3.11 are fields of homogeneous deformation denoted by the symbol $\dot{\epsilon}$ for a field of simple vertical compression and lateral expansion and by $\dot{\gamma}$ for simple shear. Since Tresca material is incompressible, volume is conserved in the plastic deformation. The lateral strain rate in the vertical compression mode must always be equal and opposite to the vertical strain rate (Fig. 3.11a).

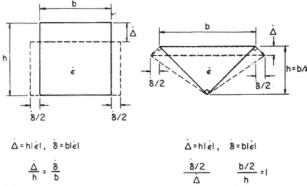

$\Delta = h|\dot{\epsilon}|, \quad \delta = b|\dot{\epsilon}|$ $\Delta = h|\dot{\epsilon}|, \quad \delta = b|\dot{\epsilon}|$

$\dfrac{\dot{\Delta}}{h} = \dfrac{\delta}{b}$ $\dfrac{\delta/2}{\Delta} \quad \dfrac{b/2}{h} = 1$

(a) Simple Vertical Compression and Lateral Expansion
    Deformation

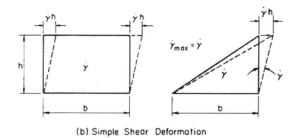

(b) Simple Shear Deformation

Fig. 3.11. Homogeneous deforming regions of Tresca material.

Dissipation of energy (Fig. 3.11)

Let us evaluate first the rate of dissipation of energy per unit volume in the field of homogeneous compression. The mode of deformation is instantaneously one of pure vertical compression and horizontal extension; the vertical strain rate $\dot{\epsilon}_y = |\dot{\epsilon}|$ and the horizontal strain rate $\dot{\epsilon}_x = -|\dot{\epsilon}|$, so the rate of dissipation of energy is equal to $\sigma_y |\dot{\epsilon}| - \sigma_x |\dot{\epsilon}|$ per unit volume, σ_y, σ_x being the corresponding principal stresses, in the plane of flow. Since the Tresca yield criterion must be satisfied in the plastically deformed region, the absolute value of the difference between σ_y and σ_x must be equal to $2k$, so:

$$D = |\dot{\epsilon}| (\sigma_y - \sigma_x) = 2k |\dot{\epsilon}| \qquad [3.43]$$

where $|\dot{\epsilon}|$ is the absolute value of the normal strain rate.

Since the flow in Fig. 3.11(b) is of simple shear, the shear stress in the simple shear flow of Tresca material must be the yield stress in pure shear, k. The rate of energy dissipation per unit volume of the simple shear flow is computed from:

$$D = k \dot{\gamma}_{max} \qquad [3.44]$$

where $\dot{\gamma}_{max}$ is the maximum rate of engineering shear strain.

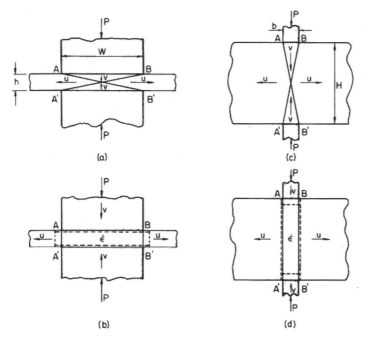

Fig. 3.12. Wide and narrow rigid punches.

Example 3.3: A strip squeezed between rigid punches

(a) Homogeneous deformation (Figs. 3.12b,d). Fig. 3.12(b,d) illustrates the use of a field of homogeneous deformation along with rigid block sliding in the well-studied problem of a strip squeezed between wide and narrow rigid punches. Certainly the field of homogeneous compression with accompanying lateral expansion in the region $A-B-B'-A'$, designated by $\dot{\epsilon}$ makes for a reasonable start. Volume is conserved in the plastic deformation; the lateral strain is equal and opposite to the vertical strain. Therefore:

$$u = v\,\frac{W}{h} \ \text{or} \ v\,\frac{b}{H} \qquad\qquad [3.45]$$

Energy is dissipated throughout the volume at the rate of the yield stress in compression ($2k$) times the strain rate in compression ($2v/h$ or $2v/H$) multiplied by the volume Wh or bH per unit dimension perpendicular to the paper (see [3.43]), making use of symmetry:

$$2k2vW \ \text{or} \ 2k2vb \qquad\qquad [3.46]$$

The rate of energy dissipation along the surfaces of discontinuity AA' and BB' is the yield stress in shear (k) multiplied by the relative velocity across the surfaces and integrated over the area of the surface or the length of the line of discontinuity of [3.1]:

$$2k \tfrac{1}{2} vh \quad \text{or} \quad 2k\tfrac{1}{2} vH \tag{3.47}$$

The relative velocity appears as $\tfrac{1}{2}v$, the average between the maximum value of v at the top and bottom of the line of discontinuity and the value of zero at the mid-height.

If the punches are smooth, there will be no dissipation at the interfaces, AB and $A'B'$. If the punches are rough, the additional rate of dissipation of energy is given by the average value of the relative velocity ($\tfrac{1}{2}u$) multiplied by the yield stress k in shear and by the total length of contact $2W$ or $2b$:

$$2k\tfrac{1}{2}uW = kvW^2/h \quad \text{or} \quad 2k\tfrac{1}{2}ub = kvb^2/H \tag{3.48}$$

For a smooth punch, the upper bounds are obtained by equating $2Pv$, the rate of work done by the external pair of forces P, to the total rate of dissipation:

$$P^u = 2kW + \tfrac{1}{2}kh \quad \text{or} \quad 2kb + \tfrac{1}{2}kH$$
$$= 2kW(1 + h/4W) \quad = 2kb(1 + H/4b) \tag{3.49}$$

For a rough punch:

$$P^u = 2kW\left(1 + \frac{h}{4W} + \frac{W}{4h}\right) \text{ or } \quad 2kb\left(1 + \frac{H}{4b} + \frac{b}{4H}\right) \tag{3.50}$$

It is no accident that the applied *stresses* $P/2kW$ or $P/2kb$ turn out to be equal for the rough punch when $W = H$ and $h = b$ as drawn. All-around hydrostatic pressure in the rectangular region $A-B-B'-A'$ has no effect on yield of a Tresca material governed by shear alone. Furthermore, expansion in the long direction and contraction in the short direction requires the same forces and dissipation as contraction in the long direction and expansion in the short direction.

(b) Rigid block sliding (Figs. 3.12a,c). The rigid block sliding pattern of Fig. 3.12(a) is similarly related to that of Fig. 3.12(c). There is no relative motion across AB, $A'B'$, AA' or BB' so that no distinction need to be made between rough and smooth punches. The result:

$$P^u = 2kW\left(\frac{W}{2h} + \frac{h}{2W}\right) \quad \text{or} \quad 2kb\left(\frac{H}{2b} + \frac{b}{2H}\right) \tag{3.51}$$

requires the computation of the relative velocity across the lines of discontinuity:

$$(v/h)\sqrt{(W^2 + h^2)} \quad \text{or} \quad (v/H)\sqrt{(H^2 + b^2)} \tag{3.52}$$

with the relation between u and v necessarily the same as before [3.45] because there is no volume change.

(c) Remarks on the solutions. Each of these simple fields of velocity with their simple discontinuities requires very little computational effort, but neither represents the actual field. In particular, the symmetry of behavior for the wide and the narrow rough punch is artificial. The region to the left of AA' and to the right of BB' in Fig. 3.12(d) will not remain rigid and so does not really correspond to the rigid punch regions above AB and below $A'B'$ in Fig. 3.12(b). As H/b becomes large, the two punches will not interact; a local solution, the indentation of a very large body loaded by a rigid punch will take over. Some upper-bound solutions to this important problem have been obtained in Example 3.1 using rigid block sliding mechanisms. The intermixing of inhomogeneous deforming regions and rigid block sliding, described in the next section, will give the "exact" indentation force, $(2 + \pi)kb$, for the semi-infinite block.

These two results for the narrow punch, together with the local solution for the semi-infinite block, are plotted in Fig. 3.13, together with the well-known slip-line solution by Hill (1950). The upper-bound solutions are not extremely close, in general, to Hill's slip-line solution which is considered to be the "exact" answer, but the diagram gives, nevertheless, a fairly clear impression of the part played by the thickness of the strip in the indentation behavior. It should be noted that when punches are rough, the $P/2kb$ vs. H/b curves in Fig. 3.13 for the narrow punch may be interpreted as the $P/2kW$ vs. W/h curves for the wide punch (solid lines only).

Before we move on to consider homogeneous deforming regions of Coulomb

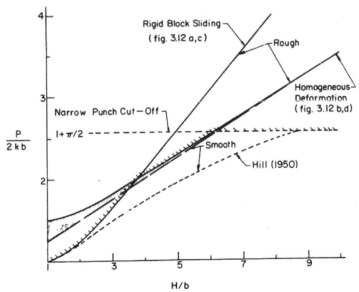

Fig. 3.13. Upper bounds for narrow punch. Note: When punches are rough, $P/2kb$ vs. H/b may be interpreted as $P/2kW$ vs. W/h (solid lines only).

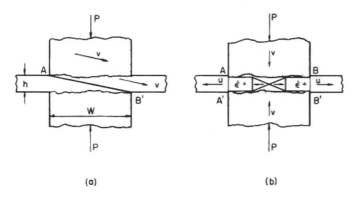

Fig. 3.14. Influence of irregularities in contact surface.

material, we note that the problem of a strip squeezed between rigid punches has been well-studied by the method of the slip-line field, a highly developed branch of continuum plasticity. The only point to be made here is that the field of homogeneous deformation is very convenient and physically satisfying as a first approximation. Interestingly enough, it does provide a rather good bound for the narrow punch as shown by the curves with hatching in Fig. 3.13. Extension of the technique of combining rigid block sliding and fields of homogeneous deformation to three dimensions is obvious. Three-dimensional bearing-capacity problems described in Chapter 7 will provide illustrative examples for such an extension.

Finally, it is worth pointing out that inhomogeneity and irregularity often dominate the picture, Fig. 3.14. This is especially true when continuum plasticity is applied on the microscale. Crude bounds obtained from block sliding on a single plane, Fig. 3.14(a), or combinations of rigid body sliding and homogeneous fields of strain rates, Fig. 3.14(b), are useful and will give a reasonably clear idea of the influence of irregularities.

3.3.2. Homogeneous deforming regions of Coulomb material

The extension of the above to include the more general case of $c - \varphi$ soils is evident. Shown in Fig. 3.15 is the counterpart of the fields of simple vertical compression and simple shear flow described previously in Fig. 3.11 for pure cohesive soils. Now a simple compression or a simple shear flow must always be accompanied by an increase in volume, i.e., volume is not conserved in the plastic deformation for $c - \varphi$ soils.

We shall proceed by re-examining the kinematic implications of the concept of the plastic flow rule or normality condition for the case of plane plastic flow. This was discussed generally in Chapter 2 and more specifically in the preceding section for the case of thin layer situations.

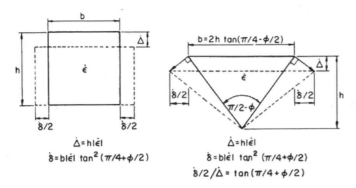

$\dot\Delta = h|\dot\epsilon|$

$\dot\delta = b|\dot\epsilon| \tan^2(\pi/4+\phi/2)$

$\dot\Delta = h|\dot\epsilon|$

$\dot\delta = b|\dot\epsilon| \tan^2(\pi/4+\phi/2)$

$\dot\delta/2 /\dot\Delta = \tan(\pi/4+\phi/2)$

(a) Simple Vertical Compression and Lateral Expansion Deformation

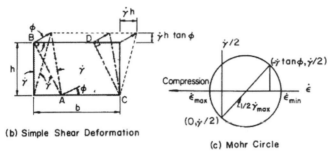

(b) Simple Shear Deformation

(c) Mohr Circle

Fig. 3.15. Homogeneous deforming regions of Coulomb material.

Let σ_{max}, σ_{min} denote the principal stresses in the plane of flow, and $\dot\epsilon_{max}$, $\dot\epsilon_{min}$ the principal plastic strain rates. Tresca's yield condition then requires that:

$$\sigma_{max} - \sigma_{min} - 2c = 0 \qquad [3.53]$$

The flow rule demands that, in an isotropic material, the principal axes of strain rate and stress coincide and that, moreover, the material is incompressible:

$$\dot\epsilon_{max} + \dot\epsilon_{min} = 0 \qquad [3.54]$$

More generally, when the internal friction angle, φ, is not zero, Coulomb's yield condition for $c - \varphi$ soils is:

$$\sigma_{max} (1 - \sin \varphi) - \sigma_{min} (1 + \sin \varphi) - 2c \cos \varphi = 0 \qquad [3.55]$$

The flow rule associated with the yield condition of [3.55] requires that:

$$\frac{\dot\epsilon_{max}}{\dot\epsilon_{min}} = - \frac{1 - \sin \varphi}{1 + \sin \varphi} = - \tan^2 (\tfrac{1}{4}\pi - \tfrac{1}{2}\varphi) \qquad [3.56]$$

or: $\dot\epsilon_{max} = - \dot\epsilon_{min} \tan^2 (45 - \tfrac{1}{2}\varphi)$ $\qquad [3.57]$

Dissipation of energy (Fig. 3.15).

Equation [3.57] implies that the field of homogeneous vertical compression deformation designated by $\dot{\epsilon} = \dot{\epsilon}_{max}$ is accompanied by the lateral tensile strain, $\dot{\epsilon}_{min} = -\dot{\epsilon}\tan^2(\frac{1}{4}\pi + \frac{1}{2}\varphi)$. Clearly, the lateral strain is greater than and opposite to the vertical strain $\dot{\epsilon}$. This is illustrated by a rectangular element and a triangular element of Fig. 3.15(a). The rate of energy dissipation, D, per unit volume of this field of homogeneous compression can now be computed from:

$$D = \sigma_{max}\,\epsilon_{max} + \sigma_{min}\,\epsilon_{min} = [\sigma_{max} - \sigma_{min}\,\tan^2(\tfrac{1}{4}\pi + \tfrac{1}{2}\varphi)]\,|\dot{\epsilon}| \qquad [3.58]$$

Since yield condition [3.55] must be satisfied in the deformation field, [3.58] is reduced to the simple form:

$$D = \frac{2c\cos\varphi}{1-\sin\varphi}\,|\dot{\epsilon}| = 2c\,|\dot{\epsilon}|\,\tan(\tfrac{1}{4}\pi + \tfrac{1}{2}\varphi) \qquad [3.59]$$

Once again it is worth pointing out that plastic dissipation is always positive, so that sign conventions on strains can be dispensed with and the vertical compressive strain, $\dot{\epsilon}$, above written in terms of absolute value.

Let us now evaluate the rate of dissipation of energy, D, per unit volume of the simple shear flow shown in Fig. 3.15(b). This simple shear can be easily visualized as a series of narrow transition layers of the type discussed in the preceding section. Each of these layers is bounded by horizontal parallel straight lines and corresponding to a relative translation of the adjacent masses of soil. The simple shear deformation designated by $\dot{\gamma}$ is accompanied by the vertical normal strain rate, $\dot{\gamma}\tan\varphi$. The rate of dissipation of energy is equal to $\tau\dot{\gamma} - \sigma\dot{\gamma}\tan\varphi$ per unit volume, τ and σ being the shear and normal stresses, respectively. Since the Coulomb yield criterion must be satisfied in the field of flow, it follows that:

$$D = c\dot{\gamma} \qquad [3.60]$$

It should be noted that the shear strain rate, $\dot{\gamma}$, shown in Fig. 3.15(b) is not the maximum shear strain rate, $\dot{\gamma}_{max}$, in the field of simple shear deformation but is related by $\dot{\gamma} = \dot{\gamma}_{max}\cos\varphi$. This is the consequence of volume expansion accompanied by plastic shearing deformation. The Mohr circle shown in Fig. 3.15(c) indicates clearly the relationship between $\dot{\gamma}$ and $\dot{\gamma}_{max}$.

The fields of simple compression deformation and simple shear deformation are of course not entirely distinct, rather they are interrelated through Mohr's circle. It follows that the internal rate of dissipation of energy for the field of simple compression deformation [3.59], should be derivable directly from the field of simple shear deformation [3.60]. This can be shown in the following way. Referring to Fig. 3.15(c), let the maximum compressive strain rate, $\dot{\epsilon}_{max}$, of the simple shear field be equal to the vertical compressive strain rate, $\dot{\epsilon}$, of the simple compressive field. From the Mohr circle and using the relation $\dot{\gamma} = \dot{\gamma}_{max}\cos\varphi$, we have:

$$\epsilon_{max} = |\dot{\epsilon}| = \tfrac{1}{2}\,\dot{\gamma}_{max} - \tfrac{1}{2}\,\dot{\gamma}\tan\varphi = \tfrac{1}{2}\,\dot{\gamma}\,\frac{1-\sin\varphi}{\cos\varphi} \qquad\qquad [3.61]$$

$$\text{or:}\quad \dot{\gamma} = 2\,|\dot{\epsilon}|\tan(\tfrac{1}{4}\pi + \tfrac{1}{2}\varphi) \qquad\qquad [3.62]$$

This relation between $\dot{\gamma}$ and $\dot{\epsilon}$ demonstrates clearly that although [3.59] and [3.60] are very different in form, they give essentially the same rate of dissipation of energy per unit volume.

Fig. 3.16 shows a number of examples of differently shaped homogeneous shearing zones. Fig. 3.16(a) can be easily visualized as a series of transition layers of the type discussed above. Each of these layers is parallel to the base AC and bounded by parallel straight lines corresponding to a relative translation of the adjacent mass of soil. Fig. 3.16(a) may also be visualized as a part of Fig. 3.15(b) as shown by the dashed and dotted lines in the figure. Fig. 3.16(b) is the half field of homogeneous deformation of Fig. 3.16(a), while a proper rigid body rotation of Fig. 3.16(b) results in the interesting field of Fig. 3.16(c).

Example 3.4: Critical height of a vertical cut with soil unable to take tension
In the laboratory, soil may exhibit the ability to resist tension. In the field,

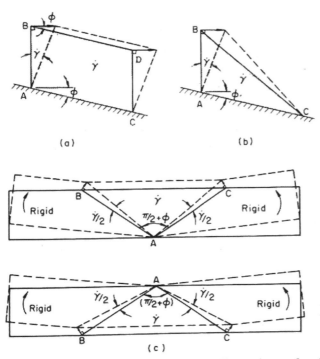

Fig. 3.16. Homogeneous shearing zone with different shapes of mode.

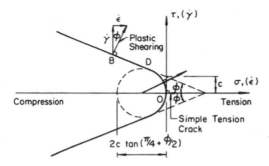

Fig. 3.17. Modified Coulomb criterion with zero-tension cut-off.

however, the presence of water or tensile cracks near the surface may destroy the tensile strength of the soil. Hence, the tensile strength of soil is not reliable and it may be neglected. This is a conservative idealization. The Coulomb yield criterion is modified by the tension cut-off as shown in Fig. 3.17, in which the requirement of zero-tension is met by the circle termination as shown (the upper half of the yield curve is *ODB*).

As the soil is unable to resist tension, the introduction of a tensile crack in a failure mechanism is permissible. No energy is dissipated in the formation of a simple tension crack; both normal and shear stress are zero on the plane of separation (see the origin in Fig. 3.17).

The rotational mechanism containing a simple-tension crack and a homogeneous shearing zone is shown in Fig. 3.18. Failure due to tipping over of the soil "slab" of thickness Δ about point A with an angular velocity ω is possible. The region

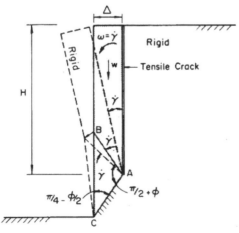

Fig. 3.18. Rotational mechanism containing a simple tension crack and the homogeneous shearing zone of Fig. 3.16 for soil unable to take tension.

$A-B-C$ of homogeneous shearing, γ, is the field shown in Fig. 3.16b which indicates that $\omega = \dot{\gamma}$. Equation [3.60] then gives $D = c\dot{\gamma} = c\omega$. The total rate of dissipation of internal energy for unit dimension perpendicular to the paper is just D times the area of the triangle ABC or:

$$(c\omega) \; [\Delta^2 \tan \left(\tfrac{1}{4}\pi + \tfrac{1}{2}\varphi\right)] \tag{3.63}$$

The rate of external work done by gravity is the weight of the soil moving downward as the "slab" rotates about A, multiplied by the velocity, which is $\omega(\tfrac{1}{2}\Delta)$:

$$W \approx \gamma\Delta H \quad \text{or} \quad \tfrac{1}{2}\gamma\Delta^2 H\omega \tag{3.64}$$

If the rate of external work is equated to the dissipation, and Δ allowed to approach zero, it yields:

$$H_{cr} = \frac{2c}{\gamma} \tan \left(\tfrac{1}{4}\pi + \tfrac{1}{2}\varphi\right) \tag{3.65}$$

This confirms Terzaghi's solution (1943) for a tensile crack extending the full height of the bank.

3.4. INTERMIXING OF INHOMOGENEOUS DEFORMING REGIONS AND RIGID BLOCK SLIDING

Two types of inhomogeneous deformation fields involving straight and curved failure lines are most frequently encountered in the applications: (1) one family of failure lines consists of concurrent straight lines, and the other of concentric circles; and (2) one family consists of concurrent straight lines, the other of logarithmic spirals. Case (1) is the familiar radial shear zone and case (2) is usually termed the logspiral zone for $c - \varphi$ soils. When such a field borders on a region which remains at rest, the border line must be a failure line. Here, we shall assume that one of the circular arcs or logarithmic spirals will be the border line.

In the following the rate of dissipation of energy for these two types of inhomogeneous fields is derived. Two alternative derivations in terms more familiar to the engineers are presented here. More formal derivation for the dissipation function corresponding to a general deformation field will be presented at the end of this Chapter.

3.4.1. Radial shear zone of Tresca material (c = k, φ = 0)

Velocity field
An approximation to this zone is given in Fig. 3.19(a) where a picture for six

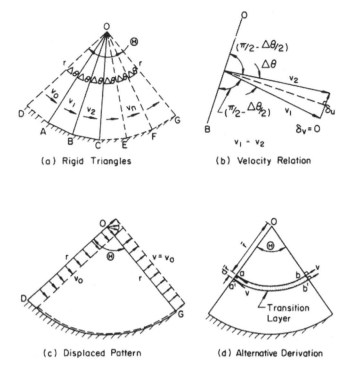

(a) Rigid Triangles

(b) Velocity Relation

(c) Displaced Pattern

(d) Alternative Derivation

Fig. 3.19. Radial shear zone for pure cohesive soil ($c = k$, $\varphi = 0$).

rigid triangles at an equal central angle $\Delta\theta$ to each other is shown. Energy dissipation takes place along the radial lines OA, OB, OC, etc. due to the discontinuity in velocity between the triangles. Energy also is dissipated on the discontinuous surface $DABCEFG$ since the material below this surface is considered at rest. Since the material must remain in contact with the surface $DABCEFG$ the triangles must move parallel to the arc surfaces. Also the rigid triangles must remain in contact with each other so that the compatible velocity diagram of Fig. 3.19(b) shows that each triangle of the mechanism must have the same speed $V_1 = V_2 = V_n = V$.

Rate of dissipation of energy

With [3.1], the rate of dissipation of energy can easily be calculated. The energy dissipation along the radial line OB, for example, is the cohesion $c = k$ multiplied by the relative velocity, δu, and the length of the line of discontinuity:

$$kr\left(2V\sin\frac{\Delta\theta}{2}\right) \qquad\qquad [3.66]$$

in which the relative velocity δu appears as $2V \sin \frac{1}{2}\Delta\theta$. Similarly, the energy dissipation along the discontinuous surface AB is:

$$k\left(2r\sin\frac{\Delta\theta}{2}\right)V \tag{3.67}$$

where the length of AB is $(2r\sin\frac{1}{2}\Delta\theta)$ and $\delta u = V$. Since the energy dissipation along the radial line OB is the same as along the arc surface AB, it is natural to expect that the total energy dissipation in the zone of radial shear, DOG, with a central angle Θ will be identical with the energy dissipated along the arc DG. This is evident since Fig. 3.19(a) becomes closer and closer to the zone of radial shear as the number of n grows. In the limit when n approaches infinity, the zone of radial shear is recovered. The total energy dissipated in the zone of radial shear is the sum of the energy dissipated along each radial line when the number n approaches infinity:

$$\lim_{n\to\infty} n\left[2krV\sin\frac{\Theta}{2n}\right] = 2krV\lim_{n\to\infty} n\sin\frac{\Theta}{2n} = kV(r\Theta) \tag{3.68}$$

where $\Delta\theta = \Theta/n$.

Indicated in Fig. 3.19(c) is the displaced pattern of the radial shear zone. This figure shows the small displacement of the field which would result if the initial velocity along OD was maintained for a short period of time. In the zone, the velocity along every radial line is constant in the direction perpendicular to the radial line. In obtaining this diagram, the material below the line DG is assumed to be at rest. The initial position of the zone is indicated by solid lines in the figure.

An alternative derivation for the dissipation of energy in the radial shear zone can also be obtained by visualizing the field of flow as a series of transition layers. Each of these transition layers is bounded by concentric circular arcs and each of these arcs is rotating as a rigid body about the center of rotation O with an angular velocity, $V/\bar{r}$, for example, for the arc surface ab or $V/(\bar{r}+d\bar{r})$ for the arc surface $a'b'$. See Fig. 3.19(d). With [3.1], the differential rate of dissipation of energy along the layer surface is found by multiplying the difference in arc length between ab and $a'b'$, i.e., $\Theta d\bar{r}$, of this layer by k times the tangential velocity, V across the layer. The total internal dissipation of energy is then found by integration over the entire radial line:

$$\int_0^r kV\Theta d\bar{r} = kVr\Theta \tag{3.69}$$

agreeing with the value obtained in [3.68].

3.4.2. Logspiral shear zone of Coulomb material (c − φ soils)

Velocity field
The extension of the above to include the more general case of a logspiral shear

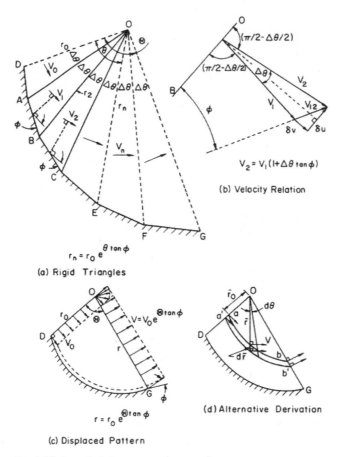

$$r_n = r_0 e^{\theta \tan \phi}$$

(a) Rigid Triangles

$$V_2 = V_1 (1 + \Delta \theta \tan \phi)$$

(b) Velocity Relation

$$r = r_0 e^{\theta \tan \phi}$$

(c) Displaced Pattern

(d) Alternative Derivation

Fig. 3.20. Logspiral shear zone of $c - \varphi$ soils.

zone for $c - \varphi$ soils is evident. Now a simple slip δu must always be accompanied by a separation $\delta v = \delta u \tan \varphi$ while there is no need for such a separation when the shear strength of a soil is due only to the cohesion. A picture of six rigid triangles at an equal angle $\Delta \theta$ to each other is shown in Fig. 3.20(a) and the corresponding compatible velocity diagram for the two typical triangles AOB and BOC is examined (Fig. 3.20b). If the central angle $\Delta \theta$ is sufficiently small, one may write:

$$V_1 = V_0 (1 + \Delta \theta \tan \varphi)$$

$$V_2 = V_1 (1 + \Delta \theta \tan \varphi)$$

$$V_n = V_{n-1} (1 + \Delta \theta \tan \varphi) \qquad \text{[3.70a]}$$

and from these relations, the velocity in the n-th triangle OEF is:

$$V_n = V_0(1 + \Delta\theta \tan \varphi)^n \qquad\qquad [3.70b]$$

where V_0 is the initial velocity.

Clearly, the logspiral shear zone is recovered as a limiting case when the number of the rigid triangles grows to infinity. Then, in the limit as $n \to \infty$, [3.70b] becomes:

$$V_0(1 + \Delta\theta \tan \varphi)^n = V_0 \left(1 + \frac{\theta \tan \varphi}{n}\right)^n \to V_0\, e^{\theta \tan \varphi} \qquad\qquad [3.71]$$

or: $V = V_0\, e^{\theta \tan \varphi}$

in which V = the velocity at any angular location, θ, along the spiral.

Fig. 3.20(c) shows the displaced position of the logspiral shear zone which would result if the soil moved with the initial velocity V_0 for a short period of time. Since a thin narrow transition layer is taken between the logspiral line DG and the material below at rest, the velocity along this line must therefore be everywhere inclined at an angle φ to the line. The velocity along every radial line of the field is constant and in the direction perpendicular to it and its magnitude is increased exponentially from the value, V_0, along the line OD to the value $V_0\exp(\Theta \tan \varphi)$ along the line OG.

Rate of dissipation of energy

With [3.19] the rate of energy dissipation along the radial line, say, OB is:

$$cr_2(V_1 \Delta\theta) \qquad\qquad [3.72]$$

in which δu appears as $V_1 \Delta\theta$. Similarly, the dissipation along the spiral surface AB is:

$$c\left(\frac{r_2\Delta\theta}{\cos \varphi}\right)(V_1 \cos \varphi) \qquad\qquad [3.73]$$

in which the length of $AB = (r_2\Delta\theta/\cos \varphi)$ and $\delta u = V_1 \cos \varphi$. Again the dissipation along a radial line is the same as along the spiral surface segment provided that the central angle $\Delta\theta$ is small. Thus, the expression for energy dissipation in the logspiral shear zone will be identical with the expression along the spiral surface which can easily be obtained by integrating [3.73] along the spiral surface $r = r_0\exp(\theta \tan \varphi)$:

$$c \int rV d\theta = c \int_0^\Theta (r_0\, e^{\theta \tan \varphi})\,(V_0\, e^{\theta \tan \varphi})\, d\theta$$

$$= \tfrac{1}{2}cV_0 r_0 \cot \varphi\, (e^{2\Theta \tan \varphi} - 1) \qquad\qquad [3.74]$$

The rate of dissipation of energy in the logspiral shear zone can *also* be determined in a very similar manner to that used above for the radial shear zone of a

Tresca material. The field of flow is now visualized as a series of transition layers bounded by concentric logarithmic spirals and each of these spirals is rotating as a rigid body about the center of rotation O with an angular velocity, $V/\bar{r}$, for the spiral surface ab or $V/(\bar{r} + d\bar{r})$ for the spiral surface $a'b'$ (Fig. 3.20d).

Clearly, from this visualization, the velocity along every spiral surface must increase exponentially along the surface. With [3.19], the differential rate of dissipation of energy between the surfaces ab and $a'b'$ is:

$$\int_0^\Theta c(V\cos\varphi) \frac{(\bar{r} + d\bar{r})\,d\theta}{\cos\varphi} - \int_0^\Theta c(V\cos\varphi) \frac{\bar{r}d\theta}{\cos\varphi} = \int_0^\Theta c(V\cos\varphi) \frac{d\bar{r}\,d\theta}{\cos\varphi} \qquad [3.75]$$

The total internal dissipation of energy is then found by integration over the entire radial line:

$$\int_0^r \int_0^\Theta c(V\cos\varphi) \frac{d\theta\,d\bar{r}}{\cos\varphi} = c \int_0^\Theta V\,r d\theta \qquad [3.76]$$

agreeing with the expression obtained in [3.74].

Example 3.5: Bearing capacity of a strip footing on $c - \varphi$ soils

In Example 3.1, we have studied the problem of bearing capacity of a strip footing resting on a pure cohesion soil (for which $\varphi = 0$). In this example the inhomogeneous fields developed above are applied to obtain the bearing capacity of a strip footing resting on a $c - \varphi$ soil. Here we attempt not only to demonstrate the use of the inhomogeneous fields described above, but also to give an indication on the relative importance of the role which the weight of the soil and the roughness of the foundation play in the load carrying capacity calculations. Some of the results will be referred to and used again in Chapters 6 and 7 when more complex two- and three-dimensional bearing capacity problems are attempted.

(a) Bearing capacity calculations neglecting weight of soil. We shall start by neglecting the weight of the soil. The two possible failure mechanisms which will give the same value for the bearing capacity of weightless soil, are illustrated in Fig. 3.21(a) and 3.22(a). They correspond to the two slip-line solutions proposed by Prandtl (1921) and by Hill (1950) for the same problem in a perfectly plastic material.

Hill mechanism (Fig. 3.22). Considering first the solution represented by Fig. 3.22(a), and since it is symmetric about the axis of the footing, we need discuss only the left-hand plastic flow regions. Plastic flow is confined to the region above the line of velocity discontinuity *OCDE*. The mechanism consists of a rigid triangular wedge, *AOC*, with base angles $\frac{1}{4}\pi + \frac{1}{2}\varphi$, a logspiral shear zone, *ACD*, of central

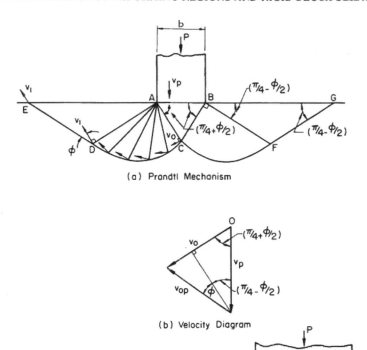

Fig. 3.21. Bearing capacity calculation based on Prandtl mechanism.

angle $\frac{1}{2}\pi$, and a wedge, ADE, with base angles $\frac{1}{4}\pi - \frac{1}{2}\varphi$. We take the downward velocity of the footing as V_p so that along AB the downward component of the velocity of the soil must be V_p. The soil below the failure line $OCDE$ remains at rest so that $OCDE$ is a line of velocity discontinuity. Thus the velocity along this line is everywhere inclined at an angle φ to the line.

Since the triangular wedge AOC moves as a rigid body in the direction perpendicular to AC and if V_0 is the velocity of the wedge, then the compatibility of velocity along AO requires that (Fig. 3.22b):

$$V_0 = V_p \sec \left(\tfrac{1}{4}\pi + \tfrac{1}{2}\varphi \right)$$ [3.77]

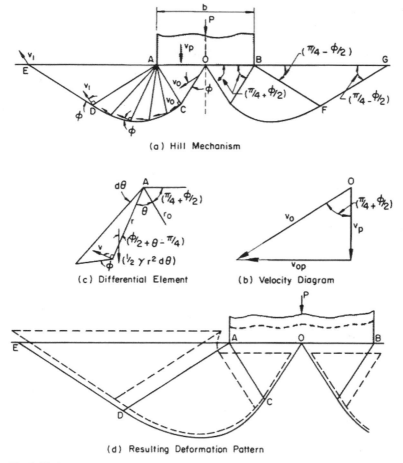

(a) Hill Mechanism

(c) Differential Element   **(b) Velocity Diagram**

(d) Resulting Deformation Pattern

Fig. 3.22. Bearing capacity calculation based on Hill mechanism.

In the logspiral shear zone ACD, we know that the velocity must increase exponentially from the value V_0 along the radial line AC to the value:

$$V_0 \, e^{(\tan \varphi)\frac{1}{2}\pi} \tag{3.78}$$

along AD. Finally, the triangular wedge ADE moves as a rigid body with the constant velocity:

$$V_1 = V_0 \, e^{(\tan \varphi)\frac{1}{2}\pi} = V_p \sec \left(\tfrac{1}{4}\pi + \tfrac{1}{2}\varphi\right) e^{(\frac{1}{2}\pi)\tan \varphi} \tag{3.79}$$

The compatible velocity diagram for the Hill mechanism is shown in Fig. 3.22(b). The small arrows in Fig. 3.22(a) represent the velocity field. Figure 3.22(d) shows the displaced position of the Hill mechanism which would result if the footing moved with the downward velocity V_p for a short period of time.

Energy is dissipated in the discontinuity surface between the material at rest and the material moving above the line $OCDE$, and also in the logspiral shear zone ACD. The rate of dissipation of energy due to the discontinuity surfaces OC and DE is equal to the areas of OC or DE multiplied by $cV_0 \cos \varphi$ or $cV_1 \cos \varphi$, respectively, since the changes in tangential velocity across these two surfaces have the constant values $V_0 \cos \varphi$ or $V_1 \cos \varphi$, respectively. The rate of dissipation of energy in the logspiral zone ACD is identical to the rate of dissipation of energy along the logspiral curve, and [3.74] can be used directly for both cases.

Equating the total rate of internal dissipation of energy to the external rate of work done by the footing load P, for half of the Hill mechanism gives (expressing all the velocities in terms of V_0):

$$\tfrac{1}{2} P V_0 \cos \left(\tfrac{1}{4}\pi + \tfrac{1}{2}\varphi\right) = c \left(V_0 \cos \varphi\right) \left[\frac{b}{4 \cos \left(\tfrac{1}{4}\pi + \tfrac{1}{2}\varphi\right)}\right]$$

$$+ c \left[V_0 \cos \varphi \, e^{\left(\tfrac{1}{2}\pi\right) \tan \varphi}\right] \left[\frac{b \, e^{\left(\tfrac{1}{2}\pi\right) \tan \varphi}}{4 \cos \left(\tfrac{1}{4}\pi + \tfrac{1}{2}\varphi\right)}\right] + \frac{c \, V_0 \, b \cot \varphi}{4 \cos \left(\tfrac{1}{4}\pi + \tfrac{1}{2}\varphi\right)} \left(e^{\pi \tan \varphi} - 1\right) \quad [3.80]$$

The last term in the equation above represents the rates of dissipation of energy in the logspiral zone ACD and along the logspiral curve CD. This term is twice the value given by [3.74]. Collecting terms of [3.80], we have:

$$\frac{P^{u}}{b} = c \cot \varphi \left[e^{\pi \tan \varphi} \tan^2 \left(\tfrac{1}{4}\pi + \tfrac{1}{2}\varphi\right) - 1\right] \quad [3.81]$$

agreeing with the solution obtained previously by the method of constructing the slip-line field.

Prandtl Mechanism (Fig. 3.21a). The Prandtl mechanism shown in Fig. 3.21(a) consists of a triangular wedge, ABC, with base angles $\tfrac{1}{4}\pi + \tfrac{1}{2}\varphi$ moving downwards as a rigid body with the velocity of the footing, V_p, a logspiral shear zone, ACD, of central angle $\tfrac{1}{2}\pi$, and a rigid wedge, ADE, with base angles $\tfrac{1}{4}\pi - \tfrac{1}{2}\varphi$. The upper-bound solution for this mechanism can be obtained in an analogous manner. In this case, the lines AC and BC in addition to the lines CDE and CFG are lines of velocity discontinuity. Referring to the left-hand side of Fig. 3.21(a), the soil below the failure line CDE remains at rest so that the velocity along this line must be everywhere inclined at an angle φ to the line. The velocity of the soil, V_0, just to the left of the discontinuity line AC is perpendicular to AC and its magnitude must be such that the change in velocity, V_{op}, across AC is inclined at an angle φ to AC. By drawing the compatibility velocity diagram shown in Fig. 3.21(b), we see that this velocity, V_0, must have the magnitude:

$$V_0 = \tfrac{1}{2} V_p \sec \left(\tfrac{1}{4}\pi + \tfrac{1}{2}\varphi\right) \quad [3.82]$$

In the logspiral shear zone ACD, the velocity increases exponentially to the value:

$$V_1 = V_0\, e^{(\frac{1}{2}\pi)\,\tan\varphi} = \tfrac{1}{2}\, V_p\, \sec\left(\tfrac{1}{4}\pi + \tfrac{1}{2}\varphi\right)\, e^{(\frac{1}{2}\pi)\,\tan\varphi} \qquad [3.83]$$

on the line AD. This triangular wedge ADE moves as a rigid body in the direction perpendicular to AD with the velocity V_1. The small arrows in Fig. 3.21(a) represent the velocity field. Figure 3.21(c) shows the displaced position of the Prandtl mechanism which would result if the footing moved with the downward velocity V_p for a short period of time.

We notice that the velocity of the soil along AE in the Prandtl mechanism is exactly half the value of the corresponding velocity in the alternative Hill mechanism and the velocities are inclined at the same angle, $\tfrac{1}{4}\pi + \tfrac{1}{2}\varphi$, to the surface AE. For a given width, AB, of the footing, however, the length of AE in Fig. 3.21(a) is twice the length of AE in Fig. 3.22(a). It follows that the volume of soil raised above the undisturbed surface after the footing has penetrated a small distance is the same in both mechanisms.

Equating internal and external rates of energy for half the Prandtl mechanism gives (expressing all the velocities in terms of V_0):

$$\tfrac{1}{2}P\, 2V_0\, \cos\left(\tfrac{1}{4}\pi + \tfrac{1}{2}\varphi\right) = c(V_0 \cos\varphi)\left[\frac{b}{2\cos\left(\tfrac{1}{4}\pi + \tfrac{1}{2}\varphi\right)}\right]$$

$$+ c\left[V_0\, e^{(\frac{1}{2}\pi)\,\tan\varphi} \cos\varphi\right]\left[\frac{b\, e^{(\frac{1}{2}\pi)\,\tan\varphi}}{2\cos\left(\tfrac{1}{4}\pi + \tfrac{1}{2}\varphi\right)}\right] + \frac{c\, V_0\, b\, \cot\varphi}{2\cos\left(\tfrac{1}{4}\pi + \tfrac{1}{2}\varphi\right)}\, (e^{\pi\tan\varphi} - 1) \quad [3.84]$$

Collecting terms gives [3.81], obtained previously for the Hill mechanism. For the particular case of Tresca material for which $c = k$ and $\varphi = 0$, the maximum bearing capacity, P, as given by [3.81] reduces to the value $(2 + \pi)kb$. This value agrees with the well-known "exact" slip-line field solution.

In the Prandtl mechanism, there is no slip between the soil and the footing, which can be considered rough, and therefore the upper-bound solution so obtained is applicable to either a smooth or a rough footing.

(b) Effect of foundation friction in weightless soil. Before we move on to consider the weight of soil in the calculations, we note that the Prandtl mechanism contains a rigid region which acts as an extension of the footing while the Hill mechanism assumes zero-friction and appreciable slip does take place. Both mechanisms give the same answer as [3.81]. Shield (1954a) has shown that by extending satisfactorily the plastic stress field associated with the Prandtl mechanism into the remaining rigid regions below the line $EDCFG$ (Fig. 3.21a), [3.81] is also a lower bound for each φ less than $75°$. This extension will be given later in section 6.6.1, Chapter 6. Therefore, the bearing capacity pressure, P/b, of [3.81] is *exact* for all possible *finite sliding friction*. This follows directly from the intuitively obvious facts that the load which produces collapse for infinite friction (no relative motion) at the

frictional interface as given by the Prandtl mechanism will produce collapse for the case of finite friction (Theorem VII, Chapter 2) and the load which is safe for zero-friction at the frictional interface as demonstrated by Shield will be safe with any values of friction (Theorem VIII, Chapter 2). The formal proof of these facts was recognized and stated in Chapter 2 as *frictional* theorems of limit analysis by Drucker (1954a). Occasionally they enable the collapse or limit load to be computed precisely for finite non-zero friction.

(c) Bearing capacity calculations considering weight of soil. The unit weight of soil has been neglected in the above calculations in order to simplify the analysis. We know that the self-weight of the soil may play an important part in many applications. Its neglect can sometimes cause misleading conclusions. In the following we attempt not only to show how the weight of the soil can be included in the upper-bound calculations, but also to give an indication that some rather different conclusions may result when the weight of the soil is neglected.

As an example, the calculation of the collapse or limit load for a *smooth* surface footing resting on a *cohesionless* soil ($c = 0$) is given as follows.

Hill Mechanism (Fig. 3.22). Considering first the Hill mechanism, Fig. 3.22(a), the rate at which work is done by the soil weight in the rigid triangular wedges is found by multiplying the area of each rigid triangular wedge by γ times the vertical component of the velocity of the rigid body and summing over all the areas in motion. Taking the left half of the symmetrical velocity field, the rate of work done in the triangular wedge under the footing AOC is:

$$\gamma(\tfrac{1}{2}r_0^2 \cos \varphi) \left[V_0 \cos (\tfrac{1}{4}\pi + \tfrac{1}{2}\varphi) \right] \tag{3.85}$$

in which r_0 denotes the length of AC, and in the triangular wedge under the free surface ADE:

$$-\gamma(\tfrac{1}{2}r_0^2 \cos \varphi \, e^{\pi \tan \varphi}) \left[V_0 \cos (\tfrac{1}{4}\pi - \tfrac{1}{2}\varphi) \, e^{(\tfrac{1}{2}\pi) \tan \varphi} \right] \tag{3.86}$$

The rate of work done in the logspiral shear zone ACD is found by considering first a differential element of the zone as shown in Fig. 3.22(c). The rate of work done by this differential element is:

$$-\gamma(\tfrac{1}{2}r^2 d\theta) \left[V \cos (\tfrac{1}{4}3\pi - \tfrac{1}{2}\varphi - \theta) \right] \tag{3.87}$$

Integrating over the entire area, we obtain:

$$\tfrac{1}{2}\gamma \int_0^{\tfrac{1}{2}\pi} r^2 V \sin (\tfrac{1}{4}\pi - \tfrac{1}{2}\varphi - \theta) \, d\theta = \tfrac{1}{2}\gamma \, r_0^2 \, V_0 \int_0^{\tfrac{1}{2}\pi} e^{3\theta \tan \varphi} \sin (\tfrac{1}{4}\pi - \tfrac{1}{2}\varphi - \theta) \, d\theta$$

$$= \frac{-\gamma \, r_0^2 \, V_0}{2(1 + 9 \tan^2 \varphi)} \left\{ [3 \tan \varphi \sin (\tfrac{1}{4}\pi + \tfrac{1}{2}\varphi) - \cos (\tfrac{1}{4}\pi + \tfrac{1}{2}\varphi)] \, e^{(\tfrac{1}{2}3\pi) \tan \varphi} \right.$$

$$+ \left. [3 \tan \varphi \cos (\tfrac{1}{4}\pi + \tfrac{1}{2}\varphi) + \sin (\tfrac{1}{4}\pi + \tfrac{1}{2}\varphi)] \right\} \tag{3.88}$$

The total rate of external work done is then obtained by adding the additional work done by the weight of the soil to the rate of work done by the footing load P:

$$\tfrac{1}{2}P\,V_0\cos\left(\tfrac{1}{4}\pi + \tfrac{1}{2}\varphi\right) \tag{3.89}$$

and by substituting b, the width of the footing, for r_0 using:

$$b = 4r_0\cos\left(\tfrac{1}{4}\pi + \tfrac{1}{2}\varphi\right) \tag{3.90}$$

and by equating the total external rate of work so obtained to zero, since the total rate of internal dissipation of energy previously calculated when the weight of soil was neglected is zero for a cohesionless soil ($c = 0$), we have:

$$P^{\mathrm{u}} = \gamma\tfrac{1}{2}b^2 N_\gamma \tag{3.91}$$

where the dimensionless bearing capacity coefficient, N_γ, is defined as:

$$N_\gamma = \tfrac{1}{4}\tan\left(\tfrac{1}{4}\pi + \tfrac{1}{2}\varphi\right)\left[\tan\left(\tfrac{1}{4}\pi + \tfrac{1}{2}\varphi\right)e^{\left(\frac{1}{2}3\pi\right)\tan\varphi} - 1\right]$$

$$+ \frac{3\sin\varphi}{1 + 8\sin^2\varphi}\left\{\left[\tan\left(\tfrac{1}{4}\pi + \tfrac{1}{2}\varphi\right) - \frac{\cot\varphi}{3}\right]e^{\left(\frac{1}{2}3\pi\right)\tan\varphi} + \tan\left(\tfrac{1}{4}\pi + \tfrac{1}{2}\varphi\right)\frac{\cot\varphi}{3} + 1\right\}$$

$$\tag{3.92}$$

This expression for N_γ is of the same form as that given by Terzaghi (1943). This expression depends only on the value of φ, and some numerical values for several values of φ are shown in Table 3.1. For comparison, values of N_γ calculated by the slip-line method of Sokolovskii (1965) are also shown and close agreement will be noted. The fact that Sokolovskii's coefficients are greater than the collapse values at large φ is to be noted and is probably due to his simplified stress field, described in chapter 4 of his book. Details of Sokolovskii's method in obtaining various slip-line field solutions are given in his book entitled "Statics of Granular Media". Further discussions on this subject will be taken up again in Chapter 6 when general bearing capacity problems of strip footing are attempted.

TABLE 3.1

Variation of N_γ with φ (Hill mechanism)

φ	N_γ				
	0	10°	20°	30°	40°
Collapse load, [3.92]	0	0.72	3.45	15.20	81.79
Sokolovskii (1965)	0	0.56	3.16	15.30	86.50

Prandtl Mechanism (Fig. 3.21). The upper-bound solution considering the weight of soil for the Prandtl mechanism can be obtained in a very similar manner to that used above for the Hill mechanism in a cohesionless soil. Further, this important bearing capacity problem will be discussed fully in Chapter 6. Here we shall not give the details. The only point to be made here is that if the Prandtl failure mechanism is used as in Fig. 3.21, the calculated collapse or limit load is exactly *twice* the value of Hill mechanism in Fig. 3.22 or [3.92]. This is in contrast to the calculated collapse loads for a cohesive weightless soil for which the Hill and Prandtl mechanisms yield the *same* value.

(d) Effect of foundation friction considering weight of soil

Hill Mechanism (Fig. 3.22). The fact that the bearing capacity of the foundation, N_γ, calculated for the condition of no slip across the foundation (Prandtl mechanism) is much greater than that for a perfectly smooth foundation (Hill mechanism) in a cohesionless soil considering the weight of soil is quite significant. It shows that foundation friction has very significant effect on the collapse load for a granular, *self-weight* soil which is in contrast to the calculated collapse load for a strongly cohesive *weightless* soil for which the Hill and Prandtl mechanisms give the same value. Therefore, if the Hill mechanism is used as in Fig. 3.22, the rate of dissipation of energy due to friction between the foundation and the soil should be taken into account in the computation of the collapse load. The rate of dissipation of energy due to friction may be computed by multiplying the discontinuity in velocity:

$$V_0 \sin \left(\tfrac{1}{4}\pi + \tfrac{1}{2}\varphi\right) \tag{3.93}$$

across the slip surface by $\tan \delta$ (δ = foundation friction angle) times normal force P acting on this surface. This rate of dissipation of energy is now equated to the previous calculated external rates of work done by the weight of soil and the footing load when the foundation base is assumed smooth. It is found that the value of N_γ is increased by the factor:

$$\frac{1}{1 - \tan \delta \, \tan \left(\tfrac{1}{4}\pi + \tfrac{1}{2}\varphi\right)} \tag{3.94}$$

over that for the smooth footing. Using $\delta = 0.2$ and $\varphi = 30°$, the foundation friction is found to increase the value of N_γ by about 53% (e.g., $N_\gamma = 23.2$). For $\varphi = 30°$ the present value of $N_\gamma = 23.2$ may be compared with the semi-graphical solution, $N_\gamma = 22.5$, obtained by Meyerhof (1951) using the conventional limit equilibrium method. The upper-bound solution based on the Hill mechanism considering the foundation friction agrees therefore within 3% in this case. The problem of shallow and deep foundations will be presented in Chapter 6.

Strictly speaking, the theorems of limit analysis as described in the preceding Chapter (section 2.6) are not applicable in general to any process in which energy is dissipated by friction. Nevertheless, for a soil, there is a strong temptation to ignore this consideration and to compute the upper bounds as discussed above. The upper bound so obtained is an upper bound for a friction material in general (Theorem X, p. 44), but lower bounds are of a less definite meaning (Drucker, 1954a). The results do provide useful information, if not the full answer.

Prandtl mechanism (Fig. 3.21). The upper-bound values of N_γ based on the Prandtl mechanism, Fig. 3.21(a) can be improved or reduced by assuming angles $(\frac{1}{4}\pi + \frac{1}{2}\varphi)$ and $(\frac{1}{4}\pi - \frac{1}{2}\varphi)$ of the chords AC and AD to be the undefined angles ξ and η, respectively. Then function $N_\gamma(\xi,\eta)$ has a minimum value when the values of ξ and η satisfy the conditions:

$$\frac{\partial N_\gamma}{\partial \xi} = 0 \quad \text{and} \quad \frac{\partial N_\gamma}{\partial \eta} = 0 \qquad\qquad [3.95]$$

Solving these equations and substituting the values of ξ and η thus obtained into function $N_\gamma(\xi,\eta)$ yields a least upper bound for the bearing capacity of the foundation based on Prandtl mechanism. Details of the minimization procedure will be presented in the following section where various techniques for obtaining critical answers are described. The solution of these equations will be taken up again there as an illustrative example.

The function $N_\gamma(\xi,\eta)$ is found to have a minimum value 26.62 for $\varphi = 30°$ when ξ and $\eta \approx 46°$ and $30°$, respectively. It is worth pointing out that the condition $\partial N_\gamma/\partial \eta = 0$ always implies $\eta = \frac{1}{4}\pi - \frac{1}{2}\varphi$.

The above minimization process for the Hill mechanism can be obtained in an analogous manner. Details of these solutions are given in Chapter 6.

3.4.3. Radial shear zone of Coulomb material (c – φ soil)

Velocity field

A circular shearing zone for a $c - \varphi$ soil with finite internal friction can be developed in a manner similar to that used above for the logspiral radial shearing zone. Consider a sector of a circle with central angle Θ to be composed to a series of n rigid triangles each of angle $\Delta\theta$, as shown in Fig. 3.23(a). The velocity vector of each triangle is directed at an angle φ to the discontinuous rigid boundary $DABCEFG$ as required by the associated flow rule idealization described earlier. Figure 3.23(b) shows the compatible velocity diagram for triangles AOB and BOC. It should be noted that the discontinuous velocity V_{12} also makes an angle φ with the line OB. This vector is shown to be composed of components δu and δv parallel and perpendicular to the discontinuity OB. Thus δu is the simple slip velocity, while δv is a separation velocity. Assuming the central angle $\Delta\theta$ is sufficiently small, we may write:

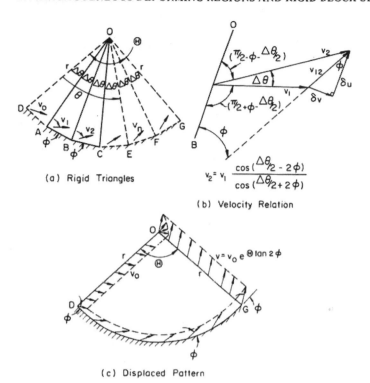

(a) Rigid Triangles

(b) Velocity Relation

(c) Displaced Pattern

Fig. 3.23. Radial shear zone for $c - \varphi$ soils.

$$V_1 = V_0 \frac{\cos \left(\frac{1}{2}\Delta\theta - 2\varphi\right)}{\cos \left(\frac{1}{2}\Delta\theta + 2\varphi\right)}$$

$$V_2 = V_1 \frac{\cos \left(\frac{1}{2}\Delta\theta - 2\varphi\right)}{\cos \left(\frac{1}{2}\Delta\theta + 2\varphi\right)}$$

.   .   .

.   .   .

.   .   .

$$V_n = \frac{V_{n-1} \cos \left(\frac{1}{2}\Delta\theta - 2\varphi\right)}{\cos \left(\frac{1}{2}\Delta\theta + 2\varphi\right)}$$ [3.96]

The velocity in the n-th triangle OEF can be expressed as:

$$V_n = V_0 \left[\frac{\cos \left(\frac{1}{2}\Delta\theta - 2\varphi\right)}{\cos \left(\frac{1}{2}\Delta\theta + 2\varphi\right)}\right]^n$$ [3.97]

where V_0 is the initial zone velocity. The circular radial shearing zone will be obtained in the limit as the number of triangles grows to infinity. Eq. [3.97] can be written as:

$$V_0 \left[\frac{\cos \left(\frac{1}{2} \Delta\theta - 2\varphi \right)}{\cos \left(\frac{1}{2} \Delta\theta + 2\varphi \right)} \right]^n = V_0 \left[\frac{\cos \left(\frac{\theta}{2n} - 2\varphi \right)}{\cos \left(\frac{\theta}{2n} + 2\varphi \right)} \right]^n = V_0 \left[\frac{1 + \tan \frac{\theta}{2n} \tan 2\varphi}{1 - \tan \frac{\theta}{2n} \tan 2\varphi} \right]^n \quad [3.98]$$

Now if $n \to \infty$ we obtain the limit:

$$\lim_{n \to \infty} \left(1 + \frac{\theta \tan 2\varphi}{n} \right)^n \to e^{\theta \tan 2\varphi} \qquad [3.99]$$

or: $V = V_0 \, e^{\theta \tan 2\varphi}$ [3.100]

where V is the velocity at any location θ along the circular arc. Eq. [3.100] is similar to the one previously derived for a log-spiral zone, [3.71]. Noting that they are different only by the factor 2.

Figure 3.23(c) shows the displaced position of the radial shear zone for the $c - \varphi$ soils which would result if the soil moved with the initial velocity V_0 for a short period of time. The velocity of any radial line of the zone is uniform. The velocity vector is no longer perpendicular to the radial line as in the previous two cases but rather makes an angle φ with a normal to the radial line. The magnitude of the velocity along the radial line is increased exponentially from the value V_0, along the line OD to the value $V_0 \exp(\Theta \tan 2\varphi)$ along the line OG.

Rate of dissipation of energy

From Fig. 3.23(a) it is clear that energy will be dissipated along every radial line and also along the boundary surfaces.

The rate of energy dissipation along a typical radial line, say OE, can be found by multiplying the cohesion c, discontinuity length r, and discontinuous tangential velocity $\delta u = V_{n-1,n} \cos \varphi$:

$$c r \, \delta u \qquad [3.101]$$

Using Fig. 3.23(b) and assuming $\Delta\theta$ to be small, we find:

$$\delta u = \frac{V_{n-1} \, \Delta\theta \, \cos \varphi}{\cos 2\varphi} \qquad [3.102]$$

and [3.101] can be written as:

$$\frac{c r V_{n-1} \, \Delta\theta \, \cos \varphi}{\cos 2\varphi} = \frac{c r V_0 \, \Delta\theta \, \cos \varphi \, e^{\theta \tan 2\varphi}}{\cos 2\varphi} \qquad [3.103]$$

Integrating over the total circular radial shear zone Θ:

$$\frac{crV_0 \cos \varphi}{\sin 2\varphi} (e^{\Theta \tan 2\varphi} - 1) \qquad\qquad [3.104]$$

Likewise, the dissipation along a typical boundary surface, CE, is given by:

$$c[2r \sin \tfrac{1}{2}\Delta\theta] \ V_{n-1} \cos \varphi \qquad\qquad [3.105]$$

which, if $\Delta\theta$ is small, becomes:

$$crV_{n-1} \Delta\theta \cos \varphi = crV_0 \Delta\theta \cos \varphi \, e^{\theta \tan 2\varphi} \qquad\qquad [3.106]$$

Integrating over the total length $DABCEFG$:

$$\frac{crV_0 \cos \varphi}{\tan 2\varphi} (e^{\Theta \tan 2\varphi} - 1) \qquad\qquad [3.107]$$

The corresponding total radial and boundary dissipation expressions for the log-spiral shearing zone are both equal and given by [3.74]. It is noted that for the circular zone case this equality does *not* occur.

Examples for intermixing of such a deforming zone with rigid block sliding are illustrated in Chapters 6 and 8. Interestingly enough, in some cases it does provide a better upper-bound solution than the familiar logspiral zone.

3.5. EVALUATION OF THE MINIMUM SOLUTION FOR AN ASSUMED MECHANISM

By upper-bound theorem of limit analysis, we know that the solution obtained from the work equation described above gives an upper bound for the collapse or limit load. We have already applied the method of differential calculus to some simple examples in order to seek the best or the most critical layout of an assumed mechanism. We shall now consider further the methods by which the most critical layout of a particular mechanism can be conveniently found. Here we shall apply the methods to some nontrivial examples.

Let us consider first the two-dimensional bearing capacity problem of concrete blocks or rock shown in Fig. 3.24(a). The strip loading P is applied on width $2a$, to a block of thickness H and width $2b$. The two-dimensional mechanism consists of a rigid wedge region, ABC, of angle 2α and a simple tension crack, CD, perpendicular to the smooth (frictionless) base. The wedge moves downward as a rigid body and displaces the surrounding material sideways. The relative velocity vector, δw, at each point along the lines of discontinuity, AC and BC is inclined at angle φ to

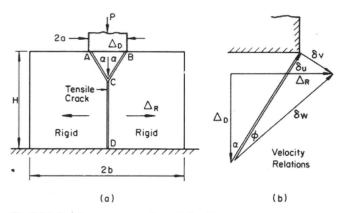

Fig. 3.24. Bearing capacity under strip loading.

these lines. The compatible velocity field with parameter α is shown in Fig. 3.24(b), from which the rate of internal dissipation of energy along the surfaces of discontinuity can be calculated. Details of the analysis for such a material are presented in Chapter 10. The upper bound is found to be a function of the tensile strength, f_t', and the compressive strength, f_c':

$$\frac{P^u}{2a} = f_t' \frac{\left(\dfrac{f_c'}{f_t'}\right)\left(\dfrac{1 - \sin \varphi}{2}\right) + \sin (\alpha + \varphi)\left(\dfrac{H}{a} \sin \alpha - \cos \alpha\right)}{\sin \alpha \cos (\alpha + \varphi)} \qquad [3.108]$$

There are two ways in which the minimum value of this algebraic equation can be found;

(1) By the use of *semi-graphical method*, the value of α for which $P/2a$ is a minimum can be determined. Essentially when using this technique, the minimum value is found by first assuming a set of numerical values for the material and dimensional constants, f_c'/f_t', φ and H/a. This is then followed by substituting several values for α into this equation, corresponding $P/2a$ values can then be formed, and a graph can be plotted and thus the approximate minimum value found.

(2) By the direct use of *differential calculus*:

$$\frac{\partial(P/2a)}{\partial \alpha} = 0 \qquad [3.109]$$

which gives the condition when the tangent to the $(P/2a)$ vs. α curve has zero slope. The critical value of α is determined from [3.109] from which the value of $P/2a$ gives a minimum.

Before we proceed to discuss the relative merits of these two algebraic techniques, it is worth noting that aside from its possible complexity of the algebra, the differential calculus method cannot always be used. The reason for this is that because of the physical conditions imposed on the variable parameters, there may not exist an absolute minimum value within the valid limits of the assumed mechanism. Such a case exists for the bearing capacity problem shown in Fig. 3.24(a). The upper-bound solution for this pattern of mechanism is obtained algebraically in terms of the variable α in [3.108]. However, if:

$$\alpha < \tan^{-1} \frac{a}{H} \qquad\qquad [3.110]$$

the geometry of the pattern changes from the general shape shown in Fig. 3.24(a) to those shown in Fig. 3.25, and the upper-bound solution [3.108] is no longer valid for this pattern. The layout shown in Fig. 3.25 is, in fact, not really a *valid* mechanism of the rigid block sliding type described above.

Since the minimum value of [3.108] can be found either semi-graphically or analytically, the best technique in general probably depends on the complexity of the algebra and the range of the material and dimensional constants with which the answer is to be sought. If the material and dimensional constants are limited to a typical set such as, for example, the combinations of $f_c'/f_t' = 10$, $\varphi = 30°$ and $H/a = 10, 20$, usually the semi-graphic technique is most suitable, especially when the equation is complex in algebra. However, if a general solution applicable to all sizes of body with various frequently occurring material constants, it is probably worthwhile to use the analytical differential calculus to obtain an algebraic solution which can be used again and again, when different sizes or material constants are used. The example shown in Fig. 3.24 is now used to demonstrate the technique of the differential calculus. This will then be followed by a separate illustration of the semi-graphic technique. Finally, an alternative to the *algebraic* technique, the *arithmetic* technique is presented.

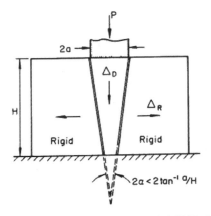

Fig. 3.25. Invalid mechanism of rigid block sliding.

3.5.1. Differential calculus method

The algebraic equation [3.108] is of the general form:

$$\frac{P}{2a} = f_t' \left[\frac{g_1(\alpha)}{g_2(\alpha)} \right]$$ [3.111]

Now this equation is a minimum when [3.109] is satisfied; but:

$$\frac{\partial}{\partial \alpha} \left(\frac{P}{2af_t'} \right) = \frac{\partial}{\partial \alpha} \left[\frac{g_1(\alpha)}{g_2(\alpha)} \right] = \frac{g_2(\alpha) \dfrac{\partial g_1(\alpha)}{\partial \alpha} - g_1(\alpha) \dfrac{\partial g_2(\alpha)}{\partial \alpha}}{[g_2(\alpha)]^2}$$ [3.112]

If the right-hand side of this equation is equated to zero and the terms rearranged, the upper bound, $P/2af_t'$, has a minimum value when:

$$\frac{g_1(\alpha)}{g_2(\alpha)} = \frac{\dfrac{\partial g_1(\alpha)}{\partial \alpha}}{\dfrac{\partial g_2(\alpha)}{\partial \alpha}}$$ [3.113]

This equation will give the critical value of α corresponding to the minimum value of $P/2af_t'$, and substitution of this value of α into [3.111] will give the most critical solution corresponding to this particular pattern of mechanism. Experience has shown, however, that the original algebraic equation [3.111] will be simplified and thus the algebraic work will be easier if one substitutes into:

$$\frac{P}{2a} = f_t' \frac{\dfrac{\partial g_1(\alpha)}{\partial \alpha}}{\dfrac{\partial g_2(\alpha)}{\partial \alpha}}$$ [3.114]

which can be seen by observing [3.113] to be equal to [3.111].

The general description of this technique is now best demonstrated by taking the specific functions $g_1(\alpha)$ and $g_2(\alpha)$ given by [3.108]:

$$\frac{g_1(\alpha)}{g_2(\alpha)} = \frac{\left(\dfrac{f_c'}{f_t'} \right) \left(\dfrac{1 - \sin \varphi}{2} \right) + \sin(\alpha + \varphi) \left(\dfrac{H}{a} \sin \alpha - \cos \alpha \right)}{\sin \alpha \cos(\alpha + \varphi)}$$

$$\frac{\dfrac{\partial g_1(\alpha)}{\partial \alpha}}{\dfrac{\partial g_2(\alpha)}{\partial \alpha}} = \frac{\sin (\alpha + \varphi)\left(\dfrac{H}{a}\cos \alpha + \sin \alpha\right) + \cos (\alpha + \varphi)\left(\dfrac{H}{a}\sin \alpha - \cos \alpha\right)}{-\sin \alpha \sin (\alpha + \varphi) + \cos \alpha \cos (\alpha + \varphi)}$$

$$= \frac{H}{a} \tan (2\alpha + \varphi) - 1 \qquad\qquad [3.115]$$

which leads to

$$(\cot \alpha - \tan \varphi)^2 \cos^2 \varphi = 1 + \frac{\dfrac{H}{a}\cos \varphi}{\left(\dfrac{f_c'}{f_t'}\right)\left[\dfrac{1 - \sin \varphi}{2}\right] - \sin \varphi} \qquad\qquad [3.116]$$

The only sensible solution of [3.116] is:

$$\cot \alpha_c = \tan \varphi + \sec \varphi \left\{1 + \frac{\dfrac{H}{a}\cos \varphi}{\left(\dfrac{f_c'}{f_t'}\right)\left[\dfrac{1 - \sin \varphi}{2}\right] - \sin \varphi}\right\}^{\frac{1}{2}} \qquad\qquad [3.117]$$

As noted previously we could substitute the critical value of α_c given by [3.117] into [3.108] to find the least upper-bound value of $P^u/2a$, but via [3.114] and [3.115] the original upper-bound equation [3.108] reduces to:

$$\frac{P^u}{2a} = f_t'\left[\frac{H}{a}\tan (2\alpha_c + \varphi) - 1\right] \qquad\qquad [3.118]$$

We should of course check that the critical value of α_c is valid for the mechanism we have postulated. Thus, from [3.110], providing:

$$\tfrac{1}{2}\pi > \alpha_c > \tan^{-1}\frac{a}{H} = \alpha_{min} \qquad\qquad [3.119]$$

The value is acceptable.

The significance of the condition [3.119] perhaps needs a little explanation. All it means is that when the H/a ratios are small, the stationary minimum value of [3.117] at which the slope tangent to the $(P/2a)$ vs. α curve is zero, occurs at a value of $\alpha_c < \alpha_{min}$. Since α must not be smaller than α_{min}, for the upper-bound solution to be valid the value of $P/2a$ we seek is that corresponding to $\alpha = \alpha_{min}$. This is a minimum value for the postulated mechanism but not a stationary minimum. This is best illustrated by taking the typical values of the material constants:

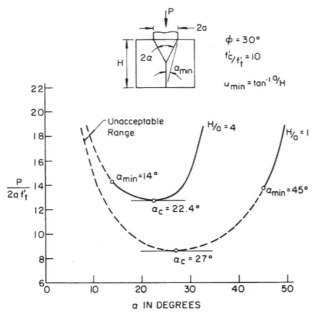

Fig. 3.26. Physically acceptable and unacceptable stationary minimum values.

$f'_c/f'_t = 10$ and $\varphi = 30°$. If we draw the curves of [3.108] when $H/a = 4$ and 1, respectively, the curves will be of the typical shape shown in Fig. 3.26. For the case of $H/a = 4$, there is a valid stationary minimum. However, when $H/a = 1$ the value of this stationary minimum is not physically acceptable.

3.5.2. Semi-graphical method

The complexity in algebra and tediousness in differentiations involved by the method of differential calculus may limit its practical use for some algebraic expressions. However, these difficulties may be considerably reduced or even eliminated altogether by the use of the semi-graphical method. In the following, this method will be demonstrated by taking the function [3.40] which has been associated with the calculation of the critical height of a vertical cut (example 3.2).

Example 3.2 (continued)

The function f given by [3.40] for example 3.2 has two variables θ_o and θ_h and one material constant φ. In order to determine the minimum value of this function by means of a semi-graphical method, we assume first a particular value of $\varphi = 20°$. Then we assign a set of values for $\theta_h = 40°$, $45°$, $50°$, ... For every value of θ_h, we substitute several values of θ_o into [3.40] and compute the corresponding values of the function f. If four or five points have been determined, the f vs. θ_o curve can be

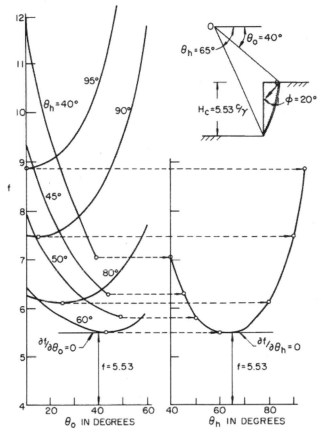

Fig. 3.27. Semi-graphic method for example 3.2.

traced easily and accurately. The f vs. θ_o curves are shown on the left-hand side of Fig. 3.27. Thus we obtain the point corresponding to the minimum value of each of these curves. Then we trace through each of these points marked by the small open circles in the figure a horizontal line to obtain the corresponding f vs. θ_h curve shown on the right-hand side of Fig. 3.27. The minimum value of f vs. θ_h curve represents the absolute minimum of the function [3.40]. The geometry corresponding to this critical value of f or height is shown as the inset in Fig. 3.27.

As far as valid values of θ_o and θ_h are concerned, the only restriction is that the value of θ_o shall not be greater than θ_h. This condition gives the minimum values for f when $\theta_h = 40°$, $45°$ and $50°$, but they are not the stationary minimum in the sense of differential calculus and hence differentation, even if attempted for in such a case, will not help us to find the value directly for these particular values of θ_h. The semi-graphic method has the further advantage that the engineer is compelled by the procedure to have at every stage of his computations and plottings a clear mental picture of what he is doing. In connection with the practical application of soil mechanics this advantage has considerable weight.

3.5.3. Combination of differential calculus method and semi-graphic method

In the preceding two examples in which the minimum solution has been found we used fully either the differential process or the graphical process, but a combination of both methods, each is most suitable in certain parts of a problem, may offer far more scope or convenience. In the following we will illustrate the use of a combination of differential calculus along with the grapical technique in the well-studied problem of bearing capacity of strip footing on $c - \varphi$ soils (example 3.5). Here we shall only consider the case of the upper-bound values of N_γ based on the Prandtl mechanism (Fig. 3.21a). It was pointed out in example 3.5 that the upper-bound solution corresponding to Prandtl mechanism in cohesionless self-weighted soil can be improved by assuming that the two angles $(\frac{1}{4}\pi + \frac{1}{2}\varphi)$ and $(\frac{1}{4}\pi - \frac{1}{2}\varphi)$ of the chords AC and AD shown in Fig. 3.21(a) to be the undefined angles ξ and η, respectively. Then function $N_\gamma(\xi,\eta)$ will have a minimum value when the values of ξ and η satisfy the conditions given by [3.95]. Referring therefore to Fig. 3.21(a) with the two undefined angles ξ and η, it is found, after some simplification, that the non-dimensional bearing capacity coefficient, N_γ is given by:

$$N_\gamma = \frac{1}{2} \frac{\cos(\xi-\varphi)}{\cos^2\xi \, \cos\varphi} \left[\frac{3\tan\varphi\cos\xi + \sin\xi}{1 + 9\tan^2\varphi} - \frac{\sin\xi\cos\xi\cos\varphi}{\cos(\xi-\varphi)} \right.$$
$$\left. + \left(\frac{\sin\eta\cos\eta\cos\varphi}{\cos(\eta+\varphi)} + \frac{3\tan\varphi\cos\eta - \sin\eta}{1 + 9\tan^2\varphi} \right) e^{3(\pi - \xi - \eta)\tan\varphi} \right] \qquad [3.120]$$

No particular difficulty exists with this equation using either of the two methods, but the main point of interest is that the differentiation of this equation with respect to the variable ξ is much more complex than that of the variable η, since η appears only in the last two terms of the above equation. The first step therefore is to differentiate [3.120] with respect to η, following the usual differential calculus procedure. It is found, after some simplification, that the condition that will give the value of η corresponding to the minimum value of N_γ is:

$$\cos 2\eta + \sin\eta\cos\eta \, [\tan(\eta+\varphi) - 3\tan\varphi] - \cos(\eta+\varphi)\cos\eta\sec\varphi = 0 \qquad [3.121]$$

independently of the variable ξ. Equation [3.121] can further be reduced to:

$$\tan\varphi\tan 2\eta = 1 \qquad [3.122]$$

the solution of which is:

$$\eta = \frac{1}{4}\pi - \frac{1}{2}\varphi \qquad [3.123]$$

TABLE 3.2

Minimum values of $N_\gamma(\varphi = 30°)$, [3.120]

ξ	15°	30°	45°	60°
N_γ	35.4	28.9	26.6	30.4

Since the complexity of algebra involved in differentiating [3.120] with respect to the variable ξ, we now adopt the semi-graphical procedure. Thus, assuming $\varphi = 30°$ and using [3.123] and by substitution of the various values of ξ into [3.120], N_γ has the values given in Table 3.2.

It is fairly obvious even without plotting that N_γ has a minimum value of approximately 26.6.

3.5.4. Arithmetic method

As an alternative to the fully or partially algebraic tehcniques described above, one can start by simply drawing several particular layouts corresponding to an assumed mechanism. The upper-bound solution is then obtained directly in terms of lengths measured from the diagram of a particular layout. Such an arithmetically obtained upper-bound solution can be repeated for several different layouts and the critical or minimum value can be picked out either visually, or from a graphical plot. This method is now demonstrated by the following example.

Example 3.6: A purely cohesive soil block with five holes

Consider a purely cohesive soil block with five holes arranged as in Fig. 3.28(a). The dimension perpendicular to the plane of the paper will be taken as unity, but all motion is supposed in the plane. The applied force P is carried through the rectangular blocks with five holes in a very elaborate pattern from the rigid top loading plate to the rigid bottom supporting base. Since this problem will be treated numerically and graphically, we take definite proportions in terms of the unity as shown. A mechanism or a velocity pattern as indicated in Fig. 3.28(a) is obtained by letting one part of the block slide on two sets of planes perpendicular to the plane of the paper and through the three holes on a 45°-plane as shown in Fig. 3.28(a). With [3.1], the internal dissipation of energy is c times the area of the surface of sliding multiplied by the discontinuity velocity, δu, or:

$$(32\sqrt{2} - 2 \times 2\sqrt{2} - 2 \times 2 - 4)c\delta u \qquad [3.124]$$

The rate at which work is done by the external force is $P\delta u/\sqrt{2}$. This leads to:

$$P^u = \sqrt{2}(28\sqrt{2} - 8)c = 44.7c \qquad [3.125]$$

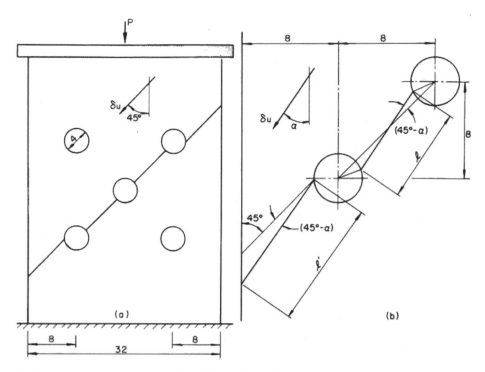

Fig. 3.28. Velocity fields for a block with five holes.

This result may be improved by changing the angle of sliding, α, slightly, as shown in Fig. 3.28(b). Although $l' + l$ is increased, $\delta u \cos \alpha$ likewise becomes larger. By trial of three different values of $\alpha = 40°$, $35°$ and $30°$, and measuring the lengths l', l directly from the diagram, P^u has the values given in Table 3.3.

From these four values of P it is observed that the minimum value is about $44.4c$. The power of the arithmetic method should never be underestimated. Mathematically it may not be as elegant as the algebraic technique, but it can help us to find the answer quickly. Here, as in the semi-graphical method, it also gives one a feeling for the analysis and the way in which the parameters should be moved. In the above example, the sliding on a $45°$-plane makes for a reasonable

TABLE 3.3

Minimum values of P^u(Fig. 3.28)

α	$45°$	$40°$	$35°$	$30°$
P^u/c	44.7	44.4	45.0	47.1

start but clearly we will stop at $\alpha = 35°$ (Table 3.3). By looking at the individual answer we can see that if we had just taken one layout between the limits $45° \geqslant \alpha \geqslant 35°$ the answer would have been within about 1½% of the minimum, because the curve is flat in this region. In short, the strict mathematical approach gives only one solution. It gives no reference to the small variations of the collapse load corresponding to the different layouts approximating to the worst possible layout.

3.5.5. Remarks on the methods

Once again it is worth pointing out that it is a fortunate feature of upper-bound limit analysis that assumed mechanisms or the selected layouts for a given problem can often be quite different from the actual mode of collapse, but the values of collapse load are often not too much in error. Fig. 3.29(c, d and e) shows a number of familiar examples of rigid block sliding separated by velocity discontinuities in the well-studied problem of bearing capacity of strip footing on a pure cohesive soil. Also indicated in the Fig. 3.29(f, g and h) are regions of homogeneous deformation denoted by the symbol $\dot{\epsilon}$ for a field of simple vertical compression and lateral expansion and by $\dot{\gamma}$ for simple shear. Although these discontinuous velocity fields are different greatly from the actual field, Fig. 3.29(a and b), and from each other, the upper bounds shown in the figure are not too much greater than the actual answer $P = (2 + \pi)kb$. These answers are obtained quickly and easily. The approximate fields are not different from each other by more than 15% in this figure.

Mechanisms involving three and even more variables do exist. However, it is considered ill-advised in many cases to attempt such problems using the direct algebraic method. A numerical process using a computer should be employed. The method of steepest descent for several variables will be used in later chapters where more complex problems are solved.

3.6. THE DISSIPATION FUNCTIONS

In the previous sections, we have studied a number of relatively simple but extremely useful deformation zones along with illustrative examples. The expressions for the rate of energy dissipation, D, per unit volume, needed for a general deforming volume, will now be given for the yield condition of Tresca, and of Coulomb, when the strain rates are continuous.

3.6.1. Tresca material (Fig. 3.30)

According to Tresca's criterion, plastic flow can occur under a constant state of

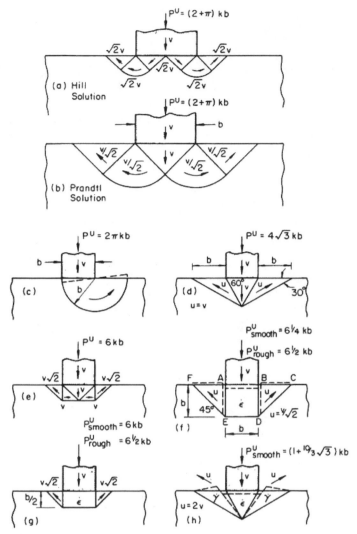

Fig. 3.29. Discontinuous velocity fields differing greatly from the slip-line fields, (a) and (b) $P = (2 + \pi)kb$, and from each other.

stress which is represented by a point on the regular hexagonal right prism equally inclined to the $\sigma_1, \sigma_2, \sigma_3$-axes or for example, by a point on the hexagon in Fig. 3.30. This figure is the intersection of the prism with a plane perpendicular to the σ_3-axis and at a distance σ_3 from the origin. Since the pure cohesive soil is isotropic, the principal axes of the plastic strain rate must coincide with the principal axes of stress and the principal components of the strain rate in the $\sigma_1, \sigma_2, \sigma_3$-directions will be denoted by $\dot{\epsilon}_1$, $\dot{\epsilon}_2$, $\dot{\epsilon}_3$, respectively. As discussed in section 2.3, Chapter 2, it is convenient to represent the plastic strain rate by a ray

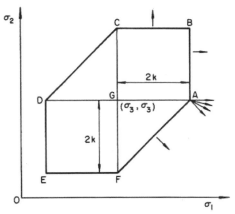

Fig.3.30. Tresca's yield condition.

or a vector with direction cosines proportional to $\dot{\epsilon}_1$, $\dot{\epsilon}_2$, $\dot{\epsilon}_3$ in the $(\sigma_1, \sigma_2, \sigma_3)$ diagram. The projection of this ray or strain vector onto the plane of Fig. 3.30 is a ray with direction cosines proportional to $\dot{\epsilon}_1$, $\dot{\epsilon}_2$. The third principal component $\dot{\epsilon}_3$, can be found when $\dot{\epsilon}_1$, $\dot{\epsilon}_2$ are known since the components satisfy the incompressibility condition for a Tresca material $(c = k, \varphi = 0)$:

$$\dot{\epsilon}_1 + \dot{\epsilon}_2 + \dot{\epsilon}_3 = 0 \qquad\qquad [3.126]$$

It follows that the projection of the ray or strain vector representing the plastic strain rate onto the plane in Fig. 3.30 determines the plastic strain rate to within an arbitrary scalar factor. In other words, the rate of dissipation of energy for a Tresca material can now be determined directly and fully from the hexagon shown in Fig. 3.30 without referring to the three-dimensional hexagonal prism.

Dissipation function

Consider now a point on the hexagon in Fig. 3.30 which does not coincide with a vertex. The concept of *normality* or *flow rule* requires that the vector representing the plastic flow which would occur under this state of stress is normal to the side of the hexagonal prism on which the point lies. The projection of the vector or ray is normal to the side of the hexagon as shown by the small arrows in Fig. 3.30. For a stress point which coincides with a vertex, the strain vector is not determined uniquely. However, the vector or ray must lie between the directions of the normals to the two sides of the hexagon which meet at the vertex. For example, at the vertex A, the vector must lie in the angular space shown by the arrows in Fig. 3.30. Although the correspondence between the plastic strain rate and the stress is not one to one, it will be shown in what follows that the internal rate of dissipation of energy is uniquely determined by the plastic strain rate. The rate of dissipation of energy is given by:

$$D = \sigma_1 \dot{\epsilon}_1 + \sigma_2 \dot{\epsilon}_2 + \sigma_3 \dot{\epsilon}_3 \qquad\qquad [3.127]$$

and in the following it will be shown that the above equation can also be reduced to the simple form:

$$D = 2k \max |\dot{\epsilon}| \qquad\qquad [3.128]$$

where $\max |\dot{\epsilon}|$ denotes the absolute value of the numerically largest principal component of the plastic strain rate.

For a point on the side AB of the hexagon, the normal to AB is parallel to the σ_1-axis so that $\epsilon_2 = 0$. Hence, $\epsilon_3 = -\epsilon_1$ from the condition of incompressibility ([3.126]). Eq. [3.127] then shows that the rate of dissipation of energy is given by:

$$D = \sigma_1 \dot{\epsilon}_1 + \sigma_3 \dot{\epsilon}_3 = (\sigma_1 - \sigma_3)\dot{\epsilon}_1 = 2k\,\dot{\epsilon}_1 \qquad\qquad [3.129]$$

since $\sigma_1 = \sigma_3 + 2k$ on AB. Now $\max |\dot{\epsilon}| = \dot{\epsilon}_1$ in this case so that $2k\dot{\epsilon}_1$ can be written in the form of [3.128]. If the stress point coincides with the vertex A, then $\sigma_1 = \sigma_3 + 2k$, $\sigma_2 = \sigma_3$ and therefore by substituting into [3.127], we have:

$$D = (\sigma_3 + 2k)\dot{\epsilon}_1 + \sigma_3 \dot{\epsilon}_2 + \sigma_3 \dot{\epsilon}_3 = 2k\dot{\epsilon}_1 \qquad\qquad [3.130]$$

on account of the incompressibility condition [3.126]. Since ϵ_1 is the absolutely largest principal component in this case, this again can be written in the form of [3.128]. In the same way it can be shown that [3.128] holds at every stress point of the hexagon.

For the particular case of plane strain, $\dot{\epsilon}_2 = -\dot{\epsilon}_1$ and $\dot{\epsilon}_3 = 0$, so that:

$$\max |\dot{\epsilon}| = \frac{|\dot{\epsilon}_1 - \dot{\epsilon}_2|}{2} = \frac{\dot{\gamma}_{max}}{2} \qquad\qquad [3.131]$$

where γ_{max} is the maximum rate of engineering shear strain. Thus for plane strain:

$$D = k\dot{\gamma}_{max} \qquad\qquad [3.132]$$

Eq. [3.128] and [3.132] for the rate of dissipations of energy were given previously in [3.43] and [3.44] for the special cases of simple plane compression and simple shear flow. This form of [3.128] was first obtained by Hodge and Prager (1951) for the special case of plane stress, $\sigma_2 = 0$ and later extended by Shield and Drucker (1953) to the present general case.

3.6.2. Coulomb material (Fig. 3.31)

(a) Volume expansion. More generally when the Coulomb yield condition is considered, the yield surface becomes a right hexagonal pyramid equally inclined to the $\sigma_1, \sigma_2, \sigma_3$-axes, and with its vertex V at the point $\sigma_1 = \sigma_2 = \sigma_3 = -c \cot \varphi$ (Fig.

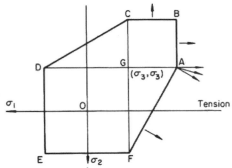

$$GA = GC = 2c \tan (\pi/4 - \phi/2) + |\sigma_3| \tan^2 (\pi/4 - \phi/2) - |\sigma_3|$$

$$GF = GD = 2c \tan (\pi/4 + \phi/2) - |\sigma_3| \tan^2 (\pi/4 + \phi/2) + |\sigma_3|$$

Fig. 3.31. Coulomb's yield condition.

2.3). Fig. 3.31 shows the section of the yield surface by the plane $\sigma_3 = $ constant $>$ $-c \cot \varphi$ where compressive stress is taken as positive. In the following we will show that the flow rule associated with the Coulomb yield criterion requires that:

$$\tan^2 (\tfrac{1}{4}\pi - \tfrac{1}{2}\varphi) \, \Sigma\dot\epsilon_t + \Sigma\dot\epsilon_c = 0 \tag{3.133}$$

where $\dot\epsilon_t$ and $\dot\epsilon_c$ denote the principal tensile and compressive components of the plastic strain rate respectively. When φ is zero the Coulomb criterion reduces to the Tresca yield criterion and [3.133] becomes the equation of incompressibility, [3.126]. For the particular case of plane plastic flow, [3.133] reduces to the equation obtained previously in [3.57].

For the faces of the pyramid, one of the principal components of the plastic strain rate is always zero, and the principal components if properly ordered are in the ratio:

$$- (1 + \sin \varphi)/(1 - \sin \varphi)/0 \tag{3.134}$$

At an edge of the pyramid (Fig. 2.3), a possible plastic strain rate is a linear combination of the components for the faces which intersect at the edge. For example, for points on the edge which passes through the point A of Fig. 3.31, the principal components $\dot\epsilon_1, \dot\epsilon_2, \dot\epsilon_3$ of the plastic strain rate can be written:

$$\dot\epsilon_1 = - (\lambda_1 + \lambda_2)(1 + \sin \varphi), \quad \dot\epsilon_2 = \lambda_1(1 - \sin \varphi), \quad \dot\epsilon_3 = \lambda_2(1 - \sin \varphi) \tag{3.135}$$

where λ_1 and λ_2 are positive scalars. Corresponding expressions apply for other stress points and it follows that for stress points on the faces and edges of the pyramid:

$$\dot\epsilon_1 + \dot\epsilon_2 + \dot\epsilon_3 = (\dot\epsilon_1 - \dot\epsilon_2 - \dot\epsilon_3) \sin \varphi \leqslant 0 \tag{3.136}$$

where $\dot\epsilon_1$ is the (only) strictly negative tensile principal component of the plastic strain rate. From the equality in [3.136], [3.133] follows where $\dot{\epsilon}_1 = \dot{\epsilon}_t$ and $\dot{\epsilon}_2 + \dot{\epsilon}_3 = \Sigma \dot{\epsilon}_c$. In the same way it can be shown that [3.133] holds at every stress point of the hexagon.

(b) Dissipation function. The rate of dissipation of energy, [3.127], for the Coulomb material has the simple form:

$$D = c \cot \varphi \, (\dot{\epsilon}_1 + \dot{\epsilon}_2 + \dot{\epsilon}_3) \tag{3.137}$$

which is seen most easily if one takes the end of a stress vector at the vertex of the pyramid. In the following, it will be shown that the rate of dissipation of energy, [3.127] or [3.137], can also be reduced to the form:

$$D = 2c \tan \left(\tfrac{1}{4}\pi + \tfrac{1}{2}\varphi \right) \Sigma \, |\dot{\epsilon}_c| \tag{3.138}$$

This equation states that the rate of dissipation of energy per unit volume is $2c \tan(\tfrac{1}{4}\pi + \tfrac{1}{2}\varphi)$ times the absolute value of the compressive strain rate or the sum of the absolute values of the compressive strain rates if two of the principals are compressive. The value $2c \tan(\tfrac{1}{4}\pi + \tfrac{1}{2}\varphi)$ is the unconfined compressive strength of the soil. The corresponding Mohr's circle for this stress state is shown by the dashed circle in Fig. 3.17.

The proof of this equation can be achieved in an analogous manner to that of Tresca material. For a point on the side AB of the hexagon, for example, the normal to AB is parallel to the σ_1-axis so that $\dot{\epsilon}_1 < 0$, $\dot{\epsilon}_2 = 0$. Hence:

$$\dot{\epsilon}_3 = - \dot{\epsilon}_1 \tan^2 \left(\tfrac{1}{4}\pi - \tfrac{1}{2}\varphi \right) \tag{3.139}$$

from the condition of volume expansion, [3.133]. Equation [3.127] then shows that the dissipation function is:

$$D = \sigma_1 \dot{\epsilon}_1 + \sigma_3 \dot{\epsilon}_3 = [\sigma_1 - \sigma_3 \tan^2 \left(\tfrac{1}{4}\pi - \tfrac{1}{2}\varphi \right)] \dot{\epsilon}_1 = - 2c \tan \left(\tfrac{1}{4}\pi - \tfrac{1}{2}\varphi \right) \dot{\epsilon}_1 \tag{3.140}$$

since: $\quad \sigma_1 = - 2c \tan \left(\tfrac{1}{4}\pi - \tfrac{1}{2}\varphi \right) + \sigma_3 \tan^2 \left(\tfrac{1}{4}\pi - \tfrac{1}{2}\varphi \right) \tag{3.141}$

on AB as shown in Fig. 3.31. Now $|\dot{\epsilon}_c| = \dot{\epsilon}_3$ in this case so that by using [3.139], [3.140] can be written in the form of [3.138]. If the stress point coincides with the vertex A, then σ_1 is still given by [3.141] and $\sigma_2 = \sigma_3$ and therefore by substituting into [3.127], we have:

$$[- 2c \tan \left(\tfrac{1}{4}\pi - \tfrac{1}{2}\varphi \right) + \sigma_3 \tan^2 \left(\tfrac{1}{4}\pi - \tfrac{1}{2}\varphi \right)] \dot{\epsilon}_1 + \sigma_3 \dot{\epsilon}_2 + \sigma_3 \dot{\epsilon}_3 = - 2c \tan \left(\tfrac{1}{4}\pi - \tfrac{1}{2}\varphi \right) \dot{\epsilon}_1 \tag{3.142}$$

on account of the fact that $\dot{\epsilon}_1 < 0$, $\dot{\epsilon}_2 > 0$, $\dot{\epsilon}_3 > 0$ and also the condition [3.133]. Since $\dot{\epsilon}_2$ and $\dot{\epsilon}_3$ are both compressive in this case, so $\Sigma |\dot{\epsilon}_c| = (\dot{\epsilon}_2 + \dot{\epsilon}_3)$ and $\dot{\epsilon}_t = \dot{\epsilon}_1$, this again can be written in the form of [3.138] making use of the condition

[3.133]. In the same way it can be shown that [3.138] holds at every stress point of the hexagon.

For the particular case of plane strain, one of the principal components of the plastic strain rate is always zero, and the other two component have the relation:

$$\dot{\epsilon}_t = - \dot{\epsilon}_c \tan^2 \left(\tfrac{1}{4}\pi + \tfrac{1}{2}\varphi\right) \tag{3.143}$$

on account of [3.133], so that the maximum rate of engineering shear strain is given by:

$$\dot{\gamma}_{max} = \dot{\epsilon}_c - \dot{\epsilon}_t = \frac{\dot{\epsilon}_c}{\cos^2 \left(\tfrac{1}{4}\pi + \tfrac{1}{2}\varphi\right)} \tag{3.144}$$

Thus for plane strain, [3.138] reduces to:

$$D = c \cos \varphi \, \dot{\gamma}_{max} \tag{3.145}$$

Equations [3.138] and [3.145] for the rate of dissipation of energy were given previously in [3.59] and [3.60] for the special cases of simple plane compression and simple shear flow of Coulomb material. This form of [3.138] was first obtained by Drucker and Prager (1952) for the special case of plane strain and later extended by Chen (1968a) to the present general case.

(c) Tension cut-offs. For a soil unable to take tension, the Coulomb yield criterion must be modified by tension cut-offs. This was demonstrated early in Fig. 3.17 (example 3.4) in which the requirement of zero-tension is met by the circle termination as shown in the figure. More generally, when the Coulomb yield surface in the $\sigma_1, \sigma_2, \sigma_3$-axes is considered (Fig. 2.3), the truncated pyramid, a section of which is shown in Fig. 3.32, can be used to derive the flow rule. For stress points which do not lie on the cut-off planes (side *DE* or *EF*), the principal

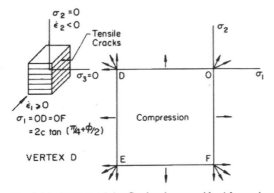

Fig. 3.32. Section of the Coulomb pyramid with tension cut-offs by the plane $\sigma_3 = 0$.

components of the plastic strain rate satisfy [3.133]. Further, the dissipation function D is again given by [3.138]. For stress points on the cut-off planes (side OD or OF and the vertex O) the left-hand side of [3.133] is negative since there is no compression strain. Tensile cracks perpendicular to each respective principal axis are permissible and the rate of dissipation of energy is zero. For stress points on the boundary of the cut-off planes (vertex D or F), a compressive strain rate is possible. In this case the dissipation function D is again given by [3.138].

Chapter 4

LIMIT ANALYSIS BY THE LOWER-BOUND METHOD

4.1. INTRODUCTION

The lower-bound method of limit analysis is different from the upper-bound method in that the *equilibrium* equation and *yield* condition instead of the work equation and failure mechanism are considered. Moreover, whereas the development of the work equation from an assumed collapse mechanism is always clear, many engineers find the construction of a plastic equilibrium stress field to be quite unrelated to physical intuition. Without physical insight there is trouble in finding effective ways to alter the stress fields when they do not give a close bound on the collapse or limit load. Often the user employs the existing stress fields from well-known texts or the more recent technical literature as a magic handbook and tries to fit his problem to the particular solutions he finds. Intuition and innovation seem discouraged by unfamiliarity and apparent complexity. Although the discontinuous fields of stress which will be drawn and discussed in this Chapter are simpler to visualize, they too are not often employed in an original manner by the design engineer. Yet, in fact, the concepts are familiar to the civil engineer in his terms and can be utilized by the designer as a working tool. In this Chapter, it is hoped that the following discussion of the stress field in plane strain problems will make this lower-bound technique of limit analysis more readily accessible to engineers.

Most of the early work on the construction of a stress field is concerned with the pure cohesive soil or Tresca material, the self-weight of the material being assumed to be insignificant. Actually, there are only limited practically important problems in soil mechanics for which this assumption is justified. Further, as a rule for $c - \varphi$ soil or Coulomb material, the stress field involves both applied forces and the self-weight of soil mass. While a number of simple stress fields of this type have been obtained during recent years, general methods allowing for the self-weight of soil have not yet been developed. Nevertheless, the lower-bound theorem and technique are presented herein because the current method is useful for Tresca material in metal plasticity. Furthermore, progress in its extension to Coulomb material in soil plasticity is anticipated in the near future.

The lower-bound theorem

We shall begin by re-examining the rules of the lower-bound theorem. As stated in the lower-bound theorem, if an equilibrium state of stress below yield can be found which satisfies the stress boundary conditions, then the loads imposed can be carried without collapse by a stable body composed of elastic—perfectly plastic material. Any such field of stress thus gives a safe or lower bound on the collapse or limit load. The stress field satisfying all these conditions is called *statically admissible stress field*. The conditions required to establish such a lower-bound solution are essentially as follows:

(1) A complete stress distribution or stress field must be found, *everywhere* satisfying the differential equation of equilibrium.

(2) The stress field at the boundary must satisfy the *stress* boundary conditions.

(3) The stress field must *nowhere* violate the yield condition.

From these rules it can be seen therefore that a lower-bound technique is based entirely on equilibrium and yield conditions but it must not, however, be confused that the limit equilibrium method or slip-line field gives a lower-bound solution. This has been discussed in Chapter 2. Once again it is worth pointing out here that in the limit equilibrium method or slip-line field, the stress state is specified only either along the slip lines or in a local plastic stress zone around the load and not everywhere in the soil mass, as required by item 1, and therefore a limit equilibrium solution or a slip-line solution does not give a complete equilibrium solution. Further, even if a complete equilibrium solution extended from the slip-line field into the rigid regions can be found, it remains to be demonstrated that such a stress distribution will not violate the yield condition in the rigid regions, as required by item 3. Hence, the slip-line field solution strictly should only be regarded as an upper-bound solution, though, in many cases, it seems most likely that it could be completed. It should also be noted that the stress distribution associated with an assumed collapse mechanism in the upper-bound calculation need *not* be in equilibrium, and is only defined in the deforming regions of the collapse mode.

Discontinuous fields

It has already been mentioned in Chapter 3 that discontinuous fields of stress and velocity may be used in applying the lower- and upper-bound theorems. The proofs given in Chapter 2 used continuous systems only for brevity. Discontinuous fields of stress are found to be especially useful in deriving lower bounds. Here, as in a discontinuous velocity situation, surfaces of stress discontinuity are clearly possible, provided the equilibrium equations of stresses are satisfied at all points of these surfaces. This stress discontinuity concept along with Mohr's circle for stress will be discussed more thoroughly in what follows. The only point to be added here is that if the stress fields are chosen for convenience to be at yield in some regions rather than below, the load so obtained may be the collapse load itself. Although such a discontinuous stress situation is useful and permissible in lower-bound calculation, it is rarely the actual state. This is in marked contrast to the velocity situation where discontinuity is not only found useful and convenient in upper-bound calculation but often is contained in actual collapse mode or mechanism. This has already been demonstrated in the preceding chapter.

4.2. MOHR'S DIAGRAM AND BASIC RELATIONS

We shall restrict our discussion here to the plane strain condition in which the dimension perpendicular to the plane of the paper is infinitely long. If a system of rectangular coordinates x, y, z with the z-axis perpendicular to the paper is chosen, then this condition requires that the velocity component v_x and v_y are independent of z, while v_z is zero. In other words, all motion is supposed in the plane of the paper so that we can take the dimension perpendicular to the paper as unity. Under these circumstances, the stress components $\tau_{yz} = \tau_{zx} = 0$ and the remaining components σ_x, σ_y, σ_z and τ_{xy} are independent of the coordinate z. By convention we take compressive stress to be positive and a tensile stress negative. This is indicated in Fig. 4.1(a), in which the stress components are shown in their positive directions.

The vanishing of the shear stresses $\tau_{yz} = \tau_{zx} = 0$ indicates that the stress component σ_z is a principal stress and the z-direction is a principal direction. Accordingly, the x- and y-directions may be chosen so as to coincide with the other two principal directions and the other two principal stresses, σ_{max} and σ_{min} must lie in the xy-plane. In the Mohr stress diagram (Fig. 4.1b) the normal stress σ and the shearing stress τ are used as coordinates. Any stress point (σ,τ) representing the normal

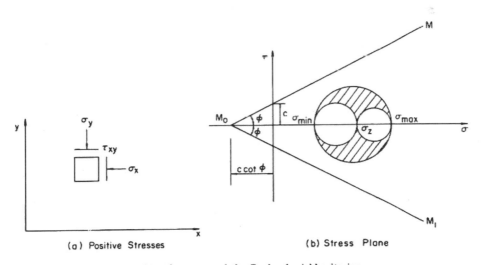

(a) Positive Stresses   (b) Stress Plane

Fig. 4.1. Mohr's representation of a stress and the Coulomb yield criterion.

and shearing stresses across any section through a point lies within the shaded area of the stress circles shown in the Mohr diagram.

Values of σ, τ satisfying the Coulomb condition are represented in Fig. 4.1(b) by points in the region to the right of the two straight lines from the point $(-c \cot \varphi, 0)$ and inclined at angles of amount φ to the σ-axis. It follows that in order not to violate the Coulomb yield condition, a state of stress σ_{max}, σ_z, σ_{min} must be such that the Mohr's circles lie within the wedge-shaped region. Yielding of the soil can occur when the largest of the circles touches the two straight lines. In the case of plane strain condition, instead of a three-Mohr circle representation, it is sufficient to consider only the stress variation in the xy-plane or the Mohr circle passing through the two points σ_{max} and σ_{min}, and instead of stresses on some element of area, we have only to consider stresses on some linear element.

From equilibrium consideration, we can express the principal normal stresses σ_{max} and σ_{min} in terms of the stress components σ_x, σ_y and τ_{xy}:

$$\left.\begin{array}{c}\sigma_{max}\\\sigma_{min}\end{array}\right\} = \tfrac{1}{2}(\sigma_x + \sigma_y) \pm [\tfrac{1}{4}(\sigma_x - \sigma_y)^2 + \tau_{xy}^2]^{1/2} \qquad [4.1]$$

and the third, σ_z, lying between them.

$$\sigma_{min} \leqslant \sigma_z \leqslant \sigma_{max} \qquad [4.2]$$

For convenience, we introduce the notations:

$$p = \tfrac{1}{2}(\sigma_{max} + \sigma_{min}), \qquad s = \tfrac{1}{2}(\sigma_{max} - \sigma_{min}) \qquad [4.3]$$

which can also be expressed in terms of the stress components as follows:

$$p = \tfrac{1}{2}(\sigma_x + \sigma_y), \qquad s = [\tfrac{1}{4}(\sigma_x - \sigma_y)^2 + \tau_{xy}^2]^{1/2} \qquad [4.4]$$

The value p represents a uniform hydrostatic pressure while the value of s represents the maximum shear value in the xy-plane. In the Mohr diagram (Fig. 4.2a) the center of the circle is seen on the σ-axis at distance p to the right of the origin; and its radius is the value of s. Consequently, the values of p and s are sufficient to draw the Mohr circle and hence the state of stress at a point.

Let the point A in Fig. 4.2(a) correspond to the surface element the normal n of which is in the positive x-direction (Fig. 4.2b). On this surface there will be applied an actual stress T forming an angle δ with the normal n and having normal and tangential components σ_x, τ_{xy}. The stress vector T may also be considered as having hydrostatic pressure p normal to the surface and maximum shear stress component s forming an angle 2α with the normal n. The intensity of the stress

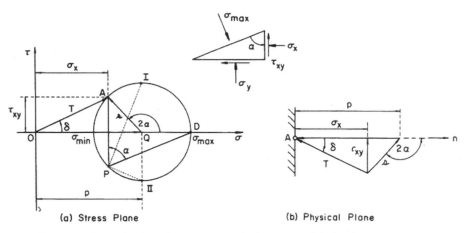

Fig. 4.2. Graphic determination of stresses at a point by means of circle of stress.

vector T on this element is then given by the length of the vector OA in Fig. 4.2(a).
The direction of T is determined by the following fact: To assume the direction of
T, the normal n must be rotated through the same angle δ but in the opposite sense
as the positive σ-axis in Fig. 4.2(a) must be rotated to assume the direction of OA.
It follows from this rule that the surface element shown in Fig. 4.2(b) is represented
on the Mohr circle by the point A with the coordinates (σ_x, τ_{xy}) (Fig. 4.2a). Note
that in the Mohr diagram sketch (see Fig. 4.2a, inset) the angle α between the
vertical plane (or x-plane) on which σ_x and τ_{xy} act and the principal plane on
which σ_{max} acts is measured positive in the *counter-clockwise* sense. Note especially
that τ_{xy} is taken as positive in the Mohr diagram when it runs in a *counter-clockwise*
sense around the free body shown.

Sum of two stress fields

In lower-bound techniques of limit analysis, one of the most important applications of the Mohr stress circle method illustrated by Fig. 4.2 is to solve the following problem: We know the directions and intensities of the principal stresses of
several uniform stress fields. We also know the Coulomb yield condition. We want
to determine whether the resultant stress field obtained from the superposition of
these individual stress fields will violate the yield condition.

In order to solve this problem we have to compute first the values of p and s of
the resultant stress field. In Fig. 4.3(a) the directions of the major principal stresses
corresponding to two different stress fields are shown by the lines $I'I'$ and $I''I''$
which intersect the horizontal axis at angles α' and α'', respectively. They intersect
each other with an angle $(\alpha'' - \alpha')$. Figures 4.3(b) and (c) show the Mohr stress
circles corresponding to the stress fields with the intensity of the principal stresses

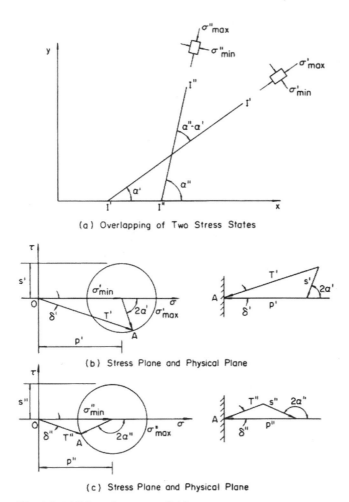

(a) Overlapping of Two Stress States

(b) Stress Plane and Physical Plane

(c) Stress Plane and Physical Plane

Fig. 4.3. Addition of two stress fields.

$\sigma'_{max}, \sigma'_{min}$ and $\sigma''_{max}, \sigma''_{min}$ respectively. Figures 4.3(b and c) also illustrate the correspondence between the points of the Mohr circles and the stresses transmitted across the vertical surface element which is perpendicular to the x-axis. The intensity of the traction T' transmitted across this vertical element shown in the right-hand side of Fig. 4.3(b) is given then by the length of the vector OA in the left-hand side of Fig. 4.3(b). The angle δ' is determined by the afore-mentioned rule. Similarly, the traction T'' and the angle δ'' of Fig. 4.3(c) are represented by the line OA and its angle of inclination to the horizontal axis. The hydrostatic pressure component p' or p'' and shear stress component s' or s'' are represented by certain vectors on these diagrams. All the constructions are obvious and do not require further explanation.

In order to solve our problem it is sufficient to remember that the resultant stress state is determined completely by the values p and s corresponding to the resultant components of the resultant traction T on any section through the point. Clearly, if the two different stress states are overlapped or added together, the stress vectors T' and T'' transmitted across this vertical element must add vectorially, or $T = T' + T''$. Since the hydrostatic pressure components p' and p'' are always perpendicular to the element, the resultant *hydrostatic* pressure component p is simply the *algebraic summation* of p' and p''. However, the resultant *shear* component s is the *vectorial summation* of s' and s''. The angle between these shear stress vectors s' and s'' is equal to *twice* their angle in the physical plane or twice the value $(\alpha'' - \alpha')$ shown in Fig. 4.3(a). Once the values of p and s are determined, the corresponding Mohr circle representing the resultant stress state on any section through point A can be constructed. If the circle of stress representing the resultant state of stress at point A does not intersect the lines of yielding M_0M and M_0M_1 (Fig. 4.1b) there is no section through point A in Fig. 4.3 which violates the stress conditions for yielding. On the other hand, if the circle of stress intersects the lines of yielding more than two points, the yield condition must be violated on some sections of point A. Hence the only circle of stress which satisfies the yield condition at point A is the circle which is tangent to the lines of yielding.

The pole of a Mohr circle

Referring now to Fig. 4.2(a), it is well known that the central angle of the arc AQD equals 2α. However, the position of point A can also be determined without laying off either α or 2α by means of the following procedure. We trace through D a line parallel to the principal section shown in the inset of Fig. 4.2(a). This line intersects the circle at point P. This point is called the *pole* of the Mohr circle. When the pole is known, the point of the circle which corresponds to a given surface element is readily found by drawing, through the pole, a line parallel to the trace of the surface element and determining the second intersection of this line with the circle. The coordinates of this intersection point represent the state of stress on the surface element. For example, the highest and lowest points of the Mohr circle (*I* and *II* in Fig. 4.2a) correspond to the surface elements across which the maximum shearing stress s and hydrostatic pressure p are transmitted. The direction of the traces of these elements is given by PI and PII in Fig. 4.2a.

Two stress conditions are sketched in Fig. 4.4. The purpose of this figure is to demonstrate a direct means of obtaining a physical feeling for states of stress without any calculation or the drawing of Mohr's circles. If σ_x or σ_y is compressive, the algebraically larger principal stress will be compressive. The direction of σ_{max} and the plane on which it acts can be determined approximately by inspection. As illustrated by the square element of Fig. 4.4(a), if $\sigma_x > \sigma_y$, the direction of σ_{max} *in*

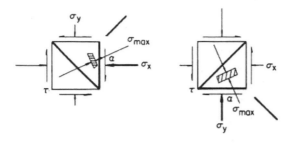

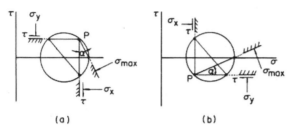

Fig. 4.4. Directions of principal stresses and planes on which they act.

the absence of τ would be the direction of the heavy arrow σ_x. The plane on which σ_{max} would be given by the heavy vertical line. If τ alone were present, $\sigma_x = \sigma_y = 0$, the direction of σ_{max} would be the heavy line sloping up to the right at $45°$ and the plane would be the heavy $45°$-diagonal of the square. Therefore this combination of σ_x, σ_y, and τ will give a σ_{max} direction (α) between the two extremes of direction as shown, and a corresponding orientation of principal plane between the two extreme plane positions. The idea and the symbolism are the same for (b). The corresponding Mohr circles and the poles are also shown in the figure.

4.3. DISCONTINUITIES IN THE STRESSES

In constructing an equilibrium distribution of stress which does not violate the yield condition, it may be advantageous to divide the soil body into several stress zones. In each zone the stress field will satisfy the equations of equilibrium and not violate the yield condition; further, the stress field will be continuous in each zone. However, the stress state on the boundary between two neighboring zones may not be identical. Here we shall exploit the possibility of discontinuity of stress between two adjacent stress elements at the boundary.

First we consider the possibility that the stress systems on two sides of a boundary or plane may be different, but yet in equilibrium. Figure 4.5(a) represents such

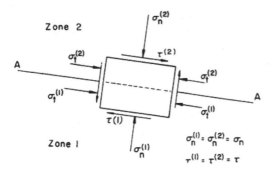

(a) Discontinuity of Stress Across A-A

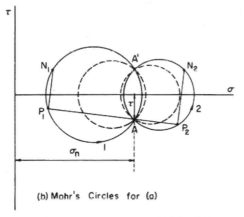

(b) Mohr's Circles for (a)

Fig. 4.5. Stress discontinuity and Mohr circles.

a boundary plane between zone 1 and zone 2. Consideration of the equilibrium of a long and narrow element containing the boundary where the state of stress on either side is denoted by the subscripts t and n, tangential and normal to the boundary respectively, shows that in the absence of self-weight of soils the normal stress σ_n, and the shear stress τ, must be continuous or:

$$\sigma_n^{(1)} = \sigma_n^{(2)} \quad \text{and} \quad \tau^{(1)} = \tau^{(2)} \tag{4.5}$$

where the superscripts denote the zones. Equilibrium, however, places no restrictions on the change of σ_t across the boundary. We conclude therefore that as far as *equilibrium* is concerned we can have a discontinuity in the σ_t-component across the boundary, although the *other* components of stress, σ_n and τ, must be continuous across the boundary from the above equation.

This situation is illustrated clearly by the Mohr circles for the regions 1 and 2 in Fig. 4.5(b). The poles of the two circles are obtained by drawing a line through A

parallel to the element AA of the line of discontinuity. The points P_1, P_2 where this line meets the circles are the poles of the circles 1 and 2, then the lines P_1N_1 and P_2N_2 give the directions of the lines normal to the line of discontinuity AA, in the regions 1 and 2, respectively. The two stress components σ_n and τ coincide at the discontinuous interface A, but the two states of stress, represented by circles 1 and 2 corresponding to two different values of σ_t, may be rather different. For given values of σ_n and τ represented by the point A in the diagram, infinitely many of such circles of various radius can be drawn through A, so as to have their centers on the σ-axis.

In order to obtain a largest lower-bound solution, it is advantageous if the soil is at the *yield point* or *plastic stress* on both sides of a boundary. In such a case,

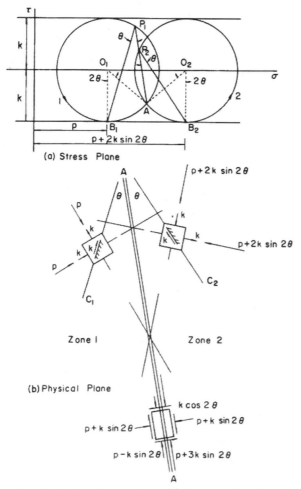

Fig. 4.6. A line of stress discontinuity separating the plastic stress fields 1 and 2 of Tresca material.

through A only two circles tangent to the yield lines can be drawn. Considering, for example, the special case of Tresca material for which $\varphi = 0$, the Mohr circles representing limiting states of stress have two parallel lines as an envelope (Fig. 4.6a). In Fig. 4.6(b), the line AA represents an element of a line of stress discontinuity separating the plastic stress state fields 1 and 2. Subscripts 1 and 2 will be used to distinguish the values which a quantity assumes on the two sides of the line. In Fig. 4.6(a), the two circles represent the two possible states of plastic stress at the discontinuity line AA. The left-hand circle corresponds to a smaller hydrostatic pressure p than the right-hand circle. For the left-hand circle, the point P_1 is the pole and the corresponding maximum shear direction AC_1 drawn in the physical plane (Fig. 4.6b) is indicated by P_1B_1 in the stress plane (Fig. 4.6a). For the right-hand circle, the pole is P_2 and the corresponding maximum shear direction AC_2 drawn in the physical plane is indicated by P_2B_2 in the stress plane. The line P_1P_2A in the stress plane is the line of stress discontinuity AA in the physical plane, separating the plastic stress regions 1 and 2. By simple geometry applied to the Mohr circles, the central angles corresponding to the arcs AB_1 and AB_2, must have the same value, say, 2θ. It follows that the discontinuity line AA must *bisect* the corresponding directions of principal shear stress on either side of the plastic stress discontinuity. In other words, the axes of principal stresses in the two plastic zones form mirror images of each other in the boundary of stress discontinuity. This fact is of considerable importance in later constructions of discontinuous stress fields for pure cohesive soils.

An illustrative example

Having described the general conditions of stress discontinuity across a surface, it is interesting to point out that such a stress discontinuity has been known for many years in some theoretical solutions. As an example, the development of the stress distribution across a slab bent by couples can be followed from elastic bending until fully plastic state occurs, as illustrated in Fig. 4.7. Fig. 4.7(a) shows the linear elastic stress distribution. As the moment is increased, elements close to the surfaces of the slab reach plastic state first, and we have the elastic–plastic situation shown in (b) with an elastic core. As the bending moment increases further toward the maximum permissible value corresponding to the limiting state configuration (c), the elastic core shrinks to a membrane, and we have the special case of a discontinuity with a saltus of $4k$ in the stress component parallel to it. This particular example represents the largest possible change or jump in stress σ_t across a discontinuity since σ_n and τ are zero in this case (Fig. 4.5a).

Here, as in the previous case of velocity discontinuity, the line of stress discontinuity can be interpreted as the limiting case of a narrow transition zone between two distinct stress fields. Because the intermediate stress state or the stresses in this

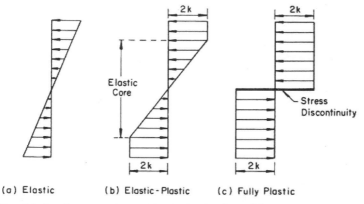

(a) Elastic   (b) Elastic-Plastic   (c) Fully Plastic

Fig. 4.7. Development of stress distribution in a bent slab.

narrow transitional zone can be expressed by several stress circles drawn through points A and A', as illustrated by the dashed line circles in Fig. 4.5(b), and in addition all these intermediate circles are always smaller than the two extreme circles marked 1 and 2 in the figure, it follows that the distribution of stress in the narrow transition zone must be in elastic state and the stress varies in a rapid but continuous manner. The stress discontinuity line in the strict mathematical sense may therefore be imagined physically as an elastic filament of infinitesimal width. If the elastic strains are neglected, the filament must be taken to be inextensible but perfectly flexible. In such case bending with large changes of curvature is permissible without violating elastic or inextensible condition in the filament because the strains still remain small or zero for finite curvature. It can therefore be concluded from this discussion that a slip or failure line cannot coincide with a line of stress discontinuity. Since a discontinuity line in the velocity field can occur only across a failure line, it follows that the velocity field must be continuous across a line of stress discontinuity. It is worth pointing out here that in the last Chapter it has been established that stresses must be continuous across the discontinuous line of velocity.

Example 3.4 (continued)

Let us now attempt a lower-bound solution of the same slope stability problem — critical height of a vertical cut with soil unable to take tension (Example 3.4, section 3.3, Chapter 3), by constructing a simple discontinuous stress field which does not violate the yield condition. The simplest possible equilibrium distribution of stress is found by having a horizontal plane of discontinuity between zones I and II and a vertical plane of discontinuity between zones II and III as shown in Fig. 4.8(a). Assuming the state of stress in zones I, II, and III to be uniaxial com-

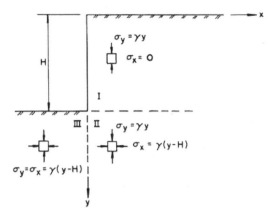

(a) An Equilibrium Solution

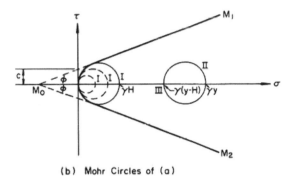

(b) Mohr Circles of (a)

Fig. 4.8. A lower-bound stress field for the stability of a vertical cut.

pression, biaxial compression, and hydrostatic compression, respectively. Figure 4.8(b) shows the corresponding Mohr circles for each zone. Coulomb yield condition with tension cut-off is satisfied when the circles representing zone I at ground level meet the yield lines M_0M_1 and M_0M_2. Therefore:

$$\tfrac{1}{2}\,\gamma H = c\,\cos\varphi + \tfrac{1}{2}\,\gamma H\,\sin\varphi \qquad\qquad [4.6]$$

Since this discontinuous stress field satisfies equilibrium everywhere in the soil mass and the boundary conditions, which in this case require both normal and shear stresses to be zero on all surfaces, and nowhere exceeds the Coulomb yield condition with zero-tension cut-off, by the lower bound theorem of limit analysis, the value H computed from [4.6] is therefore a lower bound for the critical height:

$$H_{cr} = \frac{2c}{\gamma}\,\tan\,(\tfrac{1}{4}\pi + \tfrac{1}{2}\varphi) \qquad\qquad [4.7]$$

Since this lower bound agrees with the previous upper-bound solution [3.65], the

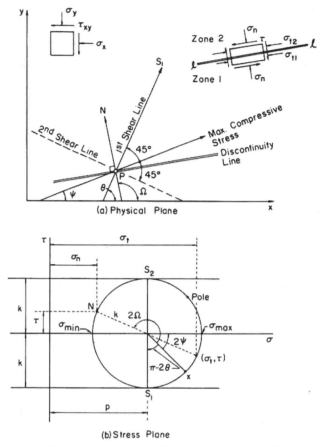

Fig. 4.9. Stress state at a point under the Tresca yield condition.

exact value of the critical height, which neglects the tensile stress of soil, is:

$$H_{cr} = \frac{2c}{\gamma} \tan\left(\tfrac{1}{4}\pi + \tfrac{1}{2}\varphi\right) \qquad\qquad [4.8]$$

It must be borne in mind however that the coincidence of upper and lower bounds provided by the velocity field, Fig. 3.18, and the stress field, Fig. 4.8(a), is by no means indicating that the two discontinuous fields are the actual state. Once again it is worth pointing out that in the limit analysis there is no theoretical restriction that the assumed stress field or velocity field need have some similarity to the actual state, although generally speaking, the closer the assumed state to the actual state is, the more realistic the resulting answer will be.

4.4. JUMP CONDITIONS AT A DISCONTINUITY SURFACE OF TRESCA MATERIAL

In the last section, a general discussion on the discontinuous characteristics of stress has been pointed out, and the main object of the following two sections is to establish certain mathematical relations on this discontinuity problem. These relations are considered useful not only with regard to applications of limit analysis but also from the viewpoint of obtaining real solutions. A discussion of stress discontinuities in a cohesive $c - \varphi$ soil will be given in the next section. The following discussion applies to a purely cohesive soil.

Let σ_x, σ_y and τ_{xy} be the normal and shearing stresses with respect to the rectangular cartesian coordinates x, y. At an arbitrary point P of the x,y-plane, let us consider the normal stress σ_n and shearing stress τ transmitted across a discontinuous surface whose orientation is defined by the angle Ω between the x-axis and the normal to the surface (Fig. 4.9a). Let us further denote the angle between the direction of σ_{max} and the x-axis by Ψ, and call this direction the *first principal direction* (Fig. 4.9a). The *second principal direction* is orthogonal to the first, and the *directions of maximum shearing stress* bisect the right angles between the principal directions. The line rotating through $45°$ in the counter-clockwise sense is called the *first direction of maximum shearing stress* which is making an angle θ with the x-axis (Fig. 4.9a). The *first shear lines* are defined as curves in the x,y-plane which everywhere have the first direction of maximum shearing stress; the *second shear lines* are their orthogonal trajectories.

With the Tresca yield condition shown in Fig. 4.9(b) as two parallel lines in the Mohr circle for the state of stress at point P shown in (a) where k is the yield stress in pure shear and p the hydrostatic pressure, it can easily be shown that:

$$\sigma_n = p + k \sin 2(\theta - \Omega)$$

$$\sigma_t = p - k \sin 2(\theta - \Omega)$$

$$\tau = k \cos 2(\theta - \Omega) \tag{4.9}$$

As already has been discussed in the last section, equilibrium requires only that the normal stress σ_n and the shearing stress τ be continuous (see right top inset, Fig. 4.9a). The normal stress σ_t transmitted across the discontinuous surface may therefore be discontinuous.

Expressed in terms of σ_n, σ_t and τ, the yield condition shown in (b) takes the form:

$$(\sigma_n - \sigma_t)^2 + 4\tau^2 = 4k^2$$

$$\text{or:} \quad \sigma_t = \sigma_n \pm 2[k^2 - \tau^2]^{1/2} \tag{4.10}$$

The jump in σ_t across the discontinuous surface is therefore given by:

$$\sigma_{t2} - \sigma_{t1} = \pm 4[k^2 - \tau^2]^{1/2} \tag{4.11}$$

If σ_n and τ are to be continuous, the jumps in p and θ must satisfy the following equations, using [4.5] and [4.9]:

$$p_1 + k \sin 2(\theta_1 - \Omega) = p_2 + k \sin 2(\theta_2 - \Omega)$$

$$\cos 2(\theta_1 - \Omega) = \cos 2(\theta_2 - \Omega) \tag{4.12}$$

where the subscripts 1 and 2 refer to the two sides of the discontinuity surface. The second equation of [4.12] gives:

$$\theta_1 + \theta_2 = 2\Omega \pm n\pi \tag{4.13}$$

when $n = 0,1,2,3...$

Substituting this into the first equation of [4.12], we obtain:

$$p_2 = p_1 + 2k \sin 2(\theta_1 - \Omega) \tag{4.14}$$

It can easily be shown by simple geometry applied to the two sides of the discontinuous surface, that [4.13] implies that, at any point of the trace of the discontinuity surface in the x,y-plane, the tangent of this trace bisects the angle formed by the tangents of the first shear lines on either side of the surface. This fact has already been demonstrated in the last section (Fig. 4.6). Making use of [4.13], we may write [4.14] in the form:

$$p_2 - p_1 = -2k \sin(\theta_2 - \theta_1) \tag{4.15}$$

which shows that the decrease of p across the discontinuous surface equals the sine of the angle between the tangents of the first shear lines on either side of the discontinuous surface.

Example 4.1: Stress fields with two stress-free surfaces

Symmetric stress field (Fig. 4.10)

Consider first the interaction of discontinuity lines and stress-free surfaces for a symmetric wedge-shape stress field shown in Fig. 4.10. The constant stress regions *ABD*, *ADE* and *AEC* are separated by two lines of stress discontinuity *AD* and *AE* which are inclined to *AB* and *AC* at an angle α, respectively to be determined. Since the stress field is symmetric about the line *AF*, only right-hand side of the field need to be considered. *AB* and *AC* are the stress-free surfaces. *FE* and *EC* are first shear lines. The different values a quantity may assume in the region *AFE* and *AEC* will be distinguished by the subscripts 1 and 2, respectively. Choosing the x and

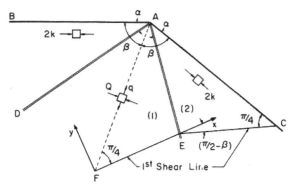

Fig. 4.10. Symmetric stress field with two stress-free surfaces.

y-axes as shown in the figure, the inclination θ of the first shear lines assumes the values:

$$\theta_1 = 0, \quad \theta_2 = -(\tfrac{1}{2}\pi - \beta) \tag{4.16}$$

The substitution of these values into the jump condition [4.13], in which Ω is put equal to $\beta - \alpha - \tfrac{1}{4}\pi$, gives the relation $\alpha = \tfrac{1}{2}\beta$.

The value of the hydrostatic pressure p in the region AEC is $p_2 = k$. The jump condition [4.15] then furnishes an equation for p_1:

$$p_1 = k - 2k \cos \beta \tag{4.17}$$

Since the Mohr stress circle for region 1 is now centered on the σ-axis at distance p_1 to the right of the origin; having k as its radius, it follows that the values of the two principal stresses are:

$$Q = p_1 + k = 2k(1 - \cos \beta)$$

$$q = p_1 - k = -2k \cos \beta \tag{4.18}$$

The negative sign in the right-hand side of the second equation indicates that the minor principal stress q in region 1 must be a tensile stress.

Unsymmetric stress field (Fig. 4.11)

A further case of such a discontinuous stress field formed by two straight stress-free surfaces OA and OD is shown in Fig. 4.11. The three regions separated by the lines of stress discontinuities OB and OC are in a plastic state of stress. The region AOB is stressed by uniaxial tension $2k$ and the region COD is stressed by uniaxial compression $2k$. The region BOC is stressed by biaxial compression–tension stress with the direction of the maximum compressive stress Q inclining at an angle δ to the y-axis. Here the angles marked α and γ and the magnitude of the principal stresses marked Q and q in the figure are to be determined.

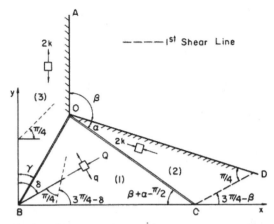

Fig. 4.11. Unsymmetric stress field with two stress-free surfaces.

The first shear line for the regions marked (1), (2) and (3) of constant stress is denoted by the respective dotted lines in the region. From this figure, we have:

$$\theta_1 = \tfrac{3}{4}\pi - \delta, \qquad \theta_2 = \tfrac{3}{4}\pi - \beta, \qquad \theta_3 = \tfrac{1}{4}\pi \tag{4.19}$$

where β is the angle between the sides AO and OD, and where the subscripts refer to the regions marked (1), (2), (3) in the figure. The angles Ω between the x-axis and the normal to the discontinuous lines OC and OB are π-α-β and π-γ, respectively. The jump condition [4.13] at the discontinuity lines OC and OB shows that:

$$\alpha = \tfrac{1}{2}\delta - \tfrac{1}{2}\beta + \tfrac{1}{4}\pi \pm \tfrac{1}{2}n\pi, \qquad \gamma = \tfrac{1}{2}\delta + \tfrac{1}{2}\pi \pm \tfrac{1}{2}n\pi \; (n = 0, 1, 2, \ldots) \tag{4.20}$$

In addition, the value of p_1 in region 1 can be deduced either from the known value of $p_2 = k$ in region 2 or $p_3 = -k$ in region 3 and the jump condition [4.15] either at the line OC or OB. Considering first the regions 1 and 2, we have:

$$p_1 = k + 2k \sin (\delta - \beta) \tag{4.21}$$

Similarly, jump condition between the regions 1 and 3 requires that:

$$p_1 = -k - 2k \cos \delta \tag{4.22}$$

Using the relations [4.20], [4.21] and [4.22] give the following additional relationship for the unknown angles α and γ:

$$\cos 2\alpha + \cos 2\gamma = 1 \tag{4.23}$$

For a given value of β, the angles α and γ may be obtained by evaluating α and γ for various δ-values. Then by substitution of the various values of α and γ so obtained into [4.23], an approximate δ-value and thus the values of α and γ can be determined. The values of Q and q can then be obtained easily from p_1 and they are found to be:

$$Q = p_1 + k = 2k[1 - \sin(\beta - \delta)]$$

$$q = p_1 - k = -2k \sin(\beta - \delta) \qquad\qquad [4.24]$$

For example, for the particular case $\beta = \frac{1}{2}\pi$, the condition of plastic equilibrium on both discontinuity lines requires the following values of angles:

$$\alpha = \gamma = \tfrac{1}{6}\pi \qquad \delta = \tfrac{1}{3}\pi \qquad\qquad [2.25a]$$

and the central region BOC is subjected to the stresses:

$$Q = k \qquad q = -k \qquad\qquad [4.25b]$$

4.5. JUMP CONDITIONS AT A DISCONTINUITY SURFACE OF COULOMB MATERIAL

The main difference between a $c - \varphi$ soil and a purely cohesive soil for which we have developed the jump condition is the form of the yield condition. The Mohr circles representing plastic states of stress have two parallel lines as an envelope for the purely cohesive soil (Fig. 4.9b) but two intersecting lines for a $c - \varphi$ soil (Fig. 4.12b). As a consequence, the two *slip* or *failure* lines are now inclined at an angle $\frac{1}{4}\pi - \frac{1}{2}\varphi$ to the direction of σ_{max}. As in the case of shear lines for a purely cohesive soil, the slip lines will be called the *first* and *second slip lines*, with the convention that the direction of the first slip line at a point is obtained from the direction of σ_{max} by a *counterclockwise* rotation of amount $\frac{1}{4}\pi - \frac{1}{2}\varphi$. The inclination of the first slip line to the x-axis will be denoted by θ (Fig. 4.12a)

In the right top inset of Fig. 4.12(a), the line ll represents an element of a line of stress discontinuity separating the plastic stress fields 1 and 2. Subscripts 1 and 2 will be used to distinguish the values which a quantity assumes on the two sides of the line. The normal to the line is inclined at an angle Ω to the x-axis and, from Fig. 4.12(b), the normal and shear stress on such a line are:

$$\sigma_n = \bar{p}[1 + \sin\varphi \sin(2\theta - 2\Omega + \varphi)] - c\cot\varphi$$

$$\tau = \bar{p}\sin\varphi \cos(2\theta - 2\Omega + \varphi) \qquad\qquad [4.26]$$

where (see Fig. 4.12b):

$$\bar{p} = p + c\cot\varphi = \frac{\sigma_{max} - \sigma_{min}}{2\sin\varphi} \geqslant 0 \qquad\qquad [4.27]$$

The equilibrium of the small rectangular element shown in the right top inset of

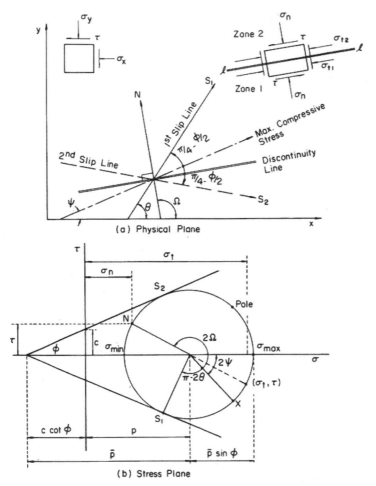

(a) Physical Plane

(b) Stress Plane

Fig. 4.12. Stress state at a point under the Coulomb yield condition.

Fig. 4.12(a) requires only that the normal and tangential components of stress σ_n and τ are continuous across the line, but the normal stress components σ_{t1} and σ_{t2} parallel to the line may be discontinuous. From [4.26], the equilibrium conditions are:

$$\bar{p}_1 [1 + \sin \varphi \sin (2\theta_1 - 2\Omega + \varphi)] = \bar{p}_2 [1 + \sin \varphi \sin (2\theta_2 - 2\Omega + \varphi)]$$

$$\bar{p}_1 \cos (2\theta_1 - 2\Omega + \varphi) = \bar{p}_2 \cos (2\theta_2 - 2\Omega + \varphi) \qquad [4.28]$$

The elimination of p_1, p_2 between these two equations, after some reduction, yields:

$$\sin (\theta_1 + \theta_2 - 2\Omega + \varphi) + \sin \varphi \cos (\theta_1 - \theta_2) = 0 \qquad [4.29]$$

provided that $\sin(\theta_1 - \theta_2) \neq 0$, that is, provided that the discontinuity is not of zero-length.

The condition [4.29] together with one of the conditions [4.28] corresponds to the jump conditions established in the previous section for a purely cohesive soil ($\varphi = 0$, $c = k$). For zero-angle of friction, condition [4.29] becomes [4.13].

Example 4.2: Wedge under unilateral pressure (Shield, 1954b)

The problem of a wedge with uniform pressure Q on one face as shown in Fig. 4.13 will be considered first and then followed by another example of loaded trapezoid stress field. In Fig. 4.13, the wedge ABD of angle β is loaded by a uniform pressure Q along AB, producing a fully plastic state of stress in the wedge region ABD. The regions ABC and BDC are regions of constant biaxial compression and uniaxial compression, respectively. The constant stress regions ABC and BCD are separated by a line of stress discontinuity BC which is inclined to AB at an angle γ to be determined. AC and CD are first slip lines. The different values a quantity may assume in the regions ABC, BCD will be distinguished by the subscripts 1 and 2, respectively. Choosing the x- and y-axes as shown in the figure, the inclination θ of the first slip lines assumes the values:

$$\theta_1 = 0, \quad \theta_2 = -(\tfrac{1}{2}\pi - \beta) \tag{4.30}$$

The substitution of these values into the jump condition [4.29], in which Ω is put equal to $\gamma - (\tfrac{1}{4}\pi - \tfrac{1}{2}\varphi)$, gives the relation:

$$\sin(\beta - 2\gamma) + \sin\varphi\sin\beta = 0 \tag{4.31}$$

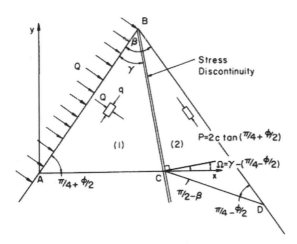

Fig. 4.13. Stress field for loaded wedge.

This equation determines the angle γ and the relevant root of this equation is found to be:

$$\gamma = \tfrac{1}{2}\beta + \tfrac{1}{2}\mu \qquad\qquad [4.32]$$

where μ is given by:

$$\sin \mu = \sin \varphi \sin \beta, \quad 0 \leqslant \mu \leqslant \tfrac{1}{2}\pi \qquad\qquad [4.33]$$

The value of $\bar{p}$ in the region BCD is determined by the condition that the state of stress in this region is uniaxial compression:

$$P = \frac{2c \cos \varphi}{1 - \sin \varphi} = 2c \tan \left(\tfrac{1}{4}\pi + \tfrac{1}{2}\varphi\right) \qquad\qquad [4.34]$$

It follows that $\bar{p}$ in this region has the value:

$$\bar{p}_2 = \frac{c \cos \varphi}{1 - \sin \varphi} + c \cot \varphi = \frac{c \cot \varphi}{1 - \sin \varphi} \qquad\qquad [4.35]$$

The second equation of [4.28] then given:

$$\bar{p}_1 = \frac{c \cot \varphi}{1 - \sin \varphi} \frac{\sin (\beta - \mu)}{\sin (\beta + \mu)} \qquad\qquad [4.36]$$

with [4.26], the normal pressure on AB can now be found and a little rearrangement gives the value:

$$Q = c \cot \varphi \left[\tan^2 \left(\tfrac{1}{4}\pi + \tfrac{1}{2}\varphi\right) \frac{\sin (\beta - \mu)}{\sin (\beta + \mu)} - 1 \right] \qquad\qquad [4.37a]$$

and with [4.27], we have:

$$q = Q - 2 \bar{p}_1 \sin \varphi = c \cot \varphi \left[\frac{\sin (\beta - \mu)}{\sin (\beta + \mu)} - 1 \right] \qquad\qquad [4.37b]$$

when $\varphi = 0$, $c = k$ (Tresca material), μ is also zero and [4.37] take the form:

$$Q = 2k(1 - \cos \beta) \quad \text{and} \quad q = -2k \cos \beta \qquad\qquad [4.38]$$

It should perhaps be pointed out that the right-half of the wedge stress field FAC shown in Fig. 4.10 is identical to that of the wedge stress field ABD considered in Fig. 4.13. Thus, the value of the normal pressure Q for the special case of Tresca material as given by [4.38], agrees with the value obtained earlier, [4.18].

For small values of β, expression [4.37a] is approximately:

$$Q = c \beta^2 \cos \varphi \qquad\qquad [4.39]$$

and it follows that for small values of β, the ratio Q/c decreases as the angle of friction φ is increased. Fig. 4.14 shows the variation of Q/c with the angle of the wedge for $\varphi = 0°$, 20° and 40°. The limiting case $\varphi = 90°$ is also shown in the diagram. When $\beta = \tfrac{1}{2}\pi$ the expression [4.37a] has the uniaxial compression value

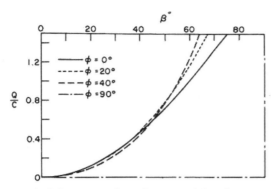

Fig. 4.14. Pressure on face of acute angled wedge.

$2c \tan \left(\frac{1}{4}\pi + \frac{1}{2}\varphi\right)$. The full lines in Fig. 4.15 show the variation in Q/c with the wedge angle β varying from $0°$ to $180°$ for friction of $0°$, $20°$ and $40°$. The broken lines are the exact solutions obtained by Prandtl (1921). Shield (1954b) has demonstrated that a velocity field can be associated with the acute angled wedge ($\beta < \frac{1}{2}\pi$) and the discontinuous stress solution [4.37] for the acute angled wedge is therefore physically admissible. However, a velocity field cannot be associated with the case of obtuse angled wedge ($\beta > \frac{1}{2}\pi$), so that the discontinuous solution [4.37] is not acceptable physically for wedge angles greater than a right angle. The pressure Q as furnished by the discontinuous solution [4.37a] for obtuse angled wedges is therefore only a lower-bound solution for the problem.

Example 4.3: Loaded trapezoid (Shield, 1954b)

A further example of discontinuous stress fields is shown in Fig. 4.16. The trapezoid *ABCD* is in a plastic state of stress due to the normal pressure Q, and Q'

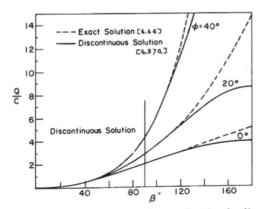

Fig. 4.15. Pressure on wedge face as given by the discontinuous and exact solutions.

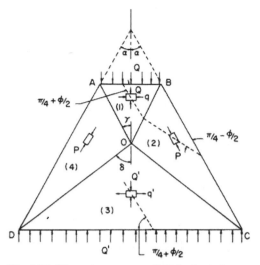

Fig. 4.16. Discontinuous stress field for loaded trapezoid.

on the parallel sides AB and CD, the sides AD and BC being free from applied stress. The lines AO, BO, CO and DO are lines of stress discontinuity separating the regions 1, 2, 3, 4 of constant stress, where the angles marked γ, δ in the figure remain to be determined. The regions ABO and DCO are regions of constant biaxial compression and biaxial compression-tension, respectively. The regions BCO and ADO are regions of uniaxial compression with the value of P given by [4.34]. The broken lines shown in the trapezoid are the first slip line for each region.

Choosing the horizontal and vertical axes as the x- and y-axes, the inclination θ of the first slip lines assumes the values:

$$\theta_1 = -(\tfrac{1}{4}\pi + \tfrac{1}{2}\varphi), \quad \theta_2 = -(\tfrac{1}{4}\pi + \tfrac{1}{2}\varphi) + \alpha, \quad \theta_3 = -(\tfrac{1}{4}\pi + \tfrac{1}{2}\varphi) \qquad [4.40]$$

where 2α is the angle between the sides AD and BC and where the subscripts refer to the regions marked 1, 2, 3 in the figure. With the values of θ, the jump condition [4.29] at the discontinuity lines BO and CO shows that:

$$\cos(2\gamma + \alpha) = \sin\varphi\cos\alpha$$

$$\cos(2\delta - \alpha) = \sin\varphi\cos\alpha \qquad\qquad\qquad\qquad [4.41]$$

and the relevant roots of these equations are:

$$\gamma = \tfrac{1}{2}\nu - \tfrac{1}{2}\alpha, \quad \delta = \tfrac{1}{2}\nu + \tfrac{1}{2}\alpha \qquad\qquad\qquad [4.42]$$

where: $\cos\nu = \sin\varphi\cos\alpha, \quad 0 \leqslant \nu \leqslant \tfrac{1}{2}\pi$ \qquad\qquad\qquad [4.43]

The values of $\bar{p}$ in the regions 1 and 3 can be deduced from the known value of $\bar{p}$

in region 2 ([4.35]) and the jump conditions [4.28] at the lines *BO* and *CO*. The pressures Q and Q' can then be obtained and it is found that:

$$Q = c \cot \varphi \left[\tan^2 \left(\tfrac{1}{4}\pi + \tfrac{1}{2}\varphi\right) \frac{\sin(\nu + \alpha)}{\sin(\nu - \alpha)} - 1 \right] \qquad [4.44]$$

$$Q' = c \cot \varphi \left[\tan^2 \left(\tfrac{1}{4}\pi + \tfrac{1}{2}\varphi\right) \frac{\sin(\nu - \alpha)}{\sin(\nu + \alpha)} - 1 \right] \qquad [4.45]$$

As the angle of friction φ tends to zero and set $c = k$ these expressions tend to the values:

$$Q = 2k(1 + \sin \alpha) \quad \text{and} \quad Q' = 2k(1 - \sin \alpha) \qquad [4.46]$$

agreeing with the values obtained by Winzer and Carrier (1948) for a Tresca material.

It can be shown (Shield, 1954b) that a velocity field cannot be associated with the stress field of Fig. 4.16. The stress field is therefore physically inadmissible. It provides only a lower-bound solution to a given problem.

4.6. DISCONTINUOUS FIELDS OF STRESS VIEWED AS PIN-CONNECTED TRUSSES – TRESCA MATERIAL

Stress discontinuities in two dimensions have been discussed thoroughly in the preceding two sections in terms of the jump conditions across the surfaces of stress discontinuity. Discontinuous fields of stress were employed years ago to give lower bounds to basic problems of interest by Bishop (1953), Shield and Drucker (1953), Brady and Drucker (1953) and many others for Tresca material. Nevertheless, the essential simplicity of the technique in obtaining a lower-bound solution for a problem is not generally recognized. Perhaps this is a result of an overidentification with exact solutions to perfectly plastic problems and the customary use of stress fields "at yield".

An alternative but not basically new point of view is discussed here. Plane strain problems, such as the one illustrated in Fig. 4.17 provide simple and instructive examples. The dimension perpendicular to the plane of the paper will be taken as unity, but all motion is supposed in the plane. The applied force P is carried through the rectangular block of Fig. 4.17(a) in a very elaborate pattern from the smooth (or rough) punch to the smooth (or rough) supporting plane. Details as well as broad principles are discussed in the original approach of Bishop (1953) and in subsequent work by Bishop et al. (1956). Suppose instead that a pin-connected truss is imagined to carry the load inside the body, Fig. 4.17(b). The forces in the members of the truss are determined directly by summation of forces at each pin as:

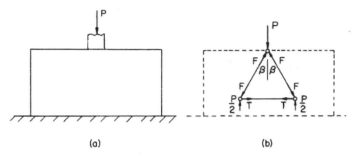

Fig. 4.17. Bearing capacity of a rectangular block. Truss action to carry the load.

$$F = \frac{P}{2 \cos \beta}, \qquad T = F \sin \beta = \tfrac{1}{2} P \tan \beta \qquad\qquad [4.47]$$

In the usual structural design, the cross-sectional area (the width in this plane problem) of each member is taken large enough to give a safe or permissible axial stress. Here the stress must be chosen at or below $2k$, where k is the yield stress in shear for Tresca material, if a lower bound on the limit load is to be found or if the safety of applying P to the block is to be determined. Once definite widths are chosen, the fields of uniaxial tensile or compressive stress will be seen to overlap. A connection problem arises somewhat as in a real truss when the members all lie in a single plane and cannot pass by each other.

Stress connections (Fig. 4.18)

Fig. 4.18 shows a somewhat more general case than Fig. 4.17 to bring out both the approach and the problem more clearly. In Fig. 4.18(a), two compressive force fields Q,S of arbitrarily chosen widths e,f overlap in ABC. They are balanced by a third compressive force R whose strip width g is determined if no additional lines of discontinuity are to be introduced. No *internal* equilibrium problems arise if Q, R, and S are in equilibrium as indicated by the force triangle in Fig. 4.18. The stress in region ABC is just the sum of a uniaxial compression Q/e and a uniaxial compression S/f at the angles pictured. The different jumps across each discontinuity BC, AC, AB in the component of normal stress parallel to the line of discontinuity can be computed, but there is no need to do so. Normal and shear stresses across each discontinuity will be continuous because overall equilibrium is satisfied and all regions of space have been accounted for. If Q/e, S/f, and R/g are each at or below yield ($2k$), the only question which arises is whether Q/e and S/f sum to a state of stress below yield. Note that there are four lines of discontinuity meeting at each point (e.g., BB', BC, BA, BB'') but the jump conditions are not those for fully plastic fields, unless by chance or choice the adjacent fields are "at yield".

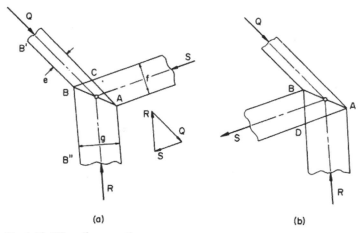

(a)   (b)

Fig. 4.18. "Truss" connections.

In Fig. 4.18(b), the force S has the same magnitude and direction as in Fig. 4.18(a), but now it is tensile instead of compressive. It overlaps with R in the region ABD before it encounters Q. However, the discontinuity line AB is the same line in (b) as in (a). Of course, the combination of a tensile stress S/f and a compressive stress R/g is likely to give a much higher maximum shear stress in ABD than is produced by the combination of the two compressive stresses in ABC.

Fig. 4.19 is a return to Fig. 4.17 with the arbitrary choice of $\beta = 30°$ and a minimum width of the inclined legs so as to give the yield value of $2k$ for the compressive stress in each:

$$F = 2kb \cos 30°, \quad P = 2F \cos 30° = 4kb \cos^2 30° = 3kb \qquad [4.48]$$

This is a proper lower bound for the limit value of P provided the state of stress in the overlap region ABC is not above yield, and further that the forces at the lower pins can be carried without violating yield. The sum of two compressive fields of $2k$ each at an angle of $60°$ has been used much before because it does have the permissible compressive principal stresses of k and $3k$. As was demonstrated in section 4.2 the general result for the sum of two stress fields in the plane is that the mean stresses add algebraically and the maximum shear stresses add as vectors with an angle between them equal to twice the angle in the physical plane. As far as T and $\frac{1}{2}P$ are concerned, no problem arises either. The picture bears an obvious resemblance to the more general Fig. 4.18(b). It is simpler because T and $\frac{1}{2}P$ are at right angles so that the overlap condition is:

$$\frac{P}{2w} + \frac{T}{h} \leqslant 2k \qquad [4.49]$$

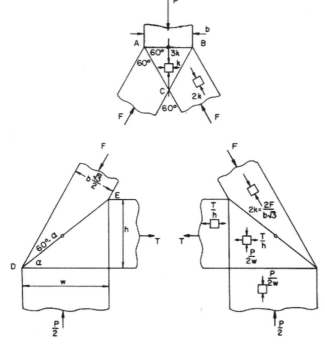

Fig. 4.19. Details of the truss of Fig. 4.17, $\beta = 30°$

This condition is met for $\alpha \geqslant 30°$. As before, there is no need to look at the jump conditions across the lines of stress discontinuity AC, DE because they are satisfied automatically.

If α is chosen as $30°$ and the picture is drawn in its most compact configuration, the familiar trapezoidal discontinuous pattern emerges, Fig. 4.20 which is the special case of Fig. 4.16 when $\varphi = 0$, $c = k$. Its very compactness, however, does tend to hide its simple meaning which is exhibited far better in Fig. 4.19.

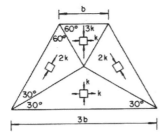

Fig. 4.20. Compact discontinuous stress fields at yield.

Addition of stress legs (Figs. 4.21, 4.22)

The truss picture of Fig. 4.17(b) clearly does not provide the maximum possible support for the load P. A vertical leg, at least, is a reasonable addition if more load is to be carried than $3kb$. Fig. 4.21(a) shows the addition of the vertical leg to the 30° legs. Again for simplicity all legs are taken at yield; the compressive stress in each is $2k$. Now, however, the overlap regions are above yield. Satisfaction of the intuitive feeling that load should be carried both vertically down and along inclined directions thus requires either a reduction of the stresses in the members of the truss, or the addition of a horizontal thrust of $2k$ as shown in Fig. 4.21(b) to balance the vertical $2k$. If an estimate of the maximum load carrying capacity is sought, the horizontal thrust is needed. Often, the problem is to decide on the safety at a given load and Fig. 4.17(b) alone may suffice.

The resemblance to the Prandtl slip-line field of classic plasticity, Fig. 4.22(a), is evident and becomes closer and closer the more the number of supporting legs chosen. A picture for nine vertical and inclined legs at 10° to each other is shown in Fig. 4.22(b). As the number of legs grows, the stress in each decreases. In the limit, the slip-line field beneath the punch is recovered exactly while the stress state away from this region is given by the overlap angle Ψ, Fig. 4.22(c).

The stress of $0.695k$ in the legs of Fig. 4.22(b) and of $2k$ in those of Fig. 4.21(b) are special cases of $4k\sin 2\theta$ in Fig. 4.23. When as in Fig. 4.22(b), $2\theta = 10°$, $\sin 2\theta = 0.1737$, when as in Fig. 4.21(b), $2\theta = 30°$, $\sin 2\theta = 0.500$. The value $4k\sin 2\theta$ represents the largest possible uniaxial compression stress which can be added to the stress field in zone 1 to produce the jump and thus the stress field in zone 2 without violating the yield condition (Fig. 4.23). The Mohr circles corresponding to zone 1 and zone 2 are shown in Fig. 4.6(a). Further discussions on the overlapping of stress fields will be given in the later part of this Chapter.

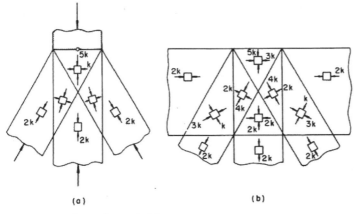

(a)   (b)

Fig. 4.21. Addition of a vertical leg.

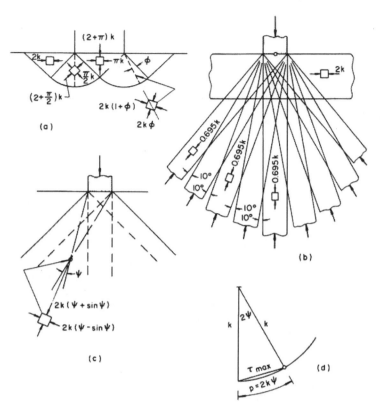

Fig. 4.22. Supporting legs give slip-line in limit. (a) Slip-line field. (b) 9 legs at 10°. (c) Limit of infinitely many legs in region $\psi \leqslant 30°$. (d) Addition of overlap fields.

Change of stress direction (Fig. 4.24)

Addition of the horizontal stress, $2k$, although necessary, does not contribute to P in the local sum of vertical forces. If the body is real, and therefore finite in extent, the force in the horizontal truss members must be turned around in the space available without violating yield. This extension of the plastic stress field under the punch was shown to be possible by Bishop (1953) who also obtained an explicit solution in a total width of $8.7b$. Infinitely many choices exist when the width available is ample as indicated schematically by the truss members of Fig. 4.24(a) and the stress patterns of Fig. 4.24(b). The more sharply the truss member directions are turned, the larger the force or stress which must be carried in tension. Fig. 4.18(b) may be thought of as the turning of a compressive force Q to give a compressive force R. The picture for half a turn of the entire picture is symmetric; it is represented by the lower portion of Fig. 4.19 where the half turn angle is 30°. Fig. 4.18 and 4.19 show that there is no symmetry requirement or

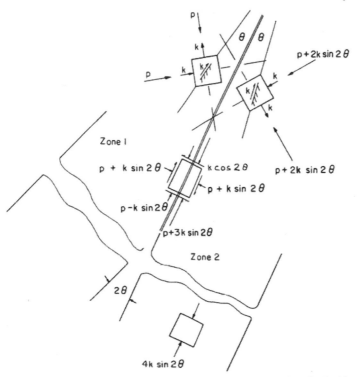

Fig. 4.23. Addition of compression Leg–Stress fields at yield on both sides of discontinuity.

need for fields of stress at yield. The geometric relations for the special choice of stress fields at yield are well known and indicated in Fig. 4.24(c). As in Fig. 4.23 the discontinuity line bisects corresponding directions of principal shear stress and so is at an angle θ to the 45°-directions, where the total angle of turn is 4θ.

A trapezoidal stress field (Fig. 4.25)

As a last example of the plane truss approach, consider the 30°-legs in a trapezoidal or wedge region, Figs. 4.19, 4.20, 4.25(a). The slip-line field solution gives the total force $P = (2 + \frac{1}{3}\pi)$ kb which is very little higher than the $3kb$ of the discontinuous field, Figs. 4.19 and 4.20. Suppose the question is asked whether a given geometric configuration will support $3kb$. For example, is the width of $3b$ in Fig. 4.20 needed? It is sufficient, but will a smaller width be enough? This is a very relevant question for the notched bar in tension which is mathematically equivalent to a wedge in compression.

The truss picture of Fig. 4.25(a) in its most compact form, Fig. 4.25(b) provides the information that $3kb$ can be carried in a total width of $2.63b$. This reduction

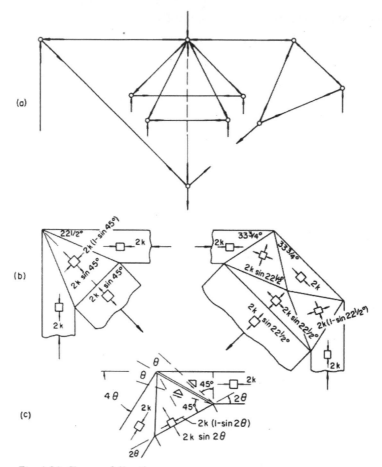

Fig. 4.24. Change of direction.

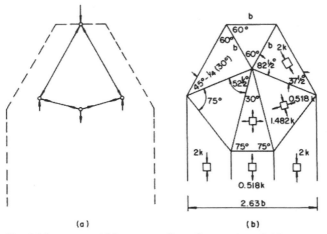

Fig. 4.25. A trapezoidal, compact discontinuous stress field.

from $3b$ with so little effort seems a worth while accomplishment but there is no reason to suppose that the minimum value has been found. Discontinuous fields rarely are real. The truss approach is an engineering tool for obtaining essential but relatively crude information quickly.

Extension of the truss concept to three dimensions is obvious and there does not compete with known solutions and established techniques. Earlier work on the punch problem for a semi-infinite domain (Shield and Drucker, 1953) provides an indication of its possible usefulness. Further discussions on the applications of this technique to some three-dimensional bearing capacity problems will be presented in Chapter 7 for soils and in Chapter 10 for rock-like materials.

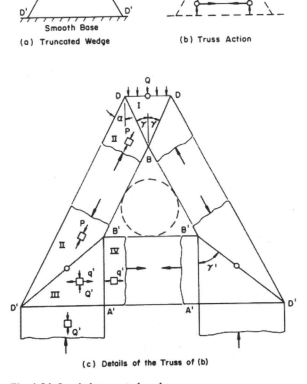

(a) Truncated Wedge

(b) Truss Action

(c) Details of the Truss of (b)

Fig. 4.26. Loaded truncated wedge.

4.7. DISCONTINUOUS FIELDS OF STRESS VIEWED AS PIN-CONNECTED TRUSSES – COULOMB MATERIALS

Truncated wedge stress fields (Fig. 4.26)

This section will continue the discussion of the techniques described in the preceding section for a pure cohesive soil to the more general case of a $c - \varphi$ soil. This extension is best illustrated by the example of constructing the stress field of a loaded truncated wedge as shown in Fig. 4.26. As in Fig. 4.17, the dimension perpendicular to the plane of the paper will be taken as unity, and all motion is supposed in the plane. The weight of the soil mass is neglected. The applied pressure Q is carried through the trapezoid region of Fig. 4.26(a) in a very elaborate pattern from the smooth (or rough) footing to the smooth (or rough) supporting rock base.

Suppose instead that a pin-connected truss is imagined to carry the load inside the body, Fig. 4.26(b). The truss action of the truncated wedge would then indicate a stress field pattern shown in Fig. 4.26(c) where the two triangles marked I, III are under a biaxial state of stress and the regions marked II, IV are under uniaxial compression P and uniaxial tension q', respectively. Four elementary stress fields are shown in Fig. 4.27(b, c, b′ and c′). The steps are self-evident in Fig. 4.27. As the field is symmetrical about the axis of applied pressure and the field of (b′) or (c′) is actually the same as in (b) or (c) except that α is negative, one only needs to discuss the representative one of Fig. 4.27(b).

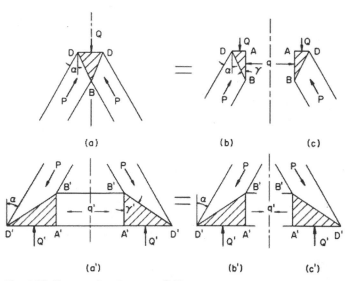

Fig. 4.27. Truncated wedge stress fields.

To obtain the highest lower bound of applied pressure Q which is supported by the horizontal stress q and inclined stress P with the inclination angle α to the vertical (**Fig. 4.28a**), it is clear that one will take the largest permissible value for the stress P. Since the state of stress in region II is uniaxial compression, the value of P is given by [4.34]. The value Q, q, and the inclination angle, γ, of line BD to the vertical are determined in terms of the known quantities φ and α from the following conditions: (1) The over-all equilibrium of forces acting on the triangular element ABD in vertical and horizontal directions: (2) the assumption that the material in the triangular element ABD is plastic. Referring now to Fig. 4.28(a), if the length of AD is taken as unity, equilibrium of forces (see Fig. 4.28b) gives:

Σ in vertical direction = 0

$$Q = \frac{P \sin (\alpha + \gamma)}{\sin \gamma} \cos \alpha \quad \text{or} \quad \tan \gamma = \frac{P \sin \alpha \cos \alpha}{Q - P \cos^2 \alpha} \qquad [4.50]$$

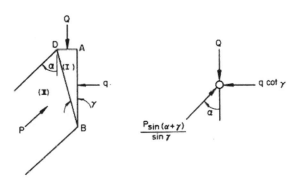

(a) Stress Field   (b) Force Equilibrium

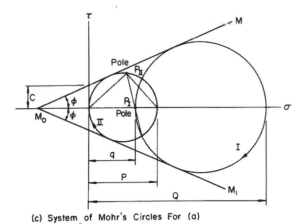

(c) System of Mohr's Circles For (a)

Fig. 4.28. The triangular stress field.

Σ in horizontal direction = 0

$$q = \frac{P \sin (\alpha + \gamma) \sin \alpha \tan \gamma}{\sin \gamma} = \frac{QP \sin^2 \alpha}{Q - P \cos^2 \alpha} \tag{4.51}$$

independent of material properties. The Coulomb yield criterion requires:

$$(q + Q) \sin \varphi + (q - Q) + 2c \cos \varphi = 0 \tag{4.52}$$

The Mohr circles for the regions marked I, II in Fig. 4.28(a) are shown in Fig. 4.28(c) where the poles of the corresponding Mohr circles are also indicated. Substituting the value of q in [4.51] into [4.52], making use of [4.34] a quadratic expression for Q is then obtained:

$$Q^2 - \frac{2 - 2 \sin \varphi + 2 \sin \varphi \sin^2 \alpha}{1 - \sin \varphi} PQ + P^2 \cos^2 \alpha = 0 \tag{4.53}$$

Taking the larger value of the relevant roots of the quadratic (i.e., positive sign):

$$Q = \frac{2c \cos \varphi [1 - \sin \varphi + \sin \varphi \sin^2 \alpha + \sin \alpha (1 - \sin^2 \varphi \cos^2 \alpha)^{1/2}]}{(1 - \sin \varphi)^2} \tag{4.54}$$

When $\varphi = 0$, $c = k$ (Tresca material), [4.54] takes the form:

$$Q = 2k(1 + \sin \alpha) \tag{4.55}$$

agreeing with the value obtained in [4.46]. To put [4.54] in a more compact form, using the relation [4.43] and after some reduction, [4.54] reduces to:

$$Q = c \cot \varphi \left[\tan^2 (\tfrac{1}{4} \pi + \tfrac{1}{2} \varphi) \frac{\sin (\nu + \alpha)}{\sin (\nu - \alpha)} - 1 \right] \tag{4.56}$$

agreeing with the value obtained in [4.44] through the application of the general "jump condition". With this value of Q, [4.50] and [4.51] give:

$$\gamma = \tfrac{1}{2}(\nu - \alpha) \tag{4.57}$$

and:

$$q = c \cot \varphi \tan \alpha \tan \tfrac{1}{2}(\nu - \alpha) \left[\tan^2 (\tfrac{1}{4} \pi + \tfrac{1}{2} \varphi) \frac{\sin (\nu + \alpha)}{\sin (\nu - \alpha)} - 1 \right] \tag{4.58}$$

As for the stress field in Fig. 4.27(b'), the value of Q', q' and γ' are exactly the same as those of Q, q and γ except for substitution of $-\alpha$ for α in the corresponding equations.

Fig. 4.16 shows the most compact configuration of the stress fields in Fig. 4.26(c) when the triangular area $BB'B'$ approaches zero. Here the uniaxial tension region IV (note: q' may not yet reach its yield value) vanishes, therefore, the whole trapezoid region $DDD'D'$ is plastic. Its very compactness, however, does tend to hide its simple meaning which is exhibited far better in Fig. 4.26(c).

It is interesting to note that the stress field of Fig. 4.26(c) is also applicable for the case where a hole (as when a flexible pipe is embedded) is present in the triangular region as shown by dotted lines in the figure.

Once again it is worth pointing out that an over-all equilibrium of forces acting on the triangular element ABD (Fig. 4.28a) demands only that the stress component σ_n normal to BD and the shear stress τ parallel to BD should be the same on both sides, but the components $\sigma_t^{(1)}$ and $\sigma_t^{(2)}$ acting parallel to BD may be different (see Fig. 4.5a). Clearly the line BD is a line of stress discontinuity. The amount $\sigma_t^{(1)}$ and $\sigma_t^{(2)}$ are restricted by the condition that the material on both sides of the line is nowhere violating the yield criterion. The results obtained by applying the jump condition [4.29] for the discontinuous line of the field in Fig. 4.28(a) are met.

Strip footing resting on a finite layer (Figs. 4.29, 4.30)

Simple variations on the above example are shown in Fig. 4.29 and Fig. 4.30; the cases referred to are as follows:

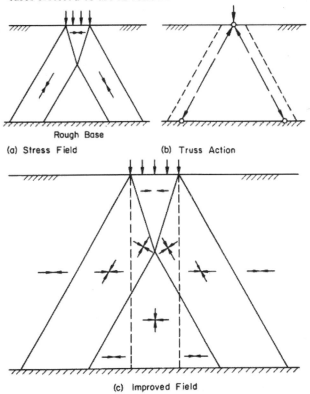

(a) Stress Field   (b) Truss Action

(c) Improved Field

Fig. 4.29. Strip foundation on a semi-infinite body resting on a perfectly rough base.

(1) *Case 1*. A strip foundation is supported by a semi-infinite mass of soil of finite thickness resting on a rough rock base (Fig. 4.29a). Physical intuition for this problem strongly suggests that the portion of the material under the foundation will behave as a truncated wedge (Fig. 4.29b). Therefore, a discontinuous stress field (Fig. 4.29a) is constructed by applying the idea of truss action. The foundation is supported by two uniaxial compression "legs" which rest on the base and the shear component of the traction between the "legs" and the base may have any value, since the base is rigid and is perfectly rough. Here the value of the lower bound can be further increased through utilizing the stress-free material more fully, that is, superimposing a vertical compression vertically below the strip foundation and a horizontal compression throughout the material (see Fig. 4.29c). The state of stress for different regions is represented by the small arrows in the figure. The details of this stress field will be presented in the later part of this section as an illustrative example.

(2) *Case 2*. In Fig. 4.30, the problem is the same as in Fig. 4.29 except that the

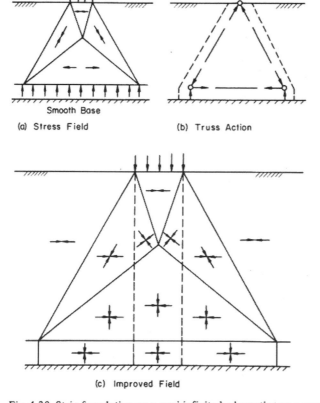

Smooth Base

(a) Stress Field   (b) Truss Action

(c) Improved Field

Fig. 4.30. Strip foundation on a semi-infinite body resting on a smooth base.

base is smooth. A stress field suggested by the truss action (Fig. 4.30b) is shown in Fig. 4.30(a), and its improvement by using the same idea as in case 1, is also shown (Fig. 4.30c).

It should perhaps be seen now that the technique of constructing stress fields aided only by ordinary mechanics of materials appeals to and makes use of the physical intuition developed by engineers. With the limit theorems, the results so obtained do provide the engineer with a quantitative feeling for his problem although the stress field need bear no resemblance to the actual state of stress according to the theorems.

Example 4.4: Wedge under unilateral pressure (Chen, 1969)

Lower-bound solutions (Fig. 4.32)

The construction of a stress field for the previous example 4.2 of wedges with uniform pressure on one face will serve as an illustration of the technique for $c - \varphi$ soils.

The discontinuous stress fields in connection with Figs. 4.27(c') and 4.27(b) can now be extended for an obtuse wedge as well as for an acute one. By rotating the stress field of Fig. 4.27(c') counterclockwise about 90°, one obtains the field of Fig. 4.31(a) and the extension of the two sides $D'A'$ and $D'B'$ of the triangular element $D'A'B'$ to infinity gives a stress field for the acute wedge (Fig. 4.31b). In the figure, the wedge $A'D'E'$ with a central angle β is loaded by a uniform pressure Q' along $A'D'$ and the line $D'B'$ is the line of stress discontinuities separating the two constant stress regions $A'D'B'$ and $B'D'E'$. The states of stress for different regions are represented by the small arrows in the figure. Since the pressure, Q' obtained previously is a function of the quantities c, φ, and α only it can be concluded that here the value of Q' is still given by [4.45].

In the same way it can be seen that Fig. 4.32(b) gives a stress field for the obtuse wedge and again the value of Q is given by [4.44] or [4.56]. By putting $\alpha = \beta - \frac{1}{2}\pi$

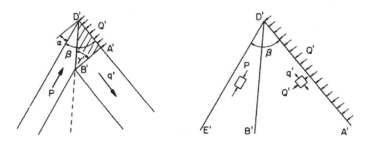

(a) STRESS FIELD OF FIG. 4.27(c')   (b) EXTENSION OF (a)

Fig. 4.31. Acute wedge $\beta \leqslant \frac{1}{2}\pi$.

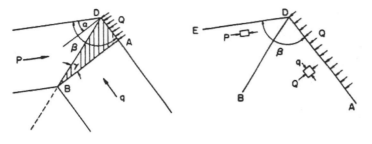

(a) STRESS FIELD OF FIG. 4.27(b)   (b) EXTENSION OF (a)

Fig. 4.32. Obtuse wedge $\beta \geqslant \frac{1}{2}\pi$.

and thus $v = \frac{1}{2}\pi - \mu$, [4.44] or [4.56] reduces to the form of [4.37a] obtained previously for the wedge problem by applying the jump conditions for the discontinuous line of the stress field. By the lower-bound theorem of limit analysis stated in the first section, Q or Q' is therefore a lower bound for the uniform critical pressure of the wedges.

Upper-bound solutions (Figs. 4.33, 4.34)

As for an upper bound of the pressure, a failure mechanism is usually indicated by physical intuition and the methods for constructing a failure mechanism in plane strain have been discussed thoroughly in the preceding Chapter. For example, for the case of an acute wedge, the simplest pattern of failure mechanism may be found by taking a plane slide as shown in Fig. 4.33. Equating external rate of work to the dissipation [3.19] gives:

$$Q'1V \cos\left(\varphi + \psi + \beta - \tfrac{1}{2}\pi\right) = cV \cos\varphi \,\frac{\sin\beta}{\sin\psi} \qquad\qquad [4.59a]$$

so that: $Q' = \dfrac{c \cos\varphi \sin\beta}{\sin\psi \sin(\psi + \beta + \varphi)}$ \qquad\qquad [4.59b]

minimizing the right-hand side gives:

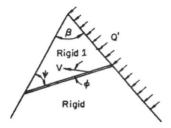

Fig. 4.33. Rigid body "slide" motion for acute wedge.

$$\psi = \tfrac{1}{2}\pi - \frac{(\beta + \varphi)}{2} \qquad [4.60]$$

and:
$$Q' \leqslant Q'^u = \frac{2c \cos\varphi \sin\beta}{1 + \cos(\beta + \varphi)} \qquad [4.61]$$

For the case of the obtuse wedge, the selected failure mechanism is shown in Fig. 4.34 which consists of two triangular regions ABC, ADE moving as a rigid body and a logarithmic spiral zone ACD. The material below the failure line $BCDE$ remains at rest so that $BCDE$ is a line of velocity discontinuity. Thus the velocity along the line must be inclined at an angle φ to the line. In the spiral zone ACD, the velocity increases exponentially from the initial velocity V_0 along the radial line AC to the velocity $V = V_0 e^{(\Theta \tan\varphi)}$ along the radial line AD as discussed in section 3.4, Chapter 3. Energy is dissipated throughout the logspiral zone as well as along the spiral surface at an equal rate as given by [3.74]. Energy is also dissipated along the surface of discontinuity BC and DE. With [3.19], the rate of dissipation of energy can easily be calculated. Take the length AC as r_0 so that the length AD is $r_0 e^{(\Theta \tan\varphi)}$. The energy dissipation along the line segments BC and DE then is given by:

$$cr_0(V_0 \cos\varphi) + c(r_0 e^{\Theta \tan\varphi})(V_0 e^{\Theta \tan\varphi} \cos\varphi) \qquad [4.62]$$

The upper bound is obtained by equating $2Qr_0\cos(\tfrac{1}{4}\pi + \tfrac{1}{2}\varphi)V_0\sin(\tfrac{1}{4}\pi - \tfrac{1}{2}\varphi)$, the rate of work done by the external pressure Q, to the total rate of dissipation:

$$Q \leqslant Q^u = c \cot\varphi \left[e^{2\Theta \tan\varphi}\tan^2(\tfrac{1}{4}\pi + \tfrac{1}{2}\varphi) - 1\right] \qquad [4.63]$$

agreeing with the value obtained by Prandtl (1921). By putting $\beta = \tfrac{1}{2}\pi + \Theta$, [4.63] can be written:

$$Q \leqslant Q^u = c \cot\varphi \left[e^{(2\beta - \pi)\tan\varphi}\tan^2(\tfrac{1}{4}\pi + \tfrac{1}{2}\varphi) - 1\right] \qquad [4.64]$$

These lower and upper bounds for the critical pressure of wedges are plotted against wedge angle β in Fig. 4.35 for $\varphi = 20°$. Further improvement of the lower-bound stress field for the case of the obtuse wedge will be given later.

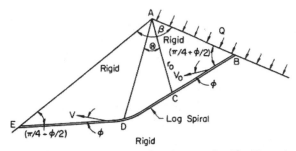

Fig. 4.34. Failure mechanism for obtuse wedge $V = V_0\exp(\Theta \tan\varphi)$.

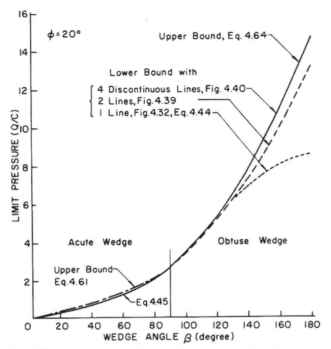

Fig. 4.35. Lower and upper bounds for a wedge under unilateral pressure, $\varphi = 20°$.

Example 4.5: Rectangular footing on $c - \varphi$ soils (Shield, 1955a)

As an example to construct the stress field shown in Fig. 4.29(c) in the two-dimensional case and also its possible extension to three-dimensional soil mechanics problems, we obtain a lower bound for the bearing capacity of a rectangular footing on the plane surface of a semi-infinite mass of $c - \varphi$ soils. The upper and lower bounds obtained in this example are applicable to either a rough or smooth footing which is centrally loaded.

Two-dimensional case (Fig. 4.36)

The stress field shown in Fig. 4.29(c) has been described earlier. The stresses acting in the various regions are given in Fig. 4.36. The field is built up in the following way. A uniaxial compression of amount P in the regions $ACFD$, $BCGE$ produces a vertical compression Q and a horizontal compression q in the region ABC, AC and BC being lines of stress discontinuity. Superimposed on this basic field are a horiztonal compression of amount R throughout the soil, and a vertical compression R in the region $ABIH$ vertically below the line of footing AB. The largest permissible value is taken for the stress R. This value is $2c \tan(\frac{1}{4}\pi + \frac{1}{2}\varphi)$ since this stress acts near the free surface in uniaxial compression state. The stress P and

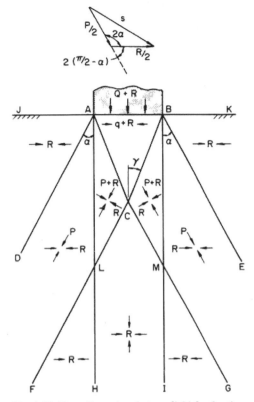

Fig. 4.36. Two-dimensional stress field for footing.

its inclination α to the vertical are chosen so that regions BCM and $BMGE$ are both at the point of yielding. Region BCM will be at yield if the uniaxial compression P and the hydrostatic pressure R satisfy the Coulomb yield condition (Fig 4.12b or [4.27]):

$$\frac{P + 2R}{2} + c \cot \varphi = \frac{P}{2 \sin \varphi} \qquad [4.65a]$$

or substituting the value of R in the equation, we have:

$$P = 2c \tan^3 \left(\tfrac{1}{4}\pi + \tfrac{1}{2}\varphi\right) \qquad [4.65b]$$

independently of the value of α. Region $BMGE$ will be at yield if the sum of the two uniaxial compressions P and R satisfy the yield condition. Since the general result for the sum of two stress fields in the plane is that the hydrostatic or mean stress add algebraically which gives the resulting hydrostatic pressure $p = (\tfrac{1}{2}P + \tfrac{1}{2}R)$, and the maximum shear stresses add as vectors with an angle between $\tfrac{1}{2}P$ and $\tfrac{1}{2}R$ equal to 2α as indicated by the shear stress diagram in Fig. 4.36. This gives the resulting maximum shear stress:

$$s^2 = (\tfrac{1}{2}P)^2 + (\tfrac{1}{2}R)^2 - 2(\tfrac{1}{2}P)(\tfrac{1}{2}R)\cos 2\alpha$$

$$= \tfrac{1}{4}(P^2 + R^2 - 2PR\cos 2\alpha) \qquad\qquad [4.66]$$

Since the resulting hydrostatic pressure p and the resulting maximum shear stress s must satisfy the Coulomb yield condition (see Fig. 4.12b):

$$s = (c\cot\varphi + p)\sin\varphi \qquad\qquad [4.67]$$

we obtain, after some simplifications, that:

$$P = \frac{4c\,(\cos 2\alpha + \sin\varphi)}{\cos\varphi\,(1 - \sin\varphi)} \qquad\qquad [4.68]$$

equating the values [4.65b] and [4.68] for P shows that α is given by:

$$\cos 2\alpha = \tfrac{1}{2}(1 + \sin^2\varphi) \qquad\qquad [4.69a]$$

which can be reduced to the simple form:

$$\sin\alpha = \frac{\cos\varphi}{2} \qquad\qquad [4.69b]$$

The compressions Q, q and the inclination γ of the stress discontinuities AC, BC to the vertical can now be determined by the equilibrium of forces in vertical and horizontal directions and also by the conditions that the material in the triangular element ABC is plastic. As described earlier in Fig. 4.28(a), the pressure is found to be:

$$Q = \tfrac{1}{2}c\tan^3(\tfrac{1}{4}\pi + \tfrac{1}{2}\varphi)\,[4 + \sin\varphi + \sin^2\varphi + (1 + \sin\varphi)(4 + \sin^2\varphi)^{1/2}] \qquad [4.70]$$

The values of $\tan\gamma$ and q are still given by [4.50] and [4.51].

Three-dimensional case (Fig. 4.37)

The extension of the two-dimensional field of Fig. 4.36 to a rectangular footing is shown in Fig. 4.37. $LMNO$ is the rectangular area of contact between the footing and soil, LO and MN being the larger sides of the rectangle. The faces of the volume $LMNOXY$ are inclined at an angle γ to the vertical; and the four prisms, two triangular and two trapezoidal, extending from the faces of the volume $LMNOXY$, are all inclined at an angle α to the vertical. Only two of the prisms are shown in Fig. 4.37. The pressure Q on the rectangular area is supported by compressions P in the four prisms. The addition of an all-round horizontal compression R throughout the soil, and a vertical compression R in the soil vertically below the rectangle $LMNO$, completes the stress field. Fig. 4.36 is then the vertical cross-section through the midpoints A, B of the longer sides of the rectangle.

Since the Coulomb yield criterion is not violated in the plane of Fig. 4.36, it only remains to check that the stress applied normal to the plane of Fig. 4.36 is intermediate of the two principal stresses in that plane, in order that the Coulomb

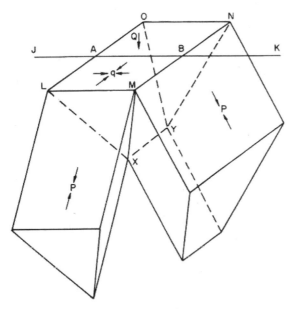

Fig. 4.37. Basis of stress field for rectangular footing.

yield criterion be not violated by the three-dimensional stress field. The verification is straightforward. We remark that the value R, which was chosen for the vertical compression added below the area of footing, is the largest which can be taken without violating the yield condition in those parts of the prisms below the area of footing.

It follows that the pressure $\bar{Q} = Q + R$, that is:

$$\bar{Q} = Q + R = \tfrac{1}{2}c \tan^3(\tfrac{1}{4}\pi + \tfrac{1}{2}\varphi)\,[4 + \sin\varphi + \sin^2\varphi + (1 + \sin\varphi)(4 + \sin^2\varphi)^{1/2}]$$

$$+ 2c\tan(\tfrac{1}{4}\pi + \tfrac{1}{2}\varphi) \quad [4.71]$$

on the area of contact can be supported by a statically admissible stress field. By the lower-bound theorem of limit analysis quoted in section 4.1, $Q + R$ is therefore a lower bound for the average footing pressure. The lower bound [4.71] is plotted against the angle of friction φ in Fig. 4.38. For comparison, the upper-bound value given by [3.81] for the bearing capacity of a very long rectangular footing (plane strain) is shown in the same diagram. The value [3.81] is an upper bound for the strip surface footing for all values of φ, and is also a lower bound for values of φ less than 75° (Shield, 1954a).

Remarks on the solutions

As a special case, the lower bound [4.71] applied to a square punch. The points

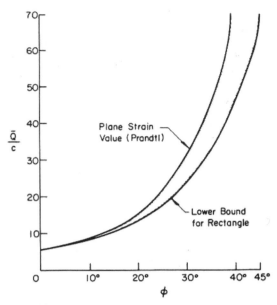

Fig. 4.38. Lower bound for the average bearing pressure for a rectangular footing and the average bearing pressure for a very long rectangle (plane strain).

X and Y in Fig. 4.37 are then coincident. Further, the stress field of Fig. 4.37 can be modified to apply to any convex area of indentation. The value [4.71] is therefore a lower bound for the average indentation pressure for any convex area.

The bound is also applied to any uniformly loaded convex area as distinct from an area loaded by a rigid footing. As remarked by Shield and Drucker (1953) for a Tresca material which will be the subject of discussion in Chapter 7, the plane strain value [3.81] is an upper bound for any uniformly loaded area. This is because the soil immediately adjacent to the boundary of the loaded area can move in a manner which is essentially a plane strain motion.

4.8. GRAPHICAL CONSTRUCTION OF DISCONTINUOUS STRESS FIELDS

Obtuse wedge with two lines of discontinuity (Fig. 4.39)

Consider again the simple problem of an obtuse wedge under unilateral pressure, as shown in Fig. 4.39(a). Let us now explore the possibility of improving the lower-bound stress field shown in Fig. 4.32(b) by using more than one line of stress discontinuity. Tentatively we replace line DB, Fig. 4.32(b), by two lines Dc and Dd. The line Dc is chosen to be perpendicular to the surface DD and the inclination of the line Dd to the surface DD is yet to be determined, as shown in Fig. 4.39(a).

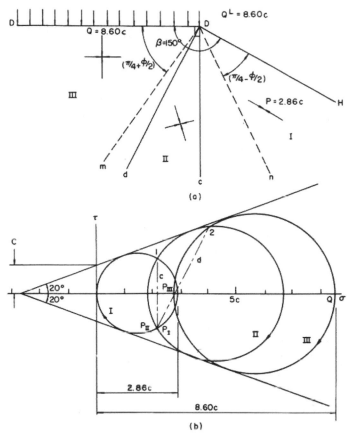

Fig. 4.39. (a) Stress field with two lines of discontinuity. (b) Graphical solution for (a).

Analytical exploration of such a stress field may become cumbersome and the graphical procedure presented herein seems preferable.

Figure 4.39(a) shows the stress system of the case and the regions marked I, II, III in (a) are shown by the Mohr circles in (b). The three Mohr circles are built up in the following way. The Mohr circle representing the uniaxial compression of amount $P = 2.86c$ in the region marked I, Fig. 4.39(a), is drawn and labeled I in Fig. 4.39(b). The point labeled P_I is the corresponding pole of the circle. Next, the line drawn through the pole P_I of the Mohr circle parallel to the discontinuous line Dc in (a) intersects in a point labeled 1 on the circle. When the point 1 is known, the Mohr circle which corresponds to the state of stress in region marked II in (a) is readily found by drawing, through the point 1, a circle tangent to the yield envelope. The second circle and its corresponding pole are labeled II and P_{II} in (b), respectively. Since the pole of the Mohr circle for the region marked III in (a) must lie on the σ-axis in (b) in order not to violate the stress boundary condition on the surface DD, it follows that the Mohr circle labeled III in (b) can be determined by the method of trial and error by assuming several inclination angles of the discontinuous line Dd to the surface DD. The correct inclination of line Dd gives the Mohr circle with its pole right on the σ-axis. This circle is labeled III in (b). Line $P_{II}\text{-}P_{III}\text{-}2$ is parallel to Dd in Fig. 4.39(a) and $P_I\text{-}1$ is parallel to Dc. The pressure on DD has the value $8.60c$ corresponding to $\beta = 150°$ and $\varphi = 20°$. The corresponding value for one line of discontinuity is $7.72c$. Clearly two lines of discontinuity are better than one from the present point of view because the last circle is seen pushed further to the right than the first and second circles.

Obtuse wedge with four lines of discontinuity (Fig. 4.40)

Let us see whether a further improvement may be made by having four lines of stress discontinuity, as shown in Fig. 4.40(a). The lines a, b, c are arbitrarily chosen

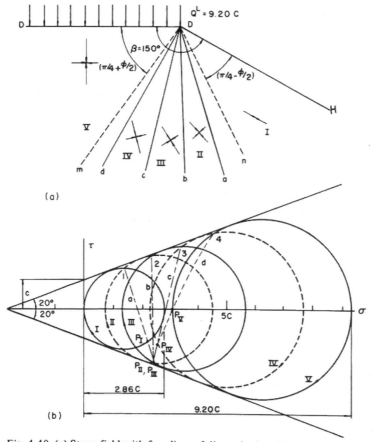

Fig. 4.40. (a) Stress field with four lines of discontinuity. (b) Graphical solution for (a).

and line d is selected to fit the boundary condition of DD. The Mohr diagram will now have five circles, disposed as in Fig. 4.40(b). All the constructions are similar to that of the two-line case and do not require further explanation. The pressure on DD now has the value $9.20c$ which is about the amount of the upper-bound value obtained previously in [4.64]. Any further increase in the pressure by introducing more discontinuous lines for this case will be of no real help as the bounds are already very close as shown in Fig. 4.35 for various values of β.

It is interesting to note that the stress field shown in Figs. 4.39 and 4.40 may be interpreted as an approximated extension of the Prandtl field of Fig. 4.34 where the continuous changing of stresses in spiral zone ACD is approximated by a few discontinuous lines. Therefore it may be concluded that Prandtl's (1921) solution as given by [4.64] gives the correct values since it may be considered as the limiting case when the numbers of discontinuous lines tend to be infinite.

In the graphical construction of the discontinuous stress fields, Fig. 4.39 and Fig. 4.40, the essential feature to be noted is that the discontinuous lines must be located inside the region bounded by the two dotted lines Dm and Dn in order to have the successive Mohr circles pushing successively further and further to the right in the Mohr diagram so that a high pressure along DD will be produced *effectively*. The lines Dm and Dn makes angles of $(\frac{1}{4}\pi + \frac{1}{2}\varphi)$ and $(\frac{1}{4}\pi - \frac{1}{2}\varphi)$ with the boundary surfaces DD and DH, respectively. The region nDm actually is the corresponding region of logarithmic spiral ACD of Fig. 4.34. The small arrow representing the principal stresses for each constant stress region shown in Fig. 4.40(a) indicate that the four discontinuous lines a, b, c, d so arranged are indeed very effective in producing a high pressure along the surface DD.

Pole trajectory (Fig. 4.41)

It is important to note that the transition from the old to the subsequently new Mohr circle as shown in Fig. 4.40(b) may be interpreted as a rolling without slippage on the lower envelope accompanied by a uniform expansion. The contact point between circle and lower envelope plays the role of both the instantaneous center of rotation and the center of expansion. This combined rolling and expansion brings the old pole, say P_{III} in Fig. 4.40(b) into a new position P_{IV}. The poles P_I, P_{II}, P_{III} ...therefore describe a *pole trajectory* when the Mohr circle rolls without slipping on the lower envelope and simultaneously expands uniformly so as to remain in contact with the upper envelope.

De Jong (1957) has shown that the shape of the pole trajectory for the case of Prandtl's slip-line field under a footing on a weightless soil is a *cycloid* as shown in Fig. 4.41. Since the stress field of Fig. 4.40(a) may be considered as the limiting case of the Prandtl's slip-line field under a finite footing when the footing size goes to infinity along one direction, it follows that the shape of the pole trajectory for

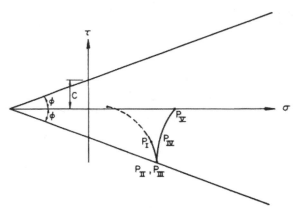

Fig. 4.41. Pole trajectory for a weightless soil.

the Mohr circles shown in Fig. 4.40(b) is still a cycloid. The points labeled P_I, P_{II}, P_{III}, P_{IV} and P_V in Fig. 4.41 are the poles of the corresponding Mohr's circles in Fig. 4.40(b).

As through a given pole two circles centered on the σ-axis can be drawn tangent to the envelope, it implies that the position of the pole is sufficient specification of the state of stress at the considered point of the physical plane in a given problem provided that it is understood whether the considered problem is one of active or passive earth pressure. This usually presents no difficulty in a physical problem. In view of the correspondence between the points of the physical plane and the poles in the stress plane, the slip-lines of the physical plane have images in the stress plane. This fact is of great value in graphical construction of slip-line fields in soil mechanics. Details of such graphical procedure have been developed by Prager (1953) for a purely cohesive soil, for which $\varphi = 0$ (Tresca material), and later extended by De Jong (1957) for the more general case of $c - \varphi$ soils accounting for the weight of the soil.

4.9. COMBINED METHOD FOR SOLVING THE PROBLEMS INVOLVING OVERLAPPING OF DISCONTINUOUS STRESS FIELDS

There are still certain difficulties, of course, which have to be solved, associated with the analytical or graphical constructions of the discontinuous stress fields. However, certain difficulties may be considerably reduced by the use of *combining* the graphical techniques with analytical calculations.

In the following, a graphical technique combined with analytical calculations for obtaining a lower-bound solution using the concept of truss action will be presented. For purposes of illustration the simple problem of a smooth flat footing indenting a weightless medium will be solved assuming plane strain conditions. The procedures for both a Tresca material and general Coulomb materials will be given. The graphical technique presented herein is based on the recognition that a state of stress can always be decomposed into two components: the mean or hydrostatic state of stress, p, and the deviatoric or pure shear state of stress, s.

4.9.1. Tresca materials ($\varphi = 0$, $c = k$)

Three stress legs (Fig. 4.42)

To begin with, the punch or bearing capacity problem for a Tresca material (a Coulomb material for which $\varphi = 0$ and $c = k$) is solved using three symmetrical legs each having a uniaxial compressive stress of $2k$ (k being the yield stress in shear) as shown in Fig. 4.42(a). Since we are trying to obtain the plastic limit load for a smooth footing, the directions of the principal stresses in region labelled (4) must be vertical and horizontal. Due to the symmetry of the arrangement of the legs, the resulting overlapping stress state in region (4) will in fact satisfy this stress bounda-

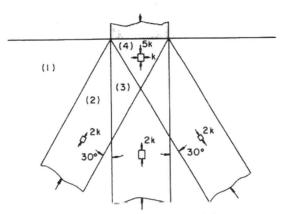

Fig. 4.42(a). Three supporting legs at 30°

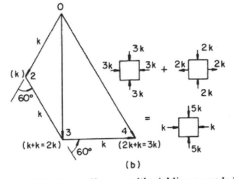

Fig. 4.42(b). Stress diagram with yielding exceeded in region 4.

ry condition. It is clear that this resulting stress field will be the result of a superposition of the three legs. The sum of various individual stress states are added such
that their hydrostatic components add algebraically while the deviatoric components add vectorially. The angle between these diviatoric stress vectors is equal to
twice their angle in the physical plane. These constructions can be validated using
the properties of the Mohr circle stress representation as was demonstrated in
section 4.2. The uniaxial compressive stress of $2k$ in each leg is decomposed into
deviatoric and hydrostatic components, each of magnitude k.

Fig. 4.42(b) gives a graphical representation of the stress states in the regions
labeled (2), (3), (4) in (a). For region (2) in (a) the magnitude of the deviatoric
stress is denoted by the vector $O-2$ in (b), while the hydrostatic stress of k is given
in parenthesis. To obtain the stress state in the overlap region, region labeled (3) in
(a), the vertical leg stress contributions are added as previously mentioned. Note
that the deviatoric component vector $2-3$ in (b) makes an angle of twice the
physical angle of $30°$. Hence, the deviatoric stress in region (3) in (a) is the vector
$O-3$ in (b) while the hydrostatic stress is given by $k + k = 2k$. Similarly, the magnitude of vector $O-4$ represents the deviatoric stress directly below the footing. The
hydrostatic stress is given by $2k + k = 3k$.

This solution however, is characterized by a violation of the yield criterion in
regions labeled (3) and (4) in (a) because the difference between the maximum and
minimum principal stresses exceeds the value $2k$ permitted by the Tresca yield criterion. If the stresses in the legs are to be kept within the value $2k$, then a horizontal leg
must also be included to balance the vertical $2k$ stress. The resulting stress diagram
and solution are shown in Fig. 4.43. Regions (3) and (4) in Fig. 4.43(a) no longer
violates the yield condition. It should be noted that Fig. 4.43(a) can alternatively be
seen to be composed of a superposition of the two symmetrical legs with a hydrostatic
region directly beneath the footing caused by the vertical and horizontal legs. The

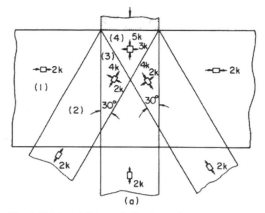

Fig. 4.43(a). Addition of a horizontal stress leg.

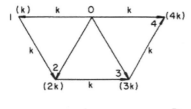

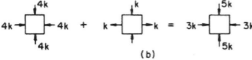

Fig. 4.43(b). Stress diagram with yield not exceeded.

addition of this hydrostatic region does not change the principal directions in regions labeled (3) and (4) in (a).

Nine stress legs (Fig. 4.44)

Fig. 4.44(a) shows the extension of the previous case to nine legs. We are first

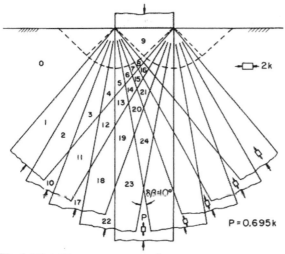

Fig. 4.44(a). Nine stress legs at $10°$

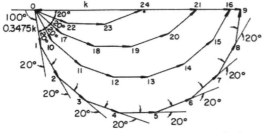

Fig. 4.44(b). Stress diagram approaches circle in limit.

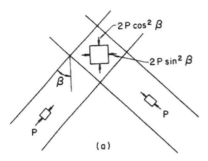

Fig. 4.45(a). Typical overlap region.

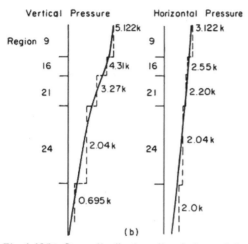

Fig. 4.45(b). Stress distributions directly beneath footing.

faced with the problem of determining the uniaxial compressive stress P in each leg such that yielding is not exceeded in the region labeled (9) in (a). The stress state in a typical symmetric overlapping region is shown in Fig. 4.45(a). Considering symmetry, there are four legs in Fig. 4.44(a) with $\beta = 10°, 20°, 30°, 40°$ where β is the angle the leg makes with the vertical. The vertical and horizontal principal stresses in region labeled (9) in Fig. 4.44(a) can then be expressed as:

$$\sigma_V = 2P[\cos^2 40° + \cos^2 30° + \cos^2 20° + \cos^2 10°] + P$$

$$\sigma_H = 2P[\sin^2 40° + \sin^2 30° + \sin^2 20° + \sin^2 10°] + 2k \qquad [4.72]$$

Using the Tresca yield condition, $\sigma_V - \sigma_H = 2k$, an expression for the compressive stress P is obtained:

$$P = \frac{2k}{(\cos 80° + \cos 60° + \cos 40° + \cos 20° + 1/2)} = 0.695k \qquad [4.73]$$

Fig. 4.44(b) shows the resulting stress diagram. For region (9) in (a) the hydrostatic pressure p and deviatoric stress s components are given as:

$$p = 9(\tfrac{1}{2}P) + k = 4.122k$$

$$s = 2k - k \quad = k \tag{4.74}$$

such that the limit load now becomes $p + s = 5.122k$. It is evident from Fig. 4.44(b) that the Tresca yield condition is satisfied in regions labeled (1)–(9) in (a) and not violated in other regions. Thus this is a statically admissible stress field and the value $5.122k$ is therefore a lower-bound solution to the collapse load. Using the stress diagram the distribution of vertical and horizontal stress directly beneath the footing (regions labeled (9), (16), (21), (24) in (a)) can also be evaluated. For example in region labeled (21) in Fig. 4.44(a) the hydrostatic pressure and deviatoric stress are (see the point labeled (21) in Fig. 4.44b):

$$p = 5(0.3475k) + k = 2.735k$$

$$s = 2 \times 0.3475k \left[\cos 40° + \cos 20° + \tfrac{1}{2}\right] - k = 0.535k \tag{4.75}$$

such that the vertical and horizontal stresses are given as:

$$\sigma_V = 2.735k + 0.535k = 3.27k$$

$$\sigma_H = 2.735k - 0.535k = 2.20k \tag{4.76}$$

Fig. 4.45(b) shows the resulting piecewise distribution for nine legs, and the continuous distribution for infinitely many legs.

Infinitely many stress legs

The above formulation can now be generalized for N legs, where N is half the number of non-vertical legs, such that:

$$N = \frac{\tfrac{1}{4}\pi}{\delta\beta} - \tfrac{1}{2} \tag{4.77}$$

where the angle $\delta\beta$ is the leg separation angle between two adjacent legs. For a Tresca material the magnitude of uniaxial compression P is given by:

$$P = 2k \bigg/ \left[\sum_{i=1}^{N} \cos(2i\delta\beta) + \tfrac{1}{2}\right] \tag{4.78}$$

Using the expression for the vertical pressure directly beneath the footing (region labeled 9 in Fig. 4.44a):

$$\sigma_V = 2P \left[\sum_{i=1}^{N} \cos^2 (i\delta\beta) + \tfrac{1}{2} \right] \tag{4.79}$$

and substituting expression [4.78] for P into expression [4.79], a general bearing expression for N legs is obtained:

$$Q = \sigma_V = 4k \ \frac{\displaystyle\sum_{i=1}^{N} \cos^2 (i\delta\beta) + \tfrac{1}{2}}{\displaystyle\sum_{i=1}^{N} \cos (2i\delta\beta) + \tfrac{1}{2}} \tag{4.80}$$

Expressing the trigonometric series terms in integral form, using [4.77]:

$$\sum_{i=1}^{N} \cos^2 (i\delta\beta) = \int_{1}^{N} \cos^2 (i\delta\beta) \, di + \epsilon_1$$

$$= \frac{1}{2\delta\beta} \left[\tfrac{1}{4}\pi - \frac{3\delta\beta}{2} + \tfrac{1}{2} \cos \delta\beta - \tfrac{1}{2} \sin 2\delta\beta \right] + \epsilon_1 \tag{4.81}$$

and:

$$\sum_{i=1}^{N} \cos (2i\delta\beta) = \int_{1}^{N} \cos (2i\delta\beta) \, di + \epsilon_2 = \frac{1}{2\delta\beta} \left[\cos \delta\beta - \sin 2\delta\beta \right] + \epsilon_2 \tag{4.82}$$

where ϵ_1, ϵ_2 are small constants. If the number of legs is now allowed to approach infinity the limiting form of [4.80] becomes:

$$Q = \lim_{\delta\beta \to 0} \left[\frac{\tfrac{1}{4}\pi - \dfrac{\delta\beta}{2} + \tfrac{1}{2} \cos \delta\beta - \tfrac{1}{2} \sin 2\delta\beta + 2\epsilon_1 \delta\beta}{\cos \delta\beta - \sin 2\delta\beta + \delta\beta + 2\epsilon_2 \delta\beta} \right] 4k \tag{4.83}$$

such that $Q = 4k(\tfrac{1}{2} + \tfrac{1}{4}\pi) = k(2 + \pi)$, which is exactly the value obtained from the Prandtl slip-line field (Fig. 4.22a) of classic theory of plasticity. For this limiting process the stress diagram of Fig. 4.44(b) would approach a semi-circle and the stress distribution directly beneath the footing would become smooth as shown by the solid lines in Fig. 4.45(b).

4.9.2. Coulomb materials (c − φ soils)

The basic difference between perfectly plastic Tresca materials ($\varphi = 0$) and the $c - \varphi$ Coulomb materials, is that for the latter the hydrostatic stress state does effect the onset of yielding and hence failure. Intuitively this would mean that the stress diagram for the footing problem no longer would resemble a circle in the limit as the number of legs approaches infinity. It will be shown that the resulting stress diagram resembles a logarithmic spiral.

Three stress legs (Fig. 4.46)

As before, the footing problem will first be solved using three legs as shown in Fig. 4.46(a). For the given problem P and α are unknowns. The value of R needed for the horizontal leg yielding is easily expressed in terms of the material constants c and φ from a simple Mohr diagram construction:

$$R = \frac{2c \cos \varphi}{1 - \sin \varphi} = 2c \tan \left(\tfrac{1}{4}\pi + \tfrac{1}{2}\varphi\right) \qquad [4.84]$$

In order to satisfy the Coulomb yield condition:

$$s = c \cos \varphi + p \sin \varphi \qquad [4.85]$$

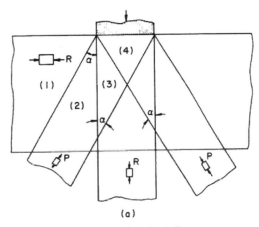

(a)

Fig. 4.46(a). Three stress legs for soil.

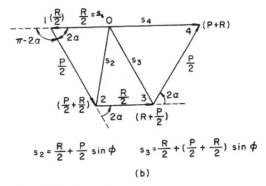

$$s_2 = \frac{R}{2} + \frac{P}{2} \sin \phi \qquad s_3 = \frac{R}{2} + \left(\frac{P}{2} + \frac{R}{2}\right) \sin \phi$$

(b)

Fig. 4.46(b). Stress diagram.

in the overlapping region, the magnitude of the deviatoric stress increment Δs in the overlapping region may be computed by using the expression relating incremental changes in hydrostatic and deviatoric stress:

$$\Delta s = \Delta p \sin \varphi \qquad \qquad [4.86]$$

For triangle O-1-2 with $\Delta p = \frac{1}{2}P$ being added to the region (1) where $s_1 = \frac{1}{2}R$, the magnitude of the resulting deviatoric stress s_2 in region (2) to produce yielding is:

$$s_2 = s_1 + \Delta s = \frac{1}{2}R + \frac{1}{2}P \sin \varphi \qquad \qquad [4.87a]$$

and similarly:

$$s_3 = \frac{1}{2}R + (\frac{1}{2}P + \frac{1}{2}R) \sin \varphi \qquad \qquad [4.87b]$$

Using the law of cosines for triangle O-1-2 and the expressions for s_2 and R, the value of P is obtained:

$$P = \frac{2R(\cos 2\alpha + \sin \varphi)}{\cos^2 \varphi} = \frac{4c(\cos 2\alpha + \sin \varphi)}{\cos \varphi (1 - \sin \varphi)} \qquad \qquad [4.88]$$

From the geometry of the stress diagram, s_3 can be equated to $\frac{1}{2}P$ such that:

$$P = \frac{R(1 + \sin \varphi)}{1 - \sin \varphi} = 2c \tan^3 (\frac{1}{4}\pi + \frac{1}{2}\varphi) \qquad \qquad [4.89]$$

Equating [4.88] and [4.89] and solving for α:

$$\cos 2\alpha = \frac{1}{2}(1 + \sin^2 \varphi) \quad \text{or} \quad \sin \alpha = \frac{\cos \varphi}{2} \qquad \qquad [4.90]$$

agreeing with [4.69b] obtained previously for example (4.5). The hydrostatic pressure p_4 and deviatoric stress s_4 in the region labeled (4) in Fig. 4.46(a) can now be obtained from Fig. 4.46(b):

$$p_4 = \frac{1}{2}R + \frac{1}{2}P + \frac{1}{2}R + \frac{1}{2}P = P + R$$

$$s_4 = (\frac{1}{2}P \cos 2\alpha + \frac{1}{2}R + \frac{1}{2}P \cos 2\alpha) - \frac{1}{2}R = P \cos 2\alpha \qquad \qquad [4.91]$$

Using the relation [4.90], the lower-bound bearing capacity value Q^l from these stresses can be computed:

$$Q^l = p_4 + s_4 = P + R + P \cos 2\alpha = P[1 + \frac{1}{2}(1 + \sin^2 \varphi)] + R \qquad \qquad [4.92a]$$

or substituting the values for P and R, we have:

$$\frac{Q^l}{c} = \tan^3 (\frac{1}{4}\pi + \frac{1}{2}\varphi) (3 + \sin^2 \varphi) + 2 \tan (\frac{1}{4}\pi + \frac{1}{2}\varphi) \qquad \qquad [4.92b]$$

For $\varphi = 20°$, $Q^l = 11.94c$. The lower-bound value obtained in [4.71] is $13.32c$ (Fig. 4.36). If the values of p_4 and s_4 obtained for region (4) are inserted into the Coulomb yield equation [4.85], it is found that yielding has not been reached, hence the low value of Q.

Nine stress legs for a semi-infinite footing (Fig. 4.47)

Fig. 4.47(a) shows a picture of nine stress legs at an equal central angle $\Delta\theta$ in the logspiral zone CAB. The magnitude of the uniaxial compressive stresses P_1, P_2, ... P_9 in each leg can be determined from the condition that each overlapping region must satisfy the Coulomb yield criterion. Using the relation [4.86], the resulting stress states in various overlapping regions satisfying the yield criterion are shown graphically in Fig. 4.47(b). If the central angle $\Delta\theta$ is sufficiently small, one may write:

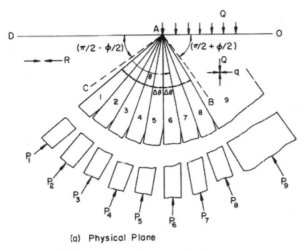

(a) Physical Plane

Fig. 4.47(a). Nine stress legs for a $c - \psi$ soil.

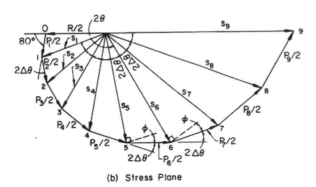

(b) Stress Plane

Fig. 4.47(b). Stress diagram of (a).

Components of shear stress, s

$$s_2 = s_1 + s_1\, 2\Delta\theta\, \tan\varphi = s_1(1 + 2\Delta\theta\, \tan\varphi)$$

$$s_3 = s_2(1 + 2\Delta\theta\, \tan\varphi) = s_1(1 + 2\Delta\theta\, \tan\varphi)^2$$

$$s_{n+1} = s_1(1 + 2\Delta\theta\, \tan\varphi)^n \qquad [4.93]$$

Components of hydrostatic pressure, p

$$p_2 = p_1 + \frac{P_2}{2} = p_1 + \frac{s_1\, 2\Delta\theta}{\cos\varphi}$$

$$p_3 = p_2 + \frac{P_3}{2} = p_1 + \frac{s_1\, 2\Delta\theta}{\cos\varphi} + \frac{s_2\, 2\Delta\theta}{\cos\varphi} = p_1 + \frac{2\Delta\theta}{\cos\varphi}(s_1 + s_2)$$

$$p_{n+1} = p_1 + \frac{2\Delta\theta}{\cos\varphi}(s_1 + s_2 + \ldots + s_n) \qquad [4.94]$$

Using [4.93], [4.94] can be written in the form:

$$p_{n+1} = p_1 + \frac{s_1\, 2\Delta\theta}{\cos\varphi}\,[1 + (1 + 2\Delta\theta\, \tan\varphi) + \ldots + (1 + 2\Delta\theta\, \tan\varphi)^{n-1}]$$

$$= p_1 + \frac{s_1\, 2\Delta\theta}{\cos\varphi}\left[\frac{1 - (1 + 2\Delta\theta\, \tan\varphi)^n}{1 - (1 + 2\Delta\theta\, \tan\varphi)}\right]$$

$$= p_1 - s_1\, \csc\varphi + s_1\, \csc\varphi\,(1 + 2\Delta\theta\, \tan\varphi)^n \qquad [4.95]$$

If the number of stress legs is now allowed to approach infinity the limiting forms of [4.93] and [4.95] become:

$$s = \lim_{n\to\infty} s_{n+1} = \lim_{n\to\infty} s_1\left(1 + \frac{2\theta\, \tan\varphi}{n}\right)^n = \tfrac{1}{2}R\, e^{2\theta\, \tan\varphi} \qquad [4.96a]$$

$$p = \lim_{n\to\infty} p_{n+1} = \lim_{n\to\infty}\left[p_1 - s_1\csc\varphi + s_1\csc\varphi\left(1 + \frac{2\theta\, \tan\varphi}{n}\right)^n\right]$$

$$= \tfrac{1}{2}R - \tfrac{1}{2}R\, \csc\varphi + \tfrac{1}{2}R\, \csc\varphi\, e^{2\theta\, \tan\varphi}$$

$$= \tfrac{1}{2}R\, \frac{1}{\sin\varphi}\,(e^{2\theta\, \tan\varphi} + \sin\varphi - 1) \qquad [4.96b]$$

where $\Delta\theta = \theta/n$ and noting that $s_1 \to \frac{1}{2}R$ and $p_1 \to \frac{1}{2}R$ in the limit as $n \to \infty$ (Fig. 4.47b). In [4.96], s and p are the values of the components of shear stress and hydrostatic pressure at any angular location θ from the first stress leg P_1 (Fig. 4.47a). From this limiting process the stress diagram of Fig. 4.47(b) would approach a semi-logspiral.

With [4.96] and [4.84], the vertical and horizontal stresses Q and q in region labeled (9) directly beneath the load (Fig. 4.47a) can be computed:

$$Q^l = (p + s)|_{\theta = \frac{1}{2}\pi} = \frac{1}{2}R\,\frac{1}{\sin\varphi}\,(e^{\pi\tan\varphi} + \sin\varphi - 1) + \frac{1}{2}R\,e^{\pi\tan\varphi}$$

$$= c\cot\varphi\,[e^{\pi\tan\varphi}\tan^2(\tfrac{1}{4}\pi + \tfrac{1}{2}\varphi) - 1] \qquad [4.97a]$$

$$q^l = (p - s)|_{\theta = \frac{1}{2}\pi} = c\cot\varphi\,(e^{\pi\tan\varphi} - 1) \qquad [4.97b]$$

The lower-bound pressure Q^l agrees with the upper-bound value obtained previously in [4.63] by putting $\Theta = \frac{1}{2}\pi$ in [4.63]. By the upper- and lower-bound theorems of limit analysis, the expression Q in [4.97a] is therefore the *correct* limit pressure for the *semi-infinite* footing problem. It will be demonstrated in section 6.6.1, Chapter 6 that the pressure Q as given in [4.97a] gives also the correct limit pressure for the case of a *finite* footing.

Chapter 5

PROGRESSIVE FAILURE OF FOOTINGS

5.1. INTRODUCTION

As has been indicated in Chapter 1, a complete elastic—plastic analysis of stress and strain in a mass of soil as the footing load is increased to failure is almost always complicated, and limit analysis is one of the methods that furnishes the load-carrying capacity of a footing on a soil in a more direct manner. Before we move on to apply the techniques of limit analysis developed in the preceding two Chapters to various stability problems in soil mechanics, we shall first, in this Chapter, present some typical elastic—plastic responses of soils to footing loads so that physical insight for this class of problems is obtained.

We present here the numerical results of a detailed study of plane strain and axisymmetric flat rigid footings bearing on an elastic—perfectly plastic soil (except section 5.8). Complete load displacement histories are presented from zero-load to failure, encompassing initial elastic behavior, contained plastic flow and collapse. We show stress distributions and zones of yielding at various load levels during continued loading. We show also velocity fields at the collapse state. Particular attention is given to the effect of the changing soil geometry on the response of the soil stratum to the footing loads. The *finite element method* and an incremental integration scheme are used to numerically solve the governing equations. Details of this finite element formulation are given in Chapter 12.

The finite element formulation was developed originally for structural analysis and it has been used for solving many complex structural problems. Although early development of the finite element method was primarily concerned with linear systems, the method has been extended to non-linear problems by many investigators (see, for example, Zienkiewicz, 1971). However, a successful formulation and solution of the large deformation problem in soil mechanics is presented here for the first time. The numerical results are described in this Chapter, but the formulation of the finite element matrix is presented in Chapter 12.

The soil is modeled here as a linear elastic—perfectly plastic material with the *extended* Von Mises yield condition [2.7] and associated flow rule (section 2.2, Chapter 2). The effect of large deformations on the response of the soil mass is included in the analysis. Both drained and undrained analyses are considered. A Von Mises model [2.6] which is a special case of the extended Von Mises model ($\alpha = 0$ in [2.7]) is used for total stress analysis of undrained clay while an extended

Von Mises (or Drucker-Prager model) is utilized for effective stress (drained) analysis of overconsolidated clay. Some elastic—plastic analyses of undrained clay strata have appeared in the literature. However, the more general case of a $c - \varphi$ soil has been treated only briefly with inconclusive results. We present here in-depth treatment of both drained and undrained cases. It will be shown here for the first time that the well known Prandtl velocity field corresponds to the true failure mode for a rough footing bearing on an extended Von Mises or Von Mises material.

The individual problems connected with the progressive failure of footings on soils will be taken up in the following sequence. First we consider in section 5.2 a notched elastic—plastic tensile specimen under plane strain condition, because this is a very relevant problem for a footing on a slope which is mathematically equivalent to a notched bar in tension. Upper- and lower-bound limit analysis solutions along with the concept of contained plastic deformation and impending plastic flow will be discussed. Then we present an elastic—plastic analysis of plane strain flat punch indentation into a block of finite dimensions (section 5.3). In sections 5.4, 5.5, 5.6 and 5.7, we consider a single, strip, surface footing bearing on a finite stratum of clay. Finally, in section 5.8, an elastic—plastic analysis of axisymmetric flat circular punch indentation into a specimen of finite dimensions is presented.

5.2. PLANE STRAIN NOTCHED TENSILE SPECIMEN (VON MISES MATERIAL)

5.2.1. Finite element solution (Davidson, 1974)

Fig. 5.1 shows one half of a tensile specimen with 90° angular notches which is supposed to be tested under conditions of plane strain. The relevant dimensions and material properties are also shown in Fig. 5.1 where Young's modulus, E, is 7000 kg/mm^2, Poisson's ratio, ν, is 0.2 and the yield stress in simple tension, σ_0, is 24.3 kg/mm^2. The material is assumed to be elastic—perfectly plastic with a Von Mises yield condition and, in terms of the extended Von Mises model used here ([2.7]), $\alpha = 0$ and $k = \sigma_0/\sqrt{3} = 14.03$ kg/mm^2. The finite element mesh used here is shown in the left-hand half of Fig. 5.1 and consists of 105 nodes and 169 triangular elements.

As the surface stress σ applied to the end sections of the specimen is gradually increased starting from zero, the specimen is first stressed in a purely elastic manner. Eventually, the stress intensity reaches the critical value k at the notch root elements of the notches, and plastic regions begin to form here. In Fig. 5.2, the applied stress σ is plotted versus the centerline displacement at the end of the specimen. The curve remains almost linear up to 18 kg/mm^2, after which it bends over quite rapidly. Zones of yielding for loads of 15 and 18 kg/mm^2 are shown in

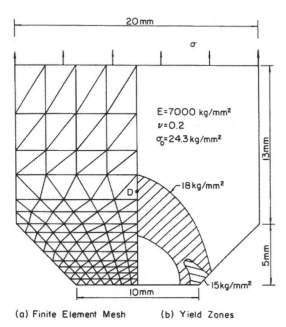

(a) Finite Element Mesh   (b) Yield Zones

Fig. 5.1. Plane strain notched tensile specimen (Von Mises' yield condition).

the right-hand half of Fig. 5.1. Yielding starts at the notch root and spreads upward and toward the centerline. While finite amounts of material are stressed plastically in this stage, large plastic deformations are ruled out by the fact that there is a central elastic strip separating the peripheral plastic regions. The overall elongation of the test specimen under this applied stress is determined primarily by the elastic properties of this strip (*contained plastic deformation*). As we keep increasing the applied stress σ, the plastic regions increase in size and eventually at 18 kg/mm^2 a

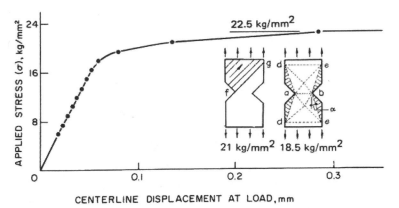

Fig. 5.2. Load displacement curve – notched tensile specimen.

bridge of plastic material is formed which extends all the way across the entire specimen. Until this point, the load displacement response is essentially linear and afterwards it becomes highly nonlinear. Finally, at 22.5 kg/mm^2, the specimen is first able to extend under this constant applied stress (*impending plastic flow*). The maximum applied stress σ_{max} = 22.5 kg/mm^2 for impending plastic flow characterizes the load carrying capacity of the specimen (*limit stress* or *load*).

Before entering upon a discussion of the slip-line solution and the determination of bounds for the load carrying capacity, let us remark that the development of the plastic regions from the elastic limit to the instant of impending plastic flow may take place in a rather unexpected manner. According to the computations, a new plastic region begins to form at the point marked D in Fig. 5.1, when the applied stress reaches approximately 17 kg/mm^2. At this instant the plastic region which developed around the notch root is not much larger than the shaded region in Fig. 5.1 which corresponds to σ = 15 kg/mm^2. The new plastic region originating at the point D spreads extremely fast; for σ = 17.5 kg/mm^2 approximately, the new plastic zone attains already the size about to merge with the plastic region which originated at the notch root. At σ = 18 kg/mm^2 the two zones of yielding bridge together and extend all the way across the entire specimen. The size of the plastic yielding corresponding to this applied stress is shown by the shaded region marked 18 kg/mm^2 in Fig. 5.1.

5.2.2. Slip-line solution

Hill's (1949) slip-line solution for this problem gives:

$$\sigma_{max} = (1 + \tfrac{1}{4}\pi)k = 1.785k \qquad\qquad\qquad\qquad [5.1]$$

For k = 14.03 kg/mm^2, the limit stress σ_{max} = 25 kg/mm^2. The limit stress 22.5 kg/mm^2 obtained by the finite element method is 10% below the slip-line solution.

One remark should be made here. Hill's slip-line solution is not *complete*. The conditions which have to be satisfied by slip-line solution in order to result in the *exact* value of the limit load have been discussed earlier in Chapter 1. In particular, it is necessary to extend the stress state in the plastic zones into the rigid zones such that the yield criterion is nowhere exceeded, then the stress field so constructed in the whole of the specimen will be a *statically admissible stress field* which enables a *lower bound* of the limit load to be obtained. As we have previously discussed in Chapter 1, this is not amenable to verification. Further, a *complete* solution requires the dissipation to be *positive everywhere in the slip-line field* which gives an *upper bound* of the limit load. This is a condition of compatibility of stress and velocity fields, and can be verified in most cases, though not always simply. In the following we will show that Hill's slip-line solution is *definitely* not a *lower* bound and it gives a poor upper-bound value for this particular case.

5.2.3. Limit analysis solutions

Lower bounds

It is not difficult to obtain a crude lower bound, by inscribing in the net area a smooth strip of width 10 mm with uniaxial stress $\sigma_0 = 24.3$ kg/mm^2; then $\sigma^l = 24.3 \times 10/20 = 12.15$ kg/mm^2 gives a lower bound for the applied stress for impending plastic flow. As Prager has shown, a substantially better estimate can be obtained by inscribing in the notched specimen the trapeziums shown by the dotted lines in the inset of Fig. 5.2. The shaded zones the stresses are zero. Here the stress on ab has the value $2k(1 + \sin \alpha)$ ([4.55]) when the angle of the wedge is 2α (see also Fig. 7.4). The region vertically below de is in a state of uniaxial tension of magnitude $2k(1 - \sin \alpha)$. The ratio of the width de/ab has the value $\tan^2(\frac{1}{4}\pi + \frac{1}{2}\alpha)$. When the angle of the wedge 2α is equal to $60°$ the pressure on ab has the value $3k$ and the width de has the value three times that of ab. The corresponding stress field for this special case is shown in Fig. 4.20.

Here the angle α must be chosen so that the tensile stress is greatest, it follows that the ratio of de/ab must set equal to 2 or α must have the value $20°$. The corresponding tensile stress is:

$$\sigma^l = 2k(1 - \sin 20°) = 1.316k = 18.5 \text{ kg/mm}^2 \qquad [5.2]$$

The value 18.5 kg/mm^2 is then an improved lower bound for the limit stress of the problem.

Upper bound

In the inset of Fig. 5.2 shows a very simple discontinuous velocity field: the shaded portion moves as a rigid body in the direction indicated by the arrow, while the unshaded portion remains at rest. It is found that the upper bound for σ^u obtained from this field assumes the minimum value $1.5k = 21$ kg/mm^2 when the line of discontinuity fg is inclined under $45°$ against the horizontal axis. Thus, even the very crude stress and velocity fields discussed in connection with the insets of Fig. 5.2 give the following bounds for the critical or limit load σ_c:

$$18.5 \text{ kg/mm}^2 = 1.32k \leqslant \sigma_c \leqslant 1.50k = 21 \text{ kg/mm}^2 \qquad [5.3]$$

For the limit load we can take the mean value $\sigma_c = 1.41k = 19.8$ kg/mm^2; this involves a maximum error of $\pm 7\%$.

Note that the limit stress 22.5 kg/mm^2 obtained by the finite element method slightly exceeds the upper-bound value 21 kg/mm^2 This may be expected because an error must be introduced when replacing a continuous medium by a finite number of elements. Further, it may be possible that the bridge of plastic material which extends across the specimen must attain some finite width before the specimen begins to extend under constant applied surface stress.

Note also that the slip-line solution for $\sigma_{max} = 1.785k$ given by Hill exceeds the upper-bound value $1.5k$. We can deduce here indirectly that a statically admissible stress distribution cannot exist outside the slip-line stress field. The slip-line solution for this problem is therefore not the *correct* solution. Although it appears to give an *upper-bound* value for the limit load, we cannot say definitely that it is a *valid* upper-bound solution.

5.2.4. Comparison with known solutions

A number of investigators have presented numerical results for the plane strain notched tensile specimen (Marcal and King, 1967; Zienkiewicz et al., 1969). The finite element mesh shown in the left-hand half of Fig. 5.1 is similar to, but not exactly the same as that used by Zienkiewicz et al. (1969). They reported a so-called lower bound for the limit load of 19.4 kg/mm^2, that is, this was the last load at which their iterative procedure converged. This solution represents a true lower bound for the discretized body since the iterative procedure utilized ensures that the discrete equations of equilibrium and the yield inequality are satisfied at the end of each increment. However, this may not represent a lower bound for the continuum since the continuum equations of equilibrium are in general not satisfied by a finite element solution.

Zones of yielding for loads of 15 and 18 kg/mm^2 as shown in the right-hand half of Fig. 5.1 are similar to those reported by Marcal and King (1967) and by Zienkiewicz et al. (1969).

5.3. PLANE STRAIN PUNCH INDENTATION OF RECTANGULAR BLOCKS (VON MISES MATERIAL)

5.3.1. Finite element solution (A.C.T. Chen, 1973)

We shall study here the elastic–plastic behavior of a finite block compressed by two equal and opposite rigid punches (Fig. 5.3a). The contact surface between the punch and the block is assumed to be perfectly rough. The geometry used in the finite element analysis is also shown in Fig. 5.3(a), where b/w is 6, and h/b is 1. The width of the punch $2w$ is 1 inch (2.54 cm) and the width of the block $2b$ is 6 inches (15.24 cm). Because of the double symmetry of the problem, only the upper-right quarter of the problem needs to be considered (Fig. 5.3b).

The material is assumed to be elastic–perfectly plastic obeying Von Mises yield condition and its associated flow rule. The material properties used for the computation are E = 10,000 ksi (68,947 MN/m^2), ν = 0.33, and σ_0 = 13 ksi (89.6 MN/m^2). These values approximate the static stress–strain curve for commercially pure aluminum 1100-0.

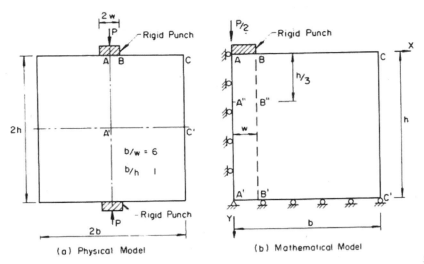

Fig. 5.3. Double punch — plane strain case.

Fig. 5.4 shows the average punch pressure as a function of the punch displacement. The pressure increases linearly first and then the curve shows a fast bend with small rate of increase in pressure beyond the point marked f on the curve. A well defined limit pressure at the point g is reached at about $5.2(\sigma_0/\sqrt{3}) = 5.2k$. The finite element mesh used here is shown in Fig. 5.5 which consists of 274 triangular

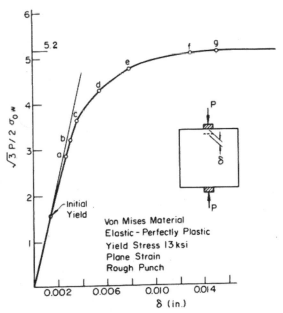

Fig. 5.4. The pressure—displacement curve for the punch indentation of a block.

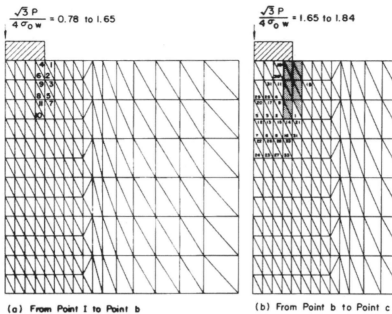

$\dfrac{\sqrt{3}\,P}{4\,\sigma_0\,w} = 0.78$ to 1.65

(a) From Point I to Point b

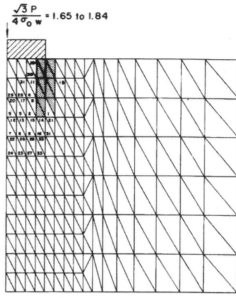

$\dfrac{\sqrt{3}\,P}{4\,\sigma_0\,w} = 1.65$ to 1.84

(b) From Point b to Point c

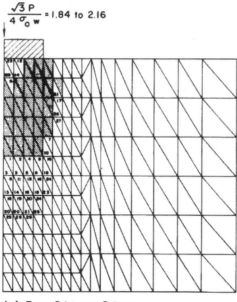

$\dfrac{\sqrt{3}\,P}{4\,\sigma_0\,w} = 1.84$ to 2.16

(c) From Point c to Point d

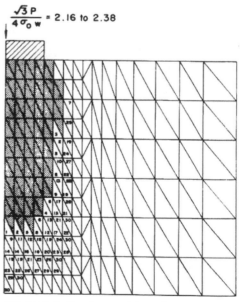

$\dfrac{\sqrt{3}\,P}{4\,\sigma_0\,w} = 2.16$ to 2.38

(d) Load further Increased from Point d to Point e

Fig. 5.5. For caption see next page.

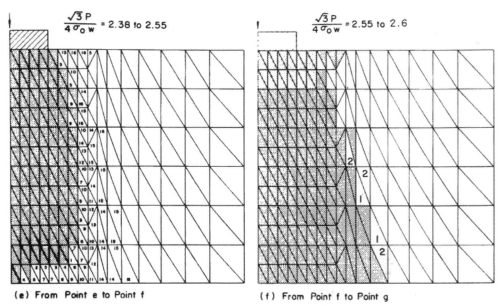

(e) From Point e to Point f   (f) From Point f to Point g

Fig. 5.5. The spreading of elastic–plastic zones of a metal block under plane strain condition.

elements. Also, Fig. 5.5 shows the development of the plastic zone, and the manner in which the plastic zone spreads as the load is continuously increased from zero. The numeral in an element indicates the order in which the element yields during the loading interval in question and the shaded area indicates elements already yielded before the loading interval. The loadings corresponding to various stages of plastification are marked on the load displacement curve in Fig. 5.4.

The initial development of the plastic zone around the punch corner (Fig. 5.5a) is seen not to affect the load displacement relationship significantly until the plastic zone has developed to a sufficient size. During the loading increment from point b to point c, the elastic zone under the punch forms a single truncated wedge and a further increase in load causes the plastic zone to spread. As soon as the plastic zone beneath the punch spreads (Fig. 5.5c), the load displacement curve begins to bend noticeably. At the point e (Fig. 5.5d), the entire region beneath the punch becomes plastic and the slope of the load displacement curve assumes an almost steady-state value. Between the points e and f, the plastic zone spreads away from the punch during the transition from contained plastic flow to unrestricted plastic flow (Fig. 5.5e). The slope of the load displacement curve then becomes almost zero and the plastic zone grows slowly in all directions as shown in Fig. 5.5(f).

The vertical stress distributions along the punch surface AB are shown in Fig. 5.6. The stress values are those at the centers of elements in contact with the rigid punch. The trend of the stress distribution is the increase of the magnitude toward the edge of the punch.

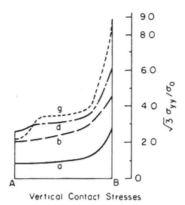

Fig. 5.6. Stress distributions along the punch surface.

5.3.2. Slip-line solution

The only slip-line solution which is relevant here is due to Hill (1950) for *smooth* punches. Hill's solution for this problem with smooth punches gives the average value 4.4k as the maximum or collapse pressure. The limit pressure 5.2k obtained by the finite element method for rough punches is about 15% above the slip-line solution for smooth punches.

5.3.3. Limit analysis solution

Lower bound

We use the "compact" discontinuous stress field shown in Fig. 7.4. Here, as in the notched tensile specimen case (Fig. 5.2), we can use two such stress fields back-to-back (i.e., mirror images in plane *de*) to fit the punches and the narrow block. The remainder of the block is regarded as stress-free. If the width of the trapezium $(2w)\tan^2(\frac{1}{4}\pi + \frac{1}{2}\alpha)$ is fitted to the width of the block 2b, we find $\tan^2(\frac{1}{4}\pi + \frac{1}{2}\alpha) = b/w = 6$, so the angle α must have the value 45.6°. The corresponding compressive stress is $2k(1 + \sin 45.6°) = 3.43k$ which gives a lower bound to our problem. The value 3.43k is also a lower bound for a block whose height to punch width ratio (h/w) is greater than 4.9.

Upper bound

Fig. 3.12(c and d) show a simple discontinuous velocity field and a field of homogeneous deformation along with rigid block sliding. The results of the narrow punch problem are shown in Fig. 3.13 together with the local solution for the semi-infinite block. For the dimensions considered here (Fig. 5.3), the least of these upper bounds is 5k. The lower and upper bounds therefore show that the critical or limiting indentation pressure for two rough punches on a square block lies within ± 18% of the value 4.22k.

Note that the limit pressure 5.2k obtained by the finite element solution slightly exceeds the upper-bound value 5k. As previously indicated, this may be due to the error introduced by replacing continuous media by a finite number of elements. It appears therefore that a *finer* finite element mesh is needed in order to capture the theoretical limit load. This crude limit analysis solution does provide a useful guide for checking the finite element results.

5.4. UNIFORM STRIP LOAD ON A SHALLOW STRATUM OF UNDRAINED CLAY (VON MISES MATERIAL)

Apparently Höeg et al. (1968) were the first to treat soil as an elastic–perfectly plastic material for the purpose of obtaining the complete load displacement response of a strip footing. A shallow layer of undrained clay, shown in Fig. 5.7, was analyzed using the finite difference-like technique of Ang and Harper (1964). The Tresca yield condition and its associated flow rule were utilized with a cohesive strength of 17.5 psi (120.6 kN/m^2), and the footing load was assumed to be uniformly distributed.

Uniform finite element mesh (Fig. 5.7)

Here we solve this same plane strain problem using the finite element method and the Von Mises yield condition. Both yield conditions should give the same limit load in terms of the value k whose precise value depends on the yield criterion. In Tresca's criterion the value k is equal to $\sigma_0/2$ while for Von Mises' it is equal to $\sigma_0/\sqrt{3}$ where σ_0 is the yield stress in simple tension test. It should be noted that the intermediate response may be different for both criteria. Two different meshes

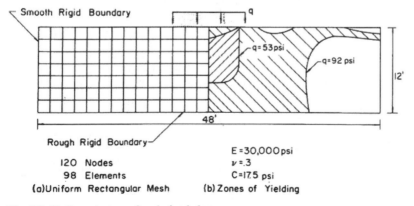

Fig. 5.7. Shallow stratum of undrained clay.

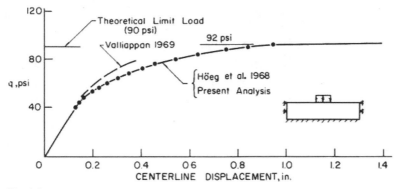

Fig. 5.8. Load–displacement curve for shallow undrained clay stratum. Uniform rectangular mesh (plane strain, Von Mises' yield condition).

are utilized here. One mesh is uniform (left-hand half of Fig. 5.7) and is similar to that used by Höeg et al. (1968), while the other is nonuniform (Fig. 5.9) and is similar to that used in the following sections for a deeper soil stratum. Boundary conditions in both meshes are identical to those used by Höeg et al. (1968). The base of the clay stratum is rigid and perfectly rough, while the vertical boundary is assumed to be rigid and perfectly smooth. The boundary conditions and dimensions used in the analysis are also shown in Fig. 5.7. The uniform mesh as shown in the left-hand half of Fig. 5.7 consists of 120 nodes and 98 rectangular elements. Each rectangle is defined by four constant strain triangles. Since the mesh is perfectly uniform we are actually considering a loading width of 10.28 ft. (3.13 m) rather than 10 ft. (3.05 m) as used by Höeg. We assume that this will make little difference in the solution and subsequent results show this to be the case.

Results for the uniform mesh are shown in Fig. 5.8 where the applied pressure is plotted versus the centerline displacement directly beneath the load. The closed circles correspond to actual computed points indicating that sixteen increments were used in the solution. This solution agrees almost point by point with that presented by Höeg. We obtained here a well defined numerical limit load of 92 psi while Höeg reported 90 psi. Both values are in remarkable agreement with the exact value of 90 psi (627 kN/m^2).

Valliappan (1969) also solved this same problem using the Von Mises yield condition and a somewhat coarser finite element mesh of 94 nodes and 150 triangular elements. As might be expected his solution, also shown in Fig. 5.8, lies above that presented here. The initial stress method (Zienkiewicz et al., 1969) was used to integrate the equations and the last load at which this iterative technique converged was 78 psi (538 kN/m^2).

Zones of yielding defined by the rectangular mesh are shown in the right-hand half of Fig. 5.7. These zones agree fairly well with those presented by Höeg for loads of 53 psi (365 kN/m^2) and 90 psi (627 kN/m^2).

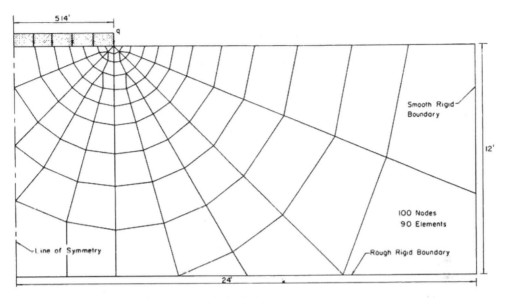

Fig. 5.9. Nonuniform mesh for shallow undrained clay stratum.

Nonuniform finite element mesh (Fig. 5.9)

A nonuniform mesh consisting of 100 nodes and 90 elements (Fig. 5.9) was also used to analyze this problem. Each quadrilateral in the nonuniform mesh was divided into four constant strain triangles. Near the edge of the footing this mesh is finer than the uniform mesh while away from the footing edge this mesh is somewhat coarser. The two meshes gave nearly identical results except near the maximum load where the nonuniform mesh overestimated the limit load by 7%.

For this particular problem in which the footing load is uniformly distributed, there is no need for such a fine mesh near the edge of the footing and the nonuniform mesh is probably too coarse away from the footing. However, comparative analyses for a rigid footing show the nonuniform mesh to be superior to the uniform mesh. Since in what follows we are primarily concerned with rigid footing and deeper soil strata, a nonuniform mesh is found to be a necessity.

5.5 RIGID STRIP FOOTING ON A ELASTIC STRATUM

5.5.1. General

We consider here a 50 feet deep soil stratum loaded by a 5 feet wide strip footing. The width of the stratum is taken to be 50 feet. The base of the soil

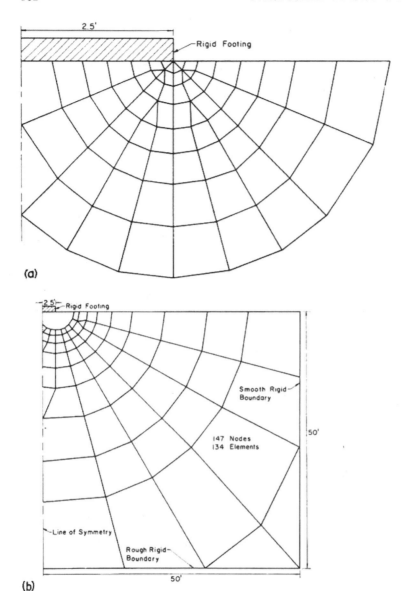

Fig. 5.10. Finite element mesh and the boundary conditions. (a) Mesh detail near footing. (b) Mesh without details near footing.

stratum is assumed to be rigid and perfectly rough, and the footing is also assumed to be rigid and perfectly rough. The boundary conditions and the dimensions used are indicated in Fig. 5.10(b).

The finite element mesh is also shown in Fig. 5.10 which consists of 147 nodes and 134 elements. The mesh is composed of a number of triangular and quadrilateral regions. The quadrilateral is further subdivided into four triangles connected to a fifth node located at the quadrilateral centroid. The stiffness for the quadrilateral region is therefore defined by four constant strain triangles. The meshes as shown in Fig. 5.10 are finest near the corner of the footing and get progressively coarser as the distance from the footing corner increases. It is not the aim here to capture the stress singularity at the corner since it is well known that it is difficult or perhaps impossible to do this with analytic finite element expansion (Sih, 1973). The aim here is simply to make the mesh fine where stress gradients are high.

Herein, an elastic stratum is treated first in order to compare the finite element solution with an existing exact solution. Subsequently we consider an elastic–plastic effective stress analysis of an overconsolidated clay (section 5.6), and finally an elastic–plastic total stress analysis of an undrained clay (section 5.7) is presented. For the elastic–plastic cases both small and large deformation analyses are considered, that is, solutions are presented with and without geometric nonlinearities included in the equilibrium equations (as described in Chapter 12).

5.5.2. Exact and finite element solutions

We compare here finite element solutions for a finite elastic stratum (Fig. 5.10) with exact solutions for an infinite halfspace (Muskhelishvili, 1963). The finite element solutions and the exact halfspace solution should be in general agreement near the footing but will not necessarily agree near the soil stratum base. Finite element stresses obtained here correspond to nodal stresses and were determined from a simple average of the stresses in all triangles adjacent to a particular node. For example, if a node is adjacent to four quadrilateral elements, each nodal stress would be an average of eight triangle stresses.

Vertical and horizontal contact stress distributions at the footing–soil interface for a Poisson's ratio of 0.3 are first compared. Both horizontal and vertical contact stresses are found to agree well with the halfspace exact solution, except near the footing corner. The vertical contact stress components agree somewhat less well than the horizontal contact stress components, with the differences ranging from 3 to 10%. Considering now the stresses below the footing centerline, the vertical stress components are found to agree remarkably well, whereas the horizontal stress components differ somewhat. This difference most likely reflects the finite stratum depth in the finite element analysis and should not be interpreted as an indicator that the present finite element mesh is not fine enough to capture the true solution. Likewise comparing the stress distributions beneath the footing corner the vertical stress components agree well while the horizontal stress components differ somewhat.

The finite element stress components are also obtained for a Poisson's ratio of 0.48. A Poisson's ratio of 0.5 is relevant to undrained total stress analysis in which clay is assumed to be incompressible. However the displacement formulation utilized herein does not admit a Poisson's ratio of 0.5 since the constitutive matrix becomes singular. Nevertheless we can attempt to approximate the incompressibility condition by using a high value of Poisson's ratio such as 0.48. It is found that we can obtain a reasonable approximation for the incompressible case by using a Poisson's ratio of 0.48.

We note here that significant developments have been made in obtaining exact and approximate solutions of problems in elasticity relevant to soil mechanics. We refer here to the book by Poulos and Davis (1974), which presents an exhaustive collection of elastic solutions for soil and rock mechanics.

5.6. RIGID STRIP FOOTING ON AN OVERCONSOLIDATED STRATUM OF INSENSITIVE CLAY (EXTENDED VON MISES MATERIAL) (DAVIDSON, 1974)

We deal here with an effective stress analysis of an overconsolidated clay. Presumably the footing load rate is such that no excess pore water pressures are generated. The material parameters utilized for the problem shown in Fig. 5.10 are:

$E = 5 \cdot 10^5$ psf (24 MN/m^2) $\varphi = 10°, 20°$ and $30°$

$\nu = 0.3$ $\gamma = 50$ pcf (801 kg/m^3)

$c = 500$ psf (24 kN/m^2) $K_o = 1.0$

where K_o is the ratio of the initial in-situ horizontal and vertical stress components. The constants α and k, which define the extended Von Mises yield criterion [2.7], are related to the $c - \varphi$ Coulomb constants through [2.8]. Only the friction angle φ is varied here with all other parameters being held constant.

In Chapter 3, section 3.4.2, we presented upper-bound solutions for smooth and rough footings bearing on a *weightless* $c - \varphi$ soil. In the following Chapter (section 6.6.1), we will show that the bearing capacity value determined by [3.81] is also a lower bound for a strip, smooth footing on a weightless $c - \varphi$ soil and therefore the correct value of the average limit pressure. We consider here the problem of a rough footing bearing on a ponderable soil, for which exact limit loads have not as yet generally been determined. We can, however, use the limit analysis solutions of Chapter 6 to estimate the limit loads to within 1–2% for the soil parameters considered here. The approximate limit loads are thus $q_o = 4350, 8260$ and $18,720$ psf (208, 396 and 896 kN/m^2) for friction angles of $\varphi = 10°, 20°$ and $30°$, respectively.

5.6.1. Load displacement curves

Load displacement curves for large and small deformation analyses are shown in

Figs. 5.11 through 5.13 for the three friction angles mentioned above. The closed and open circles represent actual computed points and the solution following each increment is plotted. None of the solutions is perfectly smooth, they all show some oscillations.

Considering first the small deformation solutions for which the original or undeformed geometry is used for the equilibrium equations, we note that the limit loads are overestimated in each case with the error increasing with increasing friction angle. For friction angles of 10, 20 and 30°, the errors are 10, 18 and 26%, respectively. Considering the entire elastic–plastic solution where the load is increased from zero to failure, we know from the above discussion that the initial part of the load displacement curve is highly accurate because the soil stratum is behaving essentially as an elastic medium. However, as the load increases we can expect that the numerical solution presented here diverges from the true solution, given the known error in the limit load. It is also apparent that as the friction angle increases, the mesh must become finer if the limit load is to be captured within a specified tolerance.

The case $\varphi = 10°$ (Fig. 5.11)

Consider now the solution obtained from a large deformation analysis for the case $\varphi = 10°$ (Fig. 5.11). In the analysis, geometric changes of the soil mass due to subsequent soil deformations are included in the equilibrium equations (Chapter 12). Except near the limit load the load displacement curve corresponding to the large deformation analysis is identical to that of the small deformation solution.

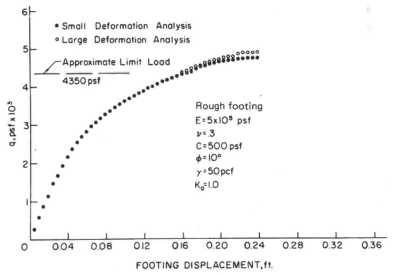

Fig. 5.11. Footing load–displacement curves, $\varphi = 10°$.

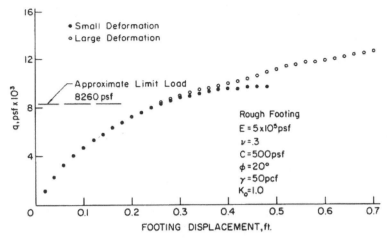

Fig. 5.12. Footing load–displacement curves, $\varphi = 20°$.

Near the limit load the two curves diverge somewhat with the large deformation curve appearing to approach a limiting load only 2% above the small deformation numerical limit load at the displacement of 0.27 feet. In this case ($\varphi = 10°$) we can thus say that the small deformation analysis is valid for all load levels up to and including the limit load. In addition the small deformation limit load is clearly a meaningful measure of the maximum bearing capacity of the footing.

The case $\varphi = 20°$ (Fig. 5.12)

Referring now to Fig. 5.12, we can see that if the friction angle is increased to 20°, the load displacement curves corresponding to the small and large deformation analyses remain essentially the same except near the small deformation limit load. Whereas the small deformation curve bends over and approaches a maximum, the large deformation curve continues to rise without any apparent limit. Although in this case the small deformation limit load is not a true measure of the maximum bearing capacity of the footing, it is nevertheless an indicator of the load level at which large increases in footing displacement can be expected for small increases in footing load.

The case $\varphi = 30°$ (Fig. 5.13)

Although there is a marked difference in the solutions by the small and large deformation analyses for the case when $\varphi = 30°$ (Fig. 5.13), the two curves are still virtually the same up to about 75% of the numerical limit load. Beyond this point the solutions diverge. There is no noticeable break in the large deformation curve, rather the curve rises smoothly past the small deformation numerical limit load. We may conclude that for this particular set of soil parameters, the small deformation

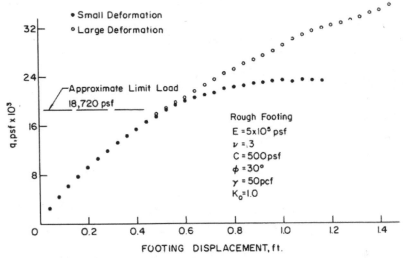

Fig. 5.13. Footing load–displacement curves, $\varphi = 30°$.

perfectly plastic limit load solution is not a meaningful measure of the bearing capacity of the footing. We might note, however, that a friction angle of $30°$ is higher than would be expected for an overconsolidated clay.

Clearly with all other parameters held fixed, an increase in the friction angle is associated with an increase in the difference between the small and large deformation solutions. In general if the elastic parameters are held fixed while the strength parameters increase, we can expect an increase in the difference between the two solutions.

Deformed surface (Fig. 5.14)

Referring to Fig. 5.12, we see that for $\varphi = 20°$, the footing has displaced almost half a foot by the time the numerical limit load has been reached. The corresponding deformed surface profile at the numerical limit load is shown in Fig. 5.14. In Fig. 5.14(a) the surface profile is drawn to scale while in Fig. 5.14(b) it is not. If the small and large deformation solutions differ significantly we would expect the deformed geometry and initial geometry of the soil stratum to differ also significantly. Fig. 5.14 clearly shows this to be the case. We can also see from this figure that soil deformation must be severe near the footing corner. In Fig. 5.14(b) we show a deformed surface profile for both small and large deformation solutions. There is a noticeable difference between the two profiles corresponding to similar load levels. The footing displacement determined from the large deformation analysis is less than that determined from the small deformation analysis. This is consistent with the load displacement curves shown in Fig. 5.12 where the large deformation solution is stiffer than the small deformation solution.

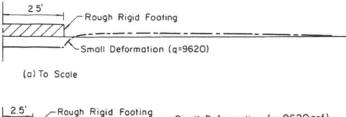

(a) To Scale

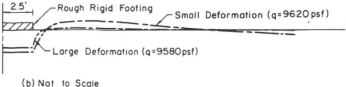

(b) Not to Scale

Fig. 5.14. Soil surface profile at the numerical limit load ($\varphi = 20°$, $c = 500$ psf or $24\,kN/m^2$).

In the remainder of this section we consider in some detail the behavior of a soil with a friction angle of $20°$ and a cohesive strength of 500 psf or ($24\,kN/m^2$). In Figs. 5.15 through 5.17 we show stress distributions at various load levels as obtained from a small deformation analysis, and in Figs. 5.18 and 5.19 stress distributions from large and small deformation analyses are compared. None of these stress distributions include the initial overburden stresses.

5.6.2. Stress distributions

Referring to Figs. 5.15 through 5.17, at a load of 1140 psf ($55\,kN/m^2$) some yielding has occurred near the footing corner but most of the soil stratum is still elastic. At $q = 4030$ psf ($193\,kN/m^2$) significant yielding has occurred and $q = 9620$ psf ($461\,kN/m^2$) corresponds to the numerical limit load.

Contact stresses (Fig. 5.15)

Considering first the contact stress at the soil–footing interface (Fig. 5.15), the stress distributions at $q = 1140$ psf ($55\,kN/m^2$) are essentially that of an elastic body. The vertical stress component, σ_y, at the footing corner is about three times that at the footing center. As the load increases and yielding spreads, the horizontal and vertical stress distributions tend to become flatter. At the numerical limit load the vertical stress, σ_y, at the corner is only about 30% greater than that at the footing center. As would be expected the shearing stress, τ_{xy}, has completely changed direction by the time the numerical limit load has been reached.

The shearing stress distribution at the limit load is nearly linear up to a peak value of about 1700 psf ($81\,kN/m^2$), after which it falls off sharply. If we define the maximum value of the mobilized friction angle between the footing base and adjacent soil to be:

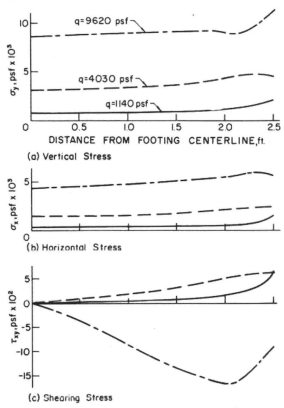

Fig. 5.15. Contact stress beneath footing ($\varphi = 20°$, $c = 500$ psf or 24 k N/m^2).

$$\delta_m = \tan^{-1}\left(\frac{\tau_{max} - c}{q_0}\right) \qquad [5.4]$$

where τ_{max} is the maximum value of the contact shearing stress, then for friction angles of 10, 20 and 30°, $\delta_m = 2°$, 7° and 11°, respectively. Thus for the material parameters studied here, a friction angle between footing base and soil of 11° is sufficient to produce an essentially perfectly rough condition. This is consistent with the results of Chen and Davidson (1973). This will be discussed further in Chapter 6.

Stress distributions below footing (Figs. 5.16 and 5.17)

In Figs. 5.16 and 5.17 horizontal and vertical stress distributions along vertical lines beneath the footing center and corner are shown. As the load increases and yielding spreads, the stress distributions change somewhat, particularly near the footing. There is a noticeable change in the shearing stress distribution beneath the footing corner as the load increases.

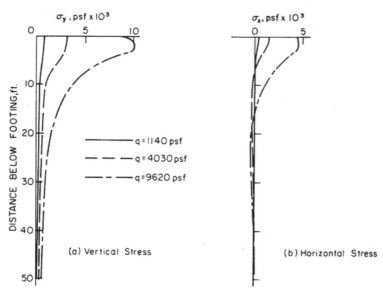

Fig. 5.16. Stress distribution below footing centerline ($\varphi = 20°$, $c = 500$ psf or 24 kN/m^2).

Comparison of small and large deformation stresses (Figs. 5.18 and 5.19)

Stress distributions at the numerical limit load are shown in Figs. 5.18 and 5.19 for small and large deformation analyses. Referring first to the contact stresses shown in Fig. 5.18, vertical and horizontal stresses differ only near the footing corner, although the shearing stresses differ all along the footing. As can be seen in Fig. 5.19, stresses beneath the footing corner differ only near the footing. At a depth of 3–4 feet (0.91–1.2 m) beneath the footing, the large and small deformation stresses are essentially the same.

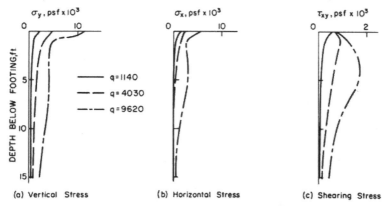

Fig. 5.17. Stress distribution below footing corner ($\varphi = 20°$, $c = 500$ psf or 24 kN/m^2).

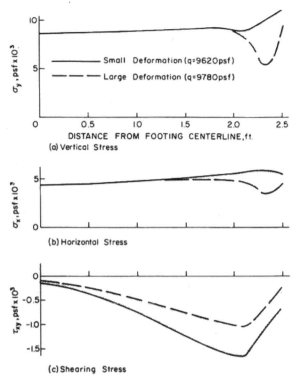

Fig. 5.18. Contact stresses beneath footing at limit load. Small and large deformation ($\varphi = 20°$, $c = 500$ psf or 24 k N/m^2).

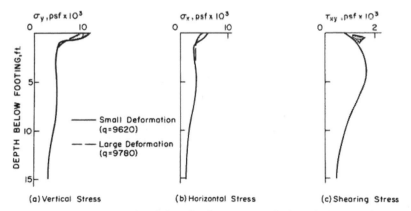

Fig. 5.19. Stress distribution below footing corner at limit load. Small and large deformation ($\varphi = 20°$, $c = 500$ psf or 24 k N/m^2).

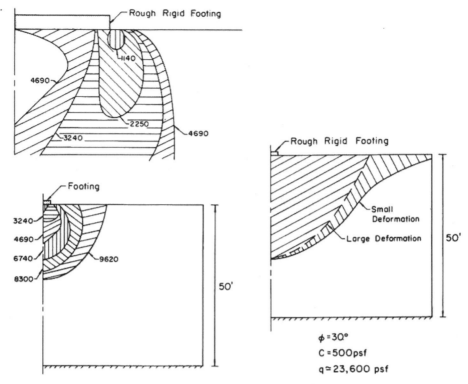

Fig. 5.20. Spread of yield zone ($\varphi = 20°$ $c = 500$ psf or 24 kN/m^2).

Fig. 5.21. Zones of yielding at the numerical limit load.

5.6.3. Yield zones

In Fig. 5.20 zones of yielding for various load levels are presented. Theoretically some yielding should occur near the footing corner for any load level since the true solution contains a singularity at the corner. In the finite element solution yielding occurred in the first increment of loading ($q = 1140$ psf or 55 kN/m^2) and a small yield zone near the footing corner can be seen in Fig. 5.20. As the load increases yielding spreads downward and toward the footing centerline. The yield zone reaches the footing centerline at a load just below 3240 psf (155 kN/m^2). The zone of yielding continues to spread outward from the footing as the load increases. In addition yielding spreads upward toward the footing until at a load of 6740 psf (323 kN/m^2) all of the soil immediately below the footing is yielded. At the numerical limit load (9620 psf or 461 kN/m^2) a significant portion of the soil stratum has yielded.

In Fig. 5.21 we show the yield zone at the numerical limit load for $\varphi = 30°$. The extent of yielding at the limit load is clearly influenced by the value of the friction angle φ. As can be seen from the figure, small and large deformation solutions give somewhat different zones of yielding at similar load levels. The large deformation analysis produces a smaller zone of yielding, and as partial explanation for this we note that as the footing punches down into the clay, an effective surcharge is created by the clay which now lies above the footing base. This surcharge should increase the hydrostatic stress component and thus increase the shear required to yield the soil.

5.6.4. Velocity fields

Rough footing (Figs. 5.22 and 5.23)

Finally in Fig. 5.22 we show the velocity field at the numerical limit load for a friction angle of $20°$. Superimposed on the figure is the outline of the Prandtl velocity field. which is only strictly applicable to weightless soils. The Prandtl velocity field has been described in Fig. 3.21, Chapter 3. However, for the particular set of soil parameters used here, the actual velocity field and the Prandtl field can be expected to be similar. It can be seen in Fig. 5.22 that the Prandtl field and the numerically determined field are indeed similar. We can clearly identify a wedge beneath the footing which moves downward with the footing. There is also an intermediate zone in which the velocity vectors are essentially perpendicular to radial lines emanating from the footing corner. The velocity magnitude can also be seen to grow as the radial line rotates counter-clockwise. A fairly well defined third zone exists which appears to be moving upward and out as a rigid body. This problem was also solved for a weightless soil (all other material parameters unchanged) and the velocity field determined at the limit load was virtually the same

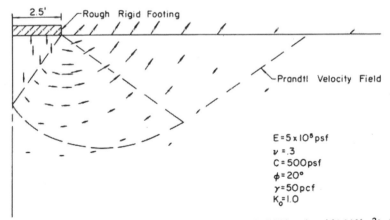

Fig. 5.22. Velocity field at the numerical limit load (9620 psf or 461 kN/m^2). Small deformation.

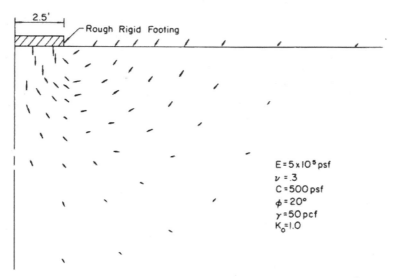

Fig. 5.23. Velocity field at the numerical limit load (9620 psf or 461 k N/m^2). Large deformation.

as that shown in Fig. 5.22. The velocity field at a similar load level and as determined from a large deformation analysis is shown in Fig. 5.23. It is no surprise that there is a distinct difference between the small and large deformation fields since the large deformation solution has yet to reach a limiting load.

Smooth footing (Fig. 5.24)
   In Fig. 5.24 the velocity field, at the numerical limit load, for a *smooth* footing

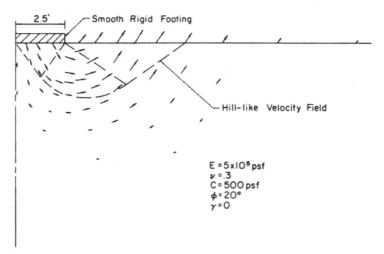

Fig. 5.24. Smooth footing velocity field at the numerical limit load.

bearing on a *weightless* $c - \varphi$ soil is shown. The outline of a Hill-like velocity field is also shown in the figure. The Hill velocity field has been described in Fig. 3.22, Chapter 3, and is characterized by two wedges beneath the footing rather than one as in the Prandtl mechanism. Each wedge makes an angle of $(\frac{1}{4}\pi + \frac{1}{2}\varphi) = 55°$ with respect to the footing base. As can be seen in Fig. 5.24, the numerically determined velocity field does not exactly correspond to the Hill field, nor does it correspond to the Prandtl field, although it contains characteristics of both fields.

5.7. RIGID STRIP FOOTING ON A STRATUM OF UNDRAINED CLAY (VON MISES MATERIAL) (DAVIDSON, 1974)

We deal here with an elastic—plastic total stress analysis of a saturated undrained clay. Presumably the load rate is such that the excess porewater pressure has no chance to dissipate, that is volumetric strain is almost zero throughout the analysis. The material parameters utilized for the problem shown in Fig. 5.10 are:

$E = 1 \cdot 10^5$ (4.8 MN/m^2) and $1 \quad 10^6$ psf (48 MN/m^2) $\varphi = 0$

$\nu = 0.48$ $\gamma = 100$ pcf (1601 kg/m^3)

$c = 1000$ psf (48 kN/m^2) $K_o = 1.0$

With $\varphi = 0$, the extended Von Mises yield function reduces to a Von Mises function.

Only Young's modulus is varied here. Two values of the ratio E/c are considered, namely 1000 and 100. This ratio in real soils is expected to range from approximately 100 to 3000 with perhaps 1000 being a typical value (D'Appolonia et al., 1971). In the following we consider first the case of a soil with a Young's modulus of $1 \cdot 10^6$ psf (48 MN/m^2) and present a fairly detailed description of the soil response to the footing load. Only limited data are presented later for the case of a soil with a Young's modulus of $1 \cdot 10^5$ psf (4.8 MN/m^2).

5.7.1. Case 1 – $E = 1 \cdot 10^6$ psf (48 MN/m^2)

Load displacement curve (Fig. 5.25)

Load displacement curves obtained from small and large deformation analyses are shown in Fig. 5.25 for a Young's modulus of $1 \cdot 10^6$ psf (48 MN/m^2). The curves are almost linear up to a load of 3600 psf (172 kN/m^2). After this point the curves bend over quite sharply and gradually approach the numerical limit loads corresponding to small and large deformation analyses. The two curves are seen to be essentially identical. Thus for $E/c = 1000$, the small deformation solution is valid for all load levels up to and including the limit load. The exact load for this problem is 5140 psf (246 kN/m^2) and the numerical limit load is 10% above this value.

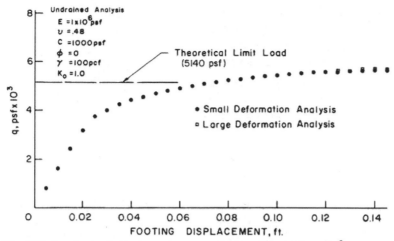

Fig. 5.25. Footing load–displacement curves, $E = 1 \cdot 10^6$ psf (48 MN/m^2).

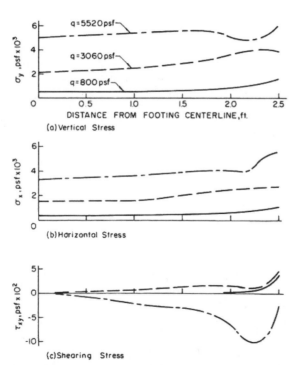

Fig. 5.26. Contact stresses beneath footing ($\varphi = 0$, $c = 1000$ psf or 48 k N/m^2).

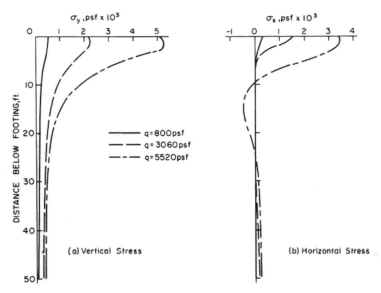

Fig. 5.27. Stress distribution below footing centerline ($\varphi = 0$, $c = 1000$ psf or 48 kN/m^2).

Stress distributions (Figs. 5.26, 5.27 and 5.28)

Stress distributions for various load levels are shown in Figs. 5.26 through 5.28. At 800 psf (38 kN/m^2) the soil stratum was essentially elastic. Considerable yielding had occurred at $q = 3060$ psf (147 kN/m^2) and $q = 5520$ psf (264 kN/m^2) corresponds to the numerical limit load. There is a marked similarity in these curves and those obtained for a $c - \varphi$ soil (Figs. 5.15 through 5.17). It is of interest to note that the maximum contact shearing stress at the limit load is equal to the cohesive strength (1000 psf or 48 kN/m^2).

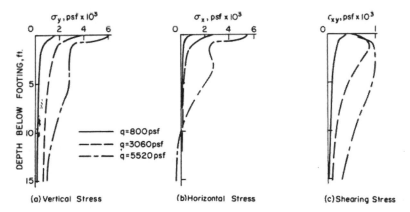

Fig. 5.28. Stress distribution below footing corner ($\varphi = 0$, $c = 1000$ psf or 48 kN/m^2).

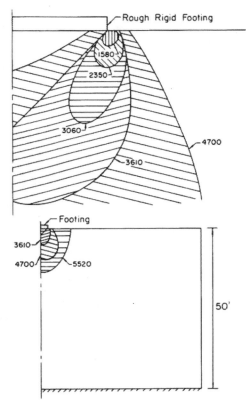

Fig. 5.29. Spread of yield zone ($\varphi = 0$, $c = 1000$ psf or 48 kN/m^2). All numbers in the figure are in psf = lb/ft^2

Zones of yielding (Fig. 5.29)

Zones of yielding for various load levels are shown in Fig. 5.29. Again yielding starts at the corner of the footing and spreads downward and toward the footing centerline. At a load of about 3610 psf (173 kN/m^2) the yield zone has just reached the footing centerline. At this point the footing and an adjacent elastic wedge (which makes a 45°-angle with the base) are separated by a band of yielded material from the remainder of the still elastic stratum. The spread of the yield zone to the centerline is coincident with the sharp break in the load displacement curve (Fig. 5.25). This kind of behavior is similar to that noted for the notched tensile specimen (section 5.2). At the numerical limit load all of the soil directly beneath the footing has yielded. At the limit load the zone of yielding in the present case ($\varphi = 0$) is considerably smaller than that of $c - \varphi$ soils (Figs. 5.20 and 5.21).

Velocity field (Figs. 5.30 and 5.31)

The corresponding velocity field at the numerical limit load is shown in Fig. 5.30

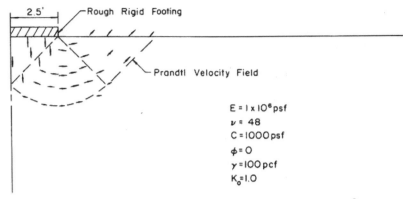

Fig. 5.30. Velocity field at the numerical limit load (5520 psf or 264 k N/m^2). Small deformation.

along with the outline of the Prandtl velocity field. The velocity field is denoted by the small arrows. The close agreement between the two fields is evident. Outside of the Prandtl field the velocity magnitudes are too small to appear in the figure. In Fig. 5.31 we show the velocity field at the limit load for a smooth footing bearing on the same material. The outline of the Hill velocity field is also superimposed on the figure. Although the numerical velocity field is similar to the Hill field, it cannot be said to be identical to the Hill field. The numerical field, in fact, appears to be a combination of the Hill and Prandtl fields. Prager and Hodge (1968) have previously suggested a combination of the Hill and Prandtl fields as a possible failure mode.

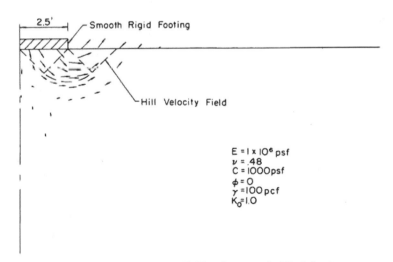

Fig. 5.31. Smooth footing velocity field at the numerical limit load.

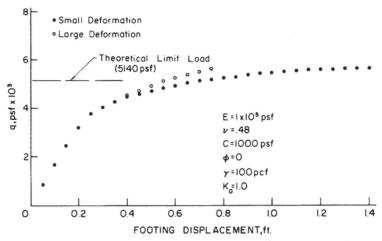

Fig. 5.32. Footing load–displacement curves, $E = 1 \cdot 10^5$ psf (4.8 MN/m²).

5.7.2. Case 2 – $E = 1 \cdot 10^5$ psf (4.8 MN/m²)

Load displacement curves for the case of a clay with a Young's modulus of $1 \cdot 10^5$ psf (4.8 MN/m²) are shown in Fig. 5.32. The shape of the small deformation curve must, of course, be identical to that for a Young's modulus of $1 \cdot 10^6$ psf (48 MN/m²). The ratio, E/c, is 100 here and we see from the figure that the small and large deformation analyses produce somewhat different results. The large deformation curve stops at a load of approximately 5600 psf (268 kN/m²). Although the analysis was continued beyond this point, the response became somewhat erratic. A

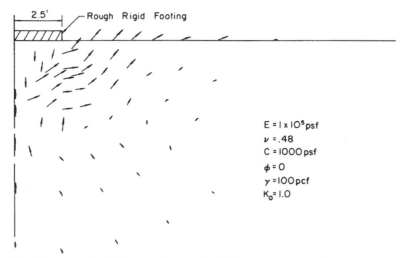

Fig. 5.33. Velocity field at maximum load (5600 psf or 268 kN/m²). Large deformation.

very curious velocity field was observed at 5600 psf (268 kN/m^2) and is shown in Fig. 5.33. A wedge beneath the footing is moving downward with the footing, and the soil adjacent to the wedge is being squeezed up and to the right. Beyond this is a region resembling a radial shear zone, and finally there is a rigid zone moving up and to the right. The velocity field is somewhat resembling the mechanism shown in Fig. 6.23(a).

In an attempt to obtain an improved solution, the problem was recomputed using half of the original increment size. The revised solution lay slightly above the original solution and a maximum load of about 5800 psf (278 kN/m^2) was obtained. At this point a solution for the linear equations could not be obtained. The problem was also solved using a reduced Poisson's ratio of 0.4 and again the solution behaved irregularly at about 5800 psf (278 kN/m^2). Thus in the context of the finite element mesh and numerical integration scheme used here, the maximum footing load is 5800 psf (278 kN/m^2). It may be the case that the velocity field shown in Fig. 5.33 corresponds to the actual failure mode for a clay with $E/c = 100$.

5.7.3. Some comments on the numerical solutions

In this section we discuss the adequacy of the increment size used in the various solutions, the accuracy of the solutions with respect to satisfaction of the discrete equilibrium equations, and finally element stresses are discussed.

Increment size

When an incremental integration scheme is utilized, there is always a question as to the adequacy of the increment size. Referring to Figs. 5.11, 5.12 and 5.13, it can be seen that about twice as many increments were used for a friction angle of 10° as were used for friction angles of 20° and 30°. In particular, 48 increments were used for $\varphi = 10°$. This problem was originally solved using about half as many increments, however, near the small deformation limit load, the large deformation solution behaved somewhat erratically. For this reason both the large and small deformation curves were recomputed using a smaller increment size. The small deformation curves were virtually the same for both increment sizes with the smaller increment size giving a 1%-reduction in the numerical limit load. The large deformation curves were the same except near the limit load where the smaller increment gave a smoother response.

As further evidence of the adequacy of the increment sizes used, the small deformation solution for $\varphi = 20°$ (Fig. 5.12), the large deformation solution for $\varphi = 30°$ (Fig. 5.13), and the small deformation undrained solution (Fig. 5.25) were all recomputed using half the original increment size. Although in all three cases the smaller increment size produced a smoother load displacement curve, damping the oscillations mentioned previously, the two solutions were essentially the same. We can thus conclude that any error in the solutions can be ascribed to the finite element discretization rather than to the integration scheme.

Accuracy of the solutions

In each increment two sets of linear simultaneous equations must be solved. As the footing load approaches the limit load we can expect these simultaneous equations to become somewhat illposed since at the limit load of the discretized body the tangent stiffness is singular. We thus need some measure of the accuracy of the linear equation solutions and herein two checks were used.

At the end of each increment the incremental displacements are substituted back into the mid-increment equations (Chapter 12) and a residual vector can be computed. Of course, for an exact solution the residual vector is identically zero. For all the solutions presented here and for each increment of those solutions, every residual vector component was less than 0.005 lb. (0.022 N).

After the last increment of every solution, overall equilibrium of the soil stratum was checked, that is, all of the external forces (including constraint forces) were summed. For all of the solutions presented here both the vertical and horizontal components of this sum were less than 0.002 lb. (0.0089 N).

Element stresses

Stresses in constant strain triangles often exhibit sharp jumps between adjacent elements. This tendency appears to be even more pronounced in elastic–plastic solutions than in elastic solutions. Stress jumps were found to be greatest in the undrained analysis where element stresses oscillated between tension and compression near the footing corner at the higher loads. It should be noted however that nodal stresses were reasonably smooth at all load levels.

5.8. RIGID CIRCULAR PUNCH ON AN ELASTIC–PLASTIC STRAIN HARDENING LAYER (ISOTROPIC HARDENING VON MISES MATERIAL)

5.8.1. Finite element solution (A.C.T. Chen, 1973)

A rigid flat circular punch indenting into a solid layer of finite dimensions, which is placed on a *smooth* rigid foundation, is shown in Fig. 5.34(a). No separation of the layer from the foundation is permitted. Thus the solution of the problem also applies to the compression of a block between two flat parallel punches similar to that shown in Fig. 5.3(a) for the case of plane strain condition. The punch is assumed to be perfectly rough and its diameter is $2w$. The diameter of the circular layer is $2R$ and its height is h. The finite element mesh and geometry used in the numerical work are shown in Fig. 5.34(b) where $R/w = 2.7, h/w = 0.6,$

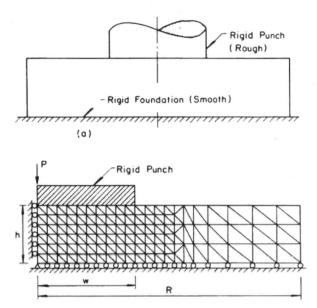

Fig. 5.34. Indentation of a circular layer by a circular punch (isotropic hardening Von Mises material).

and w is 1.0 inch (2.54 cm). The numerical solution to the problem is obtained by using the mesh shown in Fig. 5.34(b) with constant strain for each triangular ring element. The same problem has also been solved by Lee and Kobayashi (1970) using quadrilateral rings each of which is composed of four triangular ring elements with constant strain within each triangular ring element. Results corresponding to triangular ring elements and quadrilateral ring element will be compared here.

The material of the layer is assumed to be linear elastic–linear strain hardening material obeying Von Mises' criterion for initial and subsequent yielding. The material properties used for the computation are $E = 10,000$ ksi (68,948 MN/m^2), $v = 0.33$, $\sigma_0 = 13$ ksi (90 MN/m^2) and the strain hardening modulus H' used here is 20 ksi (138 MN/m^2). These values are identical to those used by Lee and Kobayashi (1970). Details of the formulation of this material model are given in Chapter 12.

Fig. 5.35 shows the average pressure as a function of punch displacement. Lee and Kobayashi's solution is indicated by small open circles in the figure. The two curves show a good agreement initially and then they deviate when the curves begin a fast bend with an increasing loading. The deviation in load is approximately 10% at the punch displacement $d/w = 0.009$. Aside from the obvious fact that the elements used for the two solutions are different (triangular rings vs. quadrilateral rings), it should be noted that Lee and Kobayashi's solution allows only one element to yield at each load increment, whereas herein several elements are allowed to yield simultaneously at each load increment.

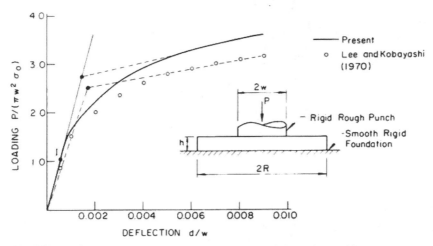

Fig. 5.35. Load−deflection curves of the circular punch indentation problem.

A unique feature of the finite element method is that it permits one to follow the path of deformation from the initial stress-free state to the collapse state. The development of the plastic zone and the manner in which the plastic zone spreads are the important aspects of the solution. Figs. 5.36 and 5.37 show the develop-

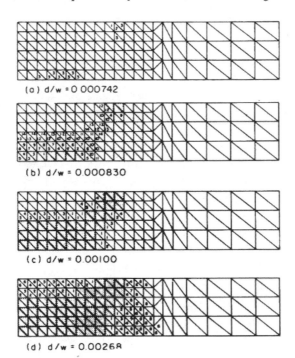

(a) d/w = 0.000742

(b) d/w = 0.000830

(c) d/w = 0.00100

(d) d/w = 0.00268

Fig. 5.36. Development of plastic zones at various stages of loading.

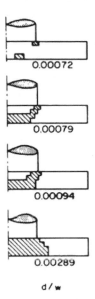

0.00072

0.00079

0.00094

0.00289

d / w

Fig. 5.37. Development of plastic zones at various stages of loading – Lee and Kobayashi (1970).

ment of the plastic zones at four different stages of loading for both present and Lee and Kobayashi's solutions. It is evident that the development of plastic zones in the present solution (Fig. 5.36) is similar to that of Lee and Kobayashi's work (Fig. 5.37).

Referring to Fig. 5.36 (or 5.37), the yielding appears first at two locations: one at the edge of the punch and the other at the central bottom of the layer as shown in Fig. 5.36(a). As soon as the plastic zone spread and two yield zones are connected (Fig. 5.36b), the load displacement curve begins to bend. This connecting plastic zone is contained by remaining elastic material, as shown in Fig. 5.36(c). The plastic zone spreads away from the punch during contained plastic deformation to almost unrestricted plastic deformation. The slope of the load displacement curve then becomes almost constant and the plastic zone grows slowly in all directions. This development of plastic zones in the circular punch case is similar to that for plane strain condition (see Fig. 5.5).

5.8.2. Limit analysis solution

The only limit analysis solution which is relevant here is given in Chapter 7 for a *smooth* punch indenting into a semi-infinite layer shown in Fig. 7.19, which is placed on a *rough* base foundation. The lower- and upper-bound solutions corresponding to this case have the average value 2.86 σ_0 ([7.36]) and 3.10 σ_0 ([7.57]), respectively. The lower and upper bounds therefore show that the limit pressure for this semi-infinite problem with a rough base foundation lies within ±5% of the value $2.98 \, \sigma_o$

Referring to Fig. 5.35, the rate of increase in pressure after passing the fast bends of the load displacement curves assumes an almost steady state. There is no noticeable break in the load displacement curves which can be used to define the limit load. Rather, the curves continue to rise without apparent limit. If we adopt an alternate definition of limit load as indicated by the small circles and dotted lines in Fig. 5.35, we obtain the values of limit pressure $2.75 \, \sigma_o$ for the present solution or $2.5 \, \sigma_o$ for Lee and Kobayashi's solution. Although this alternative definition of limit load is not the true measure of the maximum load carrying capacity of the problem, it is nevertheless a good indicator of the load level at which large increases in punch displacement can be expected for small increase in punch load.

If this crude approach is taken here, the limit analysis solution $2.98 \, \sigma_o$, which is based on the *perfect* plastic idealization for the material, provides also a meaningful answer for the case of strain-hardening material. Limit load is a matter of definition for all practical problems. The choice is not an absolute one, but is determined by the most significant features of the problem to be solved. Both crude and refined measures of load carrying capacity have their place. For design purposes, the crude may be more relevant than the refined. Yet it is always well to keep in mind that the basic experimental information on the behavior of material has been approximated so drastically.

5.9. A BRIEF HISTORICAL SKETCH

In the past few years a number of investigators have considered horizontal nonlinear clay strata subjected to vertical loads such as those transmitted by a footing. Some have treated soil as a nonlinear elastic material while others have utilized elastic—plastic models. We consider first the nonlinear elastic investigations followed by elastic—plastic investigations.

Girijavallabhan and Reese (1968) considered a circular footing bearing on an undrained clay where an isotropic nonlinear elastic model was used for the clay. Poisson's ratio was assumed to be constant and the secant value of the shear modulus was assumed to be uniquely related to the octahedral shearing strain. An iterative approach was used to solve the equilibrium equations and the finite element method was used to discretize the soil stratum. A model footing test was analyzed, and analysis and experiment were shown to agree reasonably well.

Desai and Reese (1970) also used an elastic model and the finite element method to treat circular footings bearing on an undrained clay. An incremental approach was used to integrate the equations. It was assumed that in each increment the instantaneous stiffness of the clay could be described by a constant Poisson's ratio and a tangent value of Young's modulus. The incremental material parameters were obtained directly from undrained triaxial tests. Model footing tests for a single soil layer and two soil layers were analyzed. Experiment and analysis were shown to agree well.

Desai (1971) used spline functions to numerically approximate undrained triaxial stress—strain data. An incremental integration scheme was utilized and incremental elastic parameters were determined directly from the spline functions. The finite element method was used to analyze model circular footing tests, with good results. Some comparisons were also made between footing load displacement curves obtained from spline approximations and hyperbolic approximations for undrained triaxial stress—strain curves.

Duncan and Chang (1970) used a hyperbolic representation for undrained triaxial stress—strain curves. The finite element method and a incremental integration scheme were used to analyze circular footings bearing on undrained clay.

Höeg et al. (1968) used a finite difference technique to analyze a shallow layer of undrained clay subjected to a strip load. The clay was modeled as an elastic—perfectly plastic material with a Tresca yield criterion. The numerically determined limit load and the exact limit load were shown to be identical.

Tang and Höeg (1968) utilized a linear elastic—plastic strain-hardening model developed by Christian (1966) to treat strip footings bearing on frictional materials (e.g., normally consolidated clay). The soil model is similar to the strain-hardening models proposed by Drucker et al. (1957) and Roscoe et al. (1963). A finite difference technique and an incremental integration scheme were used to solve the problem. The results were somewhat unsatisfactory with the load displacement curve having a zig-zag character. Some dynamic problems were also considered with better results.

Fernandez and Christian (1971) treated a strip footing bearing on undrained clay and both material and geometric nonlinearities were included in the formulation. A hyperbolic nonlinear elastic model and an elastic—plastic Tresca model were used to describe the clay. The finite element method and the mid-point integration rule were utilized in the solution. The results were evidently very poor, particularly for the elastic—plastic model. The load displacement curves were very irregular and the numerical limit load was far above the theoretical limit load. It is not clear if large deformations were included or excluded in the footing problem treated in the report. However for the particular soil parameters utilized in the example, the changing soil geometry should have little effect on the results.

Höeg (1972) considered a circular footing bearing on a shallow layer of undrained soft clay in which the clay was assumed to be a linear elastic—linear strain-softening material. An isotropic softening Von Mises material model was utilized. The finite element method and an incremental integration technique were used to solve the problem. For a softening stiffness equal to about 20% of the elastic stiffness, the maximum load was found to be reduced by 40%.

Finally Zienkiewicz et al. (1969) treat a uniform strip load bearing on a soil obeying the extended Von Mises yield condition and its associated flow rule. A combined iterative–incremental integration scheme in association with the finite element method was used to solve the problem. For the particular set of material properties considered, the iterative scheme failed to converge at a load which is approximately half of the theoretical limit load. No load displacement curve was presented but zones of yielding at various load levels were shown.

5.10. SUMMARY AND CONCLUSIONS

5.10.1. Summary

The primary purpose of this chapter is to present a number of example problems of progressive failure of a homogeneous soil stratum to footing loads. Soil is modeled here as a linear elastic–perfectly plastic material with the extended Von Mises yield condition and associated flow rule. The effect of large deformations on the response of the soil is also included in some of the example problems. The detailed formulation of this problem is given in Chapter 12.

In particular we considered in sections 5.5, 5.6 and 5.7 a single, strip surface footing bearing on a finite stratum of clay (loaded normally and centrally). The footing is assumed to be rigid and the interface between the footing and soil may be either smooth or rough. The base of the soil stratum is rigid and perfectly rough. A plane strain condition is assumed. Footing width and stratum depth are two of the many parameters which affect this problem. However here we chose a single set of geometric parameters while allowing the material parameters of the soil to vary.

In addition to this particular problem mentioned above, we also considered in section 5.4 a uniform strip load on a shallow layer of undrained clay as well as some additional solid mechanics problems, e.g., a notched elastic–plastic tensile specimen in section 5.2 and the indentation of finite block by two rigid rough punches in section 5.3. Finally, in section 5.8 we considered an elastic–plastic strain hardening analysis of a circular flat punch indentation into a specimen of finite dimensions.

Numerical techniques are utilized to obtain the solutions presented herein. The finite element method is used for spatial discretization while an incremental integration scheme, referred to as the mid-point rule in Chapter 12, is used to develop the complete load displacement–stress response. Constant strain triangles are used exclusively. A FORTRAN IV computer program was written to formulate and solve the governing equations. We refer here to the dissertation by Davidson (1974), which gives a brief description of this program.

The example problems considered here are clearly highly idealized. Soil is not strictly an elastic–plastic material nor are most soil strata homogeneous, and, in fact, most footings are submerged below the soil surface. Idealized problems are solved in order to gain insight, qualitative information and sometimes quantitative information with regard to real physical problems. The aim here is to look at a class of problems and to observe, through analysis, the behavior of the progressive failure of the footing–soil system. The introduction of additional parameters associated with layered soil profiles, subsurface footings and more complex soil models would confuse rather than enlighten. Limit analysis solutions of these complex two- and three-dimensional footing problems will be presented in the following two Chapters.

5.10.2. Conclusions

Limit load

It appears that a fairly fine finite element mesh is needed to capture limit loads. However the finite element method is clearly capable of predicting limit loads to within small tolerances as evidenced by the solutions for a uniform strip load on a shallow clay layer and the notched tensile specimen. In the context of the extended Von Mises model, it seems that the higher the friction angle, the finer the mesh must be in order to determine limit loads to within a specified error.

It was also demonstrated in these examples that the elastic–plastic analysis by finite element methods, combined with the determination of the limit load by limit analysis methods, is of value in obtaining a better understanding of the progressive failure problem. For any numerical work such as the finite element method used here, the particular algorithms used in a computer program as well as possible mistakes in the computer coding are all potential sources of error. In order to demonstrate that a computer program is giving reasonable and believable results, a check on the accuracy with existing closed form solutions is necessary. There exist no closed form solutions for most elastic–plastic problems. However, two extreme parts of the small displacement solution are usually known – the linear elastic solution and the limit load solution, at least for some materials. Even an approximate limit analysis solution will prove to be a powerful source for checking numerical results. Of course, we have no absolute check on the accuracy of the elastic–plastic intermediate response.

Large deformation vs. small deformation analysis

For realistic values of effective stress parameters for overconsolidated clay, changes in geometry caused by deformation of the soil are such as to affect the load displacement response only near the limit load. For reasonable values of undrained clay parameters ($E/c = 1000$), soil deformation has practically no effect on the load displacement response of a footing, even near the limit load. In such a case, small deformation analysis neglecting the changes in geometry is sufficient for an elastic–plastic analysis. For the extreme value of $E/c = 100$, clay response is affected near the limit load. For reasonable values of both drained overconsolidated parameters and undrained parameters, the small displacement limit load is a meaningful measure of the load at which footing displacements become excessive.

Clearly for practical settlement calculations a small deformation analysis is sufficient for the total and effective stress parameters considered here. Depending on the precision required, a linear analysis may, in fact, be suitable for practical settlement analysis. This is particularly true for undrained analysis where a significant portion of the load settlement curve is nearly linear.

Deformation modes at the limit state

For some time there has been a question as to the true velocity field at incipient collapse or plastic limit state for a smooth punch bearing on a perfectly plastic, weightless Von Mises or Tresca material (plane strain). The results presented here indicate that the actual field is a combination of both the Hill and Prandtl velocity fields, a possibility suggested by Prager and Hodge (1968). For a smooth punch bearing on an extended Von Mises weightless material, it was determined here that the actual failure mode contains elements of both the Prandtl and Hill velocity fields.

For both the Von Mises and extended Von Mises yield functions, we have demonstrated that the Prandtl velocity field corresponds to the actual mode of failure for a perfectly rough punch bearing on a weightless material.

Chapter 6

BEARING CAPACITY OF STRIP FOOTINGS

6.1. INTRODUCTION

The problem under consideration in this chapter is the determination of ultimate bearing capacity of a single, strip footing bearing on a plane surface of a semi-infinite mass of soil that is assumed to be elastic–perfectly plastic material. It is further assumed that the force acted on the footing is normally and centrally loaded and increased until penetration occurs as a result of plastic flow in the soil. The load required to produce the complete plastic flow or failure of the soil support is called the *critical load* or the *total ultimate bearing capacity*. The average critical load per unit of area, q_0 is called the *bearing capacity of the soil*. The value of the bearing capacity of a soil depends not only on the mechanical properties of the soil but also on the size of the loaded area, its shape, and its location with reference to the surface of the soil. In this chapter, the investigation is limited to the bearing capacity of *strip footings* on horizontal bearing areas for cohesive material with internal friction, and for special cases of purely cohesive and cohesionless materials. In the following chapter, the bearing capacity problems for three-dimensional flat square, rectangular and circular spread footings are considered.

The term *"strip footing"* is applied to a footing whose length is very long in comparison with its width. It has a uniform width which essentially gives rise to a two-dimensional *plane strain condition*. This condition simplifies the computation very considerably. This is in contrast to the term *"spread footing"* which applies to a footing whose width is approximately equal to its length, such as a square, a rectangular or a circular footing. In the following investigation, the footing is assumed to be rigid while the interface between the soil and the footing can be either smooth or rough. In most parts of this chapter, the soil is assumed to be an isotropic, homogeneous and elastic–perfectly plastic material which obeys the Coulomb yield condition and the associated flow rule. A plane strain condition is assumed in this chapter. The influence of *non-homogeneity* and *anisotropy* on the bearing capacity of footings will be discussed in section 6.8.

The limit analysis method employed herein does not consider the deformation of the soil and the solutions obtained are essentially the same as that assuming the soil to be *rigid–perfectly plastic* material. If the relative settlement or displacement restrictions on the footing are relatively unimportant, the limit analysis solutions based on rigid–plastic idealization for soil are acceptable. Otherwise, the deformation properties, i.e., stress–strain or constitutive relations of the soil must also be considered. This chapter is primarily concerned with complete failure of the footing, or its ultimate bearing capacity. This type of failure is referred to here as a *general shear failure*. On the other hand, if the mechanical properties of the soil are such that the plastic flow is preceded by a very large displacement, a complete elastic–plastic static response of soil to footing loads is required in order to estimate the settlement. This type of failure is called *local shear failure*. Some numerical results from a complete finite element solution to the problem have been discussed in Chapter 5.

In order to familiarize the reader with the limit analysis methods of computation and further to gain a better understanding of the meaning of the limit analysis and limit equilibrium solutions in soil mechanics, the individual discussions and problems connected with the bearing capacity of footings will be taken up in the following sequence. First, a brief description of the salient features of the limit analysis and limit equilibrium methods relevant to bearing capacity problems of footings will be given. Then we solve in succession the following problems: (a) computation of the ultimate bearing capacity of a footing on a *cohesive ponderable* soil ($c - \varphi - \gamma$ soil); (b) computation of the ultimate bearing capacity of a footing on a *cohesionless ponderable* soil ($\varphi - \gamma$ soil); and (c) computation of the ultimate bearing capacity of a footing on a *cohesive imponderable* (weightless) soil ($c - \varphi$ soil). Problems (b) and (c) are special cases of (a). On account of the important influence of soil weight on the bearing capacity of a footing, the special case of a footing on a cohesionless ponderable soil is therefore further discussed in section 6.5 and then on the special case of a footing on a cohesive imponderable soil in section 6.6. The main results of existing slip-line solutions are summerized in section 6.7. These slip-line solutions are frequently used in various parts of this Chapter for comparison with the theoretical limit analysis solutions.

6.2. LIMIT ANALYSIS, SLIP-LINE AND LIMIT EQUILIBRIUM METHODS

Studies of the bearing capacity of foundations under conditions of plane strain have been made by Terzaghi (1943), by Meyerhof (1951) using limit equilibrium method, by Sokolovskii (1965), by Brinch Hansen (1961) using slip-line method, by Shield (1954b), by Chen and Davidson (1973) using limit analysis method, and many others. Some of the information to be presented in this Chapter is contained, therefore, in this previous work but the relevant parts of each have not yet been compared in principle. Although all the analyses utilize the concept of *perfect plasticity*, the relation between these solutions, corresponding to different analytical methods, involves terminology and special concept that are not in common use in the field of soil mechanics. A brief description of the salient features of these methods relevant to the bearing capacity solutions of footings will therefore be given here. The discussion includes the following three methods: limit analysis, slip-line method, and limit equilibrium. A general discussion of the salient features of these methods has already been given in Chapter 1.

6.2.1. Limit equilibrium analysis (Fig. 6.1)

The comparison between these methods can best be illustrated by considering the bearing capacity of a surcharged cohesive soil in which the angle of internal friction, φ, is zero. This is illustrated, schematically, in Fig. 6.1(a). It will be assumed that there is no contact between the footing and the surcharge. The limit equilibrium solution takes as its failure mode a rotation of a semi-circular section about its own center, O, located at the corner of the footing, The distribution of normal stresses along the semi-circular surface is unknown, but the shear stress is

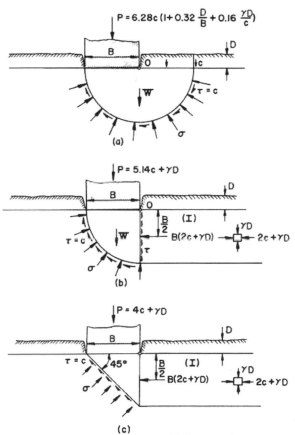

(a)

(b)

(c)

Fig. 6.1. Limit equilibrium modes of failure, $\varphi = 0$.

assumed to be equal to c, the cohesion. The summation of moments about the center of rotation, O, is taken:

$$-P\tfrac{1}{2}B + \pi BcB + BD\gamma\,\tfrac{1}{2}B + DcB = 0 \tag{6.1}$$

where the $(Dc)B$ term is due to the shearing of the surcharge material. The resultant of the weight of the rotating soil, as well as the normal stresses along the circumference, pass through the center, O, and produce no moments. The ultimate bearing capacity of the footing by limit equilibrium is then:

$$q_0 = \frac{P}{B} = 6.28c\left(1 + 0.32\,\frac{D}{B} + 0.16\,\frac{\gamma D}{c}\right) \tag{6.2}$$

Two other assumptions concerning possible surfaces of failure are given in Fig. 6.1(b and c) for purposes of comparison. The first consists in assuming the center of rotation O at the edge of the foundation, as shown in Fig. 6.1(b). The choice of this failure mode for a limit equilibrium solution would raise several questions. What is the normal and shear stress distribution along the circular shear plane? Also, what is the normal stress distribution along the vertical separation between the quarter circle and the rectangular region? The resultant of the shear stress along this line passes through O, but the normal stress resultant is not defined. A commonly accepted method in limit equilibrium analysis is to assume the quarter circle rotation produces uni-axial compression of the soil it bears against, as shown in Fig. 6.1(b). The maximum compression that may be sustained by this soil is $(2c + \gamma D)$ since $2c$ is the greatest allowable difference between principal stresses and γD corresponds to a hydrostatic pressure which is in equilibrium with the surcharge at the top surface. (The pressure due to the weight of the soil within the region is neglected.) Summing moments about O of all forces, yields:

$$q_0 = \frac{P}{B} = c(\pi + 2) + \gamma D = 5.14c + \gamma D \tag{6.3}$$

For the special case of a surface footing, $D = 0$, [6.3] reduces to $q_0 = 5.14c$ which is nearly 20% less than the value of $6.28c$ obtained in Fig. 6.1(a).

An even lower limit equilibrium solution can be produced by considering the scheme shown in Fig. 6.1(c), which assumes a triangular soil block under the punch. Equilibrium of forces shows that $q_0 = 4c + \gamma D$. As with other limit equilibrium solutions, it is not known which solutions should be chosen. If it can be shown that the stress field in Region I and the stress distribution along the failure surfaces can be extended throughout the soil mass and still satisfy equilibrium and not violate the yield condition, then this is a lower bound. Unfortunately, it is not known in the present case whether the stress field can be extended in this manner.

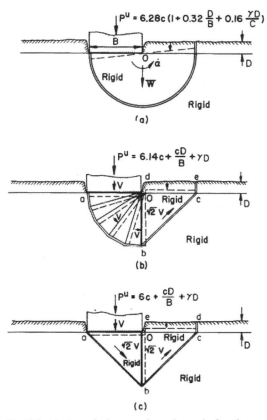

Fig. 6.2. Limit analysis upper bounds on the bearing capacity of soils, $\varphi = 0$.

6.2.2. Plastic limit analysis – Upper bounds (Fig. 6.2)

To obtain an upper bound of the same problem by limit analysis, the chosen mechanisms as shown in Fig. 6.2 are quite similar. In Fig. 6.2(a) is a rigid body rotation about O, with shearing through the surcharge, which is of depth D. If the angular velocity is $\dot\alpha$, the rate of work done by external forces is $P^u(\frac{1}{2}B)\dot\alpha - BD\gamma(\frac{1}{2}B)\dot\alpha$ while the rate of internal energy dissipation is $cB\dot\alpha \,\pi B + cB\dot\alpha \, D$. Note that the rate of work done by the weight of the body is zero since its motion is perpendicular to the direction of its own weight. Equating the rate of internal and external work will yield [6.2], the first limit equilibrium solution. The solution reduces, for the important case of $D = 0$, to $q_0 = 6.28c$.

It is apparent that since the two solutions agree, the first limit equilibrium mode has produced a solution which is an upper bound. Both solutions are applicable for smooth or rough bases, since the mechanism of the upper bound does not require relative motion between the footing and the rotating soil mass.

Different mechanisms may be employed to reduce the upper bound. One such mechanism is shown in Fig. 6.2(b). Using the upper-bound technique, the rate of external work includes terms due to the force P moving downward with velocity V, the triangular mass moving diagonally upward (a $45°$ right triangle is assumed), and the surcharge being pushed upward by the triangular soil mass, while the soil under the footing may be considered to be many rigid triangles, as in Fig. 3.19(a), sliding past one another in a zone of radial shear. The rate of external work is $P^u V - (\frac{1}{2}B^2)\gamma V - BD\gamma V + (\frac{1}{2}B^2)\gamma V$ The rate of internal energy is dissipated along the quarter circle circumference, within the quarter circle by shearing deformation (it was shown in section 3.4 that these two terms are equal), along the diagonal shear plane, by shearing through the surcharge, and by relative shearing between the quarter circle and the triangular mass (due to a relative velocity). It has the value $2cV(\frac{1}{2}\pi B) + c\sqrt{2}V\sqrt{2}B + cVD + cVB$. Equating the rate of internal and external work yields:

$$q_0^u = \frac{P^u}{B} = c\left(\pi + 3 + \frac{D}{B}\right) + \gamma D \qquad\qquad [6.4]$$

when $D = 0$, the ultimate bearing capacity is $q_0^u = 6.14c$. The improvement of the previous limit analysis solutions is only about 2%.

Figure 6.2(c) illustrates another failure mechanism. With reference to the figure, the ultimate bearing capacity is given as:

$$q_0^u = 6c + c\frac{D}{B} + \gamma D \qquad\qquad [6.5]$$

which is slightly better than the previous answer. It is apparent that the limit analysis provides the consistent upper-bound solutions while the limit equilibrium analysis cannot; when the failure modes, that can be compared, are considered in the solutions.

6.2.3. Plastic limit analysis – Lower bounds (Figs. 6.3 and 6.4)

To obtain a lower bound of the bearing capacity problem by limit analysis, choose a simple stress field, which gives $2c$ as a lower bound, as shown in Fig. 6.3(a). Since a hydrostatic pressure has no effect on plastic yielding for soils with $\varphi = 0$, the simple addition of the hydrostatic pressure $2c + \gamma D$ as shown in Fig. 6.3(b) to Fig. 6.3(a) increases the ultimate load to (Fig. 6.3c) $q_0^u = P^l/B = 4c + \gamma D$ which is an improved lower bound, but not a very good one.

Now, a little physical intuition will raise the lower bound to a very useful point. Experience has shown that the load on soil "spreads", or is carried by an even greater area, the deeper one goes (see Fig. 6.4a). Therefore, consider the stress field, shown in Fig. 6.4(b), consisting of two inclined "legs" of $2c$ each. Where they meet,

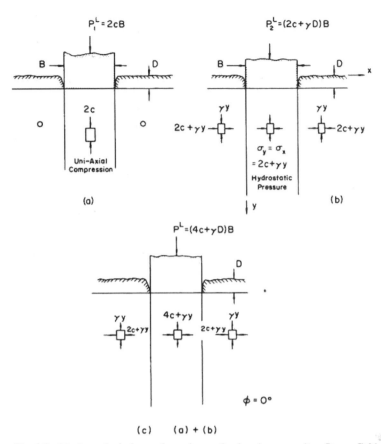

Fig. 6.3. Limit analysis lower bounds on the bearing capacity. Stress field (c) is the simple addition of the stress fields (a) and (b).

under the footing, there is a vertical component of $3c$ and horizontal component of c. If the stress field of Fig. 6.3(b) is superposed onto the field of Fig. 6.4(b), the stress field shown in Fig. 6.4(c) results and it is easily verified that the superposition of these stress fields does not violate the yield condition anywhere. Therefore, the value $q_0{}^l = 5c + \gamma D$ is a lower bound for the collapse pressure of the footing. This lower bound may be improved through a more judicious choice of stress fields. Valuable techniques for constructing lower bounds of this "truss-like" nature have been discussed in section 4.6 for Tresca material and in section 4.7 for Coulomb materials (Chapter 4). It was shown in section 4.6 that the familiar Prandtl field may be imagined physically as a load supported by infinitely many supporting legs. A picture of 9 vertical and inclined legs at $10°$ to each other is shown in Fig. 4.22(b) for a footing resting on a purely cohesive weightless soil. As the number of legs grows, the stress in each decreases. In the limit, the Prandtl field beneath the

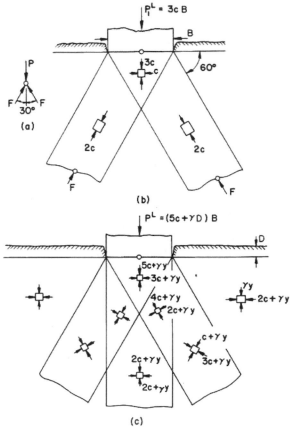

Fig. 6.4. Limit analysis lower bounds. Stress field (c) is the result of adding the stress fields in (b) and in Fig. 6.3(b).

footing is recovered. Figure 6.4(c) shows clearly that the three-legs approximation for the Prandtl field considering the weight of soil gives the answer sufficiently accurate for practical purpose.

6.2.4. Slip-line method

The application of slip-line method to this bearing capacity problem will result in a plastic equilibrium stress field around the footing; however, there is no guaran- tee that this stress field can be extended satisfactorily throughout the body, nor is it necessarily possible to associate kinematically admissible velocity fields with these stress fields. The slip-line solution for bearing capacity computation of the footing is therefore not necessarily the true solution nor is it known when it is an upper-bound or a lower-bound solution. If one employs the associated flow rule and can integrate the resulting stress—strain rate equations to yield a kinematically admissible velocity field around the footing, the slip-line solution is an upper-bound solution. If, in addition, the slip-line stress field around the footing can be extended over the entire half-space of the soil domain such that the equilibrium equations, the stress boundary conditions and the yield condition are satisfied, the slip-line solution is also a lower bound and is hence the true solution.

Slip-line solutions that have not been shown to be lower bounds are usually referred to as *incomplete solutions*. Those which have been shown to be lower bounds are referred to as *complete solutions*. The Prandtl solution (Fig. 3.21) for the bearing capacity of a surface footing resting on a cohesive weightless soil, for example, has been shown by Shield (1954a) to be complete. The few slip-line solutions for soils with weight are as yet incomplete, although it is commonly assumed, at least for smooth footings, that it will be possible to show them to be complete (for instance see Cox, 1962, page 380).

Although the slip-line method can generally be expected to give a good estimate of the correct solution, closed form solutions can only be obtained for weightless soils. The *characteristic differential equations* must be integrated numerically or graphically around the footing if soil weight is included in the analysis. To date this has only been done for the simple geometries.

6.2.5. Summary

The so-called limit equilibrium or plastic equilibrium method has traditionally been used to obtain approximate solutions for the bearing capacity of soils. Examples of this approach are the solutions of Terzaghi (1943) and Meyerhof (1951). The method can probably best be described as an approximate approach to the construction of a slip-line field and generally entails an assumed failure surface. It is necessary to make sufficient assumptions about the stress distribution within the soil domain bounded by the failure surface such that an equation of equilibrium, in terms of resultant forces, may be written for bearing capacity determination.

None of the equations of continuum mechanics are explicitly satisfied everywhere inside or outside of the failure surface. Since the stress distribution is not defined precisely everywhere inside of the assumed failure surface, one cannot say definitely that a stress distribution compatible with the assumed failure surface and satisfying equilibrium, stress boundary conditions and the yield function exists. Although the limit equilibrium technique utilizes the basic philosophy of the upper-bound theorem of limit analysis, that is, a failure surface is assumed and a least answer is sought, a limit equilibrium solution may not be an upper bound. The method basically gives no consideration to soil kinematics, and equilibrium conditions are satisfied only in a limited sense.

It is clear then that a solution obtained using the limit equilibrium method is not necessarily an upper or a lower bound. However, any upper-bound limit analysis solution will obviously be a limit equilibrium solution.

In the following computations, the upper-bound technique of limit analysis is employed to generate approximate solutions to the bearing capacity problems. The lower-bound technique of limit analysis is not considered, but the computer method presented by Lysmer (1970) can be applied and may give good lower-bound solutions.

6.3. SOIL GOVERNING PARAMETERS

6.3.1. Soil weight parameter, G

The bearing capacity of footings depends not only on the mechanical properties of the soil (cohesion c and friction angle φ), but also on the physical characteristics of the footing (width B, depth D, and roughness δ). For a Coulomb material, Cox (1962) has shown that for a smooth surface footing bearing on a soil subjected to no surcharge, the fundamental dimensionless parameters associated with the stress characteristic equations are φ and $G = \gamma B/2c$, where φ is the internal friction angle, c is the cohesive strength, γ is the unit weight of soil and B is the width of the footing. If G is small the soil behaves essentially as a cohesive weightless medium. If G is large soil weight rather than cohesion is the principal source of bearing strength. For most practical cases one can expect that φ lies in the range 0–40° and G will range from 0.1 to 1.0. These limits assume that c ranges from 500 psf (24 kN/m^2) to 1000 psf (48 kN/m^2), and that the footing width ranges from 3 to 10 feet (0.9 to 3.05 m). Two additional parameters are needed here, namely δ and D/B where δ is the friction angle between the footing base and the soil, and D is the footing depth.

It will be shown in the following section that the dimensionless bearing capacity pressure q_0/c depends only upon the angle of internal friction of the soil φ, the dimensionless *soil weight parameter G*, footing base friction angle δ, and surcharge depth ratio D/B. The ultimate bearing capacities of footings obtained in the following sections by upper-bound technique of limit analysis admit a *closed form* expression in terms of these governing parameters of the problem and the geometry of failure mechanism. Solutions for both smooth and rough footings and for surface and subsurface footings (shallow and deep) are obtained. Numerical results are presented for G lying within the range $0 \leqslant G \leqslant 10$, φ varying between 0° and 45°, and D/B ranging from 0 to 10.

6.3.2. Method of superposition

For the most part, bearing capacity of footings on soils have in the past been

calculated by a *superposition method* suggested by Terzaghi (1943), in which contributions to the bearing capacity from different soil and loading parameters are
summed. These contributions are represented by the expression:

$$q_0 = cN_c + qN_q + \gamma \tfrac{1}{2} B N_\gamma \qquad\qquad [6.6]$$

where the *bearing capacity factors*, N_c, N_q, and N_γ represent the effects due to
soil cohesion c, surface loading q, and soil unit weight γ, respectively. These parameters N are all functions of the angle of internal friction, φ. Terzaghi's quasi-
empirical method assumed that these effects are directly *superposable*, whereas the
soil behavior in the plastic range is nonlinear and thus *superposition* does not hold
for general soil bearing capacities.

Meyerhof (1951), using Terzaghi's concept, presents extensive numerical results
for shallow and deep footings by assuming failure mechanisms for the footing and
by presenting results in the form of bearing capacity factors, N.

The reason for using the simplified method (superposition method) is largely due
to the mathematical difficulties encountered when the conventional limit equilibrium method is used. Since the investigation of the influence of the weight of a soil
on the plastic equilibrium of a footing has not yet passed beyond the stage of
formulating the differential equations and integrating numerically for a given individual problem, the general bearing capacity problem can best be solved in two
stages:

The first stage is essentially based on an extension of the analytical work of
Prandtl (1920) and Reissner (1924); this assumes a *weightless* material and gives the
first part of the bearing capacity $cN_c + qN_q$ in *closed form* expressions. The second
stage is a semigraphical treatment based on an extension of the work of Ohde
(1938), or the numerical integration scheme of Sokolovskii (1965), or the graphical
method of de Jong (1957); this takes the *weight* of material into account and gives
the second part of the bearing capacity $\gamma(\tfrac{1}{2}B)N_\gamma$ in *table* or *chart* form.

It has been generally assumed that the bearing capacities obtained by Terzaghi's
method are *conservative*, and experiments on model and full-scale footings (De Beer
and Ladanyi, 1961 and De Beer and Vesic, 1958) seem to substantiate this for
cohesionless soils. However, as pointed out by Ko and Scott (1973), the angle of
internal friction as determined under an axially symmetric triaxial compression
stress is known to be several degrees *less* than that determined under plane strain
conditions for low confining pressure (Cornforth, 1964). If the φ-value obtained
under triaxial compression stress condition is used in calculating the N factors for
use in the superposition formula [6.6], a lower bearing capacity will be predicted
than if the φ-value in plane strain conditions is used. It can be seen clearly from the
plots of the N-factor in sections 6.5 and 6.6 that a few degrees increase in φ cause a
large increase in the bearing capacity factors N_γ and N_c, especially when φ is large.
Ko and Scott (1973) show that when the proper angle of internal friction is used,

the Terzaghi method of superposition may give *non-conservative* results. This, then, illustrates the importance of a proper choice of friction angle φ when comparing a theoretical solution with experimental results. If such a theoretical solution is used for practical design, then careful judgement is obviously necessary.

6.4. BEARING CAPACITY OF A STRIP FOOTING ON A GENERAL $c - \varphi - \gamma$ SOIL

The upper-bound technique of limit analysis is used herein to develop approximate solutions for the bearing capacity of a strip footing on a general cohesive soil with weight. Analytical solutions are first obtained for smooth and rough and surface and subsurface footings. Numerical results are then calculated and compared with existing slip-line and limit equilibrium solutions. The limit analysis solutions for smooth, surface footings are shown to compare favorably with slip-line solutions. Meyerhof's solutions and the limit analysis solutions for rough, subsurface footings are shown to agree remarkably well.

Failure mechanisms

Two distinct failure mechanisms, referred to here as the Prandtl and Hill mechanisms, are utilized in the analysis. In the following investigation the terms "mechanism" and "velocity field" will be used interchangeably.

6.4.1. Prandtl mechanism (Fig. 6.5)

The Prandtl mechanism, consisting of five distinct zones, is shown diagrammatically in Fig. 6.5. The wedge *abc* is translating vertically as a rigid body with the same initial downward velocity V_1 as the footing. The downward movement of the footing and wedge is accommodated by the lateral movement of the adjacent soil as indicated by the radial shear zone *bcd* and zone *bdef*. The angles ξ and η are as yet unspecified. Since the movement is symmetrical about the footing, it is only necessary to consider the movement on the right-hand side of Fig. 6.5.

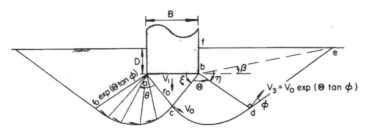

Fig. 6.5. "Prandtl" mechanism.

The radial shear zone bcd is bounded by logspiral curve cd, where the equation for the curve, in polar coordinates (r,θ), is $r = r_0 \exp(\theta \tan \varphi)$. The center of the spiral cd is at point b, and r_0 is the length of line bc. The radial shear zone bcd may be considered to be composed of a sequence of rigid triangles as shown in the left-hand side of Fig. 6.5 (or see Fig. 3.20). All the small triangles and the zone $bdef$ move as rigid bodies in directions which make an angle φ with the discontinuity lines cd and de, respectively. The velocity of each small triangle is determined by the condition that the relative velocity between the triangles in contact must have the direction which makes an angle φ to the contact surface. It is found in section 3.4, Chapter 3 that the velocity for each triangle is $V = V_0 \exp(\theta \tan \varphi)$ ([3.71]). The velocity V_3 in the zone $bdef$ is perpendicular to the radial line bd. Hence, the velocity field is continuous across line bd. Line de is constrained to be tangent to the logspiral curve at point d. Since zone $bdef$ translates as a rigid body, the zone velocity V_3 has the value $V_0 \exp(\Theta \tan \varphi)$ where $\Theta = \pi + \beta - \xi - \eta$ is the angle spanned by the radial shear zone. The velocities so determined constitute a kinematically admissible velocity field with the restriction that, $\eta > \beta$.

Rate of internal energy dissipation

Energy is dissipated at the discontinuity surfaces cd and de between the material at rest and the material in motion and at the discontinuity surface bc between adjacent rigid triangle abc and radial shear zone bcd. It is a simple matter to calculate the lengths of the lines of discontinuity. The rate of energy dissipation is then found by multiplying the length of each discontinuity line by c times the velocity difference across the line multiplied by $\cos \varphi$, and summing over all such lines ([3.19]).

Along bc. Refer now to Fig. 6.6 which shows discontinuity line bc and a super-

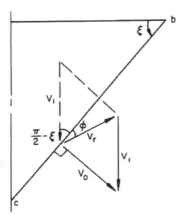

Fig. 6.6. Portion of the Prandtl mechanism with a superimposed velocity diagram.

imposed velocity diagram. Using the velocity diagram, we have:

$$V_r = \frac{V_0 \sin \xi}{\cos (\xi - \varphi)} \quad \text{and} \quad V_1 = \frac{V_0 \cos \varphi}{\cos (\xi - \varphi)} \qquad [6.7]$$

Hence the rate of energy dissipation along line bc is:

$$c V_0 r_0 \left[\frac{\sin \xi \cos \varphi}{\cos (\xi - \varphi)} \right] \qquad [6.8]$$

In radial shear zone bcd. It is found in section 3.4, [3.74], that as the number of rigid triangles shown in the left-hand side of Fig. 6.5 approaches infinity, the total rate of energy dissipation within the shear zone is:

$$\tfrac{1}{2} c V_0 r_0 \left\{ \frac{\exp [2(\pi + \beta - \xi - \eta) \tan \varphi] - 1}{\tan \varphi} \right\} \qquad [6.9]$$

Along spiral cd. It is also found in section 3.4 that the energy dissipation rate along the spiral curve is also defined by [6.9].

Along de. Since the velocity field is continuous across line bd, no energy is dissipated there. The dissipation rate along line de can be easily computed once the length of de is known and is:

$$c V_0 r_0 \left\{ \frac{\sin \eta \cos \varphi \exp [2(\pi + \beta - \xi - \eta) \tan \varphi]}{\cos (\eta + \varphi)} \right\} \qquad [6.10]$$

Rate of external work

The rate at which work is done by the soil weight is found by multiplying the area of each rigid body by γ times the vertical component of the velocity of the rigid body and summing over all the areas in motion.

Triangle region abc:

$$- \gamma V_0 r_0^2 \left[\frac{- \sin \xi \cos \xi \cos \varphi}{\cos (\xi - \varphi)} \right] = - \gamma V_0 r_0^2 h_1(\xi) \qquad [6.11]$$

where $h_1(\xi)$ has the value given in the square bracket above.

Logspiral region bcd. The external rate of work done by the soil weight in this zone can be computed by summing over the region Θ the products of differential triangle's component of vertical velocity with its weight. Referring to Fig. 6.7, the gravity force acting on a differential element is $dF = \tfrac{1}{2} \gamma r^2 d\theta$ and the differential

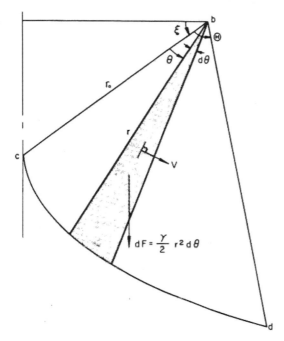

Fig. 6.7. Power calculation in radial shear zone.

rate of external work is:

$$[V \cos(\theta + \xi)] \ [\tfrac{1}{2}\gamma r^2 \, d\theta] \qquad\qquad [6.12]$$

Using the following velocity relations:

$$V = V_0 \exp(\theta \tan\varphi), \quad r = r_0 \exp(\theta \tan\varphi) \qquad\qquad [6.13]$$

The total rate of external work done by the soil weight is:

$$\tfrac{1}{2}\gamma \int_0^{\Theta} r^2 V \cos(\theta + \xi) \, d\theta = -\frac{\gamma V_0 r_0^2}{2} h_2(\xi, \eta) \qquad\qquad [6.14]$$

where:

$$h_2(\xi,\eta) =$$

$$\frac{(3 \tan\varphi \cos\xi + \sin\xi) + [3 \tan\varphi \cos(\beta - \eta) + \sin(\beta - \eta)] \ \exp[3(\pi + \beta - \xi - \eta)\tan\varphi]}{(1 + 9 \tan^2\varphi)}$$

$$[6.15]$$

Region bdef:

$$-\frac{\gamma V_0 r_0^2}{2} h_3(\xi, \eta) \qquad\qquad [6.16]$$

where:

$$h_3(\xi, \eta) = \left[\frac{\sin \eta \, \cos \varphi}{\cos (\eta + \varphi)} + \frac{\sin \beta \, \cos \beta \, \cos^2 \varphi}{\cos^2 (\eta + \varphi)}\right] \cos (\beta - \eta) \exp \left[3(\pi + \beta - \xi - \eta) \tan \varphi\right]$$

[6.17]

in which the first and second term in the square bracket represent the contributions from zone bde and zone bef, respectively. The rate of work done by the footing load is:

$$q_0 V_0 r_0 \frac{2 \cos \varphi \, \cos \xi}{\cos (\xi - \varphi)}$$

[6.18]

The solution of the Prandtl mechanism (shallow footing, $\eta > \beta$)

Equating the total rate at which work is done by the force on the foundation and the soil weight in motion to the total rate of energy dissipation along the lines of velocity discontinuity, it is found, after some simplifications, that an upper bound on the average bearing capacity pressure of the soil is:

$$\frac{q_0(\xi, \eta)}{c} = N_c(\xi, \eta) + G \, N_\gamma(\xi, \eta)$$

[6.19]

in which $G(= \gamma B/2c)$ is dimensionless soil weight parameter and the bearing capacity factors $N_c(\xi, \eta)$ and $N_\gamma(\xi, \eta)$ can be expressed in terms of the two as yet unspecified angles ξ and η under the condition $\eta > \beta$ (see Fig. 6.5):

$$N_c(\xi, \eta) = \cot \varphi \left[\frac{\cos \eta \, \cos (\xi - \varphi) \exp \{2(\pi + \beta - \xi - \eta) \tan \varphi\}}{\cos \xi \, \cos (\eta + \varphi)} - 1\right]$$

[6.20]

$$N_\gamma(\xi, \eta) =$$

$$\frac{\cos (\xi - \varphi)}{2 \cos \varphi \, \cos^2 \xi} \left[h_1(\xi) + h_2(\xi, \eta) + h_3(\xi, \eta)\right] = -\frac{\tan \xi}{2} + \frac{0.5 \cos (\xi - \varphi)}{\cos^2 \xi \, \cos \varphi (1 + 9 \tan^2 \varphi)}$$

$$\times \left[\{3 \tan \varphi \, \cos (\beta - \eta) + \sin (\beta - \eta)\} \exp \{3(\pi + \beta - \eta - \xi) \tan \varphi\} + 3 \tan \varphi \, \cos \xi + \sin \xi\right]$$

$$+ \frac{\cos (\xi - \varphi) \sin \eta \, \cos (\beta - \eta) \exp [2(\pi + \beta - \eta - \xi) \tan \varphi]}{\cos \xi \, \cos \varphi \, \sin \beta} \left(\frac{D}{B}\right)$$

$$+ \frac{2 \cos (\xi - \varphi) \cos (\beta - \eta) \exp [(\pi + \beta - \eta - \xi) \tan \varphi]}{\cos \varphi \, \tan \beta} \left(\frac{D}{B}\right)^2$$

[6.21]

where β is related to D/B, η and ξ by the transcendental equation:

$$\sin \beta \exp (\beta \tan \varphi) = \frac{2(D/B) \cos \xi \, \cos (\varphi + \eta)}{\cos \varphi \exp [(\pi - \eta - \xi) \tan \varphi]}$$

[6.22]

The best upper bound from [6.19] is found by minimizing function $q_0(\xi,\eta)/c$ with respect to variables ξ and η for the given values φ, G and D/B. The numerical solution of [6.19] and [6.22] can be obtained by the simultaneous application of the *method of steepest descent* for the optimum value of [6.19] and a *Newton-Raphson iteration* on [6.22] for angle β (see Ketter and Prawel, 1969, p. 304 and p. 467). The complete presentation of results and comparisons with various existing solutions will be discussed in later sections.

The solution of the Prandtl mechanism (surface footing)

For the special case of a surface footing for which both D/B and β are equal to zero, [6.19] reduces to:

$$
\frac{q_0(\xi,\eta)}{c} = N_c(\xi,\eta)|_{\beta=0} + G\,N_\gamma(\xi,\eta)|_{\beta=0}
$$

$$
= \cot\varphi \left[\frac{\cos\eta\,\cos(\xi-\varphi)\exp\{2(\pi-\xi-\eta)\tan\varphi\}}{\cos\xi\,\cos(\eta+\varphi)} - 1 \right]
$$

$$
+ G\left[\frac{-\tan\xi}{2} + \frac{0.5\cos(\xi-\varphi)}{\cos^2\xi\,\cos\varphi(1+9\tan^2\varphi)} \right.
$$

$$
\times \{(3\tan\varphi\cos\eta - \sin\eta)\exp[3(\pi-\eta-\xi)\tan\varphi] + 3\tan\varphi\cos\xi + \sin\xi\}
$$

$$
\left. + \frac{0.5\cos(\xi-\varphi)\sin\eta\cos\eta\exp[3(\pi-\eta-\xi)\tan\varphi]}{\cos^2\xi\,\cos(\eta+\varphi)} \right] \qquad [6.23]
$$

For a surface footing, the optimum value of η from the condition $\partial q_0/\partial\eta = 0$ is found to be, as expected:

$$
\eta = 45^\circ - \tfrac{1}{2}\varphi \qquad\qquad\qquad [6.24]
$$

Although the optimum value of ξ depends on G, the minimum values of the bearing capacities may be approximated to within 5% by using the following approximated equations:

$$
\xi = 45^\circ + \tfrac{1}{2}\varphi \qquad \text{if } G \leqslant 0.1 \qquad\qquad [6.25]
$$

$$
\xi = \varphi + 15^\circ \qquad \text{if } G > 0.1 \qquad\qquad [6.26]
$$

Since the kinematically admissible velocity field used in the Prandtl mechanism is such that no slip occurs between the footing and the soil, the upper bound obtained is applicable to either a rough or a smooth footing.

6.4.2. Hill mechanism (Fig. 6.8)

Bearing capacity calculations based on Prandtl mechanism are rigorous upper

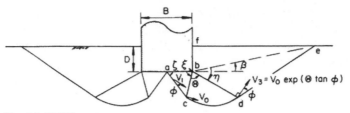

Fig. 6.8. "Hill" mechanism.

bounds for perfectly rough, moderately rough, and perfectly smooth footings. However, a better upper bound for the case of small base friction and large values of G can be obtained using the Hill mechanism shown in Fig. 6.8. Excepting the area directly below the base, the Hill mechanism closely resembles the Prandtl mechanism. Considering now the right half of the symmetric velocity field, wedge abc is translating as a rigid body with a downward velocity V_1 inclined at an angle φ to the discontinuity line ac. A portion of the Hill mechanism with a superimposed velocity diagram is shown in Fig. 6.9. Since the soil must remain in contact with the footing, the footing must move with the downward velocity $V_f = V_1 \sin(\zeta - \varphi)$. The relative velocity between the footing and soil mass is $V_r = V_1 \cos(\zeta - \varphi)$. The rest of the mechanism is similar in form to the Prandtl mechanism. Energy is

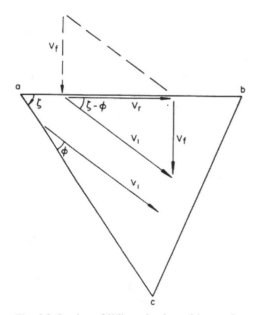

Fig. 6.9. Portion of Hill mechanism with superimposed velocity diagram.

dissipated along the lines ac, bc and de, and curve cd. Energy is also dissipated within the radial shear zone bcd.

Since the Hill mechanism admits sliding between the footing base and the adjacent soil, dissipation of energy due to friction on this surface should be taken into account in the computation of the bearing capacity of the footing. The rate of dissipation of energy due to friction may be computed by multiplying the discontinuity in velocity $V_1 \cos(\zeta - \varphi)$ across the base surface by $\tan \delta$ (δ is the friction angle between the base and the adjacent soil) times the normal force $q_0 B$ acting on this surface or $q_0 B V_1 \cos(\zeta - \varphi) \tan \delta$. The total rate of dissipation of energy is then obtained by adding this additional dissipation to all previous dissipations along discontinuity lines ac, bc, cd and de and also within the radial shear zone bcd. Since the soil is moving away from the footing wall bf, no frictional energy is dissipated along the wall (the same is true for the Prandtl mechanism).

The solution of the Hill mechanism (shallow footing, $\eta > \beta$)

Equating the external rate of work done by the force acting on the footing and the weight of soil in motion to the total internal rate of energy dissipation, it is found that the value of the upper bound on the average bearing pressure is:

$$\frac{q_0(\xi, \eta, \zeta)}{c} = \left[\frac{1}{1 - \tan \delta \, \cot(\zeta - \varphi)}\right]\left[\bar{N}_c(\xi, \eta, \zeta) + G\,\bar{N}_\gamma(\xi, \eta, \zeta)\right]$$

$$= N_c(\xi, \eta, \zeta) + G\,N_\gamma(\xi, \eta, \zeta) \qquad [6.27]$$

in which the bearing capacity factors $\bar{N}_c = [1 - \tan \delta \, \cot(\zeta - \varphi)]N_c$ and $\bar{N}_\gamma = [1 - \tan \delta \, \cot(\zeta - \varphi)]N_\gamma$ can be expressed in terms of the three as yet unspecified angles ξ, η, ζ under the condition:

$$\eta > \beta$$

$$\bar{N}_c(\xi, \eta, \zeta) = \frac{\sin \xi \cos \varphi + \sin \zeta \, \text{abs}[\cos(\xi + \zeta - \varphi)]}{\sin(\zeta + \xi)\sin(\zeta - \varphi)} +$$

$$\frac{\alpha \sin \zeta \{\exp[2(\pi + \beta - \eta - \xi)\tan \varphi] - 1\}}{\sin(\zeta + \xi)\sin \varphi \sin(\zeta - \varphi)} + \frac{\alpha \sin \eta \sin \zeta \exp[2(\pi + \beta - \eta - \xi)\tan \varphi]}{\cos(\varphi + \eta)\sin(\zeta + \xi)\sin(\zeta - \varphi)}$$

$$[6.28]$$

$$\bar{N}_\gamma\,(\xi,\eta,\zeta) = \frac{-0.5\,\sin\zeta\,\sin\xi}{\sin\,(\zeta+\xi)} + \frac{0.5\,\alpha\,\sin^2\zeta}{\sin^2\,(\zeta+\xi)\,\cos\varphi\,\sin\,(\zeta-\varphi)\,(1+9\,\tan^2\varphi)}$$

$$\times\,\{[3\,\tan\varphi\,\cos\,(\beta-\eta)+\sin\,(\beta-\eta)]\,\exp\,[3(\pi+\beta-\eta-\xi)\,\tan\varphi] +3\,\tan\varphi\,\cos\xi$$

$$+\sin\xi\}+\frac{\alpha\,\sin\zeta\,\sin\eta\,\cos\,(\beta-\eta)\,\exp\,[2(\pi+\beta-\eta-\xi)\,\tan\varphi]}{\sin\,(\zeta+\xi)\,\cos\varphi\,\sin\,(\zeta-\varphi)\,\sin\beta}\,\left(\frac{D}{B}\right)$$

$$+\frac{2\,\alpha\,\cos\,(\beta-\eta)\,\exp\,[(\pi+\beta-\eta-\xi)\,\tan\varphi]}{\tan\beta\,\cos\varphi\,\sin\,(\zeta-\varphi)}\,\left(\frac{D}{B}\right)^2 \qquad [6.29]$$

where $\alpha = \sin\,(\xi+\zeta-2\varphi)$ if $-\tfrac{1}{2}\pi+\xi+\zeta-\varphi>0$ $\qquad$ [6.30a]

otherwise $\alpha = \sin\,(\xi+\zeta)$ $\qquad$ [6.30b]

and the corresponding governing equation for β is:

$$\sin\beta\,\exp\,(\beta\,\tan\varphi) = \frac{2\,(D/B)\,\sin\,(\zeta+\xi)\,\cos\,(\varphi+\eta)}{\sin\zeta\,\cos\varphi\,\exp\,[(\pi-\eta-\xi)\,\tan\varphi]} \qquad [6.31]$$

The amplification factor $1/[1-\tan\delta\,\cot\,(\zeta-\varphi)]$ appearing in the right-hand side of [6.27] is contributed by the sliding friction. Minimization of the function $q_0(\xi,\eta,\zeta)/c$ will be discussed later.

The solution of the Hill mechanism (surface footing)

For the special case of a surface footing for which $\beta=0$, or $D/B=0$, the function $\bar{N}_c(\xi,\eta,\zeta)+G\bar{N}_\gamma(\xi,\eta,\zeta)$ in the square bracket of [6.27] reduces to:

$$\bar{N}_c(\xi,\eta,\zeta) + G\,\bar{N}_\gamma(\xi,\eta,\zeta) = \left[\frac{\sin\xi\,\cos\varphi+\sin\zeta\,\text{abs}\,[\cos\,(\xi-\zeta-\varphi)]}{\sin\,(\zeta+\xi)\,\sin\,(\zeta-\varphi)}\right.$$

$$\left.+\frac{\alpha\,\sin\zeta\,\{\exp\,[2(\pi-\eta-\xi)\,\tan\varphi]-1\}}{\sin\,(\zeta+\xi)\,\sin\varphi\,\sin\,(\zeta-\varphi)}+\frac{\alpha\,\sin\zeta\,\sin\eta\,\exp\,[2(\pi-\eta-\xi)\,\tan\varphi]}{\cos\,(\varphi+\eta)\,\sin\,(\zeta+\xi)\,\sin\,(\zeta-\varphi)}\right]$$

$$+G\left[\frac{-0.5\,\sin\zeta\,\sin\xi}{\sin\,(\zeta+\xi)}+\frac{0.5\,\alpha\,\sin^2\zeta}{\sin^2\,(\zeta+\xi)\,\cos\varphi\,\sin\,(\zeta-\varphi)\,(1+9\,\tan^2\varphi)}\right.$$

$$\times\,\{(3\,\tan\varphi\,\cos\eta-\sin\eta)\,\exp\,[3(\pi-\eta-\xi)\,\tan\varphi]+3\,\tan\varphi\,\cos\xi+\sin\xi\}$$

$$\left.+\frac{0.5\,\alpha\,\sin^2\zeta\,\sin\eta\,\cos\eta\,\exp\,[3(\pi-\eta-\xi)\,\tan\varphi]}{\sin^2\,(\zeta+\xi)\,\sin\,(\zeta-\varphi)\,\cos\,(\varphi+\eta)}\right] \qquad [6.32]$$

As was the case for the Prandtl mechanism, the optimum value of η for a surface footing is again found from the condition $\partial q_0 / \partial \eta = 0$:

$$\eta = 45° - \tfrac{1}{2}\varphi \qquad\qquad [6.33]$$

and the optimum values for the parameters ξ and ζ are found to satisfy the condition:

$$\xi + \zeta = 90° + \varphi \qquad\qquad [6.34]$$

This implies that line ac is tangent to the logspiral curve at point c (referring to Fig. 6.8). For a smooth footing for which $\delta = 0$, bearing capacities can generally be approximated to within 10% of the minimum if:

$$\xi = 45° + \tfrac{1}{2}\varphi \qquad\qquad [6.35]$$

Remarks on the solutions

Once again it is worth pointing out here that although the Hill mechanism allows explicit inclusion of finite friction between base and soil, bearing capacities computed using this mechanism are *rigorous* upper bounds only for perfectly smooth footings. This has been discussed in details in section 2.6, Chapter 2. A variant of the Hill mechanism in which sliding between soil and base is constrained was also studied. Bearing capacities computed with this alternate mechanism, which are true upper bounds for arbitrary base friction, were found to be identical with the Hill mechanism results with $\delta = \varphi$. Hence the Hill mechanism can be used to obtain upper bounds for both perfectly rough, moderately rough and perfectly smooth footings.

It is noted further that for subsurface footings, both the Prandtl and Hill mechanisms give rigorous upper bounds only if the footing wall is perfectly smooth. For cases in which the upper-bound theorem of limit analysis is not strictly applicable, the upper-bound terminology and computational process are still used. However, the most that can be said is that an approximate solution has been obtained using a rational computational procedure (section 2.6, Chapter 2).

6.4.3. Deep footings

In the preceding discussions it has been tacitly assumed that the entire area $bdef$ in Prandtl and Hill mechanisms (Figs. 6.5 and 6.8) is moving away from the footing wall bf. The Prandtl and Hill mechanisms are therefore kinematically admissible and no frictional energy is dissipated along the wall. This assumption is specified mathematically by the condition that $\eta > \beta$ which states physically that the vertical distance D between the surface of the ground and the base of the footing must be relatively small in comparison with footing width B. If this condition ($\eta > \beta$) is satisfied the bearing capacity solutions are called "*shallow footing*" solutions. In other words all solutions we have obtained in the preceding discussions are for the case of shallow footings. If the depth D is considerably greater than the width $B(\eta < \beta$, *deep footings*) the corresponding upper-bound solutions for the bearing capacity of deep footings can be obtained by substituting $\eta = \beta$ in the preceding equations [6.19] and [6.27]. The frictional energy dissipation between the footing wall and the soil located above the level of the footing base is *not* considered in such substitution. It is found after substituting $\eta = \beta$ in [6.19] and [6.27] that the average bearing capacity of a deep footing on a $c - \varphi - \gamma$ soil is:

Prandtl mechanism (deep footing)

$$
\frac{q_0(\xi)}{c} = \cot\varphi \left[\frac{\cos\beta \cos(\xi - \varphi) \exp\{2(\pi - \xi)\tan\varphi\}}{\cos\xi \cos(\beta + \varphi)} - 1 \right]
$$

$$
+ G\left[-\frac{\tan\xi}{2} + \frac{0.5 \cos(\xi - \varphi)}{(1 + 9\tan^2\varphi)\cos^2\xi \cos\varphi} \{3\tan\varphi \exp[3(\pi - \xi)\tan\varphi]\right.
$$

$$
+ 3\tan\varphi \cos\xi + \sin\xi\} + \frac{\cos(\xi - \varphi)\exp[2(\pi - \xi)\tan\varphi]}{\cos\xi \cos\varphi}\left(\frac{D}{B}\right)
$$

$$
\left. + \frac{2\cos(\xi - \varphi)\exp[(\pi - \xi)\tan\varphi]}{\cos\varphi \tan\beta}\left(\frac{D}{B}\right)^2 \right] \tag{6.36}
$$

where β is related to D/B and ξ by the transcendental equation:

$$
\frac{\sin\beta}{\cos(\beta + \varphi)} = \frac{2(D/B)\cos\xi}{\cos\varphi \exp[(\pi - \xi)\tan\varphi]} \tag{6.37}
$$

Hill mechanism (deep footing)

$$
\frac{q_0(\xi, \zeta)}{c} = \left[\frac{1}{1 - \tan\delta \cot(\zeta - \varphi)} \right] \left\{ \frac{\sin\xi \cos\varphi + \sin\zeta \, \mathrm{abs}[\cos(\xi + \zeta - \varphi)]}{\sin(\xi + \zeta)\sin(\zeta - \varphi)} \right.
$$

$$
+ \frac{\alpha \sin\zeta \{\exp[2(\pi - \xi)\tan\varphi] - 1\}}{\sin(\xi + \zeta)\sin(\zeta - \varphi)\sin\varphi} + \frac{\alpha \sin\beta \sin\zeta \exp[2(\pi - \xi)\tan\varphi]}{\cos(\beta + \varphi)\sin(\xi + \zeta)\sin(\zeta - \varphi)}
$$

$$
+ G\left[\frac{-0.5 \sin\zeta \sin\xi}{\sin(\xi + \zeta)} + \frac{0.5 \alpha \sin^2\zeta \{3\tan\varphi \exp[3(\pi - \xi)\tan\varphi] + 3\tan\varphi \cos\xi + \sin\xi}{(1 + 9\tan^2\varphi)\sin^2(\zeta + \xi)\cos\varphi \sin(\zeta - \varphi)} \right.
$$

$$
\left. \left. + \frac{\alpha \sin\zeta \exp[2(\pi - \xi)\tan\varphi]}{\sin(\xi + \zeta)\cos\varphi \sin(\zeta - \varphi)}\left(\frac{D}{B}\right) + \frac{2\alpha \exp[(\pi - \xi)\tan\varphi]}{\tan\beta \cos\varphi \sin(\zeta - \varphi)}\left(\frac{D}{B}\right)^2 \right] \right\} \tag{6.38}
$$

where α is defined in [6.30] and the corresponding governing equation for β is:

$$\frac{\sin\beta}{\cos(\beta+\varphi)} = \frac{2(D/B)\sin(\xi+\zeta)}{\sin\zeta\,\cos\varphi\,\exp\left[(\pi-\xi)\tan\varphi\right]} \tag{6.39}$$

6.4.4. Numerical solution

Once the governing material parameters (φ,G,δ) and geometric parameters (D/B) have been determined, the velocity field is uniquely defined by assigning, for example, values of the angular parameters ξ and η in the Prandtl mechanism (Fig. 6.5) and ξ, η and ζ in the Hill mechanism (Fig. 6.8). Each set of values assigned to the angular parameters is associated with a bearing capacity value that is an upper bound to the true limit load. Of course, the best upper bound is the *least* upper bound. In order to find values of the mechanism parameters that minimize the upper bound, a modified form of the *method of steepest descent* (Ketter and Prawel, 1969, p. 467) is used. The optimum mechanism parameter set is found through a sequence of incremental steps, starting with an assumed set of values. The length of the incremental parameter vector is arbitrarily assigned and its direction is defined by the negative of the function gradient. After each change in the mechanism parameter vector, the load associated with this new set is compared with the old load associated with the previous parameter set. If the new load is less than the old load, a new parameter set is computed, if not, the process is terminated or a new smaller incremental vector length is assigned.

The minimization procedure for previous equations has been programmed for a CDC 6400 computer. Incremental vector lengths of five degrees and subsequently one degree were used in the program. Optimum values so obtained have been compared to those obtained by tabulating bearing capacity versus the angular parameters and finding the minimum by hand. Excellent agreement was observed in every case checked. Optimum values were in most cases obtained within fifteen cycles. For each new set of given conditions (φ, G, D/B, δ), the minimization procedure took about one-tenth of a second of computer time.

6.4.5. Results and discussion

Charts relating bearing capacity pressure q_0/c to the various governing parameters are presented in Figs. 6.10–6.20. For the entire range of parameters considered, bearing capacity was computed using both Prandtl and Hill mechanisms. In the spirit of the upper-bound approach, the *lesser* of these two solutions is shown in the charts. Results for surface footings will be first discussed followed by a discussion of the results for shallow and deep footings.

Surface footings
Bearing capacities of surface footings are shown in Figs. 6.10 and 6.11. In

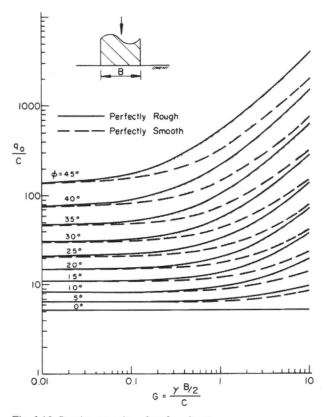

Fig. 6.10. Bearing capacity of surface footings.

Fig. 6.10 the relationship between a nondimensionalized bearing capacity, q_0/c, and the dimensionless soil weight parameter G ($= \gamma B/2c$) is shown for both perfectly smooth and perfectly rough footings, where q_0 is the average bearing pressure at failure. Soil internal friction angles ranging from zero to $45°$ are considered. The charts clearly show that base friction has little effect upon bearing capacity for small values of G. This is to be expected since Prandtl's solution for weightless soil has been shown to be independent of base friction (see example 3.5, section 3.4, p. 78). However, for relatively large values of G, base friction has a significant effect on bearing capacity. For instance, letting $G = 10$ and $\varphi = 30°$, the bearing capacity of a rough footing is approximately twice that of a smooth footing. In general the Prandtl mechanism gives the lesser upper bound for rough footings while the Hill mechanism gives the smaller bound for smooth footings. However, if G is large and $0 < \varphi \leqslant 10°$, the Hill mechanism gives a lower value than the Prandtl mechanism for rough as well as smooth footings.

The effect of base friction on the bearing capacity of footings is shown in

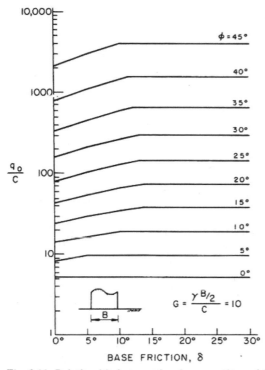

Fig. 6.11. Relationship between bearing capactity and base friction.

Fig. 6.11. The curves are characterized by a rising portion followed by a flat plateau. At the intersection of the rising portion and the plateau, base friction is just
sufficient to restrain any sliding motion between the base and the adjacent soil.
Any increase then in the base friction angle will yield no increase in bearing capacity. The results presented here indicate that a rather modest value of base friction
(less than 15° for all φ) is sufficient to create an essentially perfectly rough condition. For φ greater than or equal to 15°, the rising portion of the curve is associated
with the Hill mechanism while the flat plateau is associated with the Prandtl mechanism. For $\varphi \leqslant 10°$, the entire curve is obtained from the Hill mechanism.

Shallow and deep footings

The results for shallow and deep perfectly rough footings are presented in
Figs. 6.12–6.20. In Figs. 6.12 and 6.13 bearing capacity is plotted against depth to
breadth ratios (D/B) ranging from 0 to 1 and 0 to 10, respectively, while $G = 0$ and
10 and φ ranges from 0° to 45°.

The charts show that the increase in bearing capacity with increasing depth is
much more significant for G equal to 10 than for G equal to 0. In addition it can be
seen that the smaller the angle φ, the greater the effect of increasing depth upon

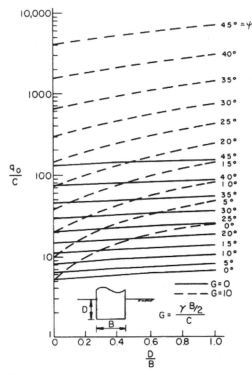

Fig. 6.12. Bearing capacity of shallow footings.

bearing capacity. For example, for $\varphi = 20°$ and $G = 10$, an increase in the depth to breadth ratio from 0 to 0.2 increases the bearing capacity by 40%. For $\varphi = 40°$ and $G = 10$, however, the bearing capacity increases by only 20% (Fig. 6.12).

In Figs. 6.14 and 6.15 bearing capacity of shallow perfectly rough footings is plotted against values of G ranging from 0.1 to 10, for five values of D/B and for φ ranging from 10° to 35°. In Figs. 6.16, 6.17 and 6.18 similar charts are presented for deep footings for depth/breadth ratios up to 5.

It was pointed out earlier that base friction can have a significant effect upon the bearing capacity of surface footings. However, the analysis presented here indicates that the significance of base friction is greatly reduced for deep footings. Fig. 6.19 and 6.20 show the relationship between bearing capacity and depth for both a perfectly smooth and a perfectly rough base for G equal to 1 and 10, respectively. The figures show clearly that as the depth increases the difference in bearing capacities of a rough and smooth footing diminishes. In fact it can be seen that at some depth the bearing capacities of a smooth and rough footing become identical.

The bearing capacity of rough footings is governed by the Prandtl mechanism. For smooth footings the Hill mechanism governs until the two curves intersect. For greater depths the Prandtl mechanism gives a smaller upper bound than does the Hill mechanism.

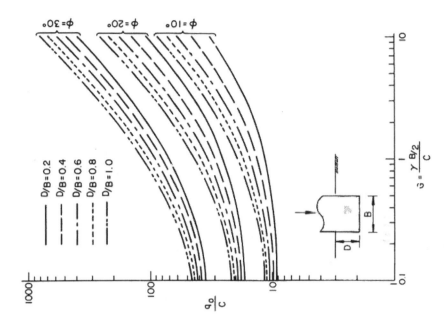

Fig. 6.14. Bearing capacity of shallow rough footings.

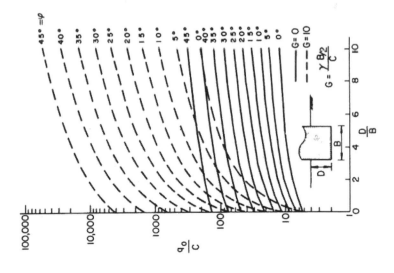

Fig. 6.13. Bearing capacity of deep footings.

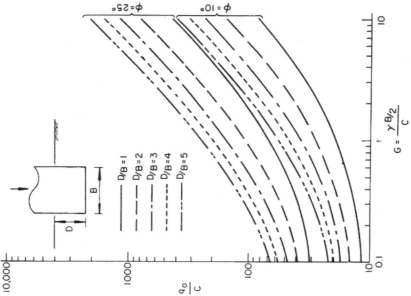

Fig. 6.16. Bearing capacity of deep rough footings.

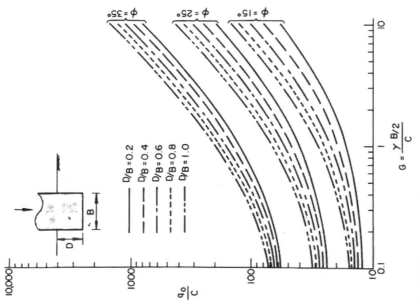

Fig. 6.15. Bearing capacity of shallow rough footings.

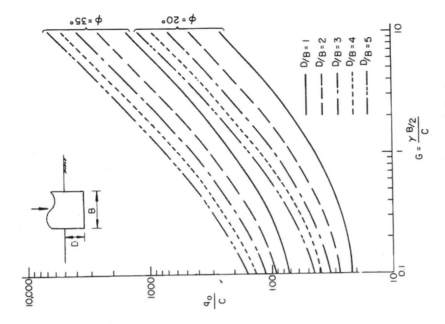

Fig. 6.18. Bearing capacity of deep rough footings.

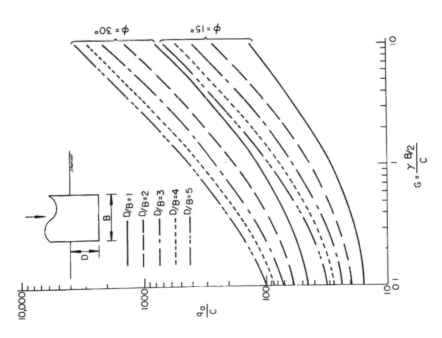

Fig. 6.17. Bearing capacity of deep rough footings.

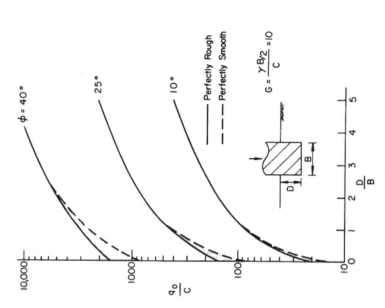

Fig. 6.20. Effect of base roughness on bearing capacity of deep footings.

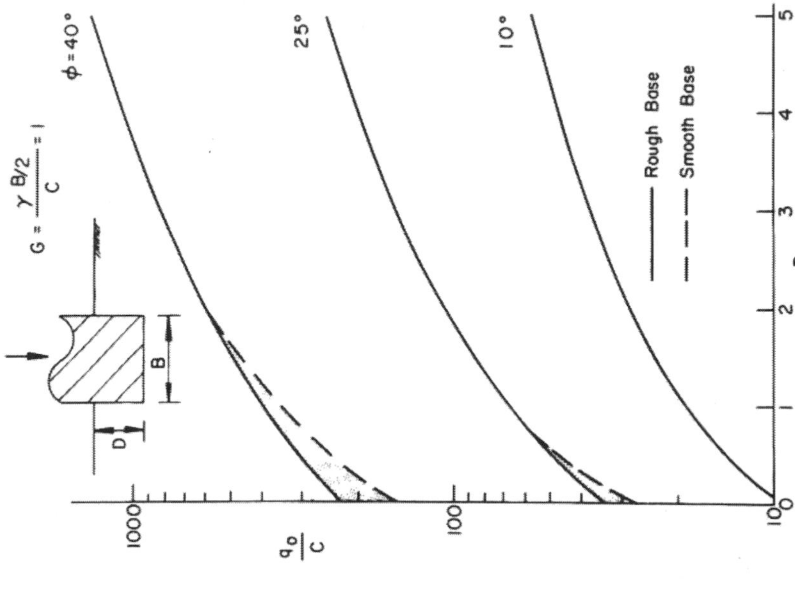

Fig. 6.19. Effect of base roughness on bearing capacity of deep footings.

6.4.6. Comparison of results with existing solutions

Since the upper-bound technique of limit analysis gives only an approximate solution to the bearing capacity problem, some measure of the accuracy of the solutions must be determined. Hence, the present solutions will first be compared to slip-line solutions, followed by a comparison with limit equilibrium solutions.

Comparison with slip-line solutions

Cox (1962) has published slip-line solutions for the bearing capacity of a *smooth surface footing*. In that work, values of G ranging from 0 to 10 and φ ranging from $0°$ to $40°$ are considered. In addition Spencer (1962) has published approximate solutions for the same problem using a *perturbation* technique, where the first term of the perturbation expansion corresponds to the solution for a weightless soil.

Cox's solutions, Spencer's solutions, and the limit analysis solutions developed here are tabulated in Table 6.1. It is noted that limit analysis gives the exact solution when the value of G is equal to zero. From examination of the tabular results it can be seen that the error associated with the limit analysis solutions increases as G and φ increase. Errors are referred to the slip-line solutions of Cox. A further discussion of limit analysis solutions for the special cases $G = \infty$ (cohesionless soil) and $G = 0$ (weightless soil) will be given in sections 6.5 and 6.6, respectively.

For G equal to 10 and φ equal to $40°$, the upper-bound limit analysis method overestimates the slip-line solution by 37%. For the value of G less than 5 and φ less than $40°$, the error can be expected to be less than 25%. As was mentioned earlier (section 6.3), however, values of G can normally be expected to range from 0.1 to 1, and within this range the maximum error in the limit analysis solutions is less than 9%.

It is of interest to compare the results of Spencer with the limit analysis solutions. The two methods give nearly identical results with the limit analysis solutions lying slightly closer to the slip-line solutions. One might interpret Spencer's technique as an upper-bound approach in which the kinematically admissible velocity field is that of a weightless soil. The velocity field used here resembles closely that of a weightless soil, however, the size of various zones (rigid body zones and radial shear zone) are varied in order to minimize the load, thus accounting for the slight improvement over Spencer's solution. Chen and Davidson (1973) found that they could duplicate all Spencer's results by using the velocity field for a weightless soil.

To date (1973) there have been no slip-line solutions published for a rough surface footing on a rough or smooth subsurface footing bearing on a *cohesive* soil with weight ($c - \varphi - \gamma$ soil). However, the success of the limit analysis approach in predicting the bearing capacity of smooth surface footings leads us to believe that it will be equally successful in predicting the bearing capacity of subsurface footings as well as rough surface footings.

TABLE 6.1.

Bearing capacities (q_0/c) of a smooth surface footing (data from Cox, 1962; Spencer, 1962)

ϕ (°)	G = 0.0			G = 0.1			G = 1.0			G = 10.0		
	Cox	Spencer	limit analysis	Cox	Spencer	limit analysis	Cox	Spencer	limit analysis	Cox	Spencer	limit analysis
0	5.14	5.14	5.14	5.14	5.14	5.14	5.14	5.14	5.14	5.14	5.14	5.14
10	8.34	8.35	8.35	8.42	8.42	8.42	9.02	9.07	9.05	13.6	—	14.4
20	14.8	14.8	14.8	15.2	15.2	15.2	17.9	18.3	18.1	37.8	—	43.4
30	30.1	30.1	30.1	31.6	31.7	31.7	42.9	45.3	44.3	127.0	—	159.0
40	75.3	75.3	75.3	83.0	83.5	83.4	139.0	157.0	151.0	574.0	—	786.0

TABLE 6.2

Limit analysis and limit equilibrium solutions for rough footings

φ (°)	Geometry and material constants						Bearing capacity, in pounds per square foot (kilonewtons per square meter)			
	c, in pounds per square foot (kilonewtons per square meter)	B, in feet (meters)	D, in feet (meters)	γ, in pounds per cubic foot (kilograms per cubic meter)	G	D/B	Limit analysis	J. Brinch Hansen (1961)	Terzaghi (1943)	Meyerhof (1951, 1955)
10	1,000 (47.9)	3 (0.914)	0	100 (1062)	0.15	0	8,560 (410.)	8,370 (401.)	8,000 (383.)	8,000 (383.)
30	500 (23.9)	10 (3.05)	0	100 (1062)	1.0	0	29,100 (1,393.)	24,100 (1,154.)	23,000 (1,101.)	23,000 (1,101.)
10	1,000 (47.9)	3 (0.914)	1.5 (0.457)	100 (1062)	0.15	0.5	10,400 (498.)	10,200 (488.)	8,380 (401.)	—
30	500 (23.9)	10 (3.05)	5 (1.52)	100 (1062)	1.0	0.5	41,800 (2,000.)	37,400 (1,791.)	32,000 (1,532.)	41,000 (1,963.)
10	1,000 (47.9)	3 (0.914)	3 (0.914)	100 (1062)	0.15	1.0	12,100 (579.)	11,100 (531.)	8,750 (419.)	—
30	500 (23.9)	10 (3.05)	10 (3.05)	100 (1062)	1.0	1.0	56,200 (2,691.)	51,100 (2,447.)	41,000 (1,963.)	54,500 (2,610.)
30	500 (23.9)	10 (3.05)	20 (6.10)	100 (1062)	1.0	2.0	88,800 (4,252.)	81,700 (3,912.)	59,000 (2,825.)	82,300 (3,940.)
30	500 (23.9)	10 (3.05)	50 (15.2)	100 (1062)	1.0	5.0	210,000 (10,055.)	183,000 (8,762.)	113,000 (5,410.)	207,000 (9,911.)

Comparison with limit equilibrium solutions

Some limit analysis and limit equilibrium solutions for surface and subsurface footings are tabulated in Table 6.2. The limit analysis solutions are for a perfectly rough footing base.

The solutions described by Terzaghi (1943) neglect the strength of soil above the footing base. Meyerhof's solutions for surface footings are obtained from Meyerhof (1955) while those for a subsurface footing are obtained from Meyerhof (1951). The conversion from D/B to Meyerhof's angle "β" (same as the angle β used here in the Prandtl and Hill geometries) was determined from a chart on p. 422 in Scott's book "Principles of Soil Mechanics" (1963). Although Meyerhof's method is equally applicable for a friction angle of $10°$ as for other friction angles, the chart mentioned above does not include friction angles less than $20°$. It is for this reason that Meyerhof's solutions for a friction angle of $10°$ are not included in the table. The solutions ascribed to Brinch Hansen were obtained from Brinch Hansen (1961) and incorporate the so-called depth factors.

Considering first the two surface footing cases, it can be seen that the limit analysis solutions exceed all the limit equilibrium solutions. For the values of φ and G under consideration, the limit analysis solutions probably lie quite close to the true solutions. The difference between the limit analysis and limit equilibrium solutions can probably be attributed to the use of Terzaghi's method of superposition in all three limit equilibrium solutions.

The solutions of Meyerhof and the limit analysis solutions agree remarkably well for subsurface footings. The two methods use somewhat similar failure mechanisms and both include the strength of soil above the footing base. It can also be seen that the solutions of Brinch Hansen agree fairly well with the limit analysis results, with the differences tending to increase with increasing depth. The Terzaghi's solutions presented here differ considerably from the other solutions. This is not surprising since the Terzaghi results do not include soil strength above the footing base.

6.4.7. Summary and conclusions

The upper-bound technique of limit analysis is used here to develop approximate solutions for the bearing capacity of cohesive soils with weight. Solutions are presented for smooth and rough and surface and subsurface footings (shallow and deep). Soil is treated as a perfectly plastic medium with the associated flow rule. The limit analysis solutions for smooth surface footings are shown to compare favorably with slip-line solutions. Meyerhof's solutions and the limit analysis solutions for rough subsurface footings are shown to agree remarkably well.

6.5. BEARING CAPACITY OF A STRIP FOOTING ON COHESIONLESS SOILS (N_γ FACTOR)

6.5.1. Surface footings

Prandtl and Hill mechanisms

In section 3.4, Chapter 3, we have shown that for a surface footing on a weightless $c - \varphi$ soil, both Prandtl and Hill mechanisms as shown in Figs. 3.21 and 3.22 respectively give the same *exact* solution, ([3.81]). However when the weight of a soil is included, available slip-line solutions indicate that the wedge-shaped zone located beneath the footing for both Prandtl and Hill mechanisms must now be curved. Furthermore, the shape of the connecting curve, *CD*, in the radial shear zone is no longer a spiral of angle φ (Figs. 3.21 and 3.22). Rather, the shape of the curve lies somewhere between a spiral of angle φ and a circular arc, as long as $\varphi \neq 0$. For a frictionless soil, $\varphi = 0$, the curve is always an arc. Fig. 6.21 shows the slip-line fields for a smooth footing as proposed by Lundgren and Mortensen (1953) (Fig. 6.21a) and by Prandtl (1921) (Fig. 6.21b). Since the mechanisms used in the previous discussions consist of only straight lines and logspiral curves, the bearing capacities obtained from such mechanisms give only approximate solutions to the bearing capacity problem of a ponderable soil. It can be shown that the mechanism as shown in Fig. 6.21(a) is the correct mechanism. This then leads to the conclusion that the bearing capacity equations obtained in the preceding section can at best give upper bounds to the correct values. Neither Hill nor Prandtl mechanisms shown in Figs. 6.5 and 6.8 can yield an exact solution to a ponderable soil.

The bearing capacity factor N_γ for a surface footing has been defined in the second bracket in [6.23] for Prandtl mechanism and [6.32] for Hill mechanism. The optimum value of the angle η for both mechanisms is found to be $\frac{1}{4}\pi - \frac{1}{2}\varphi$. For the case of Prandtl mechanism, the optimum value of ξ can be determined from the condition:

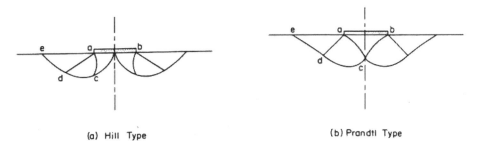

(a) Hill Type   (b) Prandtl Type

Fig. 6.21. Slip-line fields for smooth footings.

$$\left[\frac{\sin\varphi}{(3\tan\varphi\cos\xi - \sin\xi)} - \cos(\xi - \varphi)\right]\exp(-3\,\xi\tan\varphi)$$

$$= \frac{\sin\left(\frac{1}{4}\pi - \frac{1}{2}\varphi\right)}{(1 + 2\sin\varphi)\exp\left[3(\frac{3}{4}\pi + \frac{1}{2}\varphi)\tan\varphi\right]} \qquad\qquad [6.40]$$

For the case of Hill mechanism, the optimum values for the parameters ξ and ζ are found to satisfy the condition $\xi + \zeta = \frac{1}{2}\pi + \varphi$. Using the above-mentioned results for the optimum values of ξ, η, and ζ, the computations for the bearing capacity factor N_γ are tabulated numerically in Table 6.3 for φ ranging from $5°$ to $45°$. In Table 6.3 the optimum values of ξ and ζ are listed against the factor N_γ for both Prandtl and Hill mechanisms. The values ξ and ζ determine the position of the critical mechanism. For $\varphi = 30°$ the value of $N_\gamma = 26.7$ obtained from Prandtl mechanism may be compared with the limit equilibrium solutions $N_\gamma = 26$ reported by Meyerhof (1951) or $N_\gamma = 27$ by Abdul-Baki and Beik (1970). For $\varphi = 30°$ the value of $N_\gamma = 12.7$ obtained from Hill mechanism may be compared with the slip-line solutions $N_\gamma = 15.3$ reported by Sokolovskii (1965) or $N_\gamma = 14.8$ by Lundgren and Mortensen (1953). Of particular interest in Table 6.3 is the significant difference between the bearing capacity of a smooth footing determined from the Hill mechanism and the bearing capacity of a rough footing determined from the Prandtl mechanism. In all cases the Hill mechanism gives a bearing capacity slightly less than *half* of that given by the Prandtl mechanism.

Also of interest is the rapid increase in bearing capacity in Hill mechanism with increasing base friction angle δ. This is graphically demonstrated in Fig. 6.22 where the values of N_γ have been plotted against the footing base friction angle δ. The

TABLE 6.3

Values of N_γ for a surface footing on cohesionless soils

φ Friction angle (°)	Prandtl		Hill (smooth base)		
	ξ (°)	N_γ	ζ (°)	ξ (°)	N_γ
5	11	0.382	14	81	0.131
10	20	1.160	22	77	0.461
15	28	2.73	29	78	1.16
20	34	5.87	39	71	2.68
25	40	12.4	46	70	5.9
30	45	26.7	48	72	12.7
35	50	60.2	50	77	28.6
40	54	147.0	58	73	71.6
45	58	401.0	60	75	195.0

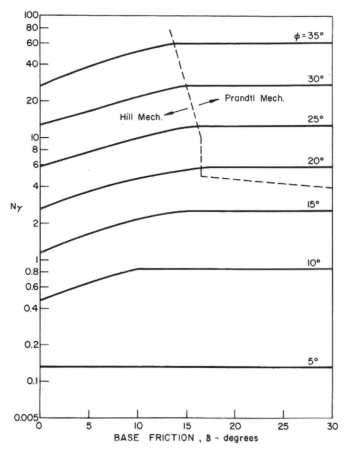

Fig. 6.22. Relationships between bearing capacity factor, N_γ, soil friction angle, φ, and base friction angle, δ, for a surface footing on cohesionless soils.

values of N_γ shown in the graph are absolute minimums obtained using either the Hill or Prandtl mechanisms. The dashed line delineates Hill mechanism controlled and Prandtl mechanism controlled regions. Some results for the Hill mechanism are presented in Table 6.4, where the values of N_γ are tabulated for two values of φ against three values of δ. The optimum values of ξ and ζ are also tabulated. For all values of φ and δ the relationship $\xi + \zeta = \frac{1}{2}\pi + \varphi$ is seen to hold under the optimum condition. This condition implies that the logspiral curve cd is tangent to line ac at point c (see Fig. 6.8). We may conclude from the above discussions that a base friction angle of approximately $17°$ is sufficient to restrain sliding between the base and the adjacent soil and hence give rise essentially to a perfectly rough base condition. It is interesting to note from Fig. 6.22 that for $\varphi \leqslant 15°$, the Hill mechanism controls for both a smooth ($\delta = 0$) and a rough ($\delta = \varphi$) footing.

TABLE 6.4

Relationship between N_γ, φ and δ for a surface footing, Hill mechanism

δ (°)	$\varphi = 20°$			$\varphi = 30°$		
	ζ (°)	ξ (°)	N_γ	ζ (°)	ξ (°)	N_γ
0	39	71	2.68	48	72	12.7
10	49	62	4.54	58	63	21.9
20	67	43	6.52	72	48	33.1

The numerical values for the bearing capacity factor N_γ of Prandtl mechanism can be approximated with an error on the safe side for $\varphi > 30°$ and on the unsafe side for $\varphi < 30°$ (not exceeding 8% for $15° < \varphi < 45°$ and not exceeding 6% for $20° < \varphi < 40°$ by the equation:

$$N_\gamma = 2[1 + e^{\pi \tan \varphi} \tan^2 (\tfrac{1}{4}\pi + \tfrac{1}{2}\varphi)] \tan \varphi \tan (\tfrac{1}{4}\pi + \tfrac{1}{5}\varphi) \qquad [6.41]$$

It should be mentioned herein that there exists in literature a great variety of proposed solutions to this problem. The differences in N_γ-values are substantial, ranging from about one-third to double the values given by [6.41]. Furthermore, as pointed out earlier, there exist the difficulties in selecting a proper value of φ for the bearing capacity computations because the plane strain value of φ may be up to 10% higher than the corresponding conventional triaxial test value. The N_γ-values according to [6.41] may therefore be considered to be sufficiently accurate for practical applications. The numerical values of [6.41] are given in Table 6.9.

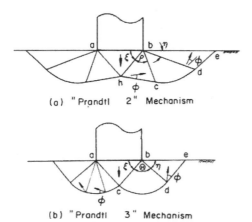

(a) "Prandtl 2" Mechanism

(b) "Prandtl 3" Mechanism

Fig. 6.23. Modified Prandtl mechanisms.

An improved Prandtl mechanism (Fig. 6.23a)

A better upper bound for the Prandtl mechanism is obtained as follows: the mechanism shown in Fig. 6.23(a) will be referred to here as "Prandtl 2". It differs from the conventional Prandtl mechanism (now referred to as "Prandtl 1") only insofar as an additional rigid body zone has been introduced (zone *bhc*). Since internal rate of dissipation of energy is zero for a cohesionless soil, it is only necessary to calculate the rates of external work done by the soil weight in triangle zones *abh*, *bhc* and *bde* and in logspiral shear zone *bcd*. When these calculations have been carried out and added to the rate at which work is done by the footing load, it is found, after some simplification, that the upper-bound value on the bearing capacity factor N_γ is:

Prandtl 2, surface footing (Fig. 6.23a)

$$N_\gamma(\xi, \rho, \eta) = -\tfrac{1}{2} \tan \xi - \frac{\sin \rho \cos(\rho - \varphi) \cos(\xi + \rho) \cos(\xi - \varphi)}{2 \cos \varphi \cos^2 \xi \cos(\rho + \varphi)}$$

$$+ \frac{\cos^2(\rho - \varphi) \cos(\xi - \varphi)}{2 \cos^2 \varphi \cos^2 \xi \cos(\rho + \varphi)(1 + 9 \tan^2 \varphi)} \{ 3 \tan \varphi \cos(\xi + \rho) + \sin(\xi + \rho)$$

$$+ [3 \tan \varphi \cos \eta - \sin \eta] \exp[3(\pi - \xi - \rho - \eta) \tan \varphi] \}$$

$$+ \frac{\cos^2(\rho - \varphi) \cos(\xi - \varphi) \sin \eta \cos \eta \exp[3(\pi - \xi - \rho - \eta) \tan \varphi]}{2 \cos \varphi \cos^2 \xi \cos(\rho + \varphi) \cos(\eta + \varphi)} \quad\quad [6.42]$$

"Prandtl 1" mechanism may be interpreted as the special case of [6.42] when $\rho = 0$. The least upper bounds obtained from [6.42] are tabulated as Prandtl 2 in Table 6.5 along with the results from "Prandtl 1" mechanism. The results for "Prandtl 1" were obtained from the mechanism shown in Fig. 6.5 with $\beta = 0$. For $\varphi = 30°$ the present value of $N_\gamma = 25$ may be compared with the previous value $N_\gamma = 26.7$ obtained from "Prandtl 1" mechanism (Fig. 6.5, $\beta = 0$). The new mechanism is seen to give a better upper bound to the problem for all values of $\varphi \leqslant 40°$.

An alternative Prandtl mechanism (Fig. 6.23b)

The mechanism which will be discussed now gives an even better upper bound for the rough footing than those obtained in the foregoing. The mechanism shown in Fig. 6.23(b) will be referred to here as "Prandtl 3". It resembles closely the conventional Prandtl mechanism, however, the shear zone, *bcd*, is now bounded by a circular arc. The velocity of any radial line of the shear zone is uniform. The velocity vector is no longer perpendicular to the radial line but rather makes an angle of φ with a normal to the radial line. The magnitude of the velocity along the

TABLE 6.5

Bearing capacity factor N_γ for surface footings

φ (°)	Rough footing, $N_\gamma = 2q_0/\gamma B$						Smooth footing, N_γ		
	Limit equilibrium Meyerhof (1951)	Slip-line Hansen and Christensen (1969)	Limit analysis				Slip-line		Limit analysis, Hill mechanism
			Prandtl 1 Fig. 6.5 $\beta = 0$	Prandtl 2 Fig. 6.23a	Prandtl 3 Fig. 6.23b		Sokolov-skii (1965)	Hansen and Christensen (1969)	
15	2.5	1.2	2.7	2.3	2.1		1.4	0.7	1.2
20	5.5	2.9	5.9	5.2	4.6		3.16	1.6	2.7
25	12.0	7.0	12.4	11.4	10.9		6.9	3.5	5.9
30	26.0	15.0	26.7	25.0	31.5		15.3	7.5	12.7
35	60.0	35.0	60.2	57.0	138.0		35.2	18.0	28.6
40	130.0	85.0	147.0	141.0	1803.0		86.5	42.0	71.6

radial line has been shown in section 3.4, Chapter 3 to increase exponentially from the initial value V_0, along the line bc to the value $V_0 \exp(\Theta \tan 2\varphi)$ along the line bd (see Fig. 6.23b). The initial and displaced patterns of such zones have been sketched in Fig. 3.23(c).

The external rate of work done by the soil weight in the circular zone bcd can be computed by dividing the region having central angle Θ into many small rigid triangles similar to that shown in Fig. 3.23(a) and by summing over the region Θ the products of each triangle's component of vertical velocity with its weight. Referring to Fig. 6.23(b) and denoting the length bc by r, this can be expressed in integral form as:

$$\tfrac{1}{2}\gamma V_0 r^2 \int_0^\Theta e^{\theta \tan 2\varphi} \cos(\xi + \theta + \varphi)\, d\theta = \frac{-\gamma V_0 r^2}{2(1 + \tan^2 2\varphi)} \; [\{\tan 2\varphi \cos(\eta - \varphi)$$

$$- \sin(\eta - \varphi)\} \exp[(\pi - \xi - \eta)\tan 2\varphi] + \tan 2\varphi \cos(\xi + \varphi) + \sin(\xi + \varphi)] \quad [6.43]$$

The rate of external work done due to self-weight in regions abc and bde is simply the vertical component of velocity in that region multiplied by the weight of the region. The total rate of work done by the soil weight is obtained by adding the rate of external work due to each region. Since a cohesionless soil is being considered here, no internal dissipation of energy occurs. Adding the total rate of external work due to soil weight to the rate at which work is done by the footing load and equating to zero, the upper-bound value on the bearing capacity factor N_γ of a surface footing is obtained.

"Prandtl 3", surface footing (Fig. 6.23b)

$$N_\gamma = -\tfrac{1}{2} \tan \xi + \frac{\cos (\xi - \varphi)}{2 \cos^2 \xi \cos 2\varphi (1 + \tan^2 2\varphi)} \{\tan 2\varphi \cos (\xi + \varphi) + \sin (\xi + \varphi)$$

$$+ [\tan 2\varphi \cos (\eta - \varphi) - \sin (\eta - \varphi)] \exp [(\pi - \xi - \eta) \tan 2\varphi] \}$$

$$+ \frac{\cos (\xi - \varphi) \tan \eta \cos (\eta - \varphi) \exp [(\pi - \xi - \eta) \tan 2\varphi]}{2 \cos^2 \xi \cos 2\varphi} \qquad [6.44]$$

The least upper bounds obtained from [6.44] are presented in "Prandtl 3" column of Table 6.5. This mechanism is seen to give better upper bounds for the footing than those obtained by "Prandtl 1" and "Prandtl 2" mechanisms for all values of $\varphi \leqslant 25°$.

Comparison of results with existing solutions (surface footing)

Slip-line solutions for perfectly rough and perfectly smooth surface footings bearing on cohesionless soils ($G = \infty$) have been presented by Lundgren and Mortensen (1953), Sokolovskii (1965), Hansen and Christensen (1969) and Graham and Stuart (1971) among others. The results of Sokolovskii, and Hansen and Christensen, as well as the limit analysis solutions obtained from Hill and various Prandtl mechanisms, are summarized in Table 6.5. The results of Meyerhof (1951) obtained by the limit equilibrium method are listed in the second column of this table.

As can be seen in Table 6.5, the limit analysis solutions compare remarkably well with Meyerhof's solutions for a rough footing and fairly well with Sokolovskii's solutions for a smooth footing but exceed the slip-line solutions of Hansen and Christensen by 50—100%. It can also be observed that the limit analysis solutions obtained from conventional Prandtl mechanism can be improved somewhat by adding additional rigid bodies and using a modified shear zone (Fig. 6.23).

Since an upper-bound solution obtained for a perfectly plastic Coulomb material with an associated flow rule is *also* an upper bound for a Coulomb material with a non-associated flow rule (Theorem X, section 2.6, p. 44), it follows that the slip-line solutions of Sokolovskii (see Table 6.5) for a smooth footing can not represent the true solution as the bearing capacity values are greater than the upper bounds obtained by limit analysis Hill mechanism (last column of Table 6.5). Likewise, the Gorbunov-Possadov solution (1965) $N_\gamma = 192$ for a rough footing bearing on a cohesionless soil with $\varphi = 40°$ cannot be correct as it is significantly greater than the value obtained from "Prandtl 1" mechanism $N_\gamma = 147$ or "Prandtl 2" mechanism $N_\gamma = 141$.

It is worth pointing out that the slip-line solutions of Hansen and Christensen are not *complete* and hence are not necessarily the true solutions. The
slip-line solutions may not, in fact, be upper bounds since they have never been
integrated to· yield a kinematically admissible velocity field. Nevertheless, these
slip-line solutions probably represent the best solutions generated to date.

6.5.2. Shallow and deep footings

Modified Prandtl and Hill mechanisms

The Prandtl and Hill mechanisms as shown in Figs. 6.5 and 6.8 will now be
modified slightly to provide a better upper bound for the bearing capacity of a deep
footing when the values of φ are small. This simple modification also gives a better
upper bound for the case of a shallow footing. Since the modification is identical
for both Prandtl and Hill mechanisms, only the Prandtl mechanism will be described in the following. The modified Prandtl mechanism is shown in the inset of
Fig. 6.24 and labeled as "Prandtl 4". It differs from the Prandtl mechanism shown
in Fig. 6.5 (also sketched in Fig. 6.24 and labeled "Prandtl 1") only insofar as an
additional discontinuous line of sliding *be* has been introduced in zone *bdef*. The
soil wedge *bef* now moves as a rigid body in the upward direction parallel to the
side *bf* of the footing. Since a cohesionless soil is being considered here, the only
internal rate of dissipation of energy occurs along the vertical surface between soil
and footing. This rate of dissipation of energy can be found by multiplying the area
of each slip surface by tan δ times the discontinuity in velocity across the length
and the normal force acting on this surface and summing over the two surfaces.
However, such an analysis presents some difficulties in obtaining solutions, since
the lateral normal force between the soil and the vertical side of the footing is
unknown. If we assume that the frictional resistance of the soil on the side of the
footing is fully mobilized, we may calculate the lateral normal force on the side of
the footing from the balance of forces on the soil wedge *bef*. This is possible
because the normal and tangential forces acting on the discontinuous line *be* must
now satisfy Coulomb yield conditions for a cohesionless soil. In the following
computations, however, this frictional dissipation of the soil on the side of the
footing is neglected.

It is a simple matter to calculate the area *bde* and area *bef* (inset, Fig. 6.24). The
rate of external work done due to this area is then found by multiplying each area
by γ times the vertical component of the velocity in that area. The total rate of
external work done due to soil weight is then obtained by adding these rates of
work to the previous calculated rates of external work for the wedge *abc* ([6.11])
and logspiral zone *bcd* ([6.14]). Adding the total external rate of work to the rate
at which work is done by the footing ([6.18]) and equating them to zero, we
obtain an upper bound for the bearing capacity factor N_{γ}.

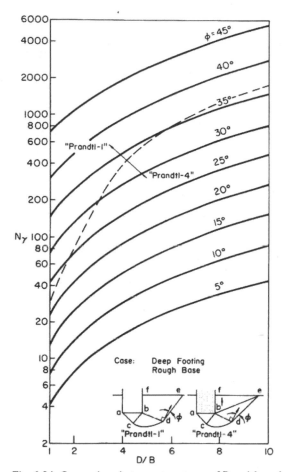

Fig. 6.24. Comparison between two types of Prandtl mechanisms.

"Prandtl 4", shallow and deep footing (Fig. 6.24)

$$N_\gamma(\xi, \eta) = \frac{\cos(\xi - \varphi)}{2 \cos\varphi \cos^2\xi} [h_1(\xi) + h_2(\xi, \eta) + h_4(\xi, \eta)] \qquad [6.45]$$

in which functions $h_1(\xi)$ and $h_2(\xi, \eta)$ have been defined in [6.11] and [6.15], respectively, and function $h_4(\xi, \eta)$ resulting from region *bdef* is expressed as:

$$h_4(\xi, \eta) =$$

$$\left[\frac{\cos\varphi \sin\eta \cos(\beta - \eta)}{\cos(\eta + \varphi)} + \frac{\cos^2\varphi \cos\beta \sin\beta}{\cos(\eta + \varphi) \cos(\beta + \varphi)} \right] \exp[3(\pi + \beta - \xi - \eta) \tan\varphi]$$

$$[6.46]$$

The Hill mechanism as shown in Fig. 6.8 can now be modified in a similar manner as the Prandtl case. Introducing the additional line of velocity discontinuity *be* and assuming the wedge *bef* moving as a rigid body in the upward direction parallel to the side *bf* of the footing, we obtain a better upper bound for the bearing capacity factor N_γ than those obtained from Fig. 6.8 when the values of φ are small. It is found that the equation for computing the bearing capacity factor $N_\gamma = \bar{N}_\gamma / [1 - \tan \delta \cot(\zeta - \varphi)]$ remains identical in its form as in the previous solution ([6.29]) except that the last term given in [6.29] must be substituted by:

$$\frac{2\,\alpha\,\cos(\eta + \varphi)\,\exp\,[(\pi + \beta - \xi - \eta)\,\tan\varphi]}{\tan\beta\,\cos\varphi\,\sin(\zeta - \varphi)\,\cos(\beta + \varphi)}\left(\frac{D}{B}\right)^2 \qquad\qquad [6.47]$$

Numerical results

For a comparison with the previous solutions obtained from conventional Prandtl (Fig. 6.5) and Hill mechanisms (Fig. 6.8), the bearing capacity factors N_γ computed from the modified mechanisms for both Prandtl and Hill types are evaluated and compared with previous numerical results. It is found that for a shallow footing with large values of φ, the conventional Prandtl and Hill mechanisms generally control the failure. For a deep footing with small values of φ, the modified Prandtl and Hill cases control the failure. For the latter case, the differences for solutions obtained using conventional type and modified type are usually very small. For the former case, however, the differences are significant but never exceed 10%.

The results of these computations are graphically represented in Fig. 6.24 for the case of Prandtl types of failure. In Fig. 6.24 the values of D/B have been plotted against the bearing capacity factor N_γ for values of φ ranging from 5° to 45°. The values of N_γ shown in the graph are the absolute minimums obtained from the two mechanisms shown in the inset of Fig. 6.24. The dashed line delineates "Prandtl 4" controlled regions. As mentioned above, in the "Prandtl 4" controlled region, the difference in N_γ-values for both mechanisms is small. In the "Prandtl 1" controlled region, the difference in N_γ is significant but never exceeds 10%.

In line with the terminology used first by Terzaghi and later by many others, the term "shallow footing" will be referred to herein as a footing whose width B is equal to or greater than the vertical distance D between the surface of the ground and the base of the footing. The results for the bearing capacity factor N_γ of a shallow footing are tabulated in Table 6.6 for three different values of base friction angle δ. All values given in the table are the absolute minimums obtained from the four mechanisms considered (two Hill types and two Prandtl types). Failure of a smooth footing is found always associated with the Hill type. As the values δ and D/B increase, the difference between the bearing capacity obtained from Hill and Prandtl types of failure decreases. When Prandtl type of failure controls, the N_γ-

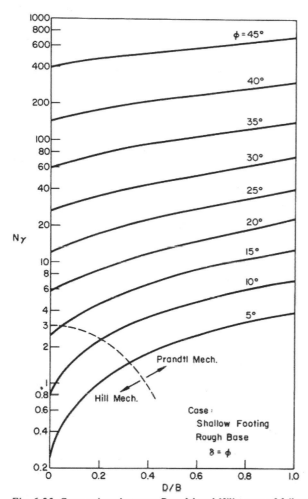

Fig. 6.25. Comparison between Prandtl and Hill types of failure for rough footing.

values in Table 6.6 are shown in parentheses. The results of this computation for the case of a rough footing are graphically represented in Fig. 6.25. In this figure the values of N_γ have been plotted against the depth ratio D/B. Prandtl type of failure controls throughout the most part of this figure. For small ratios of D/B ($\leqslant 0.3$) and small friction angles φ ($\leqslant 16°$) Hill type mechanism is seen to control the failure, as shown in the chart by the long dashed curve.

Results for the bearing capacity factor N_γ of a deep footing are shown in Fig. 6.26. In Fig. 6.26 the values of N_γ for both a rough and a smooth footing are shown. The graphs indicate clearly that for sufficiently large depth D ($>1.5B$ for $\varphi < 40°$ and $>2.5B$ for $\varphi < 45°$) the bearing capacity of a smooth footing and a rough footing are practically the same. Table 6.7 presents bearing capacity values

TABLE 6.6

Bearing capacity factor N_γ for a shallow footing on cohensionless soils

Footing base friction angle δ (°)	Depth ratio D/B	Soil internal friction angle, φ (°)								
		5	10	15	20	25	30	35	40	45
$\delta = 0$ smooth	0.1	0.508	1.10	2.21	4.26	8.41	17.1	36.5	85.9	224
	0.2	0.871	1.71	3.23	5.90	11.1	21.6	44.4	101	255
	0.3	1.23	2.33	4.21	7.64	13.9	26.2	52.8	117	288
	0.4	1.56	3.00	5.34	9.49	16.9	31.3	61.6	132	320
	0.5	1.96	3.61	6.42	11.4	20.0	36.5	70.7	149	354
	0.6	2.34	4.26	7.54	13.3	23.3	41.9	80.2	167	389
	0.7	2.71	4.92	8.68	15.2	26.7	47.6	90.0	185	426
	0.8	3.10	5.59	9.84	17.2	30.2	53.5	100	203	462
	0.9	3.48	6.28	11.0	19.2	33.8	60.0	111	222	500
	1.0	3.87	6.98	12.2	21.3	37.4	65.7	121	242	539
$\delta = 5$	0.1	0.624	1.36	2.70	5.31	10.6	21.8	47.9	115	312
	0.2	1.01	2.03	3.83	7.13	13.6	27.0	57.2	133	350
	0.3	1.40	2.69	5.00	9.05	16.7	32.4	66.9	151	389
	0.4	(1.77)	3.37	6.15	11.0	20.0	38.0	76.9	171	429
	0.5	(2.12)	4.05	7.32	13.1	23.5	43.8	87.3	190	470
	0.6	(2.48)	(4.70)	8.52	15.3	27.1	50.0	98.1	211	511
	0.7	(2.84)	(5.32)	(9.65)	15.3	30.8	56.2	109	231	554
	0.8	(3.20)	(5.95)	(10.7)	(17.2)	30.8	62.7	121	253	598
	0.9	(3.57)	(6.58)	(11.8)	(19.0)	(33.8)	(67.9)	132	275	643
	1.0	(3.93)	(7.22)	(12.9)	(20.9)	(36.9)	(73.1)	(142)	297	689
$\delta = \varphi$ rough	0.1	0.624	1.59	(3.60)	(7.28)	(14.7)	(30.7)	(67.4)	(161)	(429)
	0.2	1.01	2.30	(4.52)	(8.76)	(17.2)	(34.9)	(74.9)	(175)	(459)
	0.3	1.40	(2.89)	(5.50)	(10.3)	(19.7)	(39.2)	(82.5)	(189)	(489)
	0.4	(1.77)	(3.49)	(6.52)	(12.0)	(22.4)	(43.6)	(90.4)	(204)	(519)

TABLE 6.6 (continued)

Footing base friction angle δ (°)	Depth ratio D/B	Soil internal friction angle, φ (°)								
		5	10	15	20	25	30	35	40	45
$\delta = \varphi$ rough	0.5	(2.12)	(4.09)	(7.58)	(13.6)	(25.1)	(48.2)	(98.5)	(219)	(550)
	0.6	(2.48)	(4.70)	(8.61)	(15.4)	(27.9)	(52.9)	(107)	(235)	(582)
	0.7	(2.84)	(5.32)	(9.65)	(17.2)	(30.8)	(57.8)	(115)	(251)	(614)
	0.8	(3.20)	(5.95)	(10.7)	(19.0)	(33.8)	(62.8)	(124)	(267)	(647)
	0.9	(3.57)	(6.58)	(11.8)	(20.9)	(36.9)	(67.9)	(133)	(283)	(680)
	1.0	(3.93)	(7.22)	(12.9)	(22.8)	(40.1)	(73.1)	(142)	(300)	(714)

( ) Prandtl mechanism

TABLE 6.7

Bearing capacity factor $N\gamma$ for a deep footing on cohesiionless soils

Footing base friction angle δ (°)	Depth Ratio D/B	Soil internal friction angle, φ (°)								
		5	10	15	20	25	30	35	40	45
$\delta = 0$ smooth	1	(3.87)	(6.98)	(12.2)	(21.3)	(37.4)	(65.8)	(121)	(242)	(539)
	2	7.71	13.9	24.5	42.6	74.8	132	243	(466)	(976)
	3	11.7	21.2	37.3	64.5	113	199	359	695	1490
	4	15.9	29.1	51.5	88.8	154	273	491	932	1950
	5	20.3	37.5	66.7	115	199	351	634	1190	2450
	6	24.9	46.4	83.7	144	248	435	789	1470	2990
	7	29.6	55.8	102	174	300	576	953	1780	3570
	8	34.5	65.6	119	206	357	622	1120	2100	4180
	9	39.5	75.9	138	239	415	727	1310	2440	4840
	10	44.6	86.6	157	279	476	835	1500	2800	5530
$\delta = \varphi$ rough	1	3.94	7.22	12.9	22.8	40.1	73.1	142	300	714
	2	7.71	13.9	24.5	42.6	74.8	132	243	484	1080
	3	11.7	21.2	37.3	64.5	113	199	359	695	1490
	4	15.9	29.1	51.5	88.8	154	273	491	932	1950
	5	20.3	37.5	66.7	115	199	351	634	1190	2450
	6	24.9	46.4	83.7	144	248	435	789	1470	2990
	7	29.6	55.8	102	174	300	526	953	1780	3570
	8	34.5	65.6	117	206	357	622	1120	2100	4180
	9	39.5	75.9	138	239	415	727	1310	2440	4840
	10	44.6	86.6	157	279	476	835	1500	2800	5530

( ) Hill mechanism

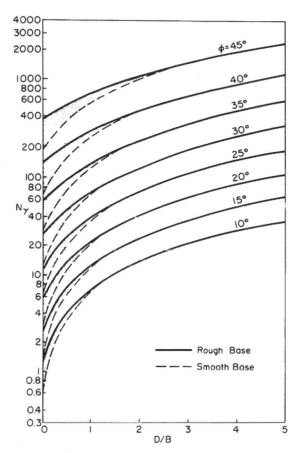

Fig. 6.26. Effect of footing base roughness on bearing capacity of deep footings.

N_γ of a deep smooth footing and a deep rough footing with D/B ratio varying from 1 to 10. Almost all values given in the table correspond to Prandtl type of failure except when $D/B < 2$ and $\delta = 0$. In such a case, Hill type mechanism may control the failure and the corresponding bearing capacity values shown in Table 6.7 are given in parentheses.

Comparison of results with existing solutions (shallow and deep footing)

Slip-line solutions for a perfectly smooth shallow footing bearing on cohesionless soils have been presented by Larkin (1968). Larkin's results are compared in Table 6.8 with some of the limit analysis solutions given in Table 6.6. The results of Abdul-Baki and Beik (1970) obtained by limit equilibrium method are also given in Table 6.8. By comparing these results, it can be observed that the limit analysis solutions compare remarkably well with Abdul-Baki and Beik's solutions for a rough footing and fairly well with Larkin's solutions for a smooth footing but

TABLE 6.8

Bearing capacity factor N_γ for shallow and deep footings

Shallow smooth footing

Depth ratio D/B	$\varphi = 30°$		$\varphi = 40°$	
	Slip-line Larkin (1968)	Limit analysis Table 6.6 δ = 0 case	Slip-line Larkin (1968)	Limit analysis Table 6.6 δ = 0 case
0.1	14	17.1	70	85.9
0.2	18	21.6	82	101
0.3	23	26.2	97	117
0.4	27	31.3	110	132
0.5	30	36.5	127	149

Deep rough footing

Depth ratio D/B	$\varphi = 30°$		$\varphi = 40°$	
	Limit equilibrium, Abdul-Baki and Beik (1970)	Limit analysis Table 6.7 δ = φ case	Limit equilibrium, Abdul-Baki and Beik (1970)	Limit analysis Table 6.7 δ = φ case
1	70	73.1	300	300
2	130	132	450	484
3	180	199	680	695
4	260	273	920	932
5	330	351	1200	1190

exceed the slip-line solutions by 13–23%. Since Abdul-Baki and Beik's solutions give very close values to those of Meyerhof (1951) when $D/B < 5$, the present limit analysis solutions should therefore yield close values to those of Meyerhof. Meyerhof's solutions are not compared here because his curves are expressed in terms of the angle β (same angle as shown in Fig. 6.5), which requires the solution of a transcendental equation.

It should be noted, however, that the slip-line solutions obtained by Larkin have not been extended throughout the soil mass in an admissible manner. Consequently, there may be other solutions which also satisfy the same boundary conditions yet give smaller bearing capacity value. Larkin's slip-line solutions are therefore not necessarily a lower bound for the bearing capacity. Also, it has not been shown that the slip-line stress fields developed by Larkin can associate with admissible velocity fields satisfying flow rule requirements. Hence, it is not known whether it is an upper bound. Nevertheless, Larkin's slip-line solutions for a smooth shallow footing probably represent a close upper bound to the problem. This is due to the fact that a velocity field has been shown (Larkin, 1972) to exist for the special case of a rough surface footing on a $c - \varphi - \gamma$ soil; it seems unlikely, since there are no new physical complications, that a velocity field associated with the slip-line stress field cannot be found for the case of a smooth shallow footing on a cohesionless soil.

6.5.3. Footings on a slope

The bearing capacity of a footing on a slope can be obtained directly from a simple modification of the solutions presented previously for a shallow footing. Taking Prandtl mechanism (Fig. 6.5) for example, when we set the area bef equal to zero, the case $-\frac{1}{2}\pi < \beta < 0$ represents the bearing capacity of a footing on a slope. The equation for computing the bearing capacity factor N_γ for this case has the following form:

Prandtl mechanism (footing on a slope, $\beta < 0$)

$$N_\gamma(\xi,\eta) = \frac{\cos(\xi - \varphi)}{2\cos\varphi\cos^2\xi}\,[h_1(\xi) + h_2(\xi,\eta)] + \frac{\cos(\xi - \varphi)\sin\eta}{2\cos(\eta + \varphi)\cos^2\xi} \qquad [6.48]$$

in which functions $h_1(\xi)$ and $h_2(\xi,\eta)$ are given in [6.11] and [6.15], respectively. The results for the case of Hill mechanism can be obtained in a similar manner.

The absolute minimum values of the factor N_γ computed from the Prandtl and Hill mechanisms for the case of a rough footing on a slope have been calculated for the slope angle β varying from 0 to $-30°$ and friction angle φ ranging from $10°$ to $45°$. The results are plotted in Fig. 6.27 within the range $-\varphi < \beta \leqslant 0$. The case $\beta = 0$ applies to a surface footing and the factor is identical to that given in

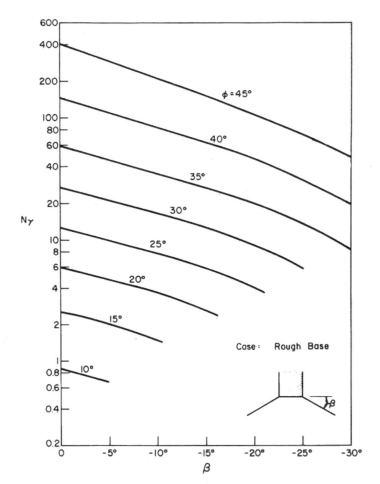

Fig. 6.27. Bearing capacity of a footing on slope.

Fig. 6.25 with $D/B = 0$. For $\varphi = 40°$ and $\beta = -30°$ the present value $N_\gamma = 19.5$ may be compared with the value $N_\gamma = 20$ obtained by Meyerhof (1951) using the method of limit equilibrium. The approximate estimate on the basis of the present upper-bound limit analysis method agrees within 2.5% in this case.

6.5.4. Some experimental observations

In the preceding bearing capacity computations, we have assumed that soil can be treated as a perfectly plastic Coulomb material for which hydrostatic pressure linearly increases the yield stress. One of the especially significant implications of this theory is the necessity of increase in volume accompanying shearing action. Further, we have seen in the preceding discussions that there exists in literature a great variety of proposed solutions to the N_γ-values based on different methods, and the differences among them are sometimes substantial. It is therefore considered advisable to check the validity of this perfect plasticity assumption experimentally and to provide some verifications of the present plasticity theory for soil under the ultimate bearing capacity conditions. After this accomplishment, the present theory can be more confidently applied to other situations such as the retaining wall, earth pressure and slope stability problems.

Experimental results of the behavior of sand under a strip footing will be discussed herein. The discussion includes the following three important items: (1) the correct N_γ-values; (2) the displacement field in the sand during the indentation of footing; and (3) the effect of footing roughness on ultimate bearing capacity.

N_γ-values

Despite an intensified experimental effort in the past on the subject of ultimate bearing capacity of a footing on sand, the question of correct N_γ-values has for some time remained unsettled, because of difficulties in selecting a correct value of the angle of internal friction of sand, φ, for the bearing capacity computations when comparing theoretical predictions with test results. Since all the theoretical analyses assume the conditions of plane strain, the plane strain value of φ should be used in the computations. Direct and indirect investigations reported in the literature indicate that under drained conditions the plane strain value of φ exceeds the value obtained from conventional triaxial tests by about $4-8°$ (Cornforth, 1964). If the φ-value obtained under triaxial compression stress conditions is used in calculating the N_γ-value, a lower bearing capacity will be predicted than if the φ-value in plane strain conditions is used. Because the N_γ factor increases rapidly with φ, the difference between the bearing capacity using triaxial compression, $\dot\varphi$, and that using a plane strain, φ, can be substantial. For example, let it be supposed, as is usually the case, that the angle φ obtained from a triaxial compression test has the value, say, $30°$. With this value of φ, we obtain $N_\gamma = 27.67$ from Table 6.9. However, if plane strain value of φ is used, this angle of friction is usually a few degrees greater than that in triaxial compression test, say, $33°$ which corresponds to $N_\gamma = 44.41$. It is seen that the increase of $3°$ in the friction angle in this case increases the bearing capacity by about 60% in the range of interest.

In view of the difficulties in selecting a proper value of angle φ, it seems unlikely that footing experiments can clearly settle the question of correct N_γ-values. Although the differences in N_γ-values among different methods may range from about one-third to double the value given in Table 6.9, this corresponds to only a few degree variations in the value of internal friction angle φ. It has generally been considered that Terzaghi's N_γ-factor (1943) gives conservative estimates of the actual bearing capacity of footings when the value of φ obtained from triaxial tests is used in the computations. However, Terzaghi's bearing capacity factor N_γ be-

TABLE 6.9

Bearing capacity factors

φ°	N_q, [6.51]	N_c, [6.49]	N_γ, [6.41]	N_q/N_c	$\tan \varphi$
0	1.000	5.140	0.000	0.185	0.000
1	1.094	5.388	0.074	0.203	0.017
2	1.197	5.636	0.156	0.212	0.035
3	1.309	5.903	0.247	0.222	0.052
4	1.433	6.188	0.350	0.232	0.070
5	1.568	6.491	0.465	0.242	0.087
6	1.716	6.815	0.595	0.252	0.105
7	1.879	7.161	0.743	0.262	0.123
8	2.058	7.530	0.909	0.273	0.141
9	2.255	7.924	1.098	0.285	0.158
10	2.472	8.347	1.313	0.296	0.176
11	2.711	8.800	1.558	0.308	0.194
12	2.974	9.287	1.837	0.320	0.213
13	3.265	9.810	2.157	0.333	0.231
14	3.586	10.37	2.522	0.346	0.249
15	3.942	10.98	2.941	0.359	0.268
16	4.336	11.63	3.423	0.373	0.287
17	4.773	12.34	3.976	0.387	0.306
18	5.259	13.11	4.614	0.401	0.325
19	5.799	13.94	5.349	0.416	0.344
20	6.401	14.84	6.198	0.431	0.364
21	7.072	15.82	7.180	0.447	0.384
22	7.823	16.89	8.318	0.463	0.404
23	8.663	18.05	9.639	0.480	0.425
24	9.605	19.33	11.18	0.497	0.445
25	10.66	20.73	12.97	0.515	0.466
26	11.86	22.26	15.05	0.533	0.488
27	13.20	23.95	17.50	0.551	0.510
28	14.72	25.81	20.36	0.570	0.532
29	16.45	27.87	23.72	0.590	0.554
30	18.41	30.15	27.67	0.610	0.577
31	20.64	32.68	32.34	0.631	0.601
32	23.18	35.50	37.86	0.653	0.625
33	26.10	38.65	44.41	0.675	0.649
34	29.45	42.18	52.20	0.698	0.675
35	33.31	46.14	61.49	0.722	0.700
36	37.76	50.50	72.62	0.746	0.727
37	42.93	55.65	85.98	0.772	0.754
38	48.95	61.37	102.1	0.798	0.781
39	55.97	67.89	121.6	0.825	0.810
40	64.21	75.34	145.3	0.852	0.839
41	73.92	83.88	174.1	0.881	0.869
42	85.40	93.74	209.5	0.911	0.900
43	99.05	105.1	253.1	0.942	0.933
44	115.4	118.4	307.0	0.974	0.966
45	134.9	133.9	374.2	1.007	1.000
46	158.6	152.2	458.2	1.042	1.036

TABLE 6.9 (continued)

φ°	N_q, [6.51]	N_c, [6.49]	N_γ, [6.41]	N_q/N_c	$\tan\varphi$
47	187.3	173.7	564.0	1.078	1.072
48	222.4	199.3	698.2	1.115	1.111
49	265.3	230.0	869.6	1.155	1.150
50	391.2	270.0	1090	1.195	1.192
51	386.1	311.9	1358	1.238	1.235
52	470.5	366.8	1750	1.283	1.280
53	577.7	434.6	2243	1.329	1.327
54	715.4	519.0	2902	1.378	1.376
55	893.9	625.2	3790	1.430	1.428
56	1128	760.1	5001	1.484	1.483
57	1439	933.6	6674	1.541	1.540
58	1856	1150	9017	1.601	1.600
59	2426	1457	12347	1.665	1.664
60	3216	1856	17160	1.733	1.732

comes non-conservative when the plane strain angle of φ is used. The recent results of footing tests on sand reported by Ko and Davidson (1973) indicate that Soko-lovskii's solutions give a slightly conservative estimate of the smooth footing bearing values when a plane strain value of φ is used.

In the experiments carried out by Ko and Davidson (1973), the smooth footings had plate glass bottoms. It is of course questionable whether this can be classified as a *perfectly* smooth base. In view of the fact that: (1) the N_γ-values increase rapidly with a small increase in base friction angle δ (Hill mechanism, Fig. 6.22); (2) the N_γ-values obtained by Hill mechanism for the case of a smooth base ($\delta = 0$) are about 13–17% smaller than those obtained by Sokolovskii (see Table 6.5), it may be concluded from Ko and Davidson's experimental bearing capacity values that the Hill mechanism solutions as given in Table 6.5, give a rather accurate estimate of the smooth footing bearing values.

Displacement field

Sylwestrowicz (1953) constructed a special container with removable side panel made of $\frac{1}{2}$ inch thick glass. A grid system can be produced on the side surface of the material and the deformation of the grid system can therefore be observed in the glass container. By taking photographs of different stages of the footing penetration, and tracing from them the grid systems corresponding to each position of the footing, a displacement field can be constructed and compared with that of Prandtl and Hill mechanisms at the ultimate bearing capacity conditions.

Sylwestrowicz (1953) observed from his footing tests that up to the maximum load, there is no discontinuity layer between the sand that moves and the sand at rest. When the maximum load is reached, two symmetrical slip surfaces start to develop and portions of the sand above them move as rigid body. With the formation of the slip surface, the force required for the further movement decreases, and a discontinuity layer is formed, which starts to move under approximately constant load. At this stage of loading, the displacement field in the sand is found in good agreement with the velocity field given by Hill mechanism (Fig. 3.22d).

An increase in volume is also observed in Sylwestrowicz's experiments with the sand. The ratio of the increase of the volume of the sand displaced by the footing is approximately 2 (for an indentation of about 0.4 inch). No exact measurements were made of the areas where the change of volume took place. Only the changes of grid areas were measured. However, from these measurements and the character of the deformation of the grid, it is clear that the main change of the volume takes place in the discontinuity layer. The increase in volume during shearing along the discontinuity layer is then in good agreement with the slip requirement implied by the assumption of perfect plasticity.

Although the increase in volume in the experiments cited corresponds in a sense to the theoretical prediction (Fig. 3.22d), one should remember that the latter is computed for a very small movement of the footing while in the experiments the penetration is quite large. Also the rapid decrease in load on the footing after the maximum load is reached seems to imply that the sand behaves in an unstable manner.

Comparison of experimental displacement field with that of theoretical Hill mechanism at the maximum footing load indicates the same order of magnitude and character in the change of thickness of the discontinuity layer, and in the ratio of the distance from the footing to the step on the upper free surface to the width of the footing (according to Fig. 3.22d, this ratio is 3.2 for $\varphi = 37°$; experimental value approximately 2.7). This agreement is still more surprising in view of the following two facts concerning the displacement field or deformation pattern as given by Fig. 3.22(d): (1) the displacement field is computed for a cohesive weightless soil, while the experiments deal with cohesionless ponderable sand; and (2) the displacement field is computed on the assumption of very small displacements while the experiments deal with quite appreciable deformations.

Footing roughness

In the experiments reported by Ko and Davidson (1973), it is observed that the sand in footing tests with glass bottoms (almost smooth) fails according to the Hill type of mechanism; two symmetrical wedges beneath the footing are developed. At the same time, the sand in footing tests with sandpaper bottoms (almost rough) fails according to the Prandtl type of mechanism; a central wedge beneath the footing is developed. Since two rather contrasting failure mechanisms are observed in the experiments corresponding to an almost smooth and an almost rough footing, this observation leads us to believe that the two types of mechanisms used in the preceding calculations for the bearing capacity of smooth and rough footings are justified.

In the following we shall examine the effect of footing roughness on ultimate bearing capacity. On the basis of his experimental results, Meyerhof (1955) suggests that the bearing capacity of a smooth footing on the surface of a cohesionless soil should be only one-half of the capacity of a rough footing. This suggestion is substantiated by the results of limit analysis, as can be seen clearly by comparing the Prandtl 1 column with the Hill column in Table 6.5.

Since a base friction angle of no more than 17° is considered to be sufficiently large to restrain sliding between the footing base and the adjacent soil (Fig. 6.22), it seems unlikely therefore that such a small value of friction angle cannot be developed in the actual situation. Thus, a Hill type of mechanism may be fictitious, and may never be realized in an actual footing. This leads Vesic (1973) to conclude that the stress and deformation pattern under a compressed area in an actual footing is such that it always leads to the formation of a single wedge. According to this view, then, the footing roughness in a practical situation has little effect on bearing capacity as long as applied external loads remain vertical. Thus, all the bearing capacity computations should be based on the condition that the footing base is perfectly rough.

6.6. BEARING CAPACITY OF A STRIP FOOTING ON A $c - \varphi$ WEIGHTLESS SOIL (N_c AND N_q FACTORS)

The bearing capacity factor N_c for a shallow footing has been defined in [6.20] for the case of Prandtl mechanism (Fig. 6.5) and in [6.28] for the case of Hill mechanism. The absolute minimum values obtained from these two equations have been presented in Fig. 6.12 ($G = 0$ case) within practical limits of $D/B < 1$ and φ. The results show that the factor N_c increases rapidly with φ and is not very sensitive to changes of D/B. This is in contrast to the N_γ-case where the factor N_γ increases rapidly with both φ and D/B (Fig. 6.25). For $D/B > 1$, the problem of deep footing is represented and the corresponding expressions for N_c have been defined in the right-hand side of [6.36] and [6.38] with $G = 0$. The results have been plotted graphically in Fig. 6.13 ($G = 0$ case).

The case $D/B = 0$ (or $\beta = 0$) applies to a surface footing, and the optimum values of ξ, ζ and η are found to be $\frac{1}{4}\pi + \frac{1}{2}\varphi$, $\frac{1}{4}\pi + \frac{1}{2}\varphi$, and $\frac{1}{4}\pi - \frac{1}{2}\varphi$, respectively for both Prandtl and Hill mechanisms ($\delta = 0$). Substituting these optimum values of ξ, ζ and η into [6.23] or [6.32] with $G = 0$, the bearing capacity factor N_c for both Prandtl and Hill mechanisms has the same form.

Surface footing

$$N_c = \cot \varphi \, [e^{\pi \tan \varphi} \tan^2 (\tfrac{1}{4}\pi + \tfrac{1}{2}\varphi) - 1]$$
[6.49]

which is identical to the result derived previously by Prandtl (1920). The numerical values of this factor are given in Table 6.9.

When an uniform surcharge q is applied vertically on the surface of the soil, the rate of external work due to this surcharge should be included in the computation of the bearing capacity of the footing. For weightless soil ($\gamma = 0$), the total bearing capacity q_0 has the form:

$$q_0 = cN_c + qN_q$$
[6.50]

in which N_c is defined in [6.49] and the dimensionless factor N_q resulting from the surcharge load can be expressed as:

$$N_q = e^{\pi \tan \varphi} \tan^2 (\tfrac{1}{4}\pi + \tfrac{1}{2}\varphi)$$
[6.51]

which is identical to the result derived by Reissner (1924). The N_q-values according to [6.51] are given in Table 6.9.

It is worth pointing out that while the question of correct N_γ-values has for some time remained unsettled, the values in N_c and N_q as given in [6.49] and [6.51] are generally accepted as the correct or exact solution for a strip footing. The factor N_c ([6.49]) and the approximate equation N_γ as given by [6.41] may be expressed in terms of the factor N_q as:

$$N_c = \cot \varphi \, (N_q - 1)$$
[6.52]

$$N_\gamma = 2(1 + N_q) \tan \varphi \tan (\tfrac{1}{4}\pi + \tfrac{1}{5}\varphi)$$
[6.53]

The factors N_q, N_c and N_γ as defined in [6.51], [6.52] and [6.53], respectively are rather simple in form yet sufficiently accurate and may be considered to be the most reliable factors available at present. The numerical values of these factors are given in Table 6.9.

6.6.1. Complete stress field of a surface footing

The failure mechanisms or velocity fields obtained in the preceding sections are *incomplete* solutions, leading only to the conclusion that the bearing capacities obtained are upper bounds to the correct values. In the following we will outline a method of showing that the bearing capacity factor N_c of a surface footing as given by [6.49] is also a lower bound and hence is the *correct* value.

In section 4.9, Chapter 4, we have obtained a statically admissible stress field for the simpler case of a flat surface of infinite extent, loaded by pressure Q extending indefinitely in one direction, as shown in Fig. 4.47(a). We now investigate whether the stress field shown in Fig. 4.47(a) is adaptable to the problem of the finite strip footing.

The extended stress field (Fig. 6.28)

The method to be adopted here essentially follows that proposed by Shield (1954a) as one possible way of determining an acceptable modification of the stress field in the problem of a semi-infinite footing (Fig. 4.47). Briefly, this method is as follows. In Fig. 6.28, the region above the line of stress discontinuity $BGSF$, which is as yet unspecified, is composed of the stress field of two regions of constant state, OAB and DAC, and a region of radial shear $CABGSF$. These regions are identical to the regions shown in Fig. 4.47(a). The stress field below the line of stress discontinuity $BGSF$ is determined from the following conditions:

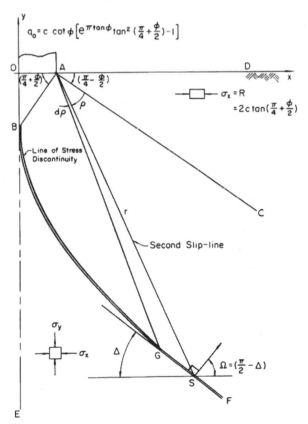

Fig. 6.28. A complete stress field for a surface footing on a $c - \varphi$ weightless soil.

(1) σ_x and σ_y are functions of y and x only, respectively, and τ_{xy} is zero, satisfying the equilibrium equations and the symmetry condition across OE.

(2) In the immediate neighborhood of the line of discontinuity, the material is in a plastic state of stress. These conditions, together with the jump conditions across the line of discontinuity discussed in section 4.5, Chapter 4, are sufficient to determine the stresses in the region $BEFSGB$ and to determine the line $BGSF$.

According to our convention in section 4.5, the first slip or failure line is obtained from the direction of σ_{max} (maximum compressive principal stress) by a counterclockwise rotation of amount $\frac{1}{4}\pi - \frac{1}{2}\varphi$. The angle of inclination of the first slip-line to the x-axis is denoted by θ (Fig. 4.12a). Across a line of stress discontinuity in a plastic stress field, the normal and tangential components of stress must be continuous across the line for equilibrium. The equilibrium condition across a line of discontinuity separating two plastic stress fields 1 and 2 are given by [4.28] and [4.29] where Ω is the inclination to the x-axis of the normal to the line at a current segment of the line. Subscripts 1 and 2 distinguish the values $\bar{p}$ and θ assumed on the two sides of the line. $\bar{p}$ is defined in [4.27].

Referring now to Fig. 6.28, the surface AD is free from traction and $\sigma_y = 0$ in the constant state region DAC. Since the whole region above the line of discontinuity $BGSF$ is in a plastic state of stress, it follows that the uniaxial compressive stress R has the value:

$$R = \frac{2c\cos\varphi}{1-\sin\varphi} = 2c\tan\left(\tfrac{1}{4}\pi + \tfrac{1}{2}\varphi\right) \qquad [6.54]$$

in the region DAC. In section 4.9, we have shown that the hydrostatic pressure p in the typical constant state element GAS in the region of radial shear $CABGSF$ has the value:

$$p = \tfrac{1}{2}R\,\frac{1}{\sin\varphi}\left[e^{2\rho\tan\varphi} - (1-\sin\varphi)\right] \qquad [6.55]$$

or using [6.54], the above equation reduces to the form:

$$p = \frac{c\cot\varphi}{1-\sin\varphi}\,e^{2\rho\tan\varphi} - c\cot\varphi \qquad [6.56]$$

where ρ is the angle defining the radial line AS (Fig. 6.28). Substituting p of [6.56] into [4.27], we obtain the value $\bar{p}$ in the constant stress element GAS:

$$\bar{p} = \frac{c\cot\varphi}{1-\sin\varphi}\,e^{2\rho\tan\varphi} \qquad [6.57]$$

Referring now to Fig. 4.47, the radial lines AB and AC are the first slip or failure line in the constant state regions OAB and DAC, respectively. Since the change of direction of maximum shear stress in the radial shear zone BAC follows the stress diagram shown in Fig. 4.47(b), it can be concluded that each radial line in this zone is also the direction of the first slip or failure line. It should be noted that if the stress field of Fig. 4.47(a) is drawn on the right-hand side of the footing as shown in Fig. 6.28, all the radial lines are now called second slip lines according to our convention.

It has been mentioned above that the material just below the line of stress discontinuity $BGSF$ is assumed to be in a plastic state of stress. We shall denote by a and b the two plastic stress fields immediately *above* and immediately *below* the line element GS, respectively. Since the stress state in the element GAS is constant and the radial line AS is a second slip line, the value of $\bar{p}$, $\bar{p}_a$, in region a at line element GS is also given by [6.57] and we have $\theta_a = \frac{1}{4}\pi - \frac{1}{2}\varphi - \rho$. In region b at line element GS, $\sigma_x > \sigma_y$, so that $\theta_b = \frac{1}{4}\pi - \frac{1}{2}\varphi$. Also the normal to the line of discontinuity element GS is inclined at an angle $\Omega = \frac{1}{2}\pi - \Delta$ to the axis, where Δ is the inclination of the line to the negative x-axis. Substitution of these values into the jump condition [4.29] gives:

$$\cos(2\Delta - \rho) = \sin\varphi \cos\rho \tag{6.58}$$

and the relevant root of this equation is:

$$\Delta = \frac{1}{4}\pi + \frac{1}{2}\rho - \frac{1}{2}\mu \tag{6.59}$$

$$\text{where: } \sin\mu = \sin\varphi \cos\rho \quad 0 \leqslant \mu \leqslant \varphi \tag{6.60}$$

with this value of Δ, the jump condition [4.28] gives:

$$\bar{p}_b = \bar{p}_a \frac{\cos(\rho + \mu)}{\cos(\rho - \mu)} \tag{6.61}$$

or:

$$\bar{p}_b = \frac{c[1 + \sin^2\varphi - 2\sin\varphi \sin(\rho + \mu)] \, e^{2\rho \tan\varphi}}{(1 - \sin\varphi) \sin\varphi \cos\varphi} \tag{6.62}$$

where the value [6.57] for $\bar{p}_a$ and [6.60] have been used.

Referring now to Fig. 4.12(b), since $2\psi = (\frac{1}{2}\pi + \varphi) - (\pi - 2\theta) = -\frac{1}{2}\pi + 2\theta + \varphi$, it can be shown from the Mohr circle that:

$$\sigma_x = \bar{p}[1 + \sin\varphi \sin(2\theta + \varphi)] - c \cot\varphi$$

$$\sigma_y = \bar{p}[1 - \sin\varphi \sin(2\theta + \varphi)] - c \cot\varphi$$

$$\tau_{xy} = \bar{p} \sin\varphi \cos(2\theta + \varphi) \tag{6.63a,b,c}$$

From [6.63], the non-zero stresses in region b at the line element GS are given by:

$$\sigma_x = \bar{p}_b(1 + \sin\varphi) - c\cot\varphi$$

$$\sigma_y = \bar{p}_b(1 - \sin\varphi) - c\cot\varphi \qquad\qquad\qquad \text{[6.64a,b]}$$

For equilibrium, σ_x and σ_y are taken to be functions of y only and x only, respectively in region $BGSFEB$. The values of σ_x and σ_y just below the line $BGSF$ are known from [6.64] and [6.62] so that σ_x and σ_y can be found at any point of the region $BGSFEB$. It can be shown that $\bar{p}_b$ is a monotonic increasing function of ρ. It follows that just below the line $BGSF$, σ_x and σ_y are increasing functions of ρ, and hence that in region $BGSFEB$, σ_x is a monotonic increasing function of y and σ_y is a monotonic decreasing function of x.

This extension of the stress field of Fig. 4.47 to the case of a *finite* footing is permissible only if the Coulomb yield condition:

$$\tfrac{1}{2}(\sigma_x + \sigma_y)\sin\varphi - [\tfrac{1}{4}(\sigma_x - \sigma_y)^2 + \tau_{xy}^2]^{1/2} + c\cos\varphi = 0 \qquad\qquad \text{[6.65]}$$

is nowhere violated in region $BGSFEB$, i.e., if the expression on the left-hand side of [6.65] is less than or equal to zero at all points in the region. Because of the monotonic character of the stresses σ_x and σ_y, the yield conditions will not be violated anywhere in the region if it is not violated at the point E at infinity on the y-axis. At the point E, σ_x has a minimum value and σ_y has a maximum value and these values are:

$$\sigma_x|_{\rho=0} = 2c\tan(\tfrac{1}{4}\pi + \tfrac{1}{2}\varphi) \qquad\qquad\qquad\qquad\qquad \text{[6.66]}$$

$$\sigma_y|_{\rho=\frac{1}{2}\pi} = c\cot\varphi\,[e^{\pi\tan\varphi}\tan^2(\tfrac{1}{4}\pi - \tfrac{1}{2}\varphi) - 1] \qquad\qquad \text{[6.67]}$$

Substituting these values into the expression on the left-hand side of [6.65] and setting the resulting expression less than or equal to zero gives, after reduction, the inequality:

$$e^{\pi\tan\varphi} \leqslant \tan^6(\tfrac{1}{4}\pi + \tfrac{1}{2}\varphi) \qquad\qquad\qquad\qquad\qquad \text{[6.68]}$$

The inequality is satisfied if the angle φ is less than an angle which lies between $75°$ and $76°$. Thus the yield condition is nowhere violated in region $BGSFEB$ if φ is less than $75°$

If the length of the line AS in Fig. 6.28 is denoted by r, then we have:

$$-dr\tan(\tfrac{1}{4}\pi - \tfrac{1}{2}\varphi + \rho - \Delta) = rd\rho \qquad\qquad\qquad\qquad \text{[6.69]}$$

or using [6.59]:

$$dr = rd\rho\cot(\tfrac{1}{2}\varphi - \tfrac{1}{2}\rho - \tfrac{1}{2}\mu) \qquad\qquad\qquad\qquad \text{[6.70]}$$

This differential equation and the condition $r = AB$ when $\rho = \tfrac{1}{2}\pi$, determine the line of discontinuity. As ρ tends to 0, the line tends asymptotically to a straight line inclined at an angle $\frac{1}{4}\pi - \frac{1}{2}\varphi$ to the negative x-axis.

Since it has been shown in section 4.9, Chapter 4 that the bearing capacity value:

$$Q = c \cot \varphi \left[e^{\pi \tan \varphi} \tan^2 \left(\tfrac{1}{4}\pi + \tfrac{1}{2}\varphi \right) - 1 \right] \qquad [6.71]$$

is a lower bound for the collapse value of the average bearing pressure Q over the *semi-infinite* footing as shown in Fig. 4.47(a) and we now have demonstrated that a statically admissible modification of this stress field into the case of a *finite* footing is found also possible, the value [6.71] or the N_c-factor given previously in [6.49] is also a lower bound for a strip, finite surface footing and therefore the correct value of the average pressure.

Remarks on the extended stress field

In view of the fact that the required construction of a statically admissible stress field has been shown to be possible for the case of a surface footing on $c - \varphi$ soils (Fig. 6.28), and, in Fig. 4.22(b) or Fig. 4.44(b), for Tresca material for which $\varphi = 0$, it seems unlikely, since there are no new physical complications, that a similar modification of the stress field shown in Fig. 4.40(a) to the case of a finite strip footing on a slope cannot be made. It follows then from the theorems of limit analysis that the bearing capacity factor N_c:

Footing on a slope ($\beta < 0$)

$$N_c = \cot \varphi \left[e^{(\pi + 2\beta) \tan \varphi} \tan^2 \left(\tfrac{1}{4}\pi + \tfrac{1}{2}\varphi \right) - 1 \right] \qquad [6.72]$$

is an upper bound ([4.63]) and also a lower bound and therefore the correct value of the bearing capacity factor N_c for a strip footing on a $c - \varphi$ slope. It is of interest to note that the case $\beta = -\frac{1}{2}\pi$ represents unconfined compression, while for $-\frac{1}{2}\pi < \beta < 0$ the bearing capacity of a footing on a slope is represented. The case $\beta = 0$ applies to a surface footing, and [6.72] is identical to that given in [6.49]. For the special case of Tresca material for which $\varphi = 0$, [6.72] reduces to the value:

$$N_c = \pi + 2 + 2\beta \qquad [6.73]$$

Although the determination of a complete solution is strictly necessary for the case of shallow footing, the fact that a complete solution has been shown to exist for the case of a surface footing would suggest that the same is also true for at least some range of the values D/B of a *very* shallow footing. Further, there is no apparent reason to suppose that the method of determining the stress field outlined above should not also be capable of application in the shallow footing case. It is concluded that the bearing capacity factor N_c obtained in this section gives not only the *correct* value for the cases of a surface footing and a footing on a slope,

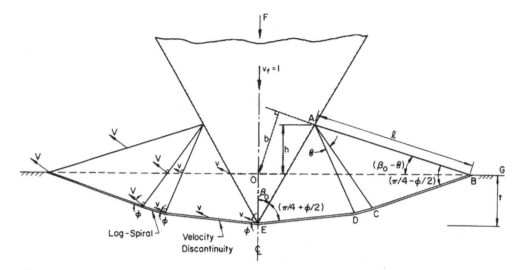

Fig. 6.29. Bearing capacity of a smooth wedge footing on weightless $c - \varphi$ soils (Shield, 1953).

but also gives a very close upper bound to the correct value for the case of a shallow footing.

6.6.2. Bearing capacity of a wedge footing

We now consider the problem of the bearing capacity of a semi-infinite mass of soil loaded by a smooth, rigid wedge footing as shown in Fig. 6.29. In this figure, as the wedge is pressed into the soil, the displaced soil will form a raised lip at each side of the wedge, and the shape of the lips must be determined as part of the solution to the problem. This is in contrast to the previous case of a flat footing where the development of the lip or the sinking in can be neglected. In the following we shall assume that the surfaces of the lips are straight and the configuration of the failure mechanism as shown in Fig. 6.29 is *geometrically similar* at each stage of the penetration. Because of these assumptions it is possible, therefore, to obtain the bearing capacity value by the limit analysis method as well as the complete history of the penetration without following the deformation step by step.

The bearing capacity value (Fig. 6.29)

In Fig. 6.29, the pattern of the failure mechanism is shown where AE is the right flank of the wedge, which is of angle $2\beta_0$. The line BG is the as yet undeformed surface of the soil and AB is the lip. This mechanism is an extension into the wedge case of a simple modification of the Hill mechanism used previously for the case of flat footing. Since the movement is symmetrical about OE, it is only necessary to consider the movement on the right-hand side of Fig. 6.29. The two triangular areas

AED and ACB move as rigid bodies in the direction making an angle φ with the lines ED and CB, respectively while ADC is a zone of logspiral of radial shear of angle Θ.

The internal dissipation of energy is due to the discontinuity surfaces $EDCB$ between the material at rest and the material in motion and due to the logspiral radial shear zone ADC. Since the velocity vector v in the triangular area ADE must make an angle φ with the line ED and also must remain in contact with the right flank of the smooth wedge, the triangle ADE in contact with the footing must move with the velocity:

$$v = v_f \sin \beta_0 \sec \left(\tfrac{1}{4}\pi + \tfrac{1}{2}\varphi\right) \tag{6.74}$$

where v_f denotes the downward velocity of the wedge footing. In the logspiral zone of radial shear the velocity increases exponentially from the value v along the radial line AD to the maximum value:

$$V = v e^{\Theta \tan \varphi} = v_f \sin \beta_0 \sec \left(\tfrac{1}{4}\pi + \tfrac{1}{2}\varphi\right) e^{\Theta \tan \varphi} \tag{6.75}$$

along the line AC. At a given instant, the region ABC is moving as a rigid body with velocity V in the direction perpendicular to AC. The velocity field in the plastic region is illustrated on the left of Fig. 6.29.

The rates of dissipation of energy are:

Along ED

$$c(v \cos \varphi)r_0 \quad \text{where} \quad r_0 = \overline{AD} \tag{6.76}$$

Along CB

$$c(V \cos \varphi)r_0 \, e^{\Theta \tan \varphi} = cv \cos \varphi \, r_0 \, e^{2\Theta \tan \varphi} \tag{6.77}$$

Along spiral DC (see [3.74])

$$\tfrac{1}{2}cvr_0 \cot \varphi \, (e^{2\Theta \tan \varphi} - 1) \tag{6.78}$$

In logspiral region ADC (see [3.74])
same as [6.78].

The external rate of work due to the pressure q_0 on the flank of the wedge is found to be:

$$[q_0 2r_0 \cos \left(\tfrac{1}{4}\pi + \tfrac{1}{2}\varphi\right)] \, [v \sin \left(\tfrac{1}{4}\pi - \tfrac{1}{2}\varphi\right)] \tag{6.79}$$

Equating the total rate of internal dissipation of energy to the rate at which work is done by the pressure q_0, it is found that the pressure q_0 on the flank of the wedge footing has the value:

$$q_0 = c \cot \varphi \, [e^{2\Theta \tan \varphi} \tan^2 \left(\tfrac{1}{4}\pi + \tfrac{1}{2}\varphi\right) - 1] \tag{6.80}$$

agreeing with the value obtained previously by Shield (1953).

For the special case of Tresca material for which $\varphi = 0$, [6.80] reduces to the simple form:

$$q_0 = 2c(1 + \Theta) \tag{6.81}$$

agreeing with the result obtained previously by Hill et al. (1947). Since we have:

$$AE = l \tan\left(\tfrac{1}{4}\pi - \tfrac{1}{2}\varphi\right) e^{-\Theta \tan\varphi} \tag{6.82}$$

where we have denoted by l the length AB of the lip, the total downward force F necessary to drive the wedge into the soil is given by:

$$F = 2\, q_0\, l \sin\beta_0 \tan\left(\tfrac{1}{4}\pi - \tfrac{1}{2}\varphi\right) e^{-\Theta \tan\varphi} \tag{6.83}$$

The history of penetration

In the above equations, the zone of spiral angle, Θ, and the length of the lip, l, are as yet unknowns. In the following we will determine these unknowns. Referring to Fig. 6.29, we denote by b the distance of O from AB, and by h the elevation of A above OB. The depth of penetration of the wedge is denoted by t and if the downward velocity of the wedge is taken as the unit of velocity ($v_f = 1$), we may take t to be the time variable.

The lip AB makes an angle $\beta_0 - \Theta$ with the undisturbed level OB and it is easily shown that we have the following expressions for l, b, h in terms of t, Θ and β_0:

$$l = \frac{t}{[e^{-\Theta \tan\varphi} \tan\left(\tfrac{1}{4}\pi - \tfrac{1}{2}\varphi\right)\cos\beta_0 - \sin(\beta_0 - \Theta)]} \tag{6.84a}$$

$$h = \frac{t \sin(\beta_0 - \Theta)}{[e^{-\Theta \tan\varphi} \tan\left(\tfrac{1}{4}\pi - \tfrac{1}{2}\varphi\right)\cos\beta_0 - \sin(\beta_0 - \Theta)]} \tag{6.84b}$$

$$b = \frac{t \sin(\beta_0 - \Theta)\,[e^{-\Theta \tan\varphi} \tan\left(\tfrac{1}{4}\pi - \tfrac{1}{2}\varphi\right)\sin\beta_0 + \cos(\beta_0 - \Theta)]}{[e^{-\Theta \tan\varphi} \tan\left(\tfrac{1}{4}\pi - \tfrac{1}{2}\varphi\right)\cos\beta_0 - \sin(\beta_0 - \Theta)]} \tag{6.84c}$$

The projection of the velocity of the lip AB on the normal to AB is ($v_f = 1$):

$$V \cos\left(\tfrac{1}{4}\pi - \tfrac{1}{2}\varphi\right) = \sin\beta_0 \tan\left(\tfrac{1}{4}\pi + \tfrac{1}{2}\varphi\right)e^{\Theta \tan\varphi} \tag{6.85}$$

while the projection of the velocity of the vertex E, which is moving downward with unit velocity, on the normal to AB is $\cos(\beta_0 - \Theta)$. Hence, the distance of E from AB increases at the constant rate which is the sum of these two projections. At a time t, i.e., since the beginning of the indentation, the distance of E from AB has therefore reached the value:

$$t\left[\sin\beta_0 \tan\left(\tfrac{1}{4}\pi + \tfrac{1}{2}\varphi\right) e^{\Theta \tan\varphi} + \cos(\beta_0 - \Theta)\right] \tag{6.86}$$

From Fig. 6.29, we see that this distance is also equal to:

$$b + t \cos(\beta_0 - \Theta) \tag{6.87}$$

and equating the expressions [6.86] and [6.87] gives:

$$b = t \sin\beta_0 \tan(\tfrac{1}{4}\pi + \tfrac{1}{2}\varphi) e^{\Theta \tan\varphi} \tag{6.88}$$

The substitution of expression [6.84c] into this equation furnishes a relation between the angles Θ, β_0, and after some reduction we obtain:

$$\cos(2\beta_0 - \Theta) = \frac{\cos\Theta \left[e^{\Theta \tan\varphi} \tan(\tfrac{1}{4}\pi + \tfrac{1}{2}\varphi) + e^{-\Theta \tan\varphi} \tan(\tfrac{1}{4}\pi - \tfrac{1}{2}\varphi) \right]}{\left[2\sin\Theta + e^{\Theta \tan\varphi} \tan(\tfrac{1}{4}\pi + \tfrac{1}{2}\varphi) + e^{-\Theta \tan\varphi} \tan(\tfrac{1}{4}\pi - \tfrac{1}{2}\varphi) \right]} \tag{6.89}$$

agreeing with the expression obtained previously by Shield (1953).

For the special case of Tresca material for which $\varphi = 0$, [6.89] reduces to the simple form:

$$\cos(2\beta_0 - \Theta) = \frac{\cos\Theta}{1 + \sin\Theta} \tag{6.90}$$

agreeing with the result obtained previously by Hill et al. (1947).

Numerical results (Fig. 6.30)
The variation of the angle Θ with the angle β_0, obtained from [6.89] and

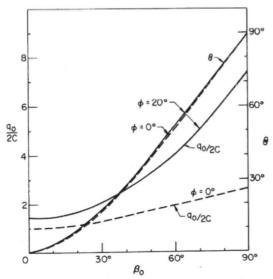

Fig. 6.30. Variation of pressure q_0 on a wedge with semi-angle β_0 and spiral angle Θ (Shield, 1953).

TABLE 6.10

Variation of angle β_0 with the angle Θ ([6.89])

Θ	$\varphi = 0°$	$\varphi = 10°$	$\varphi = 20°$	$\varphi = 30°$	$\varphi = 40°$	$\varphi = 50°$	$\varphi = 60°$
0	0	0	0	0	0	0	0
5	14.3	14.2	13.9	13.4	12.8	11.8	10.7
10	21.5	21.3	20.9	20.2	19.2	17.9	16.3
15	27.4	27.3	26.8	25.9	24.7	23.1	21.1
20	32.8	32.6	32.0	31.0	29.6	27.8	25.6
25	37.7	37.5	36.8	35.7	34.2	32.3	29.9
30	42.4	42.1	41.4	40.2	38.6	36.6	34.3
35	46.8	46.6	45.8	44.6	42.9	40.9	38.6
40	51.1	50.8	50.1	48.8	47.1	45.1	43.0
45	55.3	55.0	54.2	53.0	51.3	49.4	47.4
50	59.3	59.1	58.3	57.1	55.5	53.7	51.9
55	63.3	63.1	62.3	61.2	59.7	58.1	56.5
60	67.2	67.0	66.3	65.2	63.9	62.5	61.1
65	71.1	70.9	70.3	69.3	68.1	66.9	65.8
70	74.9	74.7	74.2	73.4	72.4	71.4	70.6
75	78.7	78.6	78.1	77.5	76.7	76.0	75.4
80	82.5	82.4	82.1	81.6	81.1	80.6	80.2
85	86.2	86.2	86.0	85.8	85.5	85.3	85.1
90	90.0	90.0	90.0	90.0	90.0	90.0	90.0

[6.90], is shown in Fig. 6.30 for $\varphi = 20°$ and $\varphi = 0°$, and is also tabulated in Table 6.10 for various values of φ. Substituting the values of Θ thus obtained into [6.80] and [6.81] the variation of q_0 is also plotted as a function of β_0 in Fig. 6.30. Once the angle Θ is known, the length of the lip l and thus the total downward driving force F of the wedge footing can be computed directly from [6.84a] and [6.83], respectively.

Summary and historical notes

The problem of the indentation of a semi-infinite medium by a smooth, rigid wedge under conditions of plane strain was first solved by Hill et al. (1947) for a perfectly plastic Tresca material. An approximate method of solution to this problem was given later by Hodge (1950). Following the work of Hill et al., Shield (1953) obtained the solution of the same problem for the more general case of Coulomb material. Paslay et al. (1968) presented a solution of this problem using a special yield surface which was believed to be particularly appropriate for rock subjected to high hydrostatic pressure such as those found during deep drilling below the earth's surface.

The essential feature of this problem is that the plastic region changes in such a way that its configuration always retains *geometrical similarity* to some initial state.

The simplest examples of this type of deformation other than the wedge footing considered above are problems of the expansion of cylindrical and spherical cavities in unbounded space, beginning from zero radius.

The problem considered here is limited to the case of a symmetric, rigid (non-deformable) wedge. Friction at the contact surface is neglected (the surface is lubricated). A number of other problems of this type of geometric similarity (oblique indentation by a rigid wedge, compression of a wedge by a rigid plane, etc.) have been studied by Hill (1950) and other authors for a perfectly plastic Tresca material.

6.7. BEARING CAPACITY DETERMINATION BY SLIP-LINE METHOD

6.7.1. Introduction

Basic concept

In deriving the slip-line solutions, the soil is assumed to be a *rigid–plastic* material in which slip or yielding occurs in plane strain when the stresses satisfy the Coulomb criterion, in the usual notation:

$$\tfrac{1}{2}(\sigma_x + \sigma_y)\sin\varphi - [\tfrac{1}{4}(\sigma_x - \sigma_y)^2 + \tau_{xy}^2]^{1/2} + c\cos\varphi = 0 \qquad [6.91]$$

This equation and the two equations of stress equilibrium:

$$\frac{\partial\sigma_x}{\partial x} + \frac{\partial\tau_{xy}}{\partial y} = 0$$

and:

$$\frac{\partial\sigma_y}{\partial y} + \frac{\partial\tau_{yx}}{\partial x} = -\gamma \qquad\qquad [6.92a,b]$$

form a hyperbolic system of equations for the determination of the stresses σ_x, σ_y and τ_{xy}. Since there are the same number of equations as unknown stress components, [6.91] and [6.92] are often sufficient to make the stresses *statically determinate* without considering a stress–strain relation of soil. Most of the bearing capacity problems considered here are statically determinate in this sense. In general, however, in many problems the boundary conditions involve rates of displacement, the stress boundary conditions above are not sufficient to make these problems statically determinate and the use of a stress–strain relation is necessary in order to obtain solutions of such problems. The term static determinacy in such cases is then misleading.

Slip-line equations

Referring now to Fig. 4.12(a), the two slip lines (or characteristic lines) are inclined at an angle $(\frac{1}{4}\pi - \frac{1}{2}\varphi)$ to the direction of the maximum compressive stress σ_{max} and will be called the first and second slip lines, with the convention that the direction of the first slip line is obtained from the direction of σ_{max} by a counter-clockwise rotation of amount $(\frac{1}{4}\pi - \frac{1}{2}\varphi)$. The angle of inclination of the first slip line to the x-axis is denoted by θ and the angle of the inclination of σ_{max} to the x-axis is denoted by ψ. The equations for the two *slip lines* are given by:

$$\frac{dy}{dx} = \tan\theta = \tan\left[\psi \pm (\tfrac{1}{4}\pi - \tfrac{1}{2}\varphi)\right] \tag{6.93}$$

Using the yielding condition [6.91], the three stress components σ_x, σ_y and τ_{xy} can be expressed in terms of two dependent variables s and ψ (compression is positive):

$$\sigma_x = (s\csc\varphi - c\cot\varphi) + s\cos 2\psi$$
$$\sigma_y = (s\csc\varphi - c\cot\varphi) - s\cos 2\psi$$
$$\tau_{xy} = s\sin 2\psi \tag{6.94a,b,c}$$

in which s is the radius of the Mohr's circle for stress. When the stress components given by [6.94] are substituted into the equilibrium equations [6.92], the variation of the two dependent variables s and ψ along the two slip-line curves [6.93] can be expressed in differential form:

$$\cot\varphi\, ds \pm 2s\, d\psi - (\cos\varphi\, dy \pm \sin\varphi\, dx)\gamma = 0 \tag{6.95}$$

which were first obtained by Kötter (1888). For the case of a weightless soil $(\gamma = 0)$, [6.95] can be integrated and reduced to the simple form:

$$\tfrac{1}{2}\cot\varphi\ln s \pm \psi = \text{constant} \tag{6.96}$$

in which ln denotes natural logarithm (or logarithm with base e).

Equations [6.95] together with the stress boundary conditions can now be used to obtain the stresses in the soil beneath a strip footing at the point of incipient failure. In the case of strip footing on a weightless soil, Prandtl (1920) obtained the closed form solution as given by [6.49] by integrating [6.96]. However, the important inclusion of soil weight γ considerably complicates the mathematical solution. Consequently, many approximate methods have been used and proposed for the bearing pressure analyses of footings. One approach is the approximate integration of the equilibrium equations [6.92] using the yield condition [6.91]. An alternative method called slip-line method is to integrate [6.95] along the slip lines. A brief review of some of this work follows.

Methods of solution

A widely used procedure of approximate integration of the governing differential equations is to use a finite difference approximation. Using such a procedure to the slip-line equations [6.93] and [6.95] Sokolovskii (1965) has obtained solutions to a number of important problems. His work has been translated and appeared in book form ("Statics of Granular Media", 1965). This finite difference approximation was used also by Cox (1962) and by Lundgren and Mortensen (1953) to obtain solutions to the surface footing problem. De Jong (1957) has discussed a graphical method for obtaining the slip-line solution for ponderable soils which is an extension of Prager's (1953) method for a Tresca material ($\varphi = 0$).

Other forms of approximate integration have also been successfully applied to the footing problem. Sokolovskii (1965) investigated certain footing problems using a solution in the form of a product of a known and an unknown function and integrated the resulting ordinary differential equations. Spencer (1962) has derived a method for obtaining a series expansion of the maximum bearing capacity in powers of a dimensionless soil weight parameter $G = \gamma B/2c$, the first-order term being derived directly from Prandtl's (1920) solution for a weightless soil. This technique is called the *perturbation method*. Also, series expansion techniques have been used by Dembicki et al. (1964).

6.7.2. Surface footing

Smooth footing with parameters G and φ (Cox, 1962)

The most important numerical results obtained by Cox (1962) are the values of the average bearing capacity pressure q_0 as a function of a dimensionless soil weight parameter, $G = \gamma B/2c$, and the angle of internal friction of the soil, φ. Table 6.11 gives values for the average bearing pressure q_0. Fig. 6.31 shows the calculated slip-line nets in the region of the stress field that has to be determined in order to obtain the stresses on the footing. The ratio AE/AB (see Fig. 6.31) gives the relative size, as G and φ vary, of the actual slip-line zone. Values of this ratio are also given in Table 6.11. This table shows that the ratio AE/AB decreases markedly as G increases, provided that φ is not zero.

Referring again to Fig. 6.31, in the case of a *weightless* soil, the slip-line net beneath the footing (region ABC) and beneath the free surface (region BDE) are straight. The 90°-fan of slip-line net BCD is an undistorted network of straight lines and logarithmic spirals. The kinematics of the failure mode associated with this stress field has been known as Hill mechanism and shown in Fig. 3.22(d). When the weight of a soil is included, it is seen from Fig. 6.31 that certain features of the slip-line field change radically. Now, the 90°-fan BCD and the region ABC beneath the footing are distorted. The lines AC and BC are curved and they are no longer equal to each other. Since the characteristics or slip lines of the stress and velocity fields coincide for a perfectly plastic Coulomb material with associated flow rule (Shield, 1953), a velocity field may be constructed if a stress field is known. This

TABLE 6.11

Values of the bearing capacity pressure, $q_0/c = N_\gamma G$ and the ratio AE/AB (see fig. 6.31). Smooth footing, Cox (1962).

φ (°)	Bearing pressure, q_0/c		$(G = \gamma B/2c)$		
	$G = 0$	$G = 0.01$	$G = 0.1$	$G = 1$	$G = 10$
0	5.14	5.14	5.14	5.14	5.14
10	8.34	8.35	8.42	9.02	13.6
20	14.83	14.87	15.2	17.9	37.8
30	30.14	30.29	31.6	42.9	127
40	75.31	76.13	83.0	139	574
	Ratio AE/AB (see Fig. 6.31)				
0	2.00	2.00	2.00	2.00	2.00
10	2.57	2.57	2.55	2.40	1.94
20	3.53	3.52	3.46	3.09	2.41
30	5.29	5.27	5.09	4.31	3.36
40	9.01	8.94	8.37	6.69	5.36

suggests that the partial stress field of Fig. 6.31 will provide some guide to the actual mode of failure. The Hill type of mechanism shown in Fig. 6.8 is clearly an approximation to the actual mode of failure for a footing on a ponderable soil. Solutions obtained from the Hill type of mechanism can therefore at best give only upper bounds to the bearing capacity pressures of a footing on a ponderable soil.

Rough footing with parameters δ and φ (Hansen and Christensen, 1969)
The bearing capacity of a strip footing on a cohesionless sand and for any

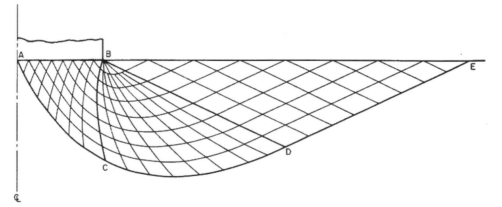

Fig. 6.31. Calculated slip-line net for a strip footing, $\varphi = 40°$, $G = 10$ (Cox, 1962).

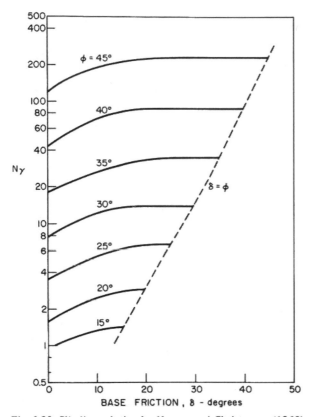

Fig. 6.32. Slip-line solution by Hansen and Christensen (1969).

value of base roughness δ has been studied by Hansen and Christensen (1969) by means of a computer program based on the method of solution used by Lundgren and Mortensen (1953). The numerical values of the bearing capacity factor N_γ obtained are shown in Fig. 6.32. For $\delta = \varphi$, the case of a perfectly rough footing is represented, while the case $\delta = 0$ represents a perfectly smooth footing. For $0 < \delta < \varphi$, the bearing capacity of a footing with finite base friction is represented. From Fig. 6.32, it can be seen that the values of N_γ decrease as the value of δ decreases, at first slowly near $\delta = \varphi$, and then at faster rate when δ decreases, until at $\delta = 0$, the N_γ takes values about half the values found for $\delta = \varphi$. For a rough base and $\varphi = 30°$, Hansen and Christensen's value of $N_\gamma = 15$ may be compared with the value $N_\gamma = 14.8$ obtained by Lundgren and Mortensen (1953).

Partial stress solutions

It must be emphasized that all the slip-line nets obtained by these authors have been derived from pure stress consideration. Therefore, they are statically admissible around the footing, but it is not shown that the stress fields around the footing can be extended *throughout* the body in a statically admissible manner. Also, the stress field around the footing may not be kinematically admissible. Any such value obtained for the bearing capacity pressure is therefore not necessarily a lower bound for the pressure nor is it known when it is an upper bound. These incomplete stress solutions obtained by integrating the slip-line equations [6.95] along the slip lines are called *partial stress solutions*.

Examples of partial stress solutions of the Hill type for smooth and rough footings are those given by Sokolovskii (1965), Cox (1962), and Hansen and Christensen (1969). Example of a partial stress solution of the Prandtl type is that given by Lundgren and Mortensen (1953). Recently, an admissible velocity field with no slip across the footing base has been obtained by Larkin (1972) for Prandtl-type partial stress fields with $\varphi = 40°$. This establishes an upper bound for a rough footing and indicates that the Lundgren and Mortensen solution is a valid upper bound. The upper-bound values of the average bearing capacity pressure q_0 obtained by Larkin (1972) are listed in Table 6.12. For comparison, available average bearing pressures obtained by Sokolovskii and Cox for perfectly smooth footings are also shown in this table. The footing roughness is seen to have little effect for a strongly cohesive soil as both solutions approach the same value when the cohesion is relatively large. This is consistent with the solution obtained previously for a weightless soil for which footing base roughness has no effect. For a weakly cohesive soil, however, the bearing pressure for a rough footing is significantly greater than that for a smooth footing.

6.7.3. Shallow footing

The bearing capacity of very shallow, perfectly *smooth* strip footings are obtained by Larkin (1968) based on the method of solution used previously by Cox (1962). Bearing capacity calculations are carried out for cohesionless soils with angle of internal friction $\varphi = 30°$ and $40°$. The results for the bearing capacity factor N_γ are given in Table 6.13. The increase of the N_γ-values with depth of burial D is seen sufficiently large to indicate that the small footing settlement prior to failure is a significant factor in a footing experiment and should be taken into account when comparing theoretical predictions with experimentally observed bearing capacity values of a *smooth* surface footing.

The bearing capacity of very shallow, *rough* strip footings has not been studied by the method of slip line. The sensitivity of the bearing capacity of rough footings to depth of burial may not be as sensitive as that of the smooth case. Since most footings in practice are probably rough, the small footing settlement prior to failure may *not* be a significant factor in practical applications where the roughness of the

TABLE 6.12

Values of the bearing capacity factor, $N_\gamma = q_0/\gamma(B/2)$, ($\varphi = 40°$; Larkin, 1972)

Footing base condition	$G = \gamma B/2c$										
	100	*40*	*20*	*18.8*	*10*	*4*	*2*	*1*	*0.4*	*0.2*	*0.1*
Rough* (no slip)	89.2	91.4	95.1	95.6	102	121	146	192	317	520	905
Smooth	45.3	47.7	51.2	–	57.4	73.2	96.5	139	261	452	830

* Rigorous upper bounds

TABLE 6.13

Values of the bearing capacity factor, $N_\gamma = q_0/\gamma(B/2)$ for a shallow footing (smooth footing, Larkin, 1968)

Depth ratio D/B	*0.005*	*0.0125*	*0.025*	*0.05*	*0.125*	*0.25*	*0.5*
$\varphi = 30°$	8.2	8.7	9.6	11	15	20	31
$\varphi = 40°$	45	47	50	56	70	90	127

footing greatly increases the value of N_γ, at least in the case of a footing on cohesionless soil.

6.8. BEARING CAPACITY OF FOOTINGS ON NONHOMOGENEOUS, ANISOTROPIC SOILS

The bearing capacity of a footing on a nonhomogeneous and/or anisotropic soil has frequently appeared in engineering practice but design data to assess the bearing capacity value of the footing are very scant. This lack of detailed information is due largely to the enormous difficulty in obtaining *exact* solutions which are needed in order to compare with the results obtained by any simplified methods. Herein, as in previous works on the bearing capacity of homogeneous and isotropic soils, the limit analysis method is apt in fact to be of greatest use. A simple example, with specially chosen angles to eliminate all calculation, is shown in Fig. 6.33. To avoid

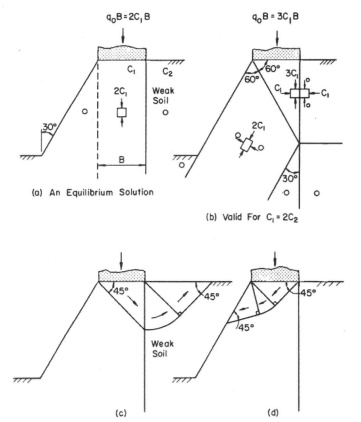

Fig. 6.33. Inhomogeneous soil with $\varphi = 0$ (Drucker, 1953).

obscuring the basic points, the internal friction angle φ is taken as zero, and the weight of the soil is considered negligible. The purpose of this simple example is to demonstrate the power of the limit analysis method in such a situation. More complex applications will then be followed.

Example 6.1: Footings on a nonhomogeneous, cohesive slope with $\varphi = 0$, Fig. 6.33 (Drucker, 1953)

The problem is to find the ultimate bearing capacity pressure q_0 of the footing shown in Fig. 6.33 for various ratios of the cohesion c_2 of the weak soil to the cohesion c_1 of the stronger soil which lies under the footing. Using the failure mechanism in the weak soil as shown in Fig. 6.33(c) which consists of a radial shear zone (Fig. 3.19) and two rigid triangles, we obtain an upper-bound solution to this problem:

$$q_0 \leqslant 2\,c_1 + (2 + \tfrac{1}{2}\pi)\,c_2 \qquad\qquad [6.97]$$

If the weak soil has no strength, $c_2 = 0$, the simple discontinuous stress field shown in Fig. 6.33(a) and the upper bound (6.97) give the obvious answer $q_0 = 2c_1$. Suppose next that $c_2 = c_1/2$. The discontinuous stress pattern of Fig. 6.33(b) which was developed previously in Fig. 4.19 coupled with the alternative failure mechanism shown in Fig. 6.33(d), give:

$$3\,c_1 \leqslant q_0 \leqslant (2 + \tfrac{1}{3}\pi)\,c_1 \qquad\qquad [6.98]$$

The first upper-bound expression [6.97] leads to a higher value than the alternative upper bound (Fig. 6.33d). Any further improvement in the values of upper and lower bounds will be of no real help as the two bounds are already very close. All bounds apply equally well to materials which cannot take tension.

6.8.1. Anisotropy and nonhomogeneity in a two-layer clay

Anisotropy (Fig. 6.34c)

The cohesion c in most clays exhibits some *anisotropy* with respect to shearing direction, and some *nonhomogeneity* with respect to depth. The variation of cohesion with respect to direction at a particular point in a soil mass has been studied by Casagrande and Carrillo (1954). It has been found that the variation of cohesion with direction approximates to the curve shown in Fig. 6.34(c). In this figure, c_h and c_v are the cohesions in the horizontal and vertical directions, respectively. The vertical cohesion c_v, for example, is determined by taking vertical samples at any site and being tested with the major principal stress applied in the same direction. The cohesion c_h and c_v may be termed as *principal cohesions*. The ratio of the principal cohesions, c_v/c_h, is approximately constant at any site. For the isotropic

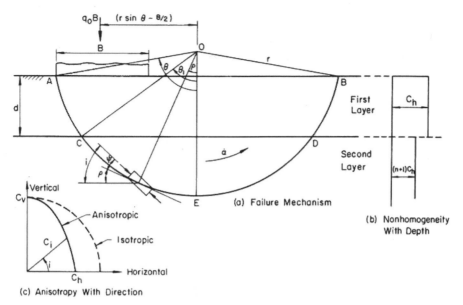

Fig. 6.34. A slip circle in a two-layer clay ($\varphi = 0$).

case, $c_v = c_h = c$ (see Fig. 6.34c). In practice, the value of the ratio c_v/c_h is found between about 0.75 and 2.0. A value of less than 1 for this ratio is obtained mostly in the case of overconsolidated clays. The angle ψ as shown in Fig. 6.34(a) is the angle between the failure plane and the direction of the major principal stress at the point under consideration. This angle, according to Lo's tests (1965), is independent of the angle of rotation of the major principal stress. The cohesion, c_i, with the major principal stress inclined at an angle i with the horizontal (Fig. 6.34c) may be written in the form:

$$c_i = c_h + (c_v - c_h) \sin^2 i \qquad\qquad [6.99]$$

This particular type of anisotropy of cohesion with direction at a point is used in the following calculations.

Nonhomogeneity (Figs. 6.34b and 6.35)

The cohesion c at any site also increases or decreases with depth. The most simple case of *nonhomogeneity* of cohesion with depth in a two-layer clay is that the horizontal cohesion c_h in each layer is constant with depth as shown in Fig. 6.34(b) where the factor n represents the relative strengths of the two layers. A more general case of nonhomogeneity of cohesion is that of a linear variation of horizontal cohesion c_h in each layer with depth z as shown diagramatically in Fig. 6.35 where λ and λ_1 denote the rate of increase or decrease of shear strength with depth. In the following computations, the case of constant cohesion in each

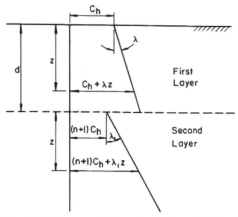

Fig. 6.35. Linear variation of cohesion with depth.

layer (Fig. 6.34b) is considered first and then followed by the case of linear variation of cohesion with depth (Fig. 6.35).

6.8.2. Bearing capacity of footings on two-layer clays

Failure mechanism

The *exact* value of the bearing capacity factor N_c for a strip footing on a homogeneous, isotropic clay ($\varphi = 0$) is $(2 + \pi) = 5.14$, according to our previous upper- and lower-bound analyses (Figs. 4.22a and b). An upper-bound value of 5.53 can be easily obtained by assuming a circular failure surface as shown in Fig. 3.3(b). Since this upper-bound value is only slightly higher than that of the exact solution, and in view of its simplicity in the mathematical analysis, this mechanism will be used in the following computations.

Case 1: Constant cohesion in each layer coupled with the anisotropy (Fig. 6.34)

For the circular mechanism with center O, angular velocity $\dot{\alpha}$, as illustrated in Fig. 6.34(a), the rate of external work is:

$$q_0 B(r \sin \theta - \tfrac{1}{2}B)\dot{\alpha} \qquad\qquad [6.100]$$

where r is the radius of the discontinuity surface, and θ the angle between the vertical line OE and the line OA. Both r and θ are as yet undefined. The rate of internal work is given by:

$$2 \int_{\theta_1}^{\theta} c_i(r\dot{\alpha})\,(rd\rho) + 2 \int_0^{\theta_1} (n+1)\,c_i(r\dot{\alpha})\,(rd\rho) \qquad\qquad [6.101]$$

in which: $\theta_1 = \cos^{-1}\left(\cos \theta + \dfrac{d}{r}\right)$ \qquad\qquad [6.102]

From Fig. 6.34(a), it can be seen that $i = \rho + \psi$ for left-hand side part of the discontinuous surface ACE and $i = \rho - \psi$ for right-hand side surface BDE. Hence, from [6.99]:

$$c_i = c_h + (c_v - c_h) \sin^2(\rho + \psi) \qquad\qquad [6.103a]$$

for part ACE, and:

$$c_i = c_h + (c_v - c_h) \sin^2(\rho - \psi). \qquad\qquad [6.103b]$$

for part BDE.

Substituting for c_i from [6.103] into [6.101] and equating the resulting expression to [6.100] give:

$$q_0 B(r \sin\theta - \tfrac{1}{2}B) = \int_{\theta_1}^{\ell} [c_h + (c_v - c_h) \sin^2(\rho + \psi)] \, r^2 d\rho$$

$$+ \int_{\theta_1}^{\theta} [c_h + (c_v - c_h) \sin^2(\rho - \psi)] r^2 d\rho$$

$$+ \int_{0}^{\theta_1} (n + 1) [c_h + (c_v - c_h) \sin^2(\rho + \psi)] r^2 d\rho$$

$$+ \int_{0}^{\theta_1} (n + 1) [c_h + (c_v - c_h) \sin^2(\rho - \psi)] r^2 d\rho \qquad [6.104]$$

Integrating and dividing both sides by $\tfrac{1}{4}Bc_v$ and simplifying, we obtain an upper-bound expression for the bearing capacity factor N_c:

$$N_c(r, \theta) = \frac{q_0}{c_v} = \frac{\left(\frac{r}{B}\right)^2}{\left(\frac{c_v}{c_h}\right)\left[\left(\frac{r}{B}\right)\sin\theta - \frac{1}{2}\right]} \left\{ 2\theta + 2n\,\theta_1 + \left[\left(\frac{c_v}{c_h}\right) - 1\right]\theta \right.$$

$$+ n\left[\left(\frac{c_v}{c_h}\right) - 1\right]\theta_1 - \frac{1}{4}\left[\left(\frac{c_v}{c_h}\right) - 1\right][\sin 2(\theta + \psi) + \sin 2 \cdot (\theta - \psi)]$$

$$\left. - \frac{1}{4}n\left[\left(\frac{c_v}{c_h}\right) - 1\right][\sin 2(\theta_1 + \psi) + \sin 2(\theta_1 - \psi)] \right\} \qquad [6.105]$$

agreeing with the results obtained previously by Reddy and Srinivasan (1967) using the method of limit equilibrium. For the special case of isotropic soil ($c_v = c_h = c$), [6.105] reduces to:

$$N_c(r, \theta) = \frac{q_0}{c} = 2\left(\frac{r}{B}\right)^2 \left\{ \frac{\theta + n\,\theta_i}{\left(\frac{r}{B}\right)\sin\theta - \frac{1}{2}} \right\}$$ [6.106]

In [6.105], r/B and θ are variables and n, c_v/c_h and ψ are parameters whose values are fixed for a given problem. The angle θ_1 defines the boundary between the first and second layer and is related to θ and d/r by [6.102].

Hence, for a least upper-bound value of $N_c(r,\theta)$ corresponding to the circle, the following conditions must be satisfied:

$$\frac{\partial N_c}{\partial\theta} = 0, \qquad \frac{\partial N_c}{\partial r} = 0$$ [6.107]

For the homogeneous and isotropic case ($n = 0$ and $c_v = c_h = c$) these two equations may be solved analytically and give a value of $N_c = 5.53$. When $n \neq 0$ and $c_v \neq c_h$, the analytical solution of [6.107] is difficult and the solution of it for any given values of c_v/c_h, d/B, and n may be found by using a computer and by substituting the solutions so obtained in [6.105]; N_c may then be obtained.

Case 2: Linear variation of cohesion in each layer (Fig. 6.35) coupled with the anisotropy (Fig. 6.34c)

Referring to Figs. 6.34(a) and 6.35, since $z = r(\cos\rho - \cos\theta)$ for the first layer and $z = r(\cos\rho - \cos\theta_1)$ for the second layer, it follows from [6.99] that the cohesion at any depth z, in a direction making an angle i with the horizontal is:

First layer

$$c_i = (c_h + \lambda z) + \left[\left(\frac{c_v}{c_h}\right)(c_h + \lambda z) - (c_h + \lambda z)\right]\sin^2 i$$

$$= [c_h + \lambda r(\cos\rho - \cos\theta)]\left[1 + \left(\frac{c_v}{c_h} - 1\right)\sin^2 i\right]$$ [6.108a]

Second layer

$$c_i = [(n+1)c_h + \lambda_1 r(\cos\rho - \cos\theta_1)]\left[1 + \left(\frac{c_v}{c_h} - 1\right)\sin^2 i\right]$$ [6.108b]

Substituting for c_i in [6.101] from [6.108], equating the resulting expression to [6.100], an upper-bound solution for N_c may be obtained as in the previous case containing the additional parameters λ and λ_1.

Considering only one layer of anisotropic clay with the principal cohesions c_h

and c_v varying linearly with depth, the expression for N_c reduces to:

$$N_c(r, \theta) = \frac{q_0}{c_v} = \frac{\left(\frac{r}{B}\right)^2}{\left(\frac{c_v}{c_h}\right)\left(\frac{r}{B} \sin \theta - \frac{1}{2}\right)} \left\{ 2\theta + \frac{1}{4}\left[\left(\frac{c_v}{c_h}\right) - 1\right][4\theta - \sin 2(\theta + \psi)\right.$$

$$- \sin 2(\theta - \psi)] + 2\left(\frac{\lambda B}{c_h}\right)\left(\frac{r}{B}\right)(\sin \theta - \theta \cos \theta) + \left(\frac{\lambda B}{c_h}\right)\left(\frac{r}{B}\right)\left(\frac{c_v}{c_h} - 1\right)\left[\frac{1}{3} \cos \psi\right.$$

$$[\sin^3 (\theta + \psi) + \sin^3 (\theta - \psi)] - \frac{1}{12} \sin \psi \, [9 \cos (\theta + \psi) - 9 \cos (\theta - \psi)$$

$$\left.\left.- \cos 3 (\theta + \psi) + \cos 3 (\theta - \psi)] - \frac{1}{4} \cos \theta \, [4\theta - \sin 2 (\theta + \psi) - \sin 2 (\theta - \psi)]\right]\right\}$$

$$[6.109]$$

As in the previous case, the expression for N_c contains two variables, r/B and θ, the other parameters being c_v/c_h, ψ, and $\lambda B/c_h$. The values of θ and r/B which give minimum N_c for any given values of c_v/c_h, ψ and $\lambda B/c_h$ may be obtained as in the previous case and N_c may be found from [6.109]. The average bearing capacity pressure is then calculated from $q_0 = c_v N_c$.

Numerical results
The minimization procedure can be programmed for a digital computer. Optimum values of N_c so obtained are presented herein. For Case 2, only the results for the problem of a footing resting on a single layer of clay are given. All the following results are based on a value of $\psi = 35°$, which is the average value found by Lo from experiments (1965).

Case 1 (Fig. 6.36)
The values of the bearing capacity factor N_c obtained from Case 1 described above are given in Fig. 6.36 for three values of $c_v/c_h = 0.8$, 1.0 and 1.2. From these curves, N_c can be determined when the values of n, c_v/c_h and d/B are known. The curves corresponding to different values of anisotropy ratio c_v/c_h are similar in shape to that for the isotropic case. The following points in connection with these curves are worth noting:

(1) When d/B is large, or $d/B = 0$ or $n = 0$ these conditions represent various homogeneous cases and the values of N_c are 6.4, 5.53 and 5.0 for $c_v/c_h = 0.8$, 1.0 and 1.2, respectively. All the curves are bounded by these extreme cases. Take the isotropic case, for example. All the curves are seen lying inside the regions bounded by the horizontal line through $N_c = 5.53$ and the straight line through the origin and the point $(N_c = 5.53, n = 0)$.

(2) When n is negative, the lower layer is weaker than the upper one. The value of N_c for any given value of d/B increases as n increases until it reaches the limiting

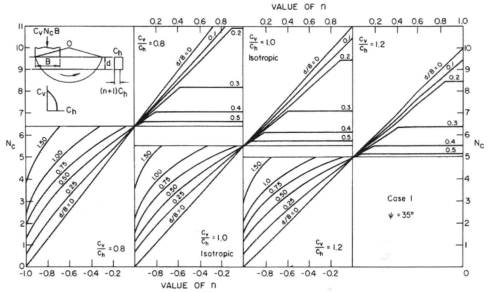

Fig. 6.36. Bearing capacity factor N_c for Case 1.

value of the horizontal boundary line and it then remains at this limiting value. At this limiting point, the slip circles lie entirely within the upper layer, and any further increase in the strength of the lower layer does not increase the bearing capacity.

(3) When n is positive, the lower layer has a greater strength than the upper one. For a particular value of d/B, the bearing capacity raises as the relative strength of the lower layer rises, but at the same time a smaller proportion of the total length of the slip surface cuts through the lower stronger layer. At a limiting point, the slip circle becomes tangential to the upper surface of the lower layer, and after this, any further increase in the strength of the lower layer will not influence the bearing capacity, as the slip surface is now lying entirely inside the upper layer. This is represented by the sudden change of the curve to the horizontal line at a certain value of n. After this point, the value of N_c for any particular value of d/B is unaltered as n is increased.

Case 2 (Fig. 6.37)

The bearing capacity factors N_c for the case of a single layer with linear variation in cohesion with depth, are presented in Fig. 6.37 for various values of $\lambda B/c_h$, which is a measure of the cohesion variation with depth, and for three values of $c_v/c_h = 0.8$, 1.0 and 1.2. When $\lambda B/c_h = 0$, the condition represents the homogeneous case and the values of N_c are 6.4, 5.53 and 5.0 for $c_v/c_h = 0.8$, 1.0 and 1.2, respectively. The value of N_c is seen to increase almost linearly as the value of $\lambda B/c_h$ increases.

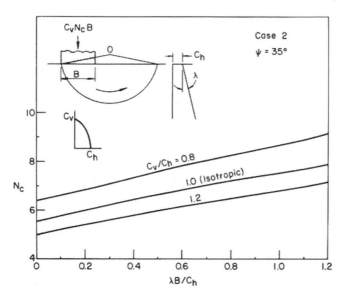

Fig. 6.37. Bearing capacity factor N_c for Case 2.

6.9. SUMMARY AND CONCLUSIONS

The upper-bound technique of limit analysis is used herein to develop approximate solutions for the bearing capacity of cohesive soils with weight. Solutions are presented for smooth and rough and surface, shallow and deep footings. Soil is treated as a perfectly plastic medium with the associated flow rule. The limit analysis solutions for smooth surface footings are shown to compare favorably with slip-line solutions. Meyerhof's solutions and the limit analysis solutions for rough, shallow and deep footings are shown to agree remarkably well.

It has been shown that the upper-bound technique of limit analysis can predict bearing capacities of cohesive ponderable soils with internal friction to within a reasonable degree of accuracy, for φ ranging from $0°$ to $40°$ and G ranging from 0 to 5. At the least, it can be said that the results compare favorably with existing limit equilibrium solutions.

The most forceful argument for the adoption of the limit analysis method is the fact that its rational basis allows it to be conveniently extended to more complex bearing capacity problems. For example, as demonstrated here, the limit analysis method can be easily adopted to the solution of layered soils. More important, the method can be used to obtain upper- and lower-bound solutions of the three-dimensional bearing capacity problem where exact solutions of the equations of plasticity are all but impossible except for the most elementary of problems. The application of the limit analysis method to some three-dimensional bearing capacity problems will be presented in the following Chapter.

Chapter 7

BEARING CAPACITY OF SQUARE, RECTANGULAR AND CIRCULAR FOOTINGS

7.1. INTRODUCTION

Limit analysis is applied in this Chapter to obtain upper and lower bounds for the average bearing capacity pressure q_0 of square, rectangular and circular footings on a purely cohesive soil ($\varphi = 0$). The individual problems connected with this application will be taken up in the following sequence. First we consider the bearing capacity of surface footings on a semi-infinite mass of soil (sections 7.2 and 7.3). Then we solve in succession the following two problems: (1) computation of the bearing capacity of footings on a finite block (sections 7.4 and 7.5); (2) computation of the bearing capacity of surface footings on a semi-infinite layer of soil resting on a rough rigid base (sections 7.6 and 7.7). When dealing with these problems the contribution of soil weight on the bearing capacity of footings will be disregarded in most cases. The influence of soil weight and the case $\varphi > 0$ on the bearing capacity of circular footing will be discussed in section 7.8.

In the following work, we are considering the bearing capacity of a perfectly plastic soil loaded normally and centrally by a flat-ended rigid footing. Except in section 7.8, the soil is assumed to be an *ideally* cohesive medium for which $\varphi = 0$ (Tresca yield condition). This condition approximates to the behavior of a frictionless cohesive soil such as clay. It is further assumed that the force on the footing is increased until penetration occurs as a result of plastic flow in the soil. The force on the end of the footing not in contact with the soil may be a concentrated force or may be distributed in any manner over the end of the footing as long as the resultant force over the footing is a normal force passing through the centroid of the contact area.

In the three-dimensional footing problems considered, the upper bounds obtained are sometimes applicable only to a smooth footing, but all the lower bounds are applicable to either rough or smooth footings.

7.2. SQUARE, RECTANGULAR AND CIRCULAR FOOTINGS ON A SEMI-INFINITE MEDIUM – LOWER BOUNDS

In this section lower bounds are obtained for the value of the average bearing

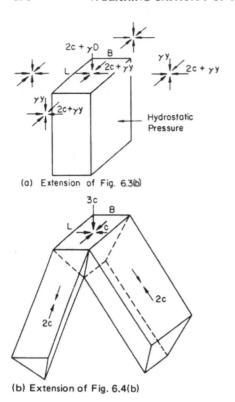

(a) Extension of Fig. 6.3(b)

(b) Extension of Fig. 6.4(b)

Fig. 7.1. Three-dimensional stress field for a bearing capacity problem, $\varphi = 0$.

capacity pressure q_0 over the rectangular area of contact which can also be used to give lower bounds for the case of circular and square footings.

A three-dimensional stress field for the rectangular footing is shown in Fig. 7.1, which is a direct extension of the two-dimensional stress field in Fig. 6.3(b) and Fig. 6.4(b). The stress field of Fig. 7.1(b) establishes $3c$ as a lower bound for the ultimate bearing pressure. It is easily verified that superposition of the stress field of Fig. 7.1(a) on that of Fig. 7.1(b) does not lead to stresses in excess of the yield limit. Thus, $5c + \gamma D$ where D is the vertical distance between the surface of the ground and the base of the footing is a lower bound for the ultimate bearing pressure in the three-dimensional rectangular or square footing.

The stress field in Fig. 7.1 can be modified, so that it applies to a circular area or any area of contact which is *convex*. This follows directly from the fact that any convex area of contact can be closely approximated by an inscribed polygon. The associated stress fields below the area of contact, which may be visualized as supporting "legs" on each side of the polygon (in a manner essentially the same as shown in Fig. 7.1), do not overlap one another as long as the contact area is convex. Since the omission of shearing stresses over the area of contact does not in itself require that the footing surface be smooth, it follows that the value $(5c + \gamma D)$ is a lower bound for either rough or smooth footings when the area is convex for which square, rectangular and circular footings are its special cases.

We now describe a stress field which provides better lower bounds than that obtained above. The improvement in the lower bounds is less than 2.84% and is therefore not of great interest insofar as numerical results are concerned. However, the result that the average bearing capacity pressure over a square footing is *not less* than the average bearing capacity pressure over a long rectangular footing (strip footing, plane strain condition), is of great interest. The stress field which can be used for this improvement is the possible extension into three dimensions of the stress field previously used (Fig. 6.28) to show the validity of the bearing capacity value:

$$q_0 = c \cot \varphi \, [e^{\pi \tan \varphi} \tan^2 (\tfrac{1}{4}\pi + \tfrac{1}{2}\varphi) - 1] \qquad [7.1]$$

to the strip surface footing on a $c - \varphi$ *weightless* soil.

The required extension of the stress field (Fig. 6.28) has been shown by Shield (1955c) to be possible for $\varphi = 0$. Since the addition of the hydrostatic pressure γy due to surcharge γD (see Fig. 6.3) will not effect the yielding of an ideally cohesive soil ($\varphi = 0$), it follows that the value $(2 + \pi)c + \gamma D = 5.14c + \gamma D$ is a lower bound for the ultimate bearing capacity pressure in the three-dimensional footing problems.

In view of the fact that the required extension of the plane strain stress field has been shown to be possible for the case $\varphi = 0$, it seems unlikely, since there are no new physical complications, that same extension cannot be found quite generally for $\varphi > 0$ at least for some range of the small values of φ. It may therefore be concluded that the bearing capacity pressure q_0 obtained from a statically admissible stress field for a strip footing problem ([7.1]) is also a lower bound for a three-dimensional footing on a $c - \varphi$ *weightless* soil.

7.3. SQUARE AND RECTANGULAR FOOTINGS ON A SEMI-INFINITE MEDIUM – UPPER BOUNDS

7.3.1. Hill mechanism (Fig. 7.2)

A simple failure mechanism for the rectangular footing is shown diagrammatically in Fig. 7.2. *lmno* is the area of footing and the downward movement of the footing is accommodated by movement of the material as indicated by small arrows in Fig. 7.2(a). In Fig. 7.2(b and c) are shown the plan and section in Fig. 7.2(a), and it can be seen that this mechanism is an extension into three dimensions of a simple modification of the two-dimensional Hill's mechanism in Fig. 3.29(a). Since the

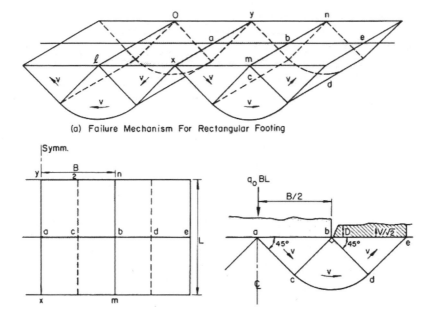

(a) Failure Mechanism For Rectangular Footing

(b) Plan View   (c) Vertical Section

Fig. 7.2. Three-dimensional Hill mechanism for a bearing capacity problem, $\varphi = 0$.

movement is symmetrical about xy, it is only necessary to consider the right-hand side of Fig. 7.2(a) for the following bearing capacity computations.

The rates of internal dissipation of energy on the discontinuous surface between the material at rest and the material in motion and in the radial shear zone are:

$$2cv\left(\frac{B}{2\sqrt{2}}L\right) + 2cv\left(\tfrac{1}{2}\pi\,\frac{B}{2\sqrt{2}}L\right) \tag{7.2}$$

in which B is footing width and L is footing length. Rates of energy dissipated on the two end surfaces are:

$$2cv\left(\frac{B}{2\sqrt{2}}\frac{B}{2\sqrt{2}}\right) + 2cv\left(\tfrac{1}{4}\pi\,\tfrac{1}{8}B^2\right) \tag{7.3}$$

The rate of energy dissipated through the surcharge is:

$$c\left(\frac{v}{\sqrt{2}}\right)(DL + DB) \tag{7.4}$$

and energy also dissipated by friction at the interface between the footing and the soil beneath the footing is:

$$\tfrac{1}{2}\,q_0\,BL\,\tan\delta\left(\frac{v}{\sqrt{2}}\right) \tag{7.5}$$

in which δ is the friction angle of footing base and q_0 is the average footing bearing pressure. The rates of external work due to footing load and soil weight are:

$$\tfrac{1}{2} q_0 BL \left(\frac{v}{\sqrt{2}} \right) - \tfrac{1}{2} \gamma \, BLD \left(\frac{v}{\sqrt{2}} \right) \qquad [7.6]$$

Equating the total rates of internal and external work yields an upper-bound solution of the bearing capacity pressure q_0:

$$q_0^u = \frac{c}{(1 - \tan \delta)} \left(5.14 + 1.26 \frac{B}{L} + 4 \frac{D}{B} \frac{D}{L} + \frac{\gamma D}{c} \right) \qquad [7.7]$$

Thus, for a square, smooth footing, for which $B = L$ and $\delta = 0$, [7.7] gives the value $[6.4c + \gamma D + 4c(D/B)^2]$ and a value of $[5.46c + \gamma D + c(D/B)^2]$ is found for a ratio $L/B = 4$. This equation tends to the value of $5.14c + \gamma D$ for rectangles whose length is great compared with their width. The failure mechanism of Fig. 7.2 and the stress field of Fig. 7.1, therefore, show that the ultimate bearing pressure for a square footing lies within 12% of the value $(5.7c + \gamma D)$ when $D/B \ll 1$.

7.3.2. Modified Hill mechanism for square footings (Fig. 7.3)

A better upper bound for a square footing is obtained as follows: In Fig. 7.3(a), *lmno* is the square area of footing which moves downward with initial velocity v. The square is divided into four equal triangles by the diagonals *ln, mo*. Taking a typical triangle *cmn*, the downward movement of the triangle is accommodated by lateral movement in the volume *cdefmn*. The volumes *dcmn* and *efmn* are tetrahedra, the points d, e being vertically below the line *cf*. The volumes *mbde* and *nbde* are two similar sections of right circular cones, the axes of which lie on *mn*. Figs. 7.3(b) and (c) show the vertical section and the plan through *cf*. Vertical sections by planes parallel to *cf* through the volume *cdefmn* are similar in shape to the section shown in Fig. 7.3(b) but are of varying size. The tetrahedral volumes *dcmn* and *efmn* move as rigid bodies in the directions parallel to *cd* and *ef*, respectively. The circular cone volume *mned* is a radial shear zone and its streamline of flow is parallel to the arc *de*. If the angle *bcd* is denoted by β, the velocity in each of these three volumes has the constant value $v\csc \beta$. The downward movement of the other three triangles is accommodated in the same way, the remainder of the material being at rest.

Energy is dissipated in the discontinuity surface between the material at rest and the material moving in the volume *cdefmn*. The rate of dissipation of energy due to this discontinuity surface is equal to the area of the surface multiplied by $cv\csc \beta$, since the change in velocity across the surface has the constant value $v\csc \beta$. It is a simple matter to calculate the three parts of the area of the surface of discontinuity. Referring now to Fig. 7.3(a), we have:

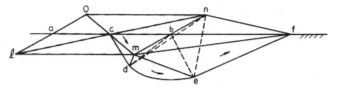

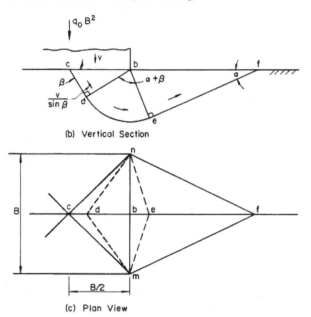

(a) Failure Mechanism for Square Footing

(b) Vertical Section

(c) Plan View

Fig. 7.3. Modified Hill mechanism for a square footing on purely cohesive soil ($\varphi = 0$) (Shield and Drucker, 1953).

Area mef = (area bef) $\dfrac{(1 + \sin^2 \beta)^{1/2}}{\sin \beta}$

$$= \tfrac{1}{8} B^2 \sin \beta \cot \alpha (1 + \sin^2 \beta)^{1/2} \qquad [7.8a]$$

where α is the angle bfe (Fig. 7.3b) and:

Area cmd = $\tfrac{1}{2}(\tfrac{1}{2}B \cos \beta)(\tfrac{1}{2}B \sin \beta) \dfrac{(1 + \sin^2 \beta)^{1/2}}{\sin \beta}$

$$= \tfrac{1}{8} B^2 \cos \beta (1 + \sin^2 \beta)^{1/2} \qquad [7.8b]$$

Area mde = $\tfrac{1}{2}(\alpha + \beta)(\tfrac{1}{2}B \sin \beta)^2 \dfrac{(1 + \sin^2 \beta)^{1/2}}{\sin \beta}$

$$= \tfrac{1}{8} B^2 (\alpha + \beta) \sin \beta (1 + \sin^2 \beta)^{1/2} \qquad [7.8c]$$

Energy is also dissipated in the circular cone volume *mned* where the material is in a state of plane strain motion (Fig. 3.19) so that expression [3.68] can be used to calculate the rate of dissipation of energy per unit thickness in the cone axis direction *mn*. Since only the radius of the cone changes with respect to the axis of the cone, it follows that the rate of dissipation of energy in the cone volume *mned* is equal to the area $mne = \frac{1}{4}B^2 \sin\beta$, multiplied by $c(v\csc\beta)(\alpha+\beta)$. When the total rate of internal dissipation of energy is equated to the external rate of work due to footing load, $q_0 B^2 v$, we obtain an upper-bound solution for the bearing capacity pressure of a smooth, square surface footing with no surcharge loading ($D = 0$):

$$q_0^u(\alpha,\beta) = c[\alpha + \beta + \sqrt{1 + \sin^2\beta}\,(\alpha + \beta + \cot\alpha + \cot\beta)] \qquad [7.9]$$

The bearing capacity function $q_0^u(\alpha,\beta)$ has the minimum value $5.80c$ when α and β are approximately $47°$ and $34°$, respectively.

7.3.3. Mechanism for rectangular footings

Further improvement in the upper bound for a square footing would require the more elaborate failure mechanism discussed by Shield and Drucker (1953). The failure mechanism which was used by Shield and Drucker provides an upper bound for the bearing capacity pressure in the rectangular footing problem and also gives a better upper bound for the square footing than those obtained in the foregoing. Their more elaborate failure mechanism is essentially an extension into a rectangular footing of a simple modification of the failure mechanism in Fig. 7.3 for the case of a square footing. In this improved mechanism the lateral movement of the material accompanied by the downward movement of the two triangles *mnc* and *col* (Fig. 7.3a) is defined by two unknown angles α and β (Fig. 7.3b) while the material accompanied by the downward movement of the other two triangles *mcl* and *cno* is now defined by another two unknown angles α_1 and β_1 which moves in the directions perpendicular to the plane through *cf*. The bearing capacity pressure q_0 for a rectangular footing can then be expressed as a function of the four unknown angles, α, β, α_1 and β_1. For rectangles for which $B/L \geqslant 0.53$, the bearing capacity function $q_0^u(\alpha,\beta,\alpha_1,\beta_1)$ has the minimum value:

$$q_0^u = c\left(5.24 + 0.47\frac{B}{L}\right) \qquad [7.10]$$

when $\alpha = 47°4'$, $\beta = 34°$, $\alpha_1 = 46°17'$ and $\beta_1 = 39°$. Thus for a smooth square footing, for which $B = L$, [7.10] gives the value $5.71c$, which is a better upper bound for the smooth, square footing than those obtained previously. For rectangles for which $B/L < 0.53$ a better upper bound than [7.10] is obtained by putting $\alpha = \beta = 45°$, $\alpha_1 = 46°17'$ and $\beta_1 = 39°$ in the function $q_0^u(\alpha,\beta,\alpha_1,\beta_1)$ to give:

$$q_0^u = c \left(5.14 + 0.66 \frac{B}{L} \right) \qquad\qquad [7.11]$$

This equation tends to the value $5.14c$ for rectangles whose length is great compared with their width ($L \gg B$), in agreement with the upper bound for the two-dimensional flat footing.

7.3.4. Uniformly loaded area

The same value of $(2 + \pi)c$ is an upper bound for any uniformly loaded area as distinguished from an area loaded by a *rigid* footing. This follows directly from the fact that kinematically admissible failure mechanisms for this problem are not restricted by the condition that the entire loaded area move as a plane surface. Therefore, the region immediately adjacent to the boundary of the area of contact can escape outward in a manner essentially that of the plane strain motion as shown in the right-half of Fig. 3.22(a) with $\varphi = 0$. Point B is then a point on the boundary; BO lies in the interior; BG is exterior. The value of $(2 + \pi)c$ is obtained by letting BO approach zero.

7.4. SQUARE AND CIRCULAR FOOTINGS ON A FINITE BLOCK – LOWER BOUNDS

Upper and lower bounds for the bearing capacity pressure have been obtained in the preceding sections for the indentation of an infinitely large block by a flat-ended rigid footing. For a square footing on a purely cohesive medium ($\varphi = 0$, or Tresca yield condition), it was shown that the average bearing pressure for smooth and rough contact surfaces lies between the values $5c$ and $5.71c$. For the case of axial symmetry which will be discussed in section 7.8, the corresponding value is $5.69c$ or $6.05c$ depending on whether the footing is smooth or perfectly rough. Herein, the problem of indentation on a block of *finite* dimensions will be described. In the following investigations, limit analysis is used to obtain upper and lower bounds of the average bearing capacity pressure during indentation of a square footing on a square block, and a circular footing on a circular cylinder. Here, as in previous cases, the material of the block and cylinder is assumed to be an elastic–perfectly plastic ideally cohesive material ($\varphi = 0$), i.e., the Tresca yield condition and its associated flow rule.

To obtain a lower bound for the average bearing pressure, the discontinuous stress pattern in Fig. 7.4 (Winzer and Carrier, 1948) is found to be useful for extension into three dimensions. The pressure on ab has the value $2c(1 + \sin \alpha)$ when the angle of the wedge is 2α. The region vertically below de is in a state of uniaxial compression of magnitude $2c(1 - \sin \alpha)$. When the angle of the wedge 2α is equal to $60°$ the pressure on ab has the value $3c$ and the corresponding stress field is shown in Fig. 4.20.

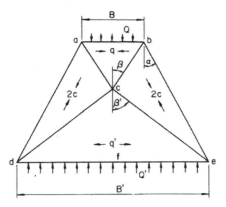

Fig. 7.4. Discontinuous stress field for loaded wedge $B' = B \tan^2 (\pi/4 + \alpha/2)$.

7.4.1. Square footings (Fig. 7.5)

The corresponding three-dimensional stress field for the square footing is shown in Fig. 7.5. *lmno* is the square area of footing. The triangular region *abc* in Fig. 7.4 becomes the pyramid-shape volume *lmnoc*. The four triangular faces of the volume are all inclined at an angle of β to the vertical. The triangular regions *bce* and *cef* in Fig. 7.4 become the volumes *mnsrc* and *rscf*, respectively. The line *rs* is parallel to the sides *mn* and *lo* of the square footing, and the rectangular face *mnsr* and triangular face *rsc* of the volume *mnsrc* are inclined at an angle of α and β' to the vertical, respectively.

The pressure $2c(1 + \sin \alpha)$ on the square area is supported by uniaxial compression of amount $2c$ in the four "legs" of material, *mnsr*, etc. This results in an all around horizontal compression of amount q in the pyramid-shape volume *lmnoc*.

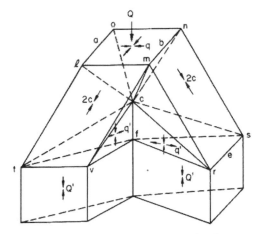

Fig. 7.5. Stress field for a three-dimensional square punch.

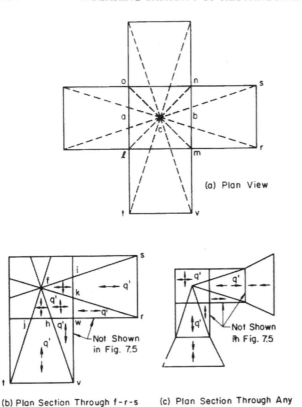

(a) Plan View

(b) Plan Section Through f-r-s

(c) Plan Section Through Any Point Between c-f

Fig. 7.6. Plan sections in Fig. 7.5.

The uniaxial compression $2c$ is supported and carried down by the biaxial state of compression–tension of amount Q', q', respectively, in the four volumes of material, $rscf$, etc. A simple tension of amount q' perpendicular to rs in volume $rwkc$ (Fig. 7.6(b), not shown in Fig. 7.5) and similarly a simple tension q' perpendicular to tv in volume $vwhc$ are added horizontally outside the volumes $rscf$ and $tvcf$, respectively. The line cw is then the intersection line of the volumes $rwkc$ and $vwhc$. The stress system in region $hwkfc$ is assumed to be composed of two equal tensions of amount q' acting in a horizontal direction, and region $kifc$ or $jhfc$ is in a state of tension–tension–compression of amount q', q', and Q', respectively. For clarity, only two of the four supporting "legs" are shown in the diagram. Fig. 7.6(a) shows the plan view through the square area of footing, and Fig. 7.6(b) shows the plane section through the lines tv and rs. Sections by plane parallel to Fig. 7.6(b), between the two points c and f, are similar in shape to the section shown in Fig. 7.6(c), but are of varying size. Fig. 7.4 is then the vertical section through the mid-points of the two parallel sides of the square.

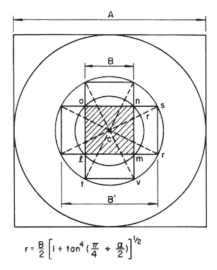

$$r = \frac{B}{2}\left[1 + \tan^4\left(\frac{\pi}{4} + \frac{\alpha}{2}\right)\right]^{1/2}$$

Fig. 7.7. Improvement of the stress field in Fig. 7.5.

The three-dimensional pattern, as shown in Fig. 7.5 and Fig. 7.6, is a statically admissible stress field for the loading of a square footing on a square block. The value $Q = 2c(1 + \sin \alpha)$ is, therefore, a lower bound for the square footing on a square block for blocks whose width, A, is greater than $\tan^2(\frac{1}{4}\pi + \frac{1}{2}\alpha)$ times the width, B, of the footing.

In order to increase the lower bound, a vertical compression of amount Q_1 is superimposed in the volume vertically below the square area of contact, increasing the pressure on the area to $2c(1 + \sin \alpha) + Q_1$. In addition, a horizontal compression of amount Q_1 is superimposed throughout the material in a cylindrical volume which circumscribes the square area of footing (Fig. 7.7).

The horizontal pressure Q_1 in the cylinder is carried by a circular tube of interior radius $B/\sqrt{2}$ and exterior radius r which is in a stress state whose circumferential stress component may be taken to be zero, and whose radial component is $BQ_1/\sqrt{2}\,r$ at the exterior face r so that the equilibrium conditions are satisfied for this tube. The second tube between the interior radius r and exterior radius $\frac{1}{2}A$ is in the state of fully plastic stress distribution, subjected to the internal pressure $BQ_1/(\sqrt{2}r) = 2c\ln(A/2r)$ (see Prager and Hodge, 1968, for example). Therefore, the average pressure on the square footing has the value $2c(1 + \sin \alpha) + Q_1$ which may be expressed as:

$$q_0^l = 2c(1 + \sin \alpha) + \sqrt{2}c\,[1 + \tan^4(\tfrac{1}{4}\pi + \tfrac{1}{2}\alpha)]^{1/2}\ln\left\{\frac{A}{B}\,[1 + \tan^4(\tfrac{1}{4}\pi + \tfrac{1}{2}\alpha)]^{-1/2}\right\}$$

$$[7.12]$$

valid for $|Q_1| \leqslant 2c$.

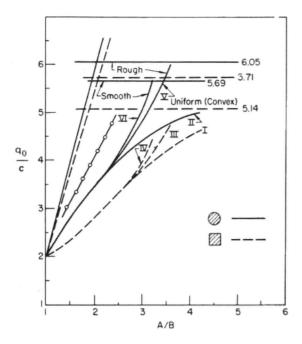

Fig. 7.8. Upper and lower bounds for a square punch on a square block and a circular punch on a circular cylinder.

The expression has a maximum value when α satisfies the condition $\partial q_0^l / \partial \alpha = 0$. This condition is:

$$\ln \frac{A}{B} = 1 - \sqrt{2} \cos \alpha \, [1 + \tan^4 (\tfrac{1}{4}\pi + \tfrac{1}{2}\alpha)]^{1/2} \cos^2 (\tfrac{1}{4}\pi + \tfrac{1}{2}\alpha) \cot^3 (\tfrac{1}{4}\pi + \tfrac{1}{2}\alpha)$$

$$+ \tfrac{1}{2} \ln [1 + \tan^4 (\tfrac{1}{4}\pi + \tfrac{1}{2}\alpha)] \qquad\qquad [7.13]$$

It can be verified that Tresca's criterion is nowhere violated in the resulting stress field, so that the stress field is statically admissible and, by the lower-bound theorem of limit analysis, the value given in [7.12] is a lower bound for the collapse value of the average bearing pressure q_0. The values of q_0^l are plotted against A/B in Fig. 7.8 (curve I).

7.4.2. Circular footings

The stress field in Fig. 7.5 can be modified, so that it applies to any area of contact which is convex. This follows directly from the fact that any convex area of contact can be closely approximated by an inscribed polygon. The associated stress fields below the area of contact, which may be visualized as supporting "legs" on each side of the polygon (in a manner essentially the same as shown in Fig. 7.5), do not overlap one another as long as the contact area is convex. In particular, the relevant formula for the average indentation pressure of a circular footing on a circular cylinder can be obtained directly from Fig. 7.7, by using an inner circular tube of interior radius $\frac{1}{2}B$ and exterior radius $r = (\frac{1}{2}B) \tan^2(\frac{1}{4}\pi + \frac{1}{2}\alpha)$. The relevant lower bound formula is found to be given by:

$$q_0^l = 2c \left\{ 1 + \sin \alpha + \tan^2(\tfrac{1}{4}\pi + \tfrac{1}{2}\alpha) \ln \left[\frac{A}{B} \tan^2(\tfrac{1}{4}\pi - \tfrac{1}{2}\alpha) \right] \right\} \qquad [7.14]$$

This expression has a maximum value when α satisfies the condition $\partial q_0^l / \partial \alpha = 0$, which yields:

$$\ln \frac{A}{B} = 1 - \cos \alpha \cos^2(\tfrac{1}{4}\pi + \tfrac{1}{2}\alpha) \cot(\tfrac{1}{4}\pi + \tfrac{1}{2}\alpha) + 2 \ln \tan(\tfrac{1}{4}\pi + \tfrac{1}{2}\alpha) \qquad [7.15]$$

Values of q_0^l are plotted against A/B in Fig. 7.8 (curve II).

For a circular cylinder for which A/B is greater than 3.59 (rough footing) or 3.20 (smooth footing), the average bearing capacity pressure becomes equal to that of a circular footing on the surface of a semi-infinite solid ($6.05c$ for a rough footing or $5.69c$ for a smooth footing, further discussions on these values will be given in section 7.8). A lower bound for the bearing capcity pressure of this limiting case by a square footing can be obtained by multiplying the circular case by $\frac{1}{4}\pi$. This follows directly from the fact that the square area of contact can be inscribed by a circular area, so that the average bearing capacity pressure over the square area is $\pi B^2 / 4B^2 = \frac{1}{4}\pi$ times the lower-bound value for a circular footing (the bearing capacity pressure outside the inscribed circle is assumed zero).

These limiting values of the lower bound are also plotted against A/B in Fig. 7.8, and are joined by a smooth curve with some of the previously obtained corresponding points (curves III, IV, V, and VI; curves III and V for rough footings and curves IV and VI for smooth footings).

7.5. SQUARE AND CIRCULAR FOOTINGS ON A FINITE BLOCK – UPPER BOUNDS

7.5.1. A symmetric mechanism for a square footing (Fig. 7.9)

A simple discontinuous velocity field for the square footing is shown diagrammatically in Fig. 7.9. *lmno* is the square area of the footing, and the initial downward velocity of this area is taken to be v_0. The downward movement of the pyramid-shape volume *lmnoc* is accommodated by the horizontal movement of the four rigid volumes *mrsnct*, etc. as indicated in Fig. 7.9(a). For clarity, only one of the four volumes is shown in the diagram. The lines *mr* and *ns* are parallel to the

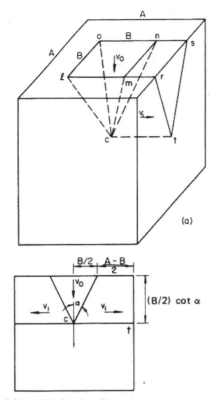

(a)

(b) Vertical Section Through c-t

Fig. 7.9. Velocity field for a square footing.

sides *lm* and *on* of the square footing. Fig. 7.9(b) shows the vertical cross-section through the mid-points of the two opposite sides of the square.

Since the velocity field is symmetrical about the two diagonal vertical planes passing through the opposite corners of the square, we need only to consider that part of the velocity field which supports the pressure on one-fourth of the footing. The internal dissipation of energy is due to the discontinuity surfaces between the material at rest, and the material in motion (surfaces *mrtc* and *nstc*), and the discontinuity surface *mnc* where the two rigid materials have a relative velocity $v_0/\cos \alpha$. The discontinuous surface *mrtc* is of area:

$$\tfrac{1}{8}B^2 \left(2\frac{A}{B} - 1\right) \cot \alpha \sec \alpha \qquad\qquad [7.16]$$

and the triangle surface *mnc* is of area $B^2/(4 \sin \alpha)$. The rate of dissipation of energy is found by multiplying the area of each discontinuity by c times the discontinuity in velocity across the surface, and summing over all the surfaces. The rate of external work is $q_0 B^2 v_0 / 4$. Equating the rate of total internal energy dissipation to the rate of external work yields:

$$q_0^u = c\left[2\frac{A}{B} - 1 + \csc\alpha\right]\sec\alpha \qquad [7.17]$$

With $A/B = 2$, the expression has the value $5.77c$ near the point $\alpha = 30°$, so that $5.77c$ is an upper bound for the collapse pressure.

7.5.2. Diagonal sliding for square and circular footings (Figs. 7.10 and 7.11)

A better upper bound for the bearing capacity pressure in the indentation of a sqaure footing on a square block can be obtained from the velocity field in which the volume *abecd* slides as a rigid body along the diagonal direction *ad* (Fig. 7.10). Equating the rate of internal energy dissipation to the rate of external work yields the upper bound:

$$q_0^u = c\left(1 + \frac{A}{B}\right)^2 \frac{(2 + \tan^2\beta)^{1/2}}{4\sin\beta} \qquad [7.18]$$

The upper bound has a minimum value at the point $\beta = 50°$, when it has the value:

$$q_0^u = 0.604c\left(1 + \frac{A}{B}\right)^2 \qquad [7.19]$$

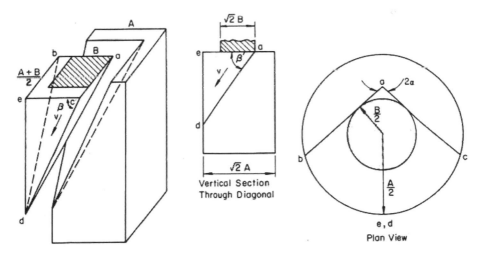

Fig. 7.10. Velocity field for a square punch on a square block.

Fig. 7.11. Modification of velocity field of Fig. 7.10 to a circular punch on a circular cylinder.

Thus, for a footing for which $A/B = 2$, [7.19] gives the value $5.44c$, so that $5.44c$ is an upper bound for the collapse pressure. The almost straight line (dashed line) in Fig. 7.8 shows the values of q_0^u plotted against A/B. It should be noted that for $A/B \leqslant 1.5$, the rigid body sliding along one side of the square footing gives upper bounds which are slightly less than those obtained by diagonal rigid body sliding as shown in Fig. 7.10.

The velocity field in Fig. 7.10 may be modified to provide an upper bound for the collapse pressure in the circular footing problem. Only the plan view of the modified velocity field is shown in Fig. 7.11. The planes abd and acd are planes of velocity discontinuity. The volume $abecd$ slides as a rigid body along the direction ad in a similar manner to that shown in Fig. 7.10. Application of the upper-bound theorem of limit analysis gives:

$$q_0^u = \frac{c}{\pi}\csc\beta'[1 + \tan^2\beta'\csc^2\alpha]^{1/2}\left[\left(\frac{A^2}{B^2} - 1\right)^{1/2} + \cot\alpha + \left(\alpha + \sin^{-1}\frac{B}{A}\right)\frac{A^2}{B^2}\right] \quad [7.20]$$

The minimum value of the upper-bound value q_0^u in [7.20] is furnished by the simultaneous conditions:

$$\frac{\partial q_0^u}{\partial\alpha} = 0, \qquad \frac{\partial q_0^u}{\partial\beta'} = 0 \qquad\qquad [7.21]$$

which leads to: $\sin\alpha = \tan^2\beta'$ $\qquad\qquad [7.22a]$

and:

$$(1 + \sin\alpha)\left(\frac{A}{B}\tan\alpha - 2\frac{B}{A}\csc 2\alpha\right) - \frac{B}{A}\cot\alpha - \frac{A}{B}\alpha = \frac{A}{B}\sin^{-1}\frac{B}{A} + \cos\left(\sin^{-1}\frac{B}{A}\right)$$

$$[7.22b]$$

The almost straight line (solid line) in Fig. 7.8 shows the values of q_0^u plotted against A/B.

There is no relative motion across the area of contact between the footing and block (Figs. 7.10 and 7.11), so that the upper bounds in [7.18] and [7.20] are applicable to both rough and smooth footings.

7.5.3. Modification of the mechanism of Fig. 7.3 (Fig. 7.12)

The velocity field used in Fig. 7.3 for the bearing capacity computation of an infinitely large block by a square footing can be modified so that it can be used to provide upper bounds for the average bearing capacity of the square footing problem. The diagram shown in Fig. 7.12 is essentially the same picture as that shown in Fig. 7.3. Application of the upper-bound theorem of limit analysis determines the upper bound:

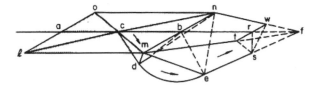

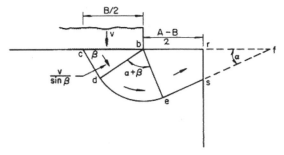

Vertical Section Through a-b

Fig. 7.12. Modification of the velocity field in Fig. 7.3 to square blocks.

$$q_0^u = c \left\{ \alpha + \beta + (1 + \sin^2 \beta)^{1/2} \left[\alpha + \beta + \cot \beta + \cot \alpha \, g(\alpha, \beta) \right] \right\} \qquad [7.23]$$

in which:

$$g(\alpha, \beta) = 1 - \left[\tan \alpha \csc \beta \left(1 - \frac{A}{B} \right) + \sec \alpha \right]^2 \qquad [7.24]$$

The function $g(\alpha,\beta)$ varies from 0 to 1. When $g = 0$; the angle α is then zero, and the lines nw, mt, and es are all parallel to the sides lm and no of the square footing. When $g = 1$; the points f and r will be coincident, and the velocity field reduces to the field shown in Fig. 7.3. For $g > 1$, the solution is independent of the ratio A/B, and the local solution computed from Fig. 7.3 will govern. Here, the upper bound has the minimum value $5.80c$ when α and β are approximately $47°$ and $34°$, respectively ($A/B \geqslant 1.767$).

It is found that the modified velocity field in Fig. 7.12 gives upper bounds which are higher than those obtained in the foregoing including the local solution $5.80c$. For example, for $A/B = 2$, [7.23] has the minimum value $6.13c$ when α and β are approximately $20°$ and $60°$, respectively. Expression [7.23] is found to be insensitive to the variables α and β, and thus the local solution $5.80c$ will be applicable for all blocks for which A/B is greater than 1.767.

7.5.4. Summary and conclusions

In this and preceding sections, limit analysis is utilized to obtain upper and lower bounds of the footing bearing pressure during indentation of a square footing on a square block, and a circular footing on a circular cylinder. Reasonably close upper and lower bounds for the average bearing pressure over the square and circular footings are obtained, and are expressed as functions of the ratio of block width to footing width (Fig. 7.8).

Suppose the question is asked, what ratio A/B is needed in order to support the force $4c(\frac{1}{4}\pi B^2)$ in the circular footing problem? The lower-bound curve of Fig. 7.8 provides the information that 2.4 is sufficient, but will a smaller ratio be enough? There is no reason to suppose that the minimum value has been found, since the upper-bound curve indicates that any ratio less than 1.5 is unsafe. This then suggests the use of average curve between upper and lower bound which is, in fact, quite close to a straight line (denoted by dash and small circle in Fig. 7.8). The average curve for the square footing is not shown, but may be determined in a similar manner to that described above.

7.6. SQUARE AND CIRCULAR FOOTINGS ON A SEMI-INFINITE LAYER – LOWER BOUNDS (SHIELD, 1955c)

Upper and lower bounds for the average bearing capacity pressure of a flat, rigid, *smooth* footing on a semi-infinite layer of elastic—perfectly plastic soil resting on a *rough*, rigid base are presented in this section. The soil is assumed to be an ideally cohesive medium ($\varphi = 0$) for which Tresca's yield criterion of constant maximum shearing stress and its associated flow rule will apply. The square footing problem is considered first and then followed by the circular footing problem.

7.6.1. Square footing

We first describe statically admissible stress fields for the two-dimensional plane strain case which are then extended to three dimensions to give lower bounds for the collapse pressure of the square footing problem.

Two-dimensional case (Fig. 7.13)

In Fig. 7.13(a), *oh* is the center-line of the footing *oa* indenting the upper surface of the layer of thickness H under condition of plane strain. A pressure $5c$ is produced on the surface of the footing by the stress system to the right of the line *af*. This stress field is the upper right portion of the stress field shown in Fig. 4.21(b) and was used previously in section 7.2 to obtain the lower bound $5c$ for the average bearing pressure in the indentation of an infinitely thick layer. The lines *ad*, *ae*, and *af* in Fig. 7.13(a) are lines of stress discontinuity and *ad*, *af* are each inclined to *ae* at an angle of 30°. Since the base is assumed to be rigid and

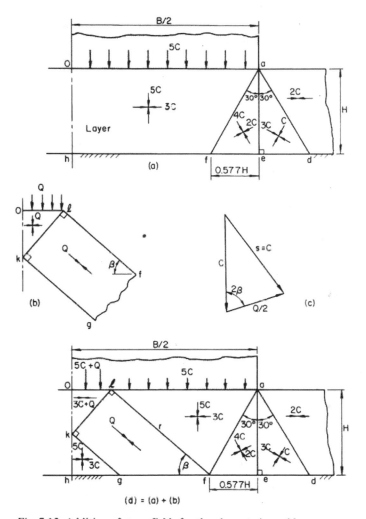

Fig. 7.13. Addition of stress fields for the plane strain problem.

perfectly rough, the stresses at the interface of the layer and the base may have any value compatible with statically admissible stress fields in the material of the layer. It follows that the value $5c$ is also a lower bound for a layer of finite thickness. The stresses in the different regions are shown in Fig. 7.13(a) by the small arrows.

In order to increase the lower bound, the pressure near the center of the footing is increased. This is done by superimposing the stress field shown in Fig. 7.13(b) onto the stress field of Fig. 7.13(a) and resulting the stress field shown in Fig. 7.13(d). In Fig. 7.13(d), the lines fl and kg are then lines of stress discontinuity inclined to the horizontal line at an angle β, and the line kl of stress discontinuity separates the stress system in $fgkl$ from the stress system in okl whose principal axes of stress are in the vertical and horizontal directions. We suppose that the regions *fgkl* and *okl* are on the point of yielding in order to increase the pressure on *ol* as much as possible. Since we know the directions and intensities of the principal stresses of the two stress fields as shown in Figs. 7.13(a and b), we now want to determine the magnitude of the principal stress Q so that the resultant stress field obtained from the superposition of the individual stress fields (a) and (b) will just be on the point of yielding. It has been shown in section 4.2, Chapter 4 that the general result for the sum of two stress fields in the plane is that the hydrostatic or mean stress add algebraically which gives the resultant hydrostatic pressure:

$$p = \frac{3c + 5c}{2} + \frac{Q}{2} = 4c + \frac{Q}{2} \qquad [7.25]$$

and the maximum shear stresses add as vectors with an angle between them equal to 2β as indicated by the shear stress diagram in Fig. 7.13(c). This gives the resultant maximum shear stress:

$$s^2 = c^2 + \left(\frac{Q}{2}\right)^2 - 2c\left(\frac{Q}{2}\right)\cos 2\beta \qquad [7.26]$$

in the overlapped region *fgkl* (Fig. 7.13d). Since hydrostatic pressure has no effect on the yielding of a Tresca material, it follows that the resultant maximum shear stress s must have the value c in order to satisfy yield condition. Equating the value s in [7.26] to c gives:

$$Q = 4c \cos 2\beta \qquad [7.27]$$

where β is the angle *gfl* in Fig. 7.13(d). It can be seen from [7.27] that the angle β must be less than $45°$ in order that Q be positive.

The pressure on *ol* now has the value $5c + Q = 5c + 4c\cos2\beta$ and the average pressure q_0^l over *oa* is found to be given by:

$$q_0^l = c \left\{ 5 + 4 \cos 2\beta \left[1 - \frac{H}{B}(1.154 + 2\cot\beta) \right] \right\} \qquad [7.28]$$

provided that $B/H > 3.154$, since $\beta < \frac{1}{4}\pi$. The stress field is statically admissible and, by lower bound theorem of limit analysis, this is a lower bound for the collapse value of the average bearing pressure.

For a given value of B/H greater than 3.154, the lower bound is found by maximizing [7.28] with respect to β. For values of B/H less than 3.154 the lower bound is $5c$. For values of B/H greater than 8 approximately, it is better to repeat this process a certain number of times, as shown in Fig. 7.14, so that the pressure distribution on the footing has more than one jump in value. By taking the lines of discontinuity *lf*, *mk*, ... in Fig. 7.14 to be inclined at an angle of $22.5°$ to the horizontal, the lower bound is found to be given by:

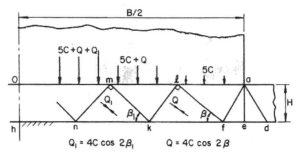

Fig. 7.14. Stress field for a wide footing (plane strain) (Shield, 1955c).

$$q_0^l = c \left[3.42 + 0.48 \left(\frac{H}{B} \right) + \frac{1}{4} \left(\frac{B}{H} \right) \right] \qquad [7.29]$$

provided that: $\quad \dfrac{B}{H} = 5.982 + 5.656n \qquad\qquad\qquad [7.30]$

where n is the number of lines of discontinuity lf, mk, ... This is a good lower bound for large values of B/H.

Three-dimensional case (Fig. 7.15)

Considering now the three-dimensional problem, a statically admissible stress field has been given in section 7.2 for the bearing capacity of an infinitely thick layer by a square footing. This stress field shows that $5c$ is a lower bound for the average bearing pressure and the bound also applies to the layer problem considered here. Here as in the preceding sections, the stress field involves a vertical compression of amount $5c$ and an all around horizontal compression of amount $3c$ for the material just below the area of contact. This state of stress extends throughout the material in the right pyramid formed by the square area of footing and four triangular faces inclined at 30° to the vertical, the vertex of the pyramid being vertically below the center of the square. In obtaining lower bounds for the layer problem we shall assume that the material of the layer which lies in the pyramid is stressed in this way. The stress near the center of the footing will be increased by a method similar to that used above in the two-dimensional plane strain problem.

Since the stress field is symmetrical about the two vertical planes passing through the mid-points of opposite sides of the square, we need only consider that part of the stress field which supports the pressure on one-quarter of the footing. The stress field is also symmetrical about the vertical planes through the diagonals of the square. In Fig. 7.15, $oabc$ is one-quarter of the area of indentation in the upper surface of the layer, o, being the center of the square area. For clarity, the points a, b, c, which are shown in the plan, are omitted from the perspective in Fig. 7.15. The triangular faces of the pyramid mentioned in the previous paragraph meet the lower surface of the layer in the lines fx, rx. The planes $lfwt$, $nrvt$ are each

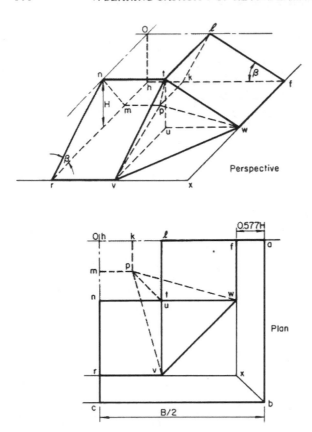

Fig. 7.15. Stress field for a square footing on a layer (Shield, 1955c).

inclined at an angle β to the horizontal, and the planes *lkpt, nmpt* are perpendicular to the planes *lfwt, nrvt*, respectively.

As mentioned above, the stress field of section 7.2 involves a vertical compression $5c$ and an all-around horizontal compression $3c$ in the material in the volume *oabcrhfx*. For the moment we assume that the material vertically above the area *hfwvr* is stress-free. An all-around horizontal compression $3c$ will be added later in this region in order to restore equilibrium. In the final stress field the stresses in the vertical plane through the line *oa* will be as shown in Fig. 7.13(a). (In Fig. 7.13(d), the line *oh* intersects the line *lk*, which is not the case in Fig. 7.15. However, if the width of the footing is such that *oh* intersects *lk* in Fig. 7.15, a simple adjustment can be made to the stress field in order that the symmetry conditions be satisfied, as they are in Fig. 7.13(d).) The minumum principal stress in region *fgkl* of Fig. 7.13(d) is $3c + 2c\cos2\beta$ so that, to satisfy the yield condition, the pressure parallel to *lt* in region *lfwtpk* of Fig. 7.15 must have this value at least.

We assume, therefore, a compression $2c\cos2\beta$ parallel to *lt* in region *lfwtpk* and similarly a compression $2c\cos 2\beta$ parallel to nt in region $nrvtpm$. The planes twp, tvp are surfaces of stress discontinuity and we assume that the stress system in region $twvp$ is composed of two equal compressions of amount S in the directions tw and tv, so that the plane tvw is stress-free. From the equilibrium conditions across twp (and tvp), it is found that S must have the value:

$$S = \frac{2c \cos 2\beta}{(1 + \sin^2 \beta) \cos^2 \beta} \qquad [7.31]$$

and that a vertical compression of amount:

$$T = 2c \cos 2\beta \sec^2 \beta \qquad [7.32]$$

must be added in the material vertically above the planes twp, tvp.

Compressions of amount $4c\cos 2\beta$ are now added in regions $lfwtpk$ and $nrvtpm$ parallel to fl and rn, respectively. Equilibrium is satisfied across the planes $lkpt$, $nmpt$ by adding a hydrostatic pressure of amount $4c\cos 2\beta$ in region $oltnmhkp$. An all-around horizontal compression $3c$ is now added throughout the material vertically above the area $hfwvr$. Vertical compressions of amount $5c$ in the region vertically above the area $hfwpvrh$ and amount $5c - T$ in the region vertically above the area $pwuv$ are added, and it can be verified that the yield condition is not violated in these regions. Above the area uvw it is found that the maximum vertical compression which can be added without violating yield is given by:

$$u(\beta) = c \left\{ 3 + 2 \left[1 - \frac{2 \cos^2 2\beta \sin^2 2\beta}{(1 + \sin^2 \beta)^2 \cos^4 \beta} \right]^{1/2} + \frac{2 \cos 2\beta(\cos 2\beta - \sin^2 \beta)}{(1 + \sin^2 \beta) \cos^2 \beta} \right\} < 5c \quad [7.33]$$

When β is less than $25°$, $u(\beta)$ has the value $5c$.

The total force on the area $oabc$ of the footing due to this stress system is the resultant of a pressure $5c$ over the area $tlabcn$ less the triangular area vertically above uvw, a pressure $5c + 4c\cos 2\beta$ over the square area $oltn$, and a pressure $u(\beta)$ over the triangular area vertically above uvw. The average pressure q_0^l on the footing is found, for $B/H > 3.154$, to be given by:

$$q_0^l = 5c + 4c \cos 2\beta \left[1 - \frac{H}{B} (1.154 + 2 \cot \beta) \right]^2 - 2 \frac{H^2}{B^2} \cot^2 \beta \, [5c - u(\beta)] \qquad [7.34]$$

Since the stress field is statically admissible, this value of q_0^l is a lower bound for the average indentation pressure. For values of B/H less than 4, maximizing [7.34] with respect to β does not give a lower bound greater than $5c$. Lower bounds for the average collapse pressure for the cases $B/H = 5, 6, 7, 8, 9, 10$ were obtained by maximizing [7.34] with respect to β. The bounds are plotted against B/H in Fig. 7.16 and joined to the line $q_0^l = 5c$ ($B/H < 4$) by a smooth curve, which is marked (i) in the figure. For values of B/H greater than 10 approximately, it is

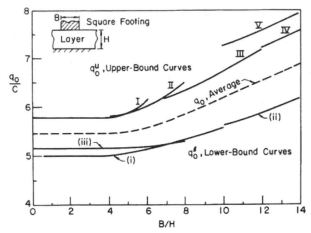

Fig. 7.16. Upper and lower bounds for the average pressure over the square footing (Shield, 1955c).

better to increase the pressure near the center of the footing by repeating this process a suitable number of times, as in the plane strain case. Lower bounds for the cases $B/H = 10$, 12, 14 were calculated by repeating the process once and maximizing with respect to the angles β, β_1 of Fig. 7.14. The results are given in Fig. 7.16, where they are joined by the curve marked (ii).

When the process of increasing the pressure towards the center is carried out n times, where the n angles β, β_1 ... are taken to be $22.5°$ the lower bound is found to be:

$$q_0^l = c\left(3.42 + \frac{1}{6}\frac{B}{H} + 3.64\frac{H}{B} - 1.04\frac{H^2}{B^2}\right) \qquad [7.35]$$

for $B/H = 5.982 + 5.656\, n$.

As previously mentioned in Section 7.2, the average bearing pressure over a square footing is *not less* than the average bearing pressure over a long rectangular footing (plane strain), it follows that the value $5.14c$ gives a better lower bound for a square footing for the case $B/H < 5$. For values of $B/H > 6$, the pressure near the center of the footing can be increased by the same method used in the stress field associated with Fig. 7.15. The relevant equations are given in Shield's paper (1955c). The results for the cases $B/H = 6$, 7, 8 are plotted in Fig. 7.16 and joined to the line $q_0^l = 5.14c$ ($B/H < 5$) by a smooth curve marked (*iii*) in the figure.

7.6.2. Circular footing (Fig. 7.17)

Attempts were made to extend the stress field of Fig. 7.13(d) to a circular area

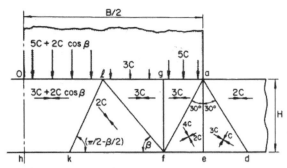

(a) Alternative Stress Field For the Plane Strain

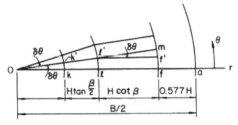

(b) Plan of (a) for a Circular Footing

Fig. 7.17. Stress field for a circular footing (Shield, 1955c).

of indentation. It proved to be a matter of some difficulty to provide the appropri-ate stresses normal to the plane of Fig. 7.13(d), i.e., the circumferential or hoop stresses, which are required by the yield criterion. An alternative two-dimensional stress field, which can be extended to the circular footing problem in a simple manner, is shown in Fig. 7.17(a). The field to the right of the line *af* produces a pressure $5c$ on the portion *ag* of the footing, as in Fig. 7.13(d). In region *fgl* the state of stress is a hydrostatic pressure of amount $3c$, and in region *fkl* a compres-sive stress $2c$ in the direction of *fl* is added to the hydrostatic pressure $3c$. The discontinuity line *lk* is inclined at an angle $\frac{1}{2}\beta$ to the vertical, where β is the angle *kfl*, and a pressure $5c + 2c\cos\beta$ is produced on *ol*. The field is statically admissible and provides lower bounds for the indentation pressure. These lower bounds are less than those obtained from the stress field of Fig. 7.13(d).

As mentioned in section 7.2, the field to the right of the line *af* in Fig. 7.17(a) can be modified so that it applies to any convex area of indentation and to a circular area in particular. The modified field lies outside the conical surface through *af*, shown in plan in Fig. 7.17(b). Vertically above the conical surface through *af*, the material is subjected to a vertical compression $5c$ and radial and circumferential compressions $3c$. The material lying between the cylindrical surface through *gf* and the conical surface through *lk* is stressed by a hydrostatic pressure $3c$. In addition, "legs" of material in this region, originating from strip elements of area on the conical surface through lk and inclined at an angle β to the horizontal, carry compressions of amount $2c$. Thus the trapezium $lkk'l'$ is a strip element of area on the conical surface through lk subtending an angle $\delta\theta$ at the axis oh and generating the "leg" $lkff'k'l'$ which carries a compression $2c$ in the direction fl. The material inside the conical surface through lk is stressed by a vertical compression $5c + 2c\cos\beta$ and equal radial and circumferential compressions of amount $3c + 2c\cos\beta$. Triangular elements of area, such as $l'f'm$ in the plan, on the area of indentation in the annulus bounded by the circles through l and g do not lie vertically above the "legs" of material. The pressure over these areas can be increased from $3c$ to $5c$ by adding a vertical compression $2c$ in the triangular prisms of material below the areas.

The field described is statically admissible since it does not violate Tresca's yield criterion and satisfies the equilibrium equation and the boundary conditions. It therefore provides a lower bound for the average collapse pressure. For layers whose thicknesses are not too large compared with the width of the footing, the process of increasing the pressure towards the center of the footing can be repeated. Thus when $B/H = 3.154 + 2.828n$, where n is an integer, the average pressure over the footing is found to be:

$$q_0^l = c\left(2.72 + \frac{1}{6}\frac{B}{H} + 6.1\frac{H}{B} - 1.76\frac{H^2}{B^2}\right) \qquad [7.36]$$

where the n angles such as angle β in Fig. 7.17(a) are taken to be $45°$.

The stress field used here for a circular footing can be modified so that it applies to any area of contact which is *convex*. It can therefore be used to provide lower bounds for any convex area of indentation.

7.7. SQUARE AND CIRCULAR FOOTINGS ON A SEMI-INFINITE LAYER – UPPER BOUNDS (SHIELD, 1955c)

7.7.1. Square footings

General discussion

Upper bounds for the bearing capacity pressure have been obtained in section 7.3 for a square footing on an infinitely thick layer. It was found that the average bearing pressure over the square area of contact cannot exceed the value $5.80c$. The kinematically admissible velocity field used in Fig. 7.3, section 7.3, to obtain the upper bound $5.80c$ was modified by Shield and Drucker (1953) to give a better upper bound $5.71c$ (see [7.10]). The mechanism shown in Fig. 7.3 involves non-zero velocities in a region of the material of the layer that extends to a depth

$(\frac{1}{2}B)\sin 34°$ below the surface of the material. It follows that $5.80c$ is also an upper bound for the average pressure in the bearing capacity calculation of a layer of finite thickness H, provided that the thickness of the layer is such that:

$$\frac{B}{H} \leqslant 2 \csc 34° = 3.58 \qquad\qquad [7.37]$$

For layers which are such that B/H is greater than 3.58, the velocity field can be modified in an obvious manner (for example by adjusting the angles α and β in Fig. 7.3) so that it lies wholly within the material of the layer and can therefore be used to provide upper bounds for the average bearing capacity pressure. The bearing capacity equation is identical to that given in [7.9]. It is found that the values $5.82c$, $5.95c$, $6.20c$ are upper bounds for the average bearing pressure when B/H has the values 4, 5, 6, respectively. These values of the upper bound q_0^u are plotted against B/H in Fig. 7.16, and are joined by a smooth curve, which is marked I in Fig. 7.16, to the line $q_0^u = 5.71c$ ($B/H < 3.58$).

For values of B/H greater than 4.6 approximately, better upper bounds are obtained by Shield (1955c) by means of velocity fields which are somewhat similar to the velocity field used by Shield and Drucker (1953) mentioned above for the value $5.71c$. The relevant equations are given in Shield's paper, the results of Shield's solution are presented in Fig. 7.16 (curves marked II, III and IV in the figure).

A failure mechanism (Fig. 7.18)

In the following, we describe a velocity field which can be used to give reasonable upper bounds for the bearing capacity pressure for large values of B/H. In Fig. 7.18, lmno is the square area of contact on the surface of the layer and as before the initial downward velocity of this area is taken to be v. The mechanism has four vertical planes of symmetry, two passing through the mid-points of opposite sides of the square and two passing through opposite corners of the square. Because of this symmetry, it is only necessary to consider that part of the mechanism which accommodates the downward motion of the triangle cbn. The triangle deh is the vertical projection of the triangle cbn on the lower surface of the layer. The volume befghn is part of a right circular cylinder with axis bn, the volume bafgpn is a right triangular prism, and angle fab is denoted by α. In region cbnhed (and in fact in the whole of the material lying vertically below the square lmno), the material is under a simple homogeneous vertical compression deformation accompanied by a horizontal expansion which is essentially an extension into three dimensions of an obvious modification of the two-dimensional velocity field shown in Fig. 3.11(a). The vertical compressive strain rate $|\dot{\epsilon}_z| = v/H$ and the horizontal tensile strain rates $\dot{\epsilon}_x = \dot{\epsilon}_y = -|\dot{\epsilon}_z|/2 = -v/2H$ satisfy the condition of incompressi-

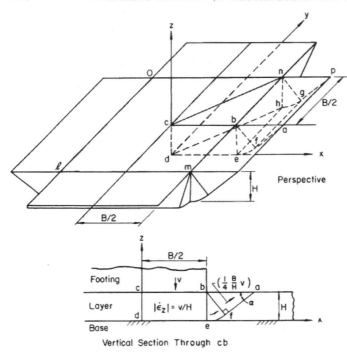

Fig. 7.18. Failure mechanism for a wide square footing (Shield, 1955c).

bility:

$$\dot{\epsilon}_x + \dot{\epsilon}_y + \dot{\epsilon}_z = 0 \qquad\qquad [7.38]$$

With expression [3.43], the rate of dissipation of energy in $\frac{1}{8}$ of the volume of this region is:

$$2c \left(\frac{B^2 H}{8}\right)\left(\frac{v}{H}\right) = \frac{c B^2 v}{4} \qquad\qquad [7.39]$$

The area *deh* is a surface of discontinuity between the material flowing in *cbnhed* and the material at rest immediately adjacent to the base. The discontinuity in velocity across this surface is given by:

$$\Delta v = [(\dot{\epsilon}_x x)^2 + (\dot{\epsilon}_y y)^2]^{1/2} = \frac{v}{2H}(x^2 + y^2)^{1/2} \qquad\qquad [7.40]$$

in which we have used the relation $\dot{\epsilon}_x = \dot{\epsilon}_y = -v/2H$. The rate of dissipation of energy in this discontinuity surface is then found by integration:

$$c \int_A \Delta v \, dA = \frac{cv}{2H} \int_0^{\frac{1}{2}B} \int_y^{\frac{1}{2}B} (x^2+y^2)^{1/2} \, dx dy = \frac{cv}{2H} \int_0^{\frac{1}{4}\pi} \int_0^{\frac{1}{2}B \sec\theta} \rho^2 d\rho d\theta = \frac{0.192 \, cvB^3}{8H}$$

$$[7.41]$$

The rectangle *behn* is also a surface of velocity discontinuity with the discontinuous velocity:

$$\Delta v = [(\dot{\epsilon}_z\, z)^2 + (\dot{\epsilon}_y\, y)^2]^{1/2} = \frac{v}{2H}\,(y^2 + 4z^2)^{1/2} \tag{7.42}$$

It follows that the rate of dissipation of energy in this surface is:

$$c \int_A \Delta v \; dA = \frac{cv}{2H} \int_0^{\frac{1}{2}B} \int_0^H (y^2 + 4z^2)^{1/2}\; dz\,dy = \frac{cvB^2}{8}\left\{\frac{1}{3}\left(1 + 16\frac{H^2}{B^2}\right)^{1/2}\right.$$

$$\left. + \frac{B}{24H}\ln\left[4\frac{H}{B} + \left(1 + 16\frac{H^2}{B^2}\right)^{1/2}\right] + \frac{8}{3}\frac{H^2}{B^2}\ln\left[\frac{B}{4H} + \left(1 + \frac{B^2}{16H^2}\right)^{1/2}\right]\right\} \tag{7.43}$$

The flow in the region *befgnh* is a radial shear plane strain motion similar to that shown in Fig. 3.19(c). The velocity in this region is parallel to the arc surface *ef* and has the constant value $|\dot{\epsilon}_x|(\frac{1}{2}B) = vB/4H$ along the arc. The rate of dissipation of energy in this radial shear zone *befgnh* has the value (using [3.69]):

$$c\left(\frac{vB}{4H}\right)\left(\frac{HB\alpha}{2}\right) = \frac{cv\alpha B^2}{8} \tag{7.44}$$

The rate of dissipation of energy in the discontinuous circular surface *efgh* is also given by [7.44]. Energy is also dissipated in the discontinuous surfaces *apgf* and *hpn*. The dissipations have the value:

$$c\left(\frac{vB}{4H}\right)\left(\frac{BH \cot \alpha}{2}\right) + c\left(\frac{vB}{4H}\right)\left(\frac{\alpha H^2}{2}\right) + c\left(\frac{vB}{4H}\right)\left(\frac{H^2 \cot \alpha}{2}\right)$$

$$= \frac{cvB^2}{8}\left[\cot \alpha + \alpha\left(\frac{H}{B}\right) + \left(\frac{H}{B}\right)\cot \alpha\right] \tag{7.45}$$

Equating the total rate of internal dissipation of energy to the rate of external work done by the footing $v(q_0B^2/8)$ yields an upper bound value for the average bearing capacity pressure of a wide square footing

$$q_0^u(\alpha) = c\left\{2 + 2\alpha + \cot \alpha + \frac{H}{B}\,(\alpha + \cot \alpha) + 0.192\frac{B}{H} + \frac{1}{3}\left(1 + 16\frac{H^2}{B^2}\right)^{1/2}\right.$$

$$\left. + \frac{B}{24H}\ln\left[4\frac{H}{B} + \left(1 + 16\frac{H^2}{B^2}\right)^{1/2}\right] + \frac{8}{3}\frac{H^2}{B^2}\ln\left[\frac{B}{4H} + \left(1 + \frac{B^2}{16H^2}\right)^{1/2}\right]\right\} \tag{7.46}$$

The bearing capacity function $q_0^u(\alpha)$ has a minimum value with respect to α when:

$$\sin \alpha = \frac{(B + H)^{1/2}}{(2B + H)^{1/2}} \qquad [7.47]$$

The upper bounds q_0^u, given by the minimum values of [7.46], are plotted in Fig. 7.16 for values of B/H between 10 and 14 (curve marked V in the figure), in order to compare the bounds with those obtained above.

For very large values of B/H, the value of α given by [7.47] tends to $45°$ Neglecting the first and higher powers of H/B and taking $\alpha = \frac{1}{4}\pi$ in [7.46], we find that for large values of B/H, the upper bound is given by:

$$q_0^u = c\left(5.07 + 0.192 \frac{B}{H}\right) \qquad [7.48]$$

For very large values of B/H, the difference between the upper bound [7.48] and the lower bound [7.35] is 15% of the lower bound, and the true value of the average bearing pressure is given by $4.25c + 0.179c\, B/H$ to within $\pm 7\%$.

Upper bounds for the average bearing capacity pressure under a *rectangular* footing could be obtained by modifying the velocity fields of this section.

7.7.2. Circular footings (Fig. 7.19)

Figure 7.19 shows a simple velocity field which satisfies the kinematical boundary conditions of a circular footing on a layer. The field is axi-symmetrical about the z-axis which passes through the center of the footing. The material lies vertically below the area of the footing in a region of homogeneous vertical compression accompanied by a horizontal expansion. The vertical compressive strain rate $|\dot{\epsilon}_z| = v/H$ is accommodated by the radial and circumferential expansions $\dot{\epsilon}_r = \dot{\epsilon}_\theta$ $= -|\dot{\epsilon}_z|/2 = -v/2H$, where v is the downward velocity of the footing. These components represent an incompressible plastic deformation. The radial and vertical components v_r and v_z of velocity in the region $bcod$ in Fig. 7.19 have the value $|v_r| = vr/2H$ and $|v_z| = vz/H$. With expression [3.43], the rate of dissipation of energy in the region $bcod$ has the value:

$$2c\left(\frac{v}{H}\right)\left(\frac{\pi B^2 H}{4}\right) = \frac{cv\pi B^2}{2} \qquad [7.49]$$

In the remainder of the velocity field, region abd, the flow is parallel to the conical surface ad, which is a surface of velocity discontinuity. The continuity of the normal velocity across the discontinuous surface bd together with the incom-

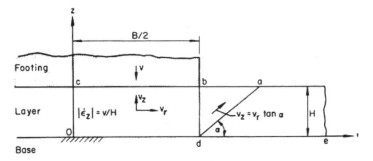

Fig. 7.19. Failure mechanism for a circular footing (Shield, 1955c).

pressibility condition determine the velocity components in region abd to be given by $|v_r| = vB^2/8rH$ and $|v_z| = |v_r| \tan \alpha$, where α is the angle ade. The plastic strain rates associated with this velocity field can be computed directly from the strain rate— velocity compatibility relations:

$$|\epsilon_r| = \left| \frac{\partial v_r}{\partial r} \right| = \frac{vB^2}{8H} \left(\frac{1}{r^2} \right)$$

$$|\epsilon_\theta| = \frac{|v_r|}{r} = \frac{vB^2}{8H} \left(\frac{1}{r^2} \right)$$

$$\epsilon_z = \frac{\partial v_z}{\partial z} = 0$$

$$\gamma_{rz} = \left| \frac{\partial v_r}{\partial z} + \frac{\partial v_z}{\partial r} \right| = \frac{vB^2}{8H} \left(\frac{1}{r^2} \right) \tan \alpha \qquad [7.50]$$

The absolute value of the numerically greatest principal component of the plastic strain rate is found from [7.50] to be:

$$\max |\dot{\epsilon}| = \frac{vB^2}{16H} (1 + \sec \alpha) \frac{1}{r^2} \qquad [7.51]$$

with the expression [3.128], the rate of dissipation of energy in region abd has the value:

$$2c \int_V \max |\dot{\epsilon}| \, dV = \frac{cvB^2}{8H} (1 + \sec \alpha) \int_{\frac{1}{2}B}^{(\frac{1}{2}B + H \cot \alpha)} \left(\frac{1}{r^2} \right) \left[H - \left(r - \frac{B}{2} \right) \tan \alpha \right] 2\pi r dr$$

$$= \frac{cv\pi B^2}{4} (1 + \sec \alpha) \left[\left(1 + \frac{B}{2H} \tan \alpha \right) \ln \left(1 + \frac{2H}{B} \cot \alpha \right) - 1 \right]$$

$$[7.52]$$

The rate of energy dissipation along the surfaces of discontinuity bd, da and od is the cohesion c multiplied by the relative velocity integrated over the area of the surface of discontinuity:

Vertical plane bd

$$c \int_0^H \left(\frac{vz}{H} + \frac{vB}{4H} \tan \alpha \right) \pi B dz = \frac{cv\pi B^2}{4} \left(\frac{2H}{B} + \tan \alpha \right) \qquad [7.53]$$

Inclined plane da

$$c \int_{\frac{1}{2}B}^{(\frac{1}{2}B+H \cot \alpha)} \left(\frac{vB^2}{8Hr} \right) \sec \alpha \, (2\pi r \sec \alpha) \, dr = \frac{cv\pi B^2}{4} \sec \alpha \csc \alpha \qquad [7.54]$$

Horizontal plane od

$$c \int_0^{\frac{1}{2}B} \left(\frac{vr}{2H} \right) 2\pi r dr = \frac{cv\pi B^2}{4} \left(\frac{1}{6} \frac{B}{H} \right) \qquad [7.55]$$

For a smooth circular footing, the upper bound is obtained by equating $(\pi B^2/4)q_0 v$, the rate of work done by the footing pressure q_0, to the total rate of dissipation:

$$q_0^u(\alpha) = c \left\{ 2 + \frac{2H}{B} + \frac{1}{6} \frac{B}{H} + \csc \alpha \sec \alpha + \tan \alpha + (1 + \sec \alpha) \left[\left(1 + \frac{B}{2H} \tan \alpha \right) \right. \right.$$

$$\left. \left. \ln \left(1 + \frac{2H}{B} \cot \alpha \right) - 1 \right] \right\} \qquad [7.56]$$

Upper bounds q_0^u are obtained by minimizing [7.56] with respect to α for particular values of B/H. For $B/H = 14$, for example, the minimum value of $q_0^u(\alpha)$ is near $\alpha = 36°$ where it has the value 7.51. If the first and higher powers of H/B are neglected in [7.56] for $q_0^u(\alpha)$ we obtain:

$$q_0^u(\alpha) = \left(2 + \frac{1}{6} \frac{B}{H} + \csc \alpha \sec \alpha + \tan \alpha \right) c$$

This expression has a minimum value for $\alpha = 35°16'$ so that, for large values of B/H:

$$q_0^u = c \left(4.83 + \frac{1}{6} \frac{B}{H} \right) \qquad [7.57]$$

is an upper bound for the collapse value of the average bearing capacity pressure. The percentage difference between the upper bound [7.57] and the lower bound [7.36] tends to zero as B/H tends to infinity.

7.8. BEARING CAPACITY OF CIRCULAR FOOTINGS BY SLIP-LINE METHOD

7.8.1. Basic concept and slip-line equations

In section 6.7, Chapter 6, the slip-line solutions for the bearing capacity of a strip footing have been described. In the case of a strip footing, there is no deformation in the long direction of the footing. This implies that the long direction is a principal stress direction with the corresponding principal stress having a value intermediate to the other two principal stresses. The equilibrium equations simplify to two equations and three unknown stress components ([6.92]). The Coulomb yield criterion [6.91] provides the third equation necessary to render the system *statically determinate*. The value of the intermediate principal stress is indeterminate but unnecessary for the computation of bearing capacity pressure. For the case

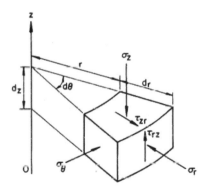

(a) Axi-Symmetrical Stress State

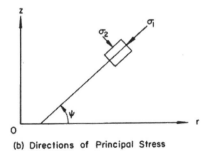

(b) Directions of Principal Stress

Fig. 7.20. Stress components in a circular footing problem.

of a circular footing, the only non-vanishing stress components in a stress distribution which is axially symmetric about the z-axis are, in the usual notation, σ_r, σ_θ, σ_z, τ_{rz} (Fig. 7.20a). The circumferential stress σ_θ is a principal stress. The other principal stress σ_1, σ_2 ($\sigma_1 \geqslant \sigma_2$, compression is positive) in the meridional planes being given by (Fig. 7.20b):

$$\sigma_1 = \tfrac{1}{2}(\sigma_r + \sigma_z) + [\tfrac{1}{4}(\sigma_r - \sigma_z)^2 + \tau_{rz}^2]^{1/2}$$

$$\sigma_2 = \tfrac{1}{2}(\sigma_r + \sigma_z) - [\tfrac{1}{4}(\sigma_r - \sigma_z)^2 + \tau_{rz}^2]^{1/2} \qquad \text{[7.58a,b]}$$

There are a number of plastic stress states permitting an axially symmetric plastic flow. It will be *assumed* herein that the circumferential stress σ_θ is equal to one of the other two principal stresses. This was hypothesized by Haar and Von Karman (1909). A detailed study of Shield (1955b) and Cox et al. (1961) has pointed out that stress distributions agreeing with the *Haar and Von Karman hypothesis* are quite general and therefore, are likely to be applicable to a number of problems. For definiteness, the circumferential stress σ_θ is taken equal to the algebraically smaller of the other two principal stresses, i.e., σ_2.

Haar and Von Karman's hypothesis

$$\sigma_\theta = \sigma_2 < \sigma_1 \qquad \text{[7.59]}$$

The *Coulomb yield criterion* then requires the principal stresses σ_1, σ_2 in the meridional planes to satisfy:

$$\sigma_1 - \sigma_2 = 2c \cos\varphi + (\sigma_1 + \sigma_2) \sin\varphi \qquad \text{[7.60]}$$

The *equations of equilibrium* take the forms:

$$\frac{\partial \sigma_r}{\partial r} + \frac{\partial \tau_{zr}}{\partial z} + \frac{n}{r}(\sigma_r - \sigma_\theta) = 0$$

$$\frac{\partial \tau_{zr}}{\partial r} + \frac{\partial \sigma_z}{\partial z} + n\frac{\tau_{zr}}{r} + \gamma = 0 \qquad \text{[7.61a,b]}$$

in which:

$$n = \begin{cases} 0 \text{ for a strip footing} \\ 1 \text{ for a circular footing} \end{cases}$$

Equations [7.59], [7.60] and [7.61] provide four equations for the determination of the four stress components so that in a sense the problem is *"statically determinate"*. Using the Coulomb yield criterion [7.60] and the Haar and Von Karman hypothesis [7.59], the four stress components can be expressed in terms of two dependent variables s and ψ (compression is positive):

$$\sigma_r = (s \csc \varphi - c \cot \varphi) + s \cos 2\psi$$

$$\sigma_z = (s \csc \varphi - c \cot \varphi) - s \cos 2\psi$$

$$\tau_{zr} = s \sin 2\psi$$

$$\sigma_\theta = (s \csc \varphi - c \cot \varphi) - s \qquad\qquad [7.62a,b,c,d]$$

in which ψ is the inclination from the r-axis of the larger principal stress direction (Fig. 7.20b), i.e., ψ specifies the orientation of the principal axis σ_1, and s is the radius of the Mohr's circle for stress. The stress components for the case of a strip footing follow immediately from [7.62] if the principal stress σ_θ is replaced by any value intermediate to the other two principal stresses.

When the stress components as given by [7.62] are substituted into the equilibrium equations [7.61], two, first-order, partial differential equations result. These equations are hyperbolic in nature with orthogonal characteristics in the (r,z)-plane which coincide with the slip lines. There are two distinct families of characteristics or slip lines given by:

$$\frac{dz}{dr} = \tan \left[\psi \pm \left(\tfrac{1}{4}\pi - \tfrac{1}{2}\varphi\right)\right] \qquad\qquad [7.63]$$

The variation of the dependent variables s and ψ along the characteristic curves can be expressed in differential form:

$$\cot \varphi \, ds \pm 2s \, d\psi - (\cos \varphi \, dz \pm \sin \varphi \, dr) \, \gamma + n \frac{s}{r} \left[\cos \varphi \, dr \pm (1 - \sin \varphi) \, dz\right] = 0 \qquad [7.64]$$

For the case of the strip footing ($n = 0$), [7.64] becomes the well-known Kötter's (1888) equations (see [6.95]).

Here, as in the strip footing case, [7.64] together with the stress boundary conditions can be solved by applying finite difference techniques. Shield (1955b) has presented such a solution for a circular *smooth* footing on a purely cohesive weightless soil and Eason and Shield (1960) extend the same solution to the case of a perfectly *rough* footing. Similarly, Berezantsev (1952), Cox et al. (1961), and Cox (1962) have solved a number of cases of the circular surface footing problems including both internal soil friction φ and soil weight γ. Recently, Larkin (1968) has solved the case of a circular footing buried at a very shallow depth in cohesionless soils.

In the following presentation the individual cases connected with the bearing capacity of circular footings will be taken up in the following sequence. First we consider the bearing capacity of a circular footing on a *purely cohesive, weightless* soil. Then we present in succession the following cases: (1) a circular footing on a $c - \varphi$ *weightless* soil; (2) a circular footing on a $c - \varphi$ *ponderable* soil; and (3) a circular footing buried at a very *shallow* depth in *cohesionless* soils.

7.8.2. Circular footings on a purely cohesive, weightless soil ($\varphi = 0$, $\gamma = 0$)

Smooth footing (Fig. 7.21a)

Shield (1955b) has presented a *complete* solution to the problem of the bearing capacity of a *smooth*, circular flat-ended rigid footing on a purely cohesive, weightless soil (Tresca material). The partial plastic stress field near the circular footing is obtained by numerical integration of [7.64]. These differential equations are replaced by finite difference equations and a slip-line field can be constructed (Fig. 7.21a). The calculated pressure distribution over the footing from the slip-line net of Fig. 7.21(a) is shown graphically in Fig. 7.22 (curve marked smooth). The average pressure is found to be $5.69c$, and referring now to Fig. 7.21(a) the distance of the point b from the origin is found to be $1.58(\frac{1}{2}B)$.

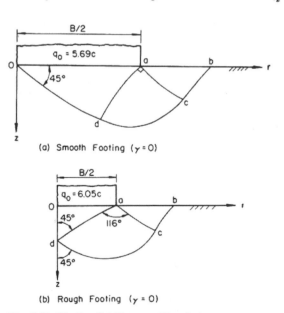

(a) Smooth Footing ($\gamma = 0$)

(b) Rough Footing ($\gamma = 0$)

Fig. 7.21. Slip-line field in a meridional plane by a circular footing.

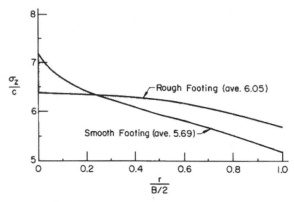

Fig. 7.22. Footing pressure distributions over a circular footing.

It can be shown that the stress field in the region *oabcd* (Fig. 7.21a) can be extended into the rest of the body without violating the yield condition. The method used by Shield is an application of the theorems of Bishop (1953) on the admissibility of incomplete solutions, and parallel that used by Bishop to show the validity of the Prandtl stress solution for the corresponding plane strain strip footing problem. The extended stress field is found to confine in a region of width $3.20(\frac{1}{2}B)$ from the center of the footing and of depth $3.36(\frac{1}{2}B)$ from the surface of the footing. The extended stress field in this region is a statically admissible stress field and the lower-bound theorem of the limit analysis shows that the value $5.69c$ is a lower bound for the average bearing capacity pressure q_0.

Shield (1955b) has shown that a kinematically admissible velocity field can also be associated with the partial stress field in the region *oabcd* of Fig. 7.21(a). The value $5.69c$ is therefore an upper bound for the average bearing capacity pressure also, so that the *exact* value of the average pressure, q_0, has been determined. Further, a theorem due to Hill (1951) states that where deformation is actually occuring the stress field is *unique*. Also a theorem due to Bishop et al. (1956) states that if any region of a complete solution is necessarily rigid, then it must be rigid in *all* complete solutions. Application of these theorems shows that the partial stress field of Fig. 7.21(a) in region *oabcd* is the *actual* plastic stress field and plastic deformation can only occur in region *oabcd* of Fig. 7.21(a).

The closely allied problem of the bearing capacity of a perfectly *rough*, circular, flat-end rigid footing on a purely cohesive weightless soil has been investigated by Eason and Shield (1960). Also, these problems are both of importance in testing the hardness of metals (see, for example, Tabor, 1951).

Rough footing (Fig. 7.21b)

The same numerical procedure adopted in the smooth footing case can also be used here for determining the partial plastic stress field for a *rough* circular footing.

The calculated pressure distribution over the footing has been reported by Eason and Shield (1960) and is also graphically shown in Fig. 7.22. The average pressure q_0 over the rough footing is found to be $6.05c$. With the footing radius oa as $\frac{1}{2}B$, the distance ob, in the slip-line net of Fig. 7.21(b), is found to be $1.88(\frac{1}{2}B)$ and the distance od is found to be $0.57(\frac{1}{2}B)$. The angle of the fan is found to be $116°$. A coefficient of friction may be defined along oa by writing:

$$\mu = \tan \delta = \frac{\tau_{rz}}{\sigma_z} \qquad [7.65]$$

The maximum value of μ along oa is found to occur at footing edge a and to have the value 0.139. Thus, the solution so obtained holds for a rough footing provided that the coefficient of friction between the material and the footing exceeds 0.139 which corresponds approximately to the friction angle $\delta = 8°$.

As in the case of the smooth footing, the partial plastic stress field (Fig. 7.21b) has been extended into the rest of the body. The method of extension closely follows that used in the previous smooth footing case in which theorems developed by Bishop (1953) for the case of plane strain were applied. The extended stress field is found to confine in a region of width $3.59(\frac{1}{2}B)$ from the center of the footing and of depth $3.65(\frac{1}{2}B)$ from the surface of the footing. The stress field obtained in this region is statically admissible and the value $6.05c$ is a lower bound for the average pressure q_0 by the limit analysis theorems. A kinematically admissible velocity can also be constructed by numerical integration to associate with the partial plastic stress field in the region $oabcd$ (Fig. 7.21b). The value 6.05 is thus the actual bearing pressure for a rough footing, as it is also an upper bound. Following the same arguments as in the smooth footing case, the partial plastic stress field of Fig. 7.21(b) in region $oabcd$ is the actual plastic stress field, and plastic deformation can only occur in region $oabcd$ of Fig. 7.21(b).

7.8.3. Circular footings on a $c - \varphi$ weightless soil $(\gamma = 0)$

Following Shield (1955b), Cox et al. (1961) obtain a *complete* solution for the bearing capacity problem of a flat-ended, *smooth*, rigid circular footing on a $c - \varphi$ *weightless* soil. Their numerical procedure is based upon the approximation of [7.64] by finite-difference equation. Details of the finite-difference equations are described in their paper.

Table 7.1 gives values of the three-dimensional average bearing capacity pressure $(q_0/c)_3$, the maximum pressure on the footing $(\sigma_z)_{max}$ which occurs at the center of the footing, and also the ratio ob/oa (Fig. 7.23), all as functions of the angle of internal friction φ for $\varphi = 0°$ $(5°)$ $40°$. For purpose of comparison, Table 7.1 also shows values of the two-dimensional average bearing capacity pressure $(q_0/c)_2$ for

TABLE 7.1

Bearing capacity of a smooth circular footing on a $c-\varphi$ weightless soil (Cox, et al., 1961).

$\varphi\ (°)$	Circular footing			Strip footing
	$(N_c)_3 = (q_0/c)_3$	$(\sigma_z/c)_{max}$	ob/oa	$(N_c)_2 = (q_0/c)_2$
0	5.69	7.1	1.58	5.14
5	7.44	9.7	1.71	6.49
10	9.98	14	1.88	8.34
15	13.9	22	2.09	11.0
20	20.1	34	2.37	14.8
25	30.5	59	2.73	20.7
30	49.3	110	3.21	30.1
35	85.8	210	3.89	46.1
40	164	430	4.86	75.3

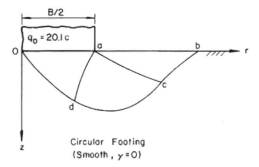

Fig. 7.23. Slip-line net for $\varphi = 20°$ (Cox et al., 1961).

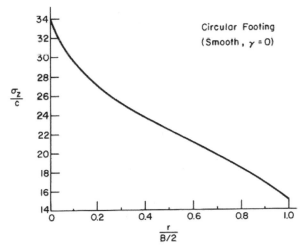

Fig. 7.24. Pressure exerted on the soil by the circular footing ($\varphi = 20°$) (Cox et al., 1961).

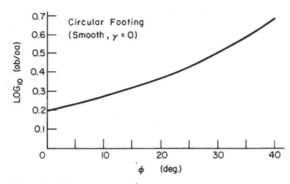

Fig. 7.25. Variation of ob/oa with φ (Cox et al., 1961).

the problem of a strip footing (see Table 6.11). Fig. 7.23 shows the calculated net of slip-lines, Fig. 7.24 shows the pressure distribution, σ_z, on the face of the footing, for the particular case when $\varphi = 20°$. For the other values of φ considered in Table 7.1, the slip-line nets exhibit the same geometrical features as for $\varphi = 20°$, the major difference being one of scale. This difference in scale is exemplified by the ob/oa which, as shown in Fig. 7.25, increases markedly with φ.

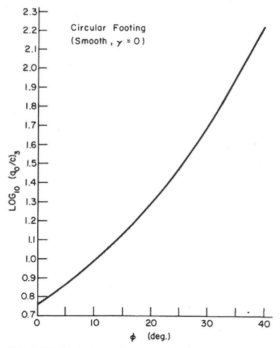

Fig. 7.26. Variation of average bearing pressure with angle of internal friction (Cox et al., 1961).

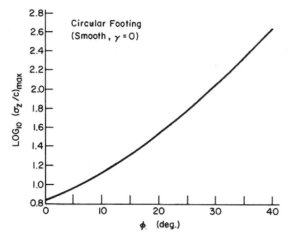

Fig. 7.27. Variation of maximum pressure exerted on the soil with angle of internal friction (Cox et al., 1961).

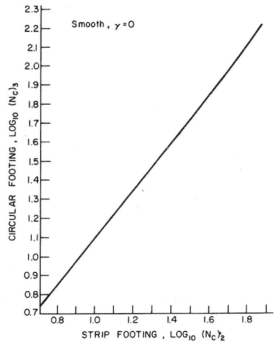

Fig. 7.28. Variation of the bearing capacity factor of a circular footing vs. the bearing capacity factor of a strip footing (Cox et al., 1961).

In Figs. 7.26 and 7.27 the quantities q_0 and $(\sigma_z)_{\max}$ are plotted against φ and in Fig. 7.28 the bearing capacity factor $(N_c)_3 = (q_0/c)_3$ for a three-dimensional circular footing is plotted against the corresponding bearing capacity factor $(N_c)_2 = (q_0/c)_2$ for a two-dimensional strip footing. Fig. 7.28 shows that, as a simple approximation, $\log_{10}(N_c)_3$ may be taken as a linear function of $\log_{10}(N_c)_2$ over the range of φ considered. Referring to Fig. 7.28, this linear function has the form:

$$\log_{10}(N_c)_3 = 1.25 \log_{10}(N_c)_2 - 0.135 \qquad [7.66]$$

or more conveniently:

$$(N_c)_3 = 0.733 \, (N_c)_2^{1.25} \qquad [7.67]$$

For a wider comparison with other studies on bearing capacity, it is convenient to introduce the shape factor $\lambda_c = (N_c)_3/(N_c)_2$ for a circular footing. Using [7.67], the shape factor λ_c for a smooth circular footing has the value:

$$\lambda_c = 0.733 \, N_c^{1/4} \qquad [7.68]$$

where $N_c = (N_c)_2$ denotes the well-known bearing capacity factor for a strip footing. The values of N_c and thus λ_c depend only on the internal friction angle φ.

A velocity field associated with the partial stress field (Fig. 7.23) has been constructed by Cox et al. (1961) for $\varphi = 20°$ and by Shield (1955b) in the preceding case for $\varphi = 0°$. Since there is no reason a priori to suppose that it cannot be similarly constructed for all the values of φ of interest here, it follows then from the upper-bound theorem of limit analysis that the average values of the bearing capacity pressure $(q_0)_3$ given in Table 7.1 are upper bounds to the *correct* values.

It is now necessary to prove that the values $(q_0)_3$ in Table 7.1 are also lower bounds and hence are the *correct* values. Thus it must be shown that the partial plastic stress field can be extended throughout the entire rigid region, without violating the conditions of equilibrium and Coulomb yield condition. Only in this way is a complete solution derived.

The method used by Shield (1955b) in the preceding case for $\varphi = 0$ has been extended by Cox et al. (1961) for the more general case $\varphi > 0$. They obtain an extended stress field for $\varphi = 20°$. In view of the fact that the required extension of the stress field has been shown to be possible for $\varphi = 20°$, and, by Shield (1955b), for $\varphi = 0$, it seems unlikely, since there are no new physical complications that a complete solution cannot be found quite generally for $0° \leqslant \varphi \leqslant 40°$. It is therefore concluded that the bearing pressures obtained in Table 7.1 are also lower bounds and hence are equal to the *correct* values. In addition, it can be concluded that the partial stress fields that were obtained in *oabcd* (Fig. 7.23) are the *actual* ones.

TABLE 7.2

Values of the average bearing capacity pressure, q_0/c for a smooth circular footing on a $c-\varphi$ ponderable soil (Cox, 1962)

φ (°)	$G = \gamma B/2c$				
	0	10^{-2}	10^{-1}	1	10
0	5.69	5.69	5.69	5.69	5.69
10	9.98	9.99	10.0	10.4	13.8
20	20.1	20.1	20.3	22.4	38.8
30	49.3	49.4	50.5	60.6	141
40	164	165	173	237	754

TABLE 7.3

Values of the ratio ob/oa (see Fig. 7.29) for a smooth circular footing on a $c-\varphi$ ponderable soil (Cox, 1962)

φ (°)	$G = \gamma B/2c$				
	0	10^{-2}	10^{-1}	1	10
0	1.58	1.58	1.58	1.58	1.58
10	1.88	1.88	1.88	1.82	1.63
20	2.37	2.37	2.35	2.24	1.94
30	3.21	3.21	3.18	2.96	2.55
40	4.86	4.86	4.76	4.32	3.77

7.8.4. Circular footings on a $c - \varphi$ ponderable soil

The bearing capacity computations presented in section 7.8.3 for a *weightless* soil loaded by a smooth circular footing have been extended by Cox (1962) to include the effect of soil weight. Results have been obtained and expressed in terms of the dimensionless soil weight parameter, $G = \gamma B/2c$, and the internal friction angle φ, for the average bearing capacity pressure q_0. Table 7.2 gives the values of q_0 as a function of G and φ. The numerical results show clearly the strong dependence of the average bearing pressure upon soil weight, for frictional soils. Fig. 7.29 shows the calculated slip-line net in the region of the partial plastic stress field $oabcd$ for $G = 10$ and $\varphi = 40°$. Table 7.3 gives the values of the ratio ob/oa and shows that this ratio decreases markedly as G increases, provided that φ is not zero.

Although the determination of a *complete* solution is strictly necessary, the fact

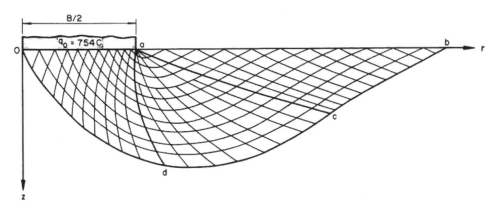

Fig. 7.29. Slip-line net for a smooth circular footing, $\varphi = 40°$, $G = 10$ (Cox, 1962).

that a complete solution has been shown to exist for an *weightless* soil (section 7.8.3) would suggest that the same is also true for a *ponderable* soil. The bearing capacity pressure q_0 presented in Table 7.2 may therefore be considered to be the *actual* values.

7.8.5. Circular shallow footings on a cohesionless soil (c = 0)

The bearing capacities of a very *shallow, smooth* circular footing on a *cohesionless* soil have been obtained by Larkin (1968) by integration of the slip-line equations [7.64]. The equations are integrated numerically by 'a finite difference approximation based on the procedure developed by Cox et al. (1961). Calculations based on partial plastic stress field near the footing are carried out for cohesionless soils with angles of internal friction of $\varphi = 30°$ and $40°$. The results are tabulated in Table 7.4 for various ratios of depth of burial D/B. It can be seen from this table that the bearing capacity increases significantly with quite small increases in depth of burial. As an example, for the case $\varphi = 30°$ the bearing capacity of a smooth circular footing at the surface can be increased 100% by initially burying the

TABLE 7.4

Values of the average bearing capacity pressure $q_0/\gamma(B/2)$ for a smooth circular footing (Larkin, 1968)

D/B	0.005	0.0125	0.025	0.05	0.125	0.25	0.5
$\varphi = 30°$	7.8	8.5	9.7	11.8	17.7	27.4	49
$\varphi = 40°$	53	57	63	73	102	151	258

TABLE 7.5

Values of the bearing capacity factors N_γ and N_q and the shape factors λ_γ and λ_q for a smooth circular footing (Larkin, 1968)

Factors	N_γ	N_q	λ_γ	λ_q
$\varphi = 30°$	7.8	45	0.94	0.91
$\varphi = 40°$	43.5	167	1.18	1.23

footing at a depth of 0.09 of the footing diameter (values of the bearing pressure for $D/B = 0$ and $D/B = 0.09$ are extrapolated graphically from Table 7.4). For $\varphi = 40°$, the 100% increase in bearing capacity can be effected by an initial burial at a depth of 0.13 of the footing diameter. This would indicate that the small footing settlement which accompanies the loading up to the point of failure, significantly increases the actual bearing capacity. This fact suggests that the small footing settlement prior to failure in footing tests may be a significant factor in explaining the large discrepancies between theoretical and experimentally observed bearing capacities.

For a wider comparison with other studies on bearing capacity, it is convenient to fit the widely used Terzaghi's equation:

$$(q_0)_2 = \gamma \tfrac{1}{2} B N_\gamma + \gamma D N_q \qquad [7.69]$$

for a two-dimensional strip footing and:

$$(q_0)_3 = \lambda_\gamma \gamma \tfrac{1}{2} B N_\gamma + \lambda_q \gamma D N_q \qquad [7.70]$$

for a three-dimensional circular footing, to the values given in Table 7.4. In [7.69] and [7.70], N_γ and N_q are the well-known bearing capacity factors for a *strip* footing. In [7.70], λ_γ and λ_q are shape factors for a *circular* footing similar to the shape factor λ_c defined in section 7.8.3. The values of N_γ, N_q and thus λ_γ and λ_q depend only on φ. Equation [7.70] is now used to fit the values in Table 7.4 by the *method of least squares*. The maximum deviation is found to be less than 3%. The resulting values for N_γ, N_q, λ_γ and λ_q are given in Table 7.5.

7.8.6. Summary and conclusions

The boundary conditions together with the slip-line equations [7.64] discussed herein suitably define the bearing capacity problem of a circular footing on a $c - \varphi - \gamma$ soil. The *complete* solution of this problem is obtained in three distinct stages. First, once suitable stress boundary conditions have been specified the stress components are calculated near the footing. This local stress field around the footing is called *partial plastic stress field* which is sufficient to determine the stresses or bearing capacity pressure on the circular footing. The solution so obtained is merely a solution and limit analysis shows that the bearing pressure so obtained is *neither* an upper bound nor a lower bound to the correct value.

Secondly, with suitable kinematical boundary conditions specified, a velocity field in this partial stress region near the footing may be constructed and being zero elsewhere. Provided that this can be done in a kinematically admissible manner, the solution so far obtained is then termed *incomplete*, and limit analysis only shows that the bearing pressure so obtained is *an upper bound* to the correct value.

Thirdly, the partial plastic stress field is, if possible, extended into the entire rigid region in such a way as to satisfy the conditions of equilibrium and yield provided this can also be done in a satisfactory manner, the solution is now termed *complete* and limit analysis shows that the bearing capacity obtained previously is also a *lower bound* to, and hence is identical with, the *correct* value. Further, a theorem due to Hill (see Bishop, 1953) shows that the stress field in the deforming region (partial stress field region) is *unique* and thus gives the actual stress field.

The discussion and the applications herein show that the *hypothesis* proposed by Haar and Von Karman are of very fundamental significance in the solution of axially symmetric problems considered here. This hypothesis states that the circumferential stress is equal to one of the principal stresses in the meridional planes during plastic deformation. From this hypothesis, the stress and velocity fields are *hyperbolic* with coincident families of characteristics or slip-lines, and the stress field is *statically determinate* under appropriate boundary conditions. This striking situation makes for such *simplicity* in the mathematical and associated computational investigations. It is interesting to note that when Tresca's yield criterion, obtaining as a special case of Coulomb criterion when $\varphi = 0$, is replaced by Von Mises's yield criterion, the governing equations for stress and velocity fields are found to have *no real* characteristics or slip-lines (Hill, 1950; Parsons, 1956). Thus, a marked degree of mathematical simplicity is achieved here through the adoption of Coulomb's yield criterion and Haar and Von Karman's hypothesis.

In the solutions of the bearing capacity problem of a circular footing on a $c - \varphi$ soil, it has been found that the bearing capacity pressure increases markedly with the angle of internal friction φ. Although the details of the solution of various cases considered vary with the values φ and γ, the basic features remain unaltered. This fact suggests that the somewhat tedious construction of *complete* (as distinct from partial or incomplete ones) solutions may well be omitted in proceeding to solve problems of a similar type. However, if such a procedure is followed, then careful judgement is obviously necessary.

Chapter 8

ACTIVE AND PASSIVE EARTH PRESSURES

8.1. INTRODUCTION

The determination of the lateral earth pressure of a fill on a retaining wall when frictional forces act on the back of the wall is one of the classical stability problems in soil mechanics. In the same way as before, this problem can quite conveniently be solved by limit analysis. In the following sections, limit analysis is first applied to obtain upper and lower bounds for the lateral earth pressure against a vertical standing smooth retaining wall. It is shown by the coincidence of upper and lower bounds that the well-known Coulomb solution (assuming plane sliding wedge) for frictionless walls are exact. Upper-bound technique of limit analysis is then applied to obtain the solutions for the general lateral earth pressure problems including the effect of wall friction.

8.1.1. Active and passive pressures (Fig. 8.1)

Before we attempt to find the pressure in the rear face of a retaining wall, we note that lateral earth pressure can be divided into active earth pressure and passive earth pressure as illustrated in Fig. 8.1(a), which shows a particular apparatus consisting of a large bin with a movable end section. By filling the bin with sand, a lateral pressure is developed against the end section which simulates the wall. This wall is constructed so that it can be held in a fixed position or moved inward or outward. A horizontal force P_n normal to the wall must be applied to this wall in order to keep the apparatus in equilibrium in its initial position. Since the wall can have two directions of motion, into the bank or away from the bank, passive and active earth pressures are developed.

If the wall is initially at rest and held by a force $P = P_o$, it is apparent that for a cohesionless soil, as the force P is reduced, the wall will be forced outward due to the weight of the soil. As P is gradually reduced, the soil undergoes first elastic deformation, then elastic–plastic deformation (contained plastic flow, where the elastic and plastic strains are of the same order) and finally, uncontained plastic flow and thus defines the active collapse load, P_{an}. Fig. 8.1(b) shows a load displacement curve depicting the behavior of the soil under active and passive earth pressure. The points marked P_o, P_{pn} and P_{an} represent the wall force at rest, at passive collapse, and at active collapse, respectively. The subscripts p, a and n

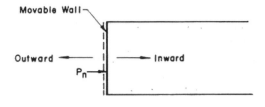

(a) Vertical Section Through Bin

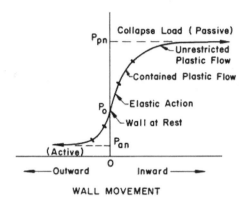

WALL MOVEMENT

(b) Load Displacement Relationship

Fig. 8.1. Results of retaining wall tests.

indicate passive, active and normal components of the force P, respectively. If the limit load theorems are applied to obtain bounds on the collapse loads, the stress (equilibrium) solutions will give force points which fall within the range $P_{pn}-P_{an}$. The velocity solutions will give force points falling outside the range of $P_{pn}-P_{an}$. When the active collapse load is sought, stress solutions give upper bounds (numerically) and velocity solutions give lower bounds.

The explanation for this lies in the definition of the upper and lower bounds. The lower bound is a load for which the structure will not fail, if equilibrium and the yield criterion are satisfied. The upper bound is a load for which the structure will fail when the rate of external work equals or exceeds the rate of internal energy dissipation in a geometrically admissible velocity field. It is seen in Fig. 8.1(b) that the loads for which the wall will not continually displace (fail) lie below the passive earth pressure and above the active earth pressure. Likewise, the upper bounds lie outside this region, above the passive earth pressure and below the active earth pressure. To approach the true active earth pressure, one must maximize the upper bound and minimize the lower bound.

If the force P is increased from P_o, displacement occurs and failure (continuous displacement under constant load) is the result as the passive earth pressure P_{pn}, is approached. The active and passive definitions are derived from the role the back-fill material plays in the two cases. In the active earth pressure case, the failure is due to the soil's weight overcoming the internal friction and pressure on the wall, that is, the soil is playing an active role. In the passive earth pressure case the failure is due to the pressure on the wall overcoming the soil's weight and internal friction, hence, the soil plays a passive role.

8.1.2. Limit equilibrium approach (Fig. 8.2)

Smooth wall ($\delta = 0$)

First, examine the conventional limit equilibrium approach, introduced by Coulomb (1776). The soil is assumed, in the active earth pressure case, to fail by sliding along a plane inclined at an angle Ω to the vertical, as illustrated in Fig. 8.2. Since the wedge is in a state of equilibrium, the force polygon indicated in Fig. 8.2 must close. The force P is the reaction along the back of the wall (the angle of wall friction, δ, is zero); F is the reaction along the face RS, at an angle φ to the normal to RS; and W is the weight of the soil wedge. The wedge angle, $\Omega = \frac{1}{4}\pi - \frac{1}{2}\varphi$, is chosen so that the sliding wedge encounters the least resistance, and the solution for a cohesionless soil ($c = 0$) is found to be:

$$P_{an} = \tfrac{1}{2}\,\gamma H^2 \tan^2\left(\tfrac{1}{4}\pi - \tfrac{1}{2}\varphi\right) \tag{8.1}$$

Rough wall ($\delta \neq 0$)

The Coulomb method can also be applied to the case where the wall is rough with an angle of friction δ. When particularized to a cohesionless soil ($c = 0$) these overall equilibrium considerations lead to the following expression for the active thrust:

$$P_{an} = \frac{\tfrac{1}{2}\gamma H^2 \tan\Omega \cos(\Omega + \varphi)\cos\delta}{\sin(\Omega + \varphi + \delta)} \tag{8.2}$$

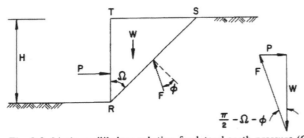

Fig. 8.2. Limit equilibrium solution for lateral earth pressure (Coulomb's solution).

which attains a maximum at the wedge angle:

$$\tan \Omega = \tan (\varphi + \delta) \left[\sqrt{1 + \cot \varphi \cot (\varphi + \delta)} - 1 \right] \qquad \qquad [8.3]$$

The values of P_{an} in [8.2] corresponding to the critical angles Ω given by [8.3] are given in table 5.2 of Heyman's book (1972). The solution may be generalized to include walls inclined at an angle α to the vertical; and backfill inclined at an angle β to the horizontal (see for example, Terzaghi, 1943; or Terzaghi and Peck, 1967).

8.2. COULOMB'S SOLUTION OF VERTICAL RETAINING WALL PROBLEMS

8.2.1. Smooth wall with a flat backfill (Fig. 8.3; $\delta = 0$, $c = 0$)

We shall attempt now to examine the limit analysis solution for a smooth vertical wall with a flat backfill of cohesionless material, as shown in Fig. 8.3. The active earth pressure corresponds to an outward motion of the wall, caused by the weight of the soil. Therefore, the soil weight is assumed to act as a wedge sliding down and outward, against the wall RT. The mechanism is two rigid bodies with plane sliding surfaces. This is a compatible velocity field for an upper-bound solution, since the velocity V of the soil mass RST is at an angle φ to the discontinuity surface RS. The compatible velocity relations are shown by solid lines in Fig. 8.3. The rate of external work is in two parts, the rate of work done by the soil weight moving downward and the rate of work done by the force P moving horizontally:

$$(\tfrac{1}{2} \gamma H^2 \tan \Omega) \left[V \cos (\Omega + \varphi) \right] - PV \sin (\Omega + \varphi) \qquad \qquad [8.4]$$

The rate of internal dissipation consists of the energy dissipated along RT and along RS (assume a frictionless wall, which implies no energy dissipated along RT). The total rate of energy dissipation is $cV\cos \varphi (H\sec \Omega)$ which, for a cohesionless soil ($c = 0$), is zero. Equating the rate of internal and external work yields:

$$P = \tfrac{1}{2} \gamma H^2 \tan \Omega \cot (\Omega + \varphi) \qquad \qquad [8.5]$$

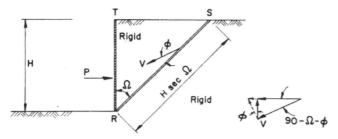

Fig. 8.3. Limit analysis velocity field for active earth pressure.

The limit load or the collapse load is obtained by maximizing Ω which yields:

$$\Omega = \tfrac{1}{4}\pi - \tfrac{1}{2}\varphi \qquad\qquad\qquad [8.6]$$

and: $P_{an} = \tfrac{1}{2}\gamma H^2 \tan^2(\tfrac{1}{4}\pi - \tfrac{1}{2}\varphi) \qquad\qquad [8.7]$

which agrees with [8.1].

For the passive earth pressure case, the mechanism is quite similar to the active earth pressure case except that since the motion is upward, the velocity, V, of the soil wedge is reversed, although still directed at an angle φ to the discontinuity surface. The rate of internal and external work are found in the same manner as for the active earth pressure, and the upper bound is then minimized to obtain the passive earth pressure:

$$\Omega = \tfrac{1}{4}\pi + \tfrac{1}{2}\varphi \qquad\qquad\qquad [8.8]$$

and: $P_{pn} = \tfrac{1}{2}\gamma H^2 \tan^2(\tfrac{1}{4}\pi + \tfrac{1}{2}\varphi) \qquad\qquad [8.9]$

To obtain lower bounds, Fig. 8.4 shows a discontinuous equilibrium solution

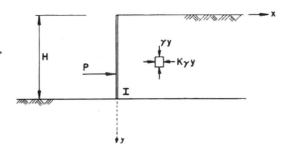

(a) An Equilibrium Solution

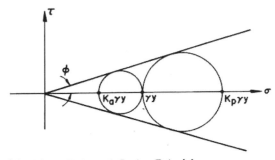

(b) Mohr Circles of Region I in (a)

Fig. 8.4. Limit analysis stress field for lateral earth pressure.

(Finn, 1967) in which K is a parameter which is chosen such that the Coulomb yield criterion will be satisfied in region I. The Mohr circles in the figure show clearly that two extreme values of K are possible which furnish the needed lower-bound solutions of the lateral earth pressure problem. From the figure, one obtains:

$$K = K_a = \tan^2 \left(\tfrac{1}{4} \pi - \tfrac{1}{2} \varphi \right) \qquad\qquad [8.10a]$$

for passive earth pressure:

$$K = K_p = \tan^2 \left(\tfrac{1}{4} \pi + \tfrac{1}{2} \varphi \right) \qquad\qquad [8.10b]$$

Equilibrium is then determined:

$$P = \int_0^H K \gamma y \, dy \qquad\qquad [8.11]$$

The results indicate that the upper and lower bounds for the active earth pressure and the passive earth pressure, respectively, agree, indicating the exact solution.

8.2.2. Rough wall with an inclined backfill (Fig. 8.6, $\delta < \varphi$)

The case considered above is a special case of the general problem, where β is the backfill angle, δ the wall friction angle ($\delta < \varphi$), and α the wall angle as shown in Fig. 8.5. All may assume different values, as well as φ (the angle of internal friction), and c (the cohesion). This general Coulomb solution will be treated in the following section. Here we consider the case of a vertical wall ($\alpha = 90°$) as shown in Fig. 8.6. The compatible velocity components for this case are also shown in Fig. 8.6. The failure plane RS is of length $H \cos \beta / \cos(\Omega + \beta)$ and the wedge is of weight:

$$W = \tfrac{1}{2} \gamma H^2 \frac{\sin \Omega \, \cos \beta}{\cos (\Omega + \beta)} \qquad\qquad [8.12]$$

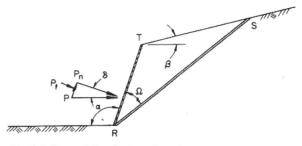

Fig. 8.5. Plane sliding for lateral earth pressure problems in general.

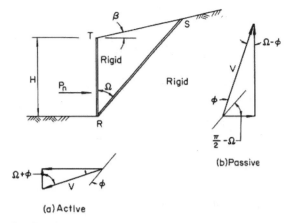

Fig. 8.6. Active and passive lateral earth pressures for a vertical wall.

The rate of external work is:
for the active case:

$$WV \cos(\Omega + \varphi) - P_{an} V \sin(\Omega + \varphi) \qquad [8.13a]$$

for the passive case:

$$- WV \cos(\Omega - \varphi) + P_{pn} V \sin(\Omega - \varphi) \qquad [8.13b]$$

For either case, the rate of energy dissipated internally along RS is:

$$cV \cos\varphi \, \frac{H \cos\beta}{\cos(\Omega + \beta)} \qquad [8.14]$$

The energy dissipated by sliding friction between the wall and the wedge is:
for the active case:

$$P_{an} \tan\delta \; V \cos(\Omega + \varphi) \qquad [8.15]$$

for the passive case:

$$P_{pn} \tan\delta \; V \cos(\Omega - \varphi) \qquad [8.16]$$

where P_{an} and P_{pn} denote the normal to the wall components of the active and passive earth pressures, respectively.

Equating the rate of internal energy dissipation and the rate of external work yields:

$$P_{an} = \frac{\cos\beta}{\cos(\Omega + \beta)} \; \frac{\frac{1}{2}\gamma H^2 \sin\Omega \cos(\Omega + \varphi) - cH \cos\varphi}{\sin(\Omega + \varphi) + \tan\delta \, \cos(\Omega + \varphi)} \qquad [8.17]$$

and:

$$P_{pn} = \frac{\cos \beta}{\cos (\Omega + \beta)} \frac{\frac{1}{2} \gamma H^2 \sin \Omega \cos (\Omega - \varphi) + cH \cos \varphi}{\sin (\Omega - \varphi) - \tan \delta \cos (\Omega - \varphi)} \qquad [8.18]$$

These equations are valid for a vertical wall, $\delta < \varphi$ (otherwise there is shearing of the soil instead of wall friction, further discussions on this point will be given later), and the assumption of a plane for a discontinuity surface. They may be easily rewritten to include α. For the special case of a rough wall with a flat cohesionless backfill ($\beta = 0$, $c = 0$), [8.17] reduces to [8.2].

The best choice, so as to maximize or minimize P, of the angle of the failure plane, Ω, may not be easily found for the general solution given by [8.17] and [8.18]. For special cases, such as discussed above ($c = \delta = \beta = 0$), it may be explicitly found or a transcendental equation, which it must satisfy, may be determined. For example, for the particular case, when $c = \beta = 0$ and $\delta < \varphi$, the condition for the best choice of angle Ω in [8.17] is when $dP_{an}/d\Omega = 0$:

$$1 + \tan \delta \cot (\Omega + \varphi) = \frac{\sin 2 \Omega}{\sin 2 (\Omega + \varphi)} \qquad [8.19a]$$

which can be solved for Ω in terms of given values of φ and δ as given by [8.3]. Alternatively, one can solve for δ in terms of Ω and φ by assuming any value of Ω. Hence:

$$\tan \delta = \frac{- \sin \varphi \cos (2\Omega + \varphi)}{\cos^2 (\Omega + \varphi)} \qquad [8.19b]$$

since $\delta \geqslant 0$, this requires $\Omega \geqslant \frac{1}{4}\pi - \frac{1}{2}\varphi$. For the case $\delta = 0$, the best choice of Ω is $\frac{1}{4}\pi - \frac{1}{2}\varphi$.

For the special case when $c = \beta = 0$ and $\delta = \varphi$, the active pressure, $P_a = P_{an} \sec \delta$ of [8.17] reduces to the simple form:

$$P_a = \frac{1}{2} \gamma H^2 \frac{\cos \varphi}{(1 + \sqrt{2} \sin \varphi)^2} \qquad [8.20]$$

For the general case, however, the easiest procedure is to hold constant the values of the different parameters, and increment Ω until the maximum $P_{an} = \frac{1}{2}\gamma H^2 K_{an}$ or the minimum $P_{pn} = \frac{1}{2}\gamma H^2 K_{pn}$ is reached. This procedure is suitable for digital computation. Table 8.1 is a comparison of Coulomb's solution and this upper-bound limit analysis solution for a vertical wall with cohesionless material. It may be seen that agreement is good.

When the angle Ω is arbitrarily chosen equal to the value $\frac{1}{4}\pi - \frac{1}{2}\varphi$ and $\beta = 0$, the total active collapse pressure P_a of [8.17] reduces to the form:

$$P_a = \frac{1}{1 + \tan \delta \tan (\frac{1}{4}\pi - \frac{1}{2}\varphi)} [\frac{1}{2} \gamma H^2 \tan^2 (\frac{1}{4}\pi - \frac{1}{2}\varphi) - 2cH \tan (\frac{1}{4}\pi - \frac{1}{2}\varphi)] \qquad [8.21]$$

TABLE 8.1

Comparison of lateral earth pressure solutions by methods of limit equilibrium and limit analysis (see Fig. 8.6)

Angle of internal friction φ (°)	Wall friction angle δ (°)	Backfill angle β (°)	Active earth pressure $K_a = K_{an} \sec \delta$		Passive earth pressure $K_p = K_{pn} \sec \delta$	
			Limit equilibrium	Limit analysis	Limit equilibrium	Limit analysis
10	0	0	0.704	0.704	1.42	1.42
	10	0	0.634	0.635	1.65	1.73
20	0	0	0.490	0.490	2.04	2.04
		10	0.569	0.566	2.59	2.59
	10	•0	0.426	0.446	2.52	2.64
		10	0.507	0.531	3.53	3.70
	20	0	0.350	0.426	2.93	3.52
		10	0.430	0.516	4.62	5.59
30	0	0	0.333	0.333	3.00	3.00
		10	0.374	0.372	4.08	4.09
		20	0.441	0.439	5.74	5.78
	10	0	0.290	0.307	3.96	4.15
		10	0.334	0.350	6.03	6.31
		20	0.401	0.420	9.94	10.41
	20	0	0.247	0.297	5.06	6.15
		10	0.282	0.338	9.05	10.91
		20	0.334	0.413	19.40	23.37
40	0	0	0.217	0.217	4.60	4.60
		10	0.238	0.237	6.84	6.85
		20	0.266	0.266	11.06	11.10
	10	0	0.195	0.204	6.63	6.92
		10	0.215	0.223	11.53	12.08
		20	0.242	0.253	24.00	25.36

For the particular case of cohesionless soil, in which $c = 0$, the value of $P_a/(\frac{1}{2}\gamma H^2)$ depends solely on the values of the angles φ and δ. For $\varphi = \delta = 30°$, the difference between the total active earth pressure $P_a/(\frac{1}{2}\gamma H^2)$ corresponding to [8.19b] where $\Omega = 35.8°$, and the arbitrary choice of $\Omega = \frac{1}{4}\pi - \frac{1}{2}\varphi$ is less than 3%. In connection with practical problems, this error is insignificant. With decreasing values of δ, the error decreases further until for $\delta = 0$, the approximate solution is exact.

The passive earth pressure case as given by [8.18] may be treated in a similar manner.

8.2.3. Rough wall with a flat backfill (Fig. 8.3, $\delta \geqslant \varphi$)

For the case $\delta \geqslant \varphi$, Coulomb shearing instead of frictional sliding along the interface provides an alternative kinematically admissible choice for a solution.

Now the relative velocity between the soil wedge and the wall is inclined at a constant angle φ to the vertical slip surface. The energy dissipated along the wall is given by [3.19]. As an illustration, considering the particular case when the backfill is horizontal (Fig. 8.3), the compatible velocity relations for the active earth pressure case are shown by the dotted lines in Fig. 8.3 and a straightforward calculation gives:

$$P_{an} = \frac{\frac{1}{2}\gamma H^2 \sin \Omega \cos (\Omega + \varphi) - cH \cos \varphi - cH \cos \Omega \cos (\Omega + \varphi)}{\cos \Omega \left[\sin (\Omega + \varphi) + \tan \varphi \cos (\Omega + \varphi) \right]} \qquad [8.22]$$

By comparing the terms on the right-hand side of [8.22] with that of [8.17] when $\beta = 0$, and noting that when the active collapse load is sought, velocity solutions give lower bounds, the equation that gives the maximum value of P_{an} is summarized as follows:

Equations governing the active collapse load

	$c = \beta = 0, \varphi \neq 0$	$\beta = 0, c \neq 0, \varphi \neq 0$
$\delta < \varphi$	[8.17] governs	[8.17] governs
$\delta = \varphi$	[8.17] or [8.22]	[8.17] governs
$\delta > \varphi$	[8.22] governs	[8.17] or [8.22] governs depending on the magnitude of c and φ

The frictional sliding case is seen to govern in most cases.

8.2.4. Remarks on the solutions

The kinematic arguments which lead to [8.17] and [8.22] are very different. For [8.17], we assume the same kind of block sliding mechanism used for a smooth wall, i.e., the relative velocity vector between the rough wall and the soil mass is parallel to the wall (Fig. 8.3). For [8.22], we assume a "gap" opens up between the rough wall and the soil mass in such a way that the relative velocity vector makes an angle φ to the wall (dashed line, Fig. 8.3). This "gap" or "separation" is a result of the use of the normal flow rule with Coulomb's yield criterion. Hence, the velocity field for the latter case is kinematically admissible and [8.22] gives a *definite lower bound* to the actual thrust P_{an} at collapse. However, as previously discussed in section 2.6, Chapter 2, the conventional limit theorems cannot be applied to any process in which energy is dissipated by friction, so it would appear that [8.17] derived from this situation cannot be interpreted in terms of strict *bounds*.

In the following we can demonstrate that in fact, in the frictional case also, [8.17] or [8.18] can be interpreted in terms of upper bounds by using the finite frictional theorem (Theorem IX) given in section 2.6, Chapter 2. According to this theorem, the limit load for an assemblage of bodies with frictional interfaces is bounded above by the limit load for the same assemblage *cemented* at the interfaces by a *cohesionless* soil of friction angle $\Phi = \tan^{-1}\mu = \delta$ where μ is the coefficient of friction at the interfaces. By analogy with plastic shearing at an interface, we now assume a mechanism with a gap which opens up between the rough wall and the soil mass in such a way that the relative velocity vector makes an angle $\Phi = \delta$ to the wall. In other words, the ratio of the relative normal and tangential velocity components at the interface is $\tan \delta$.

Since we assume the interface is "*cemented*" together with a cohesionless soil, no energy will be dissipated at the interface. The only rate of internal dissipation of energy is along RS (Fig. 8.3 or more generally, Fig. 8.6) which is given by [8.14]. Equating the rate of work done by the thrust P_{an} which has the value $P_{an} V \sin(\delta + \Omega + \varphi)/\cos \delta$, and the self-weight in RST to the rate of energy dissipation along RS gives:

$$\tfrac{1}{2} \gamma H^2 \frac{\sin \Omega \cos \beta}{\cos (\Omega + \beta)} V \cos (\Omega + \varphi) - P_{an} \frac{V \sin(\delta + \Omega + \varphi)}{\cos \delta} = cV \cos \varphi \frac{H \cos \beta}{\cos (\Omega + \beta)} \quad [8.23]$$

which simplifies to [8.17].

Note that the upper-bound theorem yields overestimates of the active load — in this case the self-weight — but a lower bound to the resisting thrust P_{an}. Hence, [8.17] now gives a definite *lower bound* to the actual thrust and is therefore an unsafe estimate. The value given in [8.17] corresponding to the critical value of Ω is hence the best lower bound calculated from this type of collapse mechanism.

This type of *dilatant velocity discontinuities* on the interface or boundary was first used by Heyman (1972) to obtain [8.2]. However, Heyman dismisses this kinematic argument on the grounds that there is no physical justification for such a dilatant velocity discontinuity occurring between the soil and a rigid boundary. Collins (1973) interpreted this kinematic argument from the viewpoint of *virtual displacement* and the principle of *maximum plastic work* (see sections 2.3 and 2.4, Chapter 2) and concluded that [8.2] is a lower bound just as in the smooth case. Mroz and Drescher (1969) have also used these dilatant velocity discontinuities on the boundary or interface to obtain estimates of the pressures exerted by flowing bulk solids on the walls of bins and hoppers and on moving plates. However, it does not seem obvious, a priori, that this kinematic argument should lead to a definite bound as opposed to mere estimates.

8.3. COULOMB'S SOLUTION OF GENERAL RETAINING WALL PROBLEMS (FIG. 8.7a)

In the following sections only the upper-bound technique of limit analysis is applied to obtain the upper bounds for the active and passive earth pressures acting

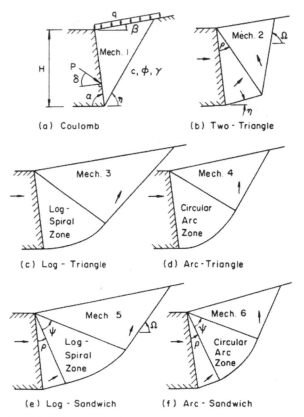

Fig. 8.7. Failure mechanisms.

on a rigid retaining wall using various failure mechanisms. A total of six different
failure mechanisms are shown in Fig. 8.7. The mechanism to be considered in this
section is the Coulomb mechanism (Fig. 8.7a). Since the solution procedure for this
case has already been described in detail in the preceding section for the case of a
vertical retaining wall, only the necessary equations and final results are recorded
here for the sake of brevity.

It has long been recognized that Coulomb's solutions greatly overestimate the
passive pressure exerted on a relatively rough wall for high soil friction angles, i.e.,
$\varphi = 30\text{--}40°$. Indeed for these situations the actual failure surface is far from
straight but is curved. This is reflected in the four failure mechanisms as shown in
Fig. 8.7(c to f). The logspiral shear zone as shown in Fig. 3.20 is used here in
mechanisms 3 and 5 of Figs. 8.7(c and e) and the circular radial shear zone for
$c - \varphi$ soils as shown in Fig. 3.23 is used in mechanisms (4) and (6) of Figs. 8.7(d
and f). Since the solution procedure for these four cases is identical a detailed
formulation will only be given for the log-sandwich mechanism (Mech. 5) and a

cohesionless soil. The necessary equations for the inclusion of cohesion and surcharge loading of this case and the solutions for other three cases are however recorded here and the subsequent sections, while the two-triangle mechanism (Mech. 2) is given in the following section.

8.3.1. Method of superposition

Before we move to find the pressure in the rear face of a retaining wall inclined at an angle α from a fill with a surcharge q, we note that if a failure mechanism is described by n independent parameters, the active and passive pressure acting on a rigid wall can be expressed as:

$$P_a = \max \left[\tfrac{1}{2} \gamma H^2 K_{a\gamma}(\theta_1, \theta_2, \ldots \theta_n) + qHK_{aq}(\theta_1, \theta_2, \ldots \theta_n)\right.$$

$$\left. + cHK_{ac}(\theta_1, \theta_2, \ldots \theta_n)\right] \qquad [8.24a]$$

$$P_p = \min \left[\tfrac{1}{2} \gamma H^2 K_{p\gamma}(\theta_1, \theta_2, \ldots \theta_n) + qHK_{pq}(\theta_1, \theta_2, \ldots \theta_n)\right.$$

$$\left. + cHK_{pc}(\theta_1, \theta_2, \ldots \theta_n)\right] \qquad [8.24b]$$

where K_γ, K_q, and K_c are pure numbers representing the effects of weight, surcharge, and cohesion, respectively. For simplicity, [8.24] can be represented approximately by the linear equation:

$$P_a = \tfrac{1}{2} \gamma H^2 \max \left[K_{a\gamma}(\theta_1, \theta_2, \ldots \theta_n)\right] + qH \max \left[K_{aq}(\theta_1, \theta_2, \ldots \theta_n)\right]$$

$$+ cH \max \left[K_{ac}(\theta_1, \theta_2, \ldots \theta_n)\right] \qquad [8.25a]$$

$$P_p = \tfrac{1}{2} \gamma H^2 \min \left[K_{p\gamma}(\theta_1, \theta_2, \ldots \theta_n)\right] + qH \min \left[K_{pq}(\theta_1, \theta_2, \ldots \theta_n)\right]$$

$$+ cH \min \left[K_{pc}(\theta_1, \theta_2, \ldots \theta_n)\right] \qquad [8.25b]$$

The value in the first term of [8.25a] or [8.25b] represents the earth pressure which the cohesionless soil mass can exert on the retaining wall if the surcharge q is equal to zero ($c = 0$ and $q = 0$), the second term is the earth pressure due exclusively to the surcharge q ($\gamma = 0$ and $c = 0$), and the third term is the earth pressure which the weightless soil mass can exert on the wall if the surcharge q is equal to zero. From the results of some numerical computations we know that the corresponding critical load P_a (or P_p) defined by [8.24], is only slightly smaller (or greater) than the sum of three terms defined by [8.25]. The functions K_γ, K_q and K_c corresponding to the six mechanisms shown in Fig. 8.7 will be derived first in what follows and then compared in the later sections.

8.3.2. Coulomb mechanism (Fig. 8.7a)

The Coulomb mechanism is completely described by the parameter η as shown in Fig. 8.7(a). The results for the coefficients of earth pressure due to weight, cohesion, and surcharge (K_γ, K_c, K_q) are given below. For simplicity we shall call a wall with a friction angle δ less than φ a *smooth wall* and a wall with δ greater than or equal to φ a *rough wall*. For the particular cases when $\delta = 0$ and $\delta = \varphi$, we shall call these conditions "*perfectly smooth*" and "*perfectly rough*", respectively.

For a smooth wall ($\delta < \varphi$):
For passive case use lower signs

$$\begin{Bmatrix} K_{a\gamma} \\ K_{p\gamma} \end{Bmatrix} = \frac{\mp \sec \delta}{\mp \sin \alpha + \tan \delta \cos \alpha - \dfrac{\tan \delta \sin (\eta \mp \varphi)}{\sin (\alpha + \eta \mp \varphi)}} \left[\frac{\sin (\alpha + \eta) \sin (\alpha + \beta) \sin (\eta \mp \varphi)}{\sin \alpha \sin (\eta - \beta) \sin (\alpha + \eta \mp \varphi)} \right]$$

$$[8.26]$$

$$\begin{Bmatrix} K_{ac} \\ K_{pc} \end{Bmatrix} = \frac{\sec \delta}{\mp \sin \alpha + \tan \delta \cos \alpha - \dfrac{\tan \delta \sin (\eta \mp \varphi)}{\sin (\alpha + \eta \mp \varphi)}} \left[\frac{\cos \varphi \sin (\alpha + \beta)}{\sin (\eta - \beta) \sin (\alpha + \eta \mp \varphi)} \right] \quad [8.27]$$

$$\begin{Bmatrix} K_{aq} \\ K_{pq} \end{Bmatrix} = \frac{\mp \sec \delta}{\mp \sin \alpha + \tan \delta \cos \alpha - \dfrac{\tan \delta \sin (\eta \mp \varphi)}{\sin (\alpha + \eta \mp \varphi)}} \left[\frac{\sin (\alpha + \eta) \sin (\eta \mp \varphi)}{\sin (\eta - \beta) \sin (\alpha + \eta \mp \varphi)} \right] \quad [8.28]$$

For a rough wall ($\delta \geqslant \varphi$):
For passive case use lower signs

$$\begin{Bmatrix} K_{a\gamma} \\ K_{p\gamma} \end{Bmatrix} = \frac{\mp \sec \delta}{\mp \sin \alpha + \tan \delta \cos \alpha} \left[\frac{\sin (\alpha + \eta) \sin (\alpha + \beta) \sin (\alpha \mp \varphi) \sin (\eta \mp \varphi)}{\sin^2 \alpha \sin (\eta - \beta) \sin (\eta + \alpha \mp 2\varphi)} \right] \quad [8.29]$$

$$\begin{Bmatrix} K_{ac} \\ K_{pc} \end{Bmatrix} = \frac{\sec \delta}{\mp \sin \alpha + \tan \delta \cos \alpha} \left[\frac{\cos \varphi \sin (\alpha + \beta) \sin (\alpha \mp \varphi)}{\sin \alpha \sin (\eta - \beta) \sin (\eta + \alpha \mp 2\varphi)} \right.$$

$$\left. + \frac{\sin (\eta \mp \varphi) \cos \varphi}{\sin \alpha \sin (\eta + \alpha \mp 2\varphi)} \right] \qquad\qquad [8.30]$$

$$\begin{Bmatrix} K_{aq} \\ K_{pq} \end{Bmatrix} = \frac{\mp \sec \delta}{\mp \sin \alpha + \tan \delta \cos \alpha} \left[\frac{\sin (\alpha + \eta) \sin (\eta \mp \varphi) \sin (\alpha \mp \varphi)}{\sin \alpha \sin (\eta - \beta) \sin (\eta + \alpha \mp 2\varphi)} \right] \qquad [8.31]$$

8.4. TWO-TRIANGLE MECHANISM (FIG. 8.8)

The two-triangle mechanism consists of two rigid sliding blocks and is complete-

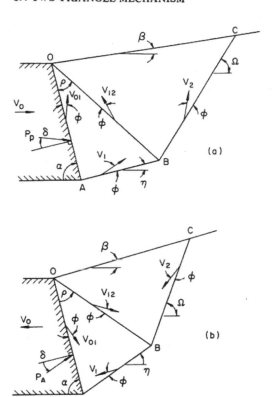

Fig. 8.8. Two-triangle mechanism.

ly described by three parameters (ρ, η, Ω). The velocity fields for both the passive and active states are shown in Figs. 8.8(a) and 8.8(b), respectively. The formulations for the coefficients of earth pressure due to weight, cohesion, and surcharge (K_γ, K_c, K_q) follow.

8.4.1. For a cohesionless soil with no surcharge loading $(c = 0, q = 0)$

The rate of external work due to self-weight in any region is simply the vertical component of velocity in that region multiplied by the weight of the region (for passive case use lower signs):

Region OAB

$$\pm \frac{\frac{1}{2} \gamma H^2 V_1 \sin \rho \sin (\alpha + \eta) \sin (\eta \mp \varphi)}{\sin^2 \alpha \sin (\alpha + \eta - \rho)}$$

[8.32]

Region OBC

$$\pm \frac{\frac{1}{2}\gamma H^2 \, V_2 \sin^2(\alpha + \eta) \sin(\alpha - \rho + \beta) \sin(\alpha - \rho + \Omega) \sin(\Omega \mp \varphi)}{\sin^2 \alpha \sin^2(\alpha + \eta - \rho) \sin(\Omega - \beta)} \qquad [8.33]$$

Moving wall load

$$\begin{Bmatrix} P_{an} \\ P_{pn} \end{Bmatrix} V_0 [\mp \sin \alpha + \tan \delta \, \cos \alpha] \qquad [8.34]$$

Rate of energy dissipation

For a smooth wall ($\delta < \varphi$) the dissipation by sliding friction is given by:

$$\begin{Bmatrix} P_{an} \\ P_{pn} \end{Bmatrix} \tan \delta \, V_{01} \qquad [8.35]$$

For a rough wall ($\delta \geq \varphi$) the dissipation of a cohesionless soil is zero.

With the use of the compatible velocity diagrams all velocities in the mechanism can be expressed in terms of the wall translational velocity V_0.

For the case of smooth walls ($\delta < \varphi$)

$$V_1 = \frac{V_0 \sin \alpha}{\sin(\eta \mp \varphi + \alpha)} \qquad V_2 = \frac{V_0 \sin \alpha \sin(\alpha - \rho + \eta \mp 2\varphi)}{\sin(\eta \mp \varphi + \alpha) \sin(\alpha - \rho + \Omega \mp 2\varphi)}$$

$$V_{01} = \frac{V_0 \sin(\eta \mp \varphi)}{\sin(\eta \mp \varphi + \alpha)} \qquad V_{12} = \frac{V_0 \sin \alpha \sin(\Omega - \eta)}{\sin(\eta \mp \varphi + \alpha) \sin(\alpha - \rho + \Omega \mp 2\varphi)} \qquad [8.36]$$

For rough walls ($\delta \geq \varphi$)

$$V_1 = \frac{V_0 \sin(\alpha \mp \varphi)}{\sin(\eta + \alpha \mp 2\varphi)} \qquad V_2 = \frac{V_0 \sin(\alpha \mp \varphi) \sin(\alpha + \eta - \rho \mp 2\varphi)}{\sin(\eta + \alpha \mp 2\varphi) \sin(\alpha - \rho + \Omega \mp 2\varphi)}$$

$$V_{01} = \frac{V_0 \sin(\eta \mp \varphi)}{\sin(\eta + \alpha \mp 2\varphi)} \qquad V_{12} = \frac{V_0 \sin(\alpha \mp \varphi) \sin(\Omega - \eta)}{\sin(\eta + \alpha \mp 2\varphi) \sin(\alpha - \rho + \Omega \mp 2\varphi)} \qquad [8.37]$$

Equating the rate of external work to the rate of internal energy dissipation:

For a smooth wall ($\delta < \varphi$)

$$\begin{Bmatrix} K_{a\gamma} \\ K_{p\gamma} \end{Bmatrix} = \frac{\mp \sec \delta}{\mp \sin \alpha + \tan \delta \, \cos \alpha - \dfrac{\tan \delta \sin(\eta \mp \varphi)}{\sin(\eta \mp \varphi + \alpha)}} \left\{ \frac{\sin \rho \sin(\alpha + \eta) \sin(\eta \mp \varphi)}{\sin \alpha \sin(\alpha + \eta - \rho) \sin(\eta \mp \varphi + \alpha)} \right.$$

$$\left. + \frac{\sin^2(\alpha + \eta) \sin(\alpha - \rho + \beta) \sin(\alpha - \rho + \Omega) \sin(\Omega \mp \varphi) \sin(\alpha + \eta - \rho \mp 2\varphi)}{\sin \alpha \sin^2(\alpha + \eta - \rho) \sin(\Omega - \beta) \sin(\alpha + \eta \mp \varphi) \sin(\alpha - \rho + \Omega \mp 2\varphi)} \right\}$$

$$[8.38]$$

For a rough wall $(\delta \geqslant \varphi)$

$$\begin{Bmatrix} K_{a\gamma} \\ K_{p\gamma} \end{Bmatrix} = \frac{\mp \sec \delta}{\mp \sin \alpha + \tan \delta \, \cos \alpha} \left| \frac{\sin \rho \, \sin (\alpha + \eta) \sin (\eta \mp \varphi) \sin (\alpha \mp \varphi)}{\sin^2 \alpha \, \sin (\alpha + \eta - \rho) \sin (\eta + \alpha \mp 2\varphi)} \right.$$

$$+ \left. \frac{\sin^2 (\alpha + \eta) \sin (\alpha - \rho + \beta) \sin (\alpha - \rho + \Omega) \sin (\Omega \mp \varphi) \sin (\alpha \mp \varphi) \sin (\alpha + \eta - \rho \mp 2\varphi)}{\sin^2 \alpha \, \sin^2 (\alpha + \eta - \rho) \sin (\Omega - \beta) \sin (\eta + \alpha \mp 2\varphi) \sin (\alpha - \rho + \Omega \mp 2\varphi)} \right|$$

$$[8.39]$$

8.4.2. *For a cohesive weightless soil with no surcharge* $(\gamma = 0, q = 0)$

The following dissipation expressions are necessary:

Along the wall (OA)

For a smooth wall $(\delta < \varphi)$ the dissipation by a sliding friction is given by [8.35].
For a rough wall $(\delta \geqslant \varphi)$ the dissipation is given by:

$$\frac{cH \cos \varphi \, V_{01}}{\sin \alpha} \qquad \qquad [8.40]$$

Along AB $\quad \dfrac{cH \, V_1 \cos \varphi \, \sin \rho}{\sin \alpha \, \sin (\alpha + \eta - \rho)} \qquad \qquad [8.41]$

Along OB $\quad \dfrac{cH \, V_{12} \cos \varphi \, \sin (\alpha + \eta)}{\sin \alpha \, \sin (\alpha + \eta - \rho)} \qquad \qquad [8.42]$

Along BC $\quad \dfrac{cH \, V_2 \cos \varphi \, \sin (\alpha + \eta) \sin (\alpha - \rho + \beta)}{\sin \alpha \, \sin (\alpha + \eta - \rho) \sin (\Omega - \beta)} \qquad \qquad [8.43]$

Using the velocity relations given by [8.36] and [8.37] we obtain:

For a smooth wall $(\delta < \varphi)$

$$\begin{Bmatrix} K_{ac} \\ K_{pc} \end{Bmatrix} = \frac{\sec \delta}{\mp \sin \alpha + \tan \delta \, \cos \alpha - \dfrac{\tan \delta \, \sin (\eta \mp \varphi)}{\sin (\eta \mp \varphi + \alpha)}} \left\{ \frac{\cos \varphi \, \sin \rho}{\sin (\eta \mp \varphi + \alpha) \sin (\alpha + \eta - \rho)} \right.$$

$$+ \frac{\cos \varphi \, \sin (\alpha + \eta) \sin (\Omega - \eta)}{\sin (\alpha + \eta - \rho) \sin (\eta \mp \varphi + \alpha) \sin (\alpha - \rho + \Omega \mp 2\varphi)}$$

$$+ \left. \frac{\cos \varphi \, \sin (\alpha + \eta) \sin (\alpha - \rho + \beta) \sin (\alpha + \eta - \rho \mp 2\varphi)}{\sin (\alpha + \eta - \rho) \sin (\Omega - \beta) \sin (\eta \mp \varphi + \alpha) \sin (\alpha - \rho + \Omega \mp 2\varphi)} \right\} \qquad [8.44]$$

For a rough wall $(\delta \geqslant \varphi)$

$$\left\{\begin{matrix} K_{ac} \\ K_{pc} \end{matrix}\right\} = \frac{\sec \delta}{\mp \sin \alpha + \tan \delta \, \cos \, \alpha} \left\{\frac{\cos \varphi \sin (\eta \mp \varphi)}{\sin \alpha \sin (\eta + \alpha \mp 2\varphi)} + \right.$$

$$+ \frac{\cos \varphi \sin \rho \sin (\alpha \mp \varphi)}{\sin \alpha \sin (\alpha + \eta - \rho) \sin (\eta + \alpha \mp 2\varphi)}$$

$$+ \frac{\cos \varphi \sin (\alpha + \eta) \sin (\alpha \mp \varphi) \sin (\Omega - \eta)}{\sin \alpha \sin (\alpha + \eta - \rho) \sin (\eta + \alpha \mp 2\varphi) \sin (\alpha - \rho + \Omega \mp 2\varphi)}$$

$$\left. + \frac{\cos \varphi \sin (\alpha + \eta) \sin (\alpha - \rho + \beta) \sin (\alpha \mp \varphi) \sin (\alpha + \eta - \rho \mp 2\varphi)}{\sin \alpha \sin (\alpha + \eta - \rho) \sin (\Omega - \beta) \sin (\eta + \alpha \mp 2\varphi) \sin (\alpha - \rho + \Omega \mp 2\varphi)}\right\}$$

$$[8.45]$$

8.4.3. Surcharge loading $(\gamma = 0, c = 0)$

The following rate of external work term is necessary for a backfill uniformly loaded with a surcharge q:

$$\frac{\pm qHV_2 \sin (\alpha + \eta) \sin (\alpha - \rho + \Omega) \sin (\Omega \mp \varphi)}{\sin \alpha \sin (\alpha + \eta - \rho) \sin (\Omega - \beta)} \qquad [8.46]$$

For a smooth wall $(\delta < \varphi)$

$$\left\{\begin{matrix} K_{aq} \\ K_{pq} \end{matrix}\right\} = \frac{\mp \sec \delta}{\mp \sin \alpha + \tan \delta \, \cos \alpha - \dfrac{\tan \delta \, \sin (\eta \mp \varphi)}{\sin (\eta \mp \varphi + \alpha)}}$$

$$\left\{\frac{\sin (\alpha + \eta) \sin (\alpha - \rho + \Omega) \sin (\Omega \mp \varphi) \sin (\alpha + \eta - \rho \mp 2\varphi)}{\sin (\alpha + \eta - \rho) \sin (\Omega - \beta) \sin (\eta \mp \varphi + \alpha) \sin (\alpha - \rho + \Omega \mp 2\varphi)}\right\} \qquad [8.47]$$

For a rough wall $(\delta \geqslant \varphi)$

$$\left\{\begin{matrix} K_{aq} \\ K_{pq} \end{matrix}\right\} = \frac{\mp \sec \delta}{\mp \sin \alpha + \tan \delta \, \cos \alpha}$$

$$\left\{\frac{\sin (\alpha + \eta) \sin (\alpha \mp \varphi) \sin (\alpha - \rho + \Omega) \sin (\Omega \mp \varphi) \sin (\alpha + \eta - \rho \mp 2\varphi)}{\sin \alpha \sin (\alpha + \eta - \rho) \sin (\Omega - \beta) \sin (\eta + \alpha \mp 2\varphi) \sin (\alpha - \rho + \Omega \mp 2\varphi)}\right\}$$

$$[8.48]$$

Coulomb's solutions obtained in the preceding section may be interpreted as the special case of two-triangle mechanism when $\rho = 0$, and $\eta = \Omega$.

In order to obtain the critical active and passive wall loads, expressions for K_γ, K_c and K_q must be either maximized or minimized respectively, with regard to the mechanism parameters ρ, η and Ω. This was accomplished with the aid of an iterative technique incorporating the *method of steepest descent.* Solutions were obtained on the CDC 6400 computer. A typical solution required at most fifteen iterations or 0.1 sec. Numerical results for the six mechanisms shown in Fig. 8.7 will be presented and compared later when all the expressions of K_γ, K_c and K_q corresponding to various mechanisms are obtained.

8.5. LOGSANDWICH MECHANISM (FIGS. 8.9 AND 8.10)

Figure 8.9(a) shows a logarithmic spiral shearing zone, *OBC*, sandwiched between two rigid blocks, *OAB* and *OCD*. Since the velocities V_1 and V_3 for the rigid triangles *OAB* and *OCD* are assumed perpendicular to the radial lines *OB* and *OC*, two angular parameters ρ and ψ describe the mechanism completely. It will be shown later that for certain limited boundary situations only one parameter need

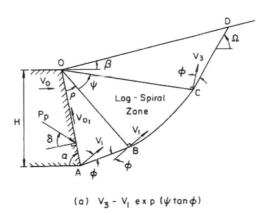

(a) $V_3 - V_l \exp(\psi \tan \phi)$

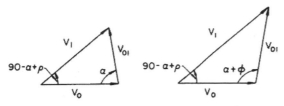

(b) Smooth Wall $\delta < \phi$ (c) Rough Wall $\delta \geq \phi$

Fig. 8.9. Passive logsandwich mechanism.

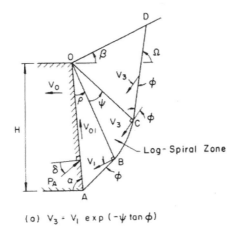

(a) $V_3 = V_1 \exp(-\psi \tan \phi)$

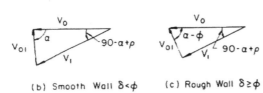

(b) Smooth Wall $\delta < \phi$ (c) Rough Wall $\delta \geq \phi$

Fig. 8.10. Active logsandwich mechanism.

be considered. This simplification reduces the complexity of the solution process. The compatible velocity diagrams corresponding to the passive pressure case are given in Figs. 8.9(b) and 8.9(c) for the smooth ($\delta < \varphi$) and rough ($\delta \geq \varphi$) wall conditions, respectively. Fig. 8.10 shows the corresponding diagrams for the active case. The wall is assumed to translate horizontally with a velocity V_0. All other velocities in the mechanism can then be expressed in terms of V_0.

8.5.1. For a cohesionless soil with no surcharge loading (c = 0, q = 0)

The rate of external work due to self-weight in any region is simply the vertical component of velocity in that region multiplied by the weight of the region (for passive case use lower signs):

Triangular region OAB

$$\pm \frac{\frac{1}{2}\gamma H^2 V_1 \sin\rho \cos(\rho \pm \varphi)\cos(\alpha - \rho)}{\sin^2\alpha \cos\varphi} \qquad [8.49]$$

Logspiral region OBC

The external rate of work done by the soil weight in this zone can be computed by summing over the region ψ the products of differential triangle's component of vertical velocity with its weight. Dividing the region OBC into small triangles similar to that shown in Fig. 3.20(a), this can be expressed in integral form as:

$$\pm \frac{1}{2} \gamma V_1 (\overline{OB})^2 \int_0^\psi \exp(\mp 3\theta \tan\varphi) \sin(\alpha - \theta - \rho) d\theta \qquad [8.50]$$

where the upper and lower signs signify the active and passive states, respectively. After integration, we obtain:

$$\pm \frac{\frac{1}{2} \gamma H^2 V_1 \cos^2(\rho \pm \varphi)}{\sin^2 \alpha \cos^2 \varphi (1 + 9 \tan^2 \varphi)} \{\cos(\alpha - \rho) [\pm 3 \tan\varphi + (\mp 3 \tan\varphi \cos\psi + \sin\psi)$$

$$\exp(\mp 3\psi \tan\varphi)] + \sin(\alpha - \rho) [1 + (\mp 3 \tan\varphi \sin\psi - \cos\psi) \exp(\mp 3\psi \tan\varphi)]\}$$

$$[8.51]$$

Triangular region OCD

$$\pm \frac{\frac{1}{2} \gamma H^2 V_1 \cos^2(\rho \pm \varphi) \sin(\alpha - \rho - \psi + \beta) \cos(\alpha - \rho - \psi) \exp(\mp 3\psi \tan\varphi)}{\sin^2 \alpha \cos\varphi \cos(\alpha - \rho - \psi + \beta \mp \varphi)}$$

$$[8.52]$$

The external work done by the components of the resultant wall load P moving in the horizontal direction with velocity V_0 is given by [8.34].

Rate of internal energy dissipation

Since a cohesionless soil is being considered, the only dissipation occurs at the soil–wall interface. For a smooth wall ($\delta < \varphi$) the dissipation by sliding friction is given by [8.35]. For a rough wall ($\delta \geqslant \varphi$) the internal dissipation of energy is 0 since $c = 0$. Using the velocity diagrams, Figs. 8.9(b) and 8.9(c), or 8.10(b) and 8.10(c), the velocities V_1 and V_{01} can be expressed in terms of the translational wall velocity V_0 for both smooth and rough cases in all the previous expressions. Equating the rate of external work to the rate of internal energy dissipation, a closed form expression for the resultant coefficients of active and passive pressures is obtained:

For a smooth wall ($\delta < \varphi$)

$$\begin{Bmatrix} K_{a\gamma} \\ K_{p\gamma} \end{Bmatrix} = \frac{\mp \sec\delta}{\mp \sin\alpha + \tan\delta \cos\alpha - \dfrac{\tan\delta \cos(\alpha - \rho)}{\cos\rho}} \left| \dfrac{\tan\rho \cos(\rho \pm \varphi) \cos(\alpha - \rho)}{\sin\alpha \cos\varphi} \right.$$

$$+ \frac{\cos^2(\rho \pm \varphi)}{\cos\rho \sin\alpha \cos^2\varphi(1 + 9\tan^2\varphi)} \Big[\cos(\alpha - \rho) [\pm 3\tan\varphi + (\mp 3\tan\varphi \cos\psi$$

$$+ \sin\psi)\exp(\mp 3\psi \tan\varphi)] + \sin(\alpha - \rho)[1 + (\pm 3\tan\varphi \sin\psi - \cos\psi)\exp(\mp 3\psi \tan\varphi)]\Big]$$

$$+ \frac{\cos^2(\rho \pm \varphi)\sin(\alpha - \rho - \psi + \beta)\cos(\alpha - \rho - \psi)\exp(\mp 3\psi \tan\varphi)}{\cos\varphi \sin\alpha \cos(\alpha - \rho - \psi \mp \varphi + \beta)\cos\rho} \Big\} \qquad [8.53]$$

For a rough wall $(\delta \geqslant \varphi)$

$$\begin{Bmatrix} K_{a\gamma} \\ K_{p\gamma} \end{Bmatrix} = \frac{\mp \sec\delta}{\mp \sin\alpha + \tan\delta \, \cos\alpha} \Big\{ \frac{\sin^2\rho \cos(\rho \pm \varphi)\cos(\alpha - \rho)\sin(\alpha \mp \varphi)}{\sin^2\alpha \cos\varphi \cos(\rho \mp \varphi)}$$

$$\mp \frac{\cos^2(\rho \pm \varphi)\sin(\alpha \mp \varphi)}{\sin^2\alpha \cos^2\varphi(1 + 9\tan^2\varphi)\cos(\rho \mp \varphi)} \Big[\cos(\alpha - \rho) [\pm 3\tan\varphi + (\mp 3\tan\varphi$$

$$\cos\psi + \sin\psi)\exp(\mp 3\psi \tan\varphi)] + \sin(\alpha - \rho)[1 + (\mp 3\tan\varphi \sin\psi - \cos\psi)$$

$$\exp(\mp 3\psi \tan\varphi)]\Big]$$

$$+ \frac{\cos^2(\rho \pm \varphi)\sin(\alpha - \rho - \psi + \beta)\cos(\alpha - \rho - \psi)\sin(\alpha \mp \varphi)\exp(\mp 3\psi \tan\varphi)}{\sin^2\alpha \cos\varphi \cos(\alpha - \rho - \psi + \beta \mp \varphi)\cos(\rho \mp \varphi)} \Big\}$$

$$[8.54]$$

8.5.2. For a cohesive weightless soil with no surcharge $(\gamma = 0, q = 0)$

For a cohesive soil, energy dissipation terms must be added for all surfaces of discontinuity as well as the shearing zone in the logspiral region (Figs. 8.9 and 8.10). Since the velocities V_1 and V_3 are perpendicular to the radial lines OB and OC, respectively, there will be no dissipation along either OB or OC due to a lack of relative movement at those surfaces. The following dissipation terms must, however, be included (for passive case use lower signs):

Along the wall (OA)

See [8.35] for a smooth wall $(\delta < \varphi)$. For a rough wall $(\delta \geqslant \varphi)$ the dissipation is given by [8.40].

Along AB $\dfrac{cH \sin\rho \, V_1}{\sin\alpha}$

$$[8.55]$$

Along CD

$$\frac{cH V_1 \cos(\rho \pm \varphi) \sin(\alpha - \rho - \psi + \beta) \exp(\mp 2\psi \tan \varphi)}{\sin \alpha \cos(\alpha - \rho - \psi \mp \varphi + \beta)}$$ [8.56]

From [3.74], the dissipation terms for shearing in the logspiral zone and along the curved discontinuity *BC* are both equal to:

$$\mp \frac{1}{2} \frac{cH V_1 \cos(\rho \pm \varphi) [\exp(\mp 2\psi \tan \varphi) - 1]}{\sin \varphi \sin \alpha}$$ [8.57]

Equating the rate of external work to the rate of internal energy dissipation:

For a smooth wall ($\delta < \varphi$)

$$\begin{Bmatrix} K_{ac} \\ K_{pc} \end{Bmatrix} = \frac{\sec \delta}{\mp \sin \alpha + \tan \delta \cos \alpha - \dfrac{\tan \delta \cos(\alpha - \rho)}{\cos \rho}}$$

$$\left\{ \tan \rho + \frac{\cos(\rho \pm \varphi) \sin(\alpha - \rho - \psi + \beta) \exp(\mp 2\psi \tan \varphi)}{\cos \rho \cos(\alpha - \rho - \psi \mp \varphi + \beta)} \right.$$

$$\left. \mp \frac{\cos(\rho \pm \varphi) [\exp(\mp 2\psi \tan \varphi) - 1]}{\sin \varphi \cos \rho} \right\}$$ [8.58]

For a rough wall ($\delta \geqslant \varphi$)

$$\begin{Bmatrix} K_{ac} \\ K_{pc} \end{Bmatrix} = \frac{\sec \delta}{\mp \sin \alpha + \tan \delta \cos \alpha} \left\{ \frac{\cos \varphi \cos(\alpha - \rho)}{\sin \alpha \cos(\rho \mp \varphi)} + \frac{\sin \rho \sin(\alpha \mp \varphi)}{\sin \alpha \cos(\rho \mp \varphi)} \right.$$

$$+ \frac{\cos(\rho \pm \varphi) \sin(\alpha - \rho - \psi + \beta) \sin(\alpha \top \varphi) \exp(\mp 2\psi \tan \varphi)}{\sin \alpha \cos(\alpha - \rho - \psi \mp \varphi + \beta) \cos(\rho \mp \varphi)}$$

$$\left. \mp \frac{\cos(\rho \pm \varphi) \sin(\alpha \mp \varphi) [\exp(\mp 2\psi \tan \varphi) - 1]}{\sin \varphi \sin \alpha \cos(\rho \mp \varphi)} \right\}$$ [8.59]

8.5.3. Surcharge loading ($\gamma = 0$, $c = 0$)

For a uniformly distributed surcharge loading q on the backfill as shown in Fig. 8.7(a), the following rate of external work term must be included:

$$\pm \frac{qH V_1 \cos(\rho \pm \varphi) \cos(\rho + \psi - \alpha) \exp(\mp 2\psi \tan \varphi)}{\sin \alpha \cos(\alpha - \rho - \psi \mp \varphi + \beta)}$$ [8.60]

Equating the rate of external work to the rate of internal energy dissipation:

For a smooth wall $(\delta < \varphi)$

$$\begin{Bmatrix} K_{aq} \\ K_{pq} \end{Bmatrix} = \frac{\mp \sec \delta}{\mp \sin \alpha + \tan \delta \, \cos \alpha - \dfrac{\tan \delta \, \cos (\alpha - \rho)}{\cos \rho}}$$

$$\left\{ \frac{\cos (\rho \pm \varphi) \, \cos (\rho + \psi - \alpha) \, \exp (\mp 2\psi \tan \varphi)}{\cos \rho \, \cos (\alpha - \rho - \psi \mp \varphi + \beta)} \right\} \qquad\qquad [8.61]$$

For a rough wall $(\delta \geqslant \varphi)$

$$\begin{Bmatrix} K_{aq} \\ K_{pq} \end{Bmatrix} = \frac{\mp \sec \delta}{\mp \sin \alpha + \tan \delta \, \cos \alpha}$$

$$\left\{ \frac{\cos (\rho \pm \varphi) \, \sin (\alpha \mp \varphi) \, \cos (\rho + \psi - \alpha) \, \exp (\mp 2\psi \tan \varphi)}{\sin \alpha \, \cos (\alpha - \rho - \psi \mp \varphi + \beta) \, \cos (\rho \mp \varphi)} \right\} \qquad\qquad [8.62]$$

In the particular case when the angle ρ of mechanism 5 (Fig. 8.7e) is zero, the logsandwich mechanism becomes the familiar logtriangle mechanism shown in Fig. 8.7(c). This logtriangle mechanism has frequently been used in the past in determining graphically the passive earth pressure of cohesive soil by the limit equilibrium method (Terzaghi, 1943). The results for the coefficients of earth pressure due to weight, cohesion, and surcharge (K_γ, K_c, K_q) corresponding to this particular case can be obtained directly as the limiting case of the logsandwich mechanism when $\rho = 0$.

The coefficients of earth pressure corresponding to the logsandwich mechanism may be improved by shifting the center of the logarithmic spiral part BC of the surface of sliding shown in Fig. 8.9 or 8.10 away from its position at the top of the wall. Two such resulting general sandwich mechanisms are shown in Fig. 8.11, which shows the logarithmic spiral whose center O is now located on the line AC. An investigation is made to the two mechanisms shown in the figure. From the numerical results obtained it is found that very little solution improvement, if any, could be expected. Such improvement is outweighed by the complexity of the optimization procedure needed, since it must now contend with various functional discontinuities resulting from unknown velocity directions V_1 and V_3 as shown in Figs. 8.11(a) and 8.11(b), respectively. Improvement beyond the present stage may perhaps be realized only for other types of simple failure mechanisms.

8.6. ARC-SANDWICH MECHANISM (FIG. 8.12)

Figure 8.12(a) shows the passive pressure case of a circular radial shearing zone,

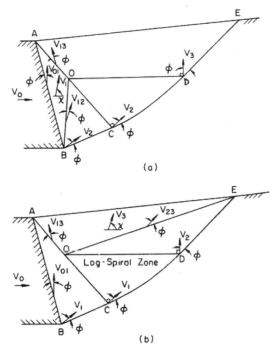

Fig. 8.11. General passive logsandwich mechanism with shifting pole.

OBC, sandwiched between two rigid triangular blocks *OAB* and *OCD.* Since the surfaces of sliding *BA* and *CD* are assumed perpendicular to the radial lines *OB* and *OC* respectively, two angular parameters ρ and ψ describe the mechanism completely. The compatible velocity diagrams corresponding to the passive pressure case are given in Figs. 8.12(b and c) for the smooth ($\delta < \varphi$) and rough ($\delta \geqslant \varphi$) wall conditions, respectively. The active pressure case can be treated in a similar manner. The formulations for the coefficients of earth pressure due to weight, cohesion, and surcharge (K_γ, K_c, K_q) follow.

8.6.1. For a cohesionless soil with no surcharge loading (c = 0, q = 0)

The rate of external work due to self-weight in any region is simply the vertical component of velocity in that region multiplied by the weight of the region (for passive case use lower signs):

Triangular region OAB

$$\pm \frac{\frac{1}{2}\gamma H^2 \, V_1 \cos\rho \sin\rho \cos(\alpha - \rho \pm \varphi)}{\sin^2\alpha} \qquad [8.63]$$

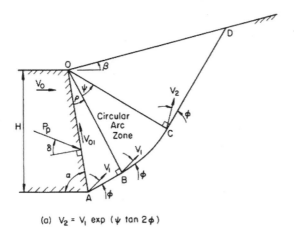

(a) $V_2 = V_1 \exp(\psi \tan 2\phi)$

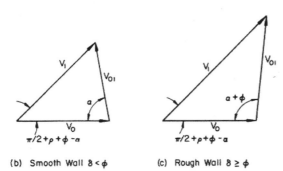

(b) Smooth Wall $\delta < \phi$ (c) Rough Wall $\delta \geq \phi$

Fig. 8.12. Passive arc-sandwich mechanism.

Triangular region OCD

$$\pm \frac{\frac{1}{2}\gamma H^2 V_2 \cos^2\rho \, \cos(\alpha - \rho - \psi \pm \varphi) \tan(\alpha - \rho - \psi + \beta)}{\sin^2\alpha} \qquad [8.64]$$

Circular-arc region OBC

The external rate of work done by the soil weight in this region can be computed by summing over the region ψ the products of differential triangle's component of vertical velocity with its weight. Dividing the region OBC into many small triangles similar to that shown in Fig. 3.23(a), this can be expressed in integral form as:

$$\pm \frac{1}{2}\gamma V_1 (\overline{OB})^2 \int_0^\psi \exp(\mp \theta \tan 2\varphi) \sin(\tfrac{1}{2}\pi - \alpha + \rho + \theta \mp \varphi) \, d\theta \qquad [8.65]$$

After integration, we obtain:

$$\pm \frac{\frac{1}{2}\gamma H^2 V_1}{\sin^2 \alpha (1 + \tan^2 2\varphi)} \{\cos(\alpha - \rho \pm \varphi)[\pm \tan 2\varphi + (\mp \tan 2\varphi \cos \psi + \sin \psi)$$

$$\exp(\mp \psi \tan 2\varphi)] + \sin(\alpha - \rho \pm \varphi)[1 + (\mp \tan 2\varphi \sin \psi - \cos \psi)\exp(\mp \psi \tan 2\varphi)]\}$$

[8.66]

The external work done by the components of the resultant wall load P moving in the horizontal direction with velocity V_0 is given by [8.34].

For a smooth wall $(\delta < \varphi)$ the dissipation by sliding friction is given by [8.35]. For a rough wall $(\delta \geqslant \varphi)$ the dissipation of a cohesionless soil is zero.

With the use of the compatible velocity diagrams all velocities in the mechanism can be expressed in terms of the wall translational velocity V_0:

For the case of smooth walls $(\delta < \varphi)$

$$V_1 = \frac{V_0 \sin \alpha}{\cos(\rho \mp \varphi)}, \qquad V_{10} = \frac{V_0 \cos(\alpha - \rho \pm \varphi)}{\cos(\rho \mp \varphi)}$$

[8.67]

For the case of rough walls $(\delta \geqslant \varphi)$

$$V_1 = \frac{V_0 \sin(\alpha \mp \varphi)}{\cos(\rho \mp 2\varphi)}, \qquad V_{10} = \frac{V_0 \cos(\alpha - \rho \pm \varphi)}{\cos(\rho \mp 2\varphi)}$$

[8.68]

and for both cases:

$$V_2 = V_1 \exp(\mp \psi \tan 2\varphi)$$

[8.69]

Equating the total rates of external work to the total rates of internal energy dissipation:

For a smooth wall $(\delta < \varphi)$

$$\begin{Bmatrix} K_{a\gamma} \\ K_{p\gamma} \end{Bmatrix} = \frac{\mp \sec \delta \csc \alpha \sec(\rho \mp \varphi) K_1}{\mp \sin \alpha + \tan \delta \cos \alpha - \dfrac{\tan \delta \cos(\alpha - \rho \pm \varphi)}{\cos(\rho \mp \varphi)}}$$

[8.70]

For a rough wall $(\delta \geqslant \varphi)$

$$\begin{Bmatrix} K_{a\gamma} \\ K_{p\gamma} \end{Bmatrix} = \frac{\mp \sec \delta \csc^2 \alpha \sin(\alpha \mp \varphi) K_1}{(\mp \sin \alpha + \tan \delta \cos \alpha)\cos(\rho \mp 2\varphi)}$$

[8.71]

in which the function $K_1(\rho, \psi)$ is defined as:

$$K_1(\rho, \psi) = \cos\rho \sin\rho \cos(\alpha - \rho \pm \varphi) + \cos^2\rho \cos(\alpha - \rho - \psi \pm \varphi) \tan(\alpha - \rho - \psi + \beta)$$

$$\exp(\mp \psi \tan 2\varphi) + \frac{1}{(1 + \tan^2 2\varphi)} \{\cos(\alpha - \rho \pm \varphi) [\pm \tan 2\varphi + (\mp \tan 2\varphi \cos\psi$$

$$+ \sin\psi) \exp(\mp \psi \tan 2\varphi)] + \sin(\alpha - \rho \pm \varphi) [1 + (\mp \tan 2\varphi \sin\psi$$

$$- \cos\varphi) \exp(\mp \psi \tan 2\varphi)] \} \qquad [8.72]$$

8.6.2. For a cohesive weightless soil with no surcharge ($\gamma = 0$, $q = 0$)

For a cohesive soil, energy dissipation terms must be added for all surfaces of discontinuity as well as the shearing zone in the circular region OBC (Fig. 8.12a). Since the velocities V_1 and V_2 are inclined at an angle $(\frac{1}{2}\pi - \varphi)$ to the radial lines OB and OC, there will be no dissipation along either OB or OC due to a lack of relative movement at those surfaces. The following dissipation terms must, however, be included (for passive case use lower signs):

Along the wall (OA)
See [8.35] for a smooth wall ($\delta < \varphi$). For a rough wall ($\delta \geqslant \varphi$) the dissipation is given by [8.40].

Along AB

$$\frac{cH\, V_1 \sin\rho \cos\varphi}{\sin\alpha} \qquad [8.73]$$

Along CD

$$\frac{cH\, V_1 \cos\rho \cos\varphi \tan(\alpha - \rho - \psi + \beta) \exp(\mp \psi \tan 2\varphi)}{\sin\alpha} \qquad [8.74]$$

Along BC
From [3.107], the dissipation along the curved discontinuity surface BC is equal to:

$$\mp \frac{cH\, V_1 \cos\rho \cos\varphi [\exp(\mp \psi \tan 2\varphi) - 1]}{\sin\alpha \tan 2\varphi} \qquad [8.75]$$

Circular region OBC
From [3.104], the total energy dissipation in the circular radial shear zone ψ is

equal to:

$$\mp \frac{cH\,V_1 \cos\rho \cos\varphi\,[\exp(\mp\,\psi\,\tan 2\varphi) - 1]}{\sin\alpha \sin 2\varphi} \qquad [8.76]$$

Equating the rate of external work to the rate of internal energy dissipation:

For a smooth wall $(\delta < \varphi)$

$$\begin{Bmatrix} K_{ac} \\ K_{pc} \end{Bmatrix} = \frac{\sec\delta\ \sec(\rho \mp \varphi)\cos\varphi\cos\rho}{\mp\sin\alpha + \tan\delta\,\cos\alpha - \dfrac{\tan\delta\,\cos(\alpha - \rho \pm \varphi)}{\cos(\rho \mp \varphi)}} \left\{ \tan\rho + \tan(\alpha - \rho - \psi + \beta) \right.$$

$$\left. \exp(\mp\,\psi\,\tan 2\varphi) \mp \frac{[\exp(\mp\,\psi\,\tan 2\varphi) - 1]\,(1 + \cos 2\varphi)}{\sin 2\varphi} \right\} \qquad [8.77]$$

For a rough wall $(\delta \geqslant \varphi)$

$$\begin{Bmatrix} K_{ac} \\ K_{pc} \end{Bmatrix} = \frac{\sec\delta\ \csc\alpha\,\sin(\alpha \mp \varphi)\,\sec(\rho \mp 2\varphi)\cos\varphi\cos\rho}{\mp\sin\alpha + \tan\delta\,\cos\alpha} \left\{ \frac{\cos(\alpha - \rho \pm \varphi)}{\cos\rho\,\sin(\alpha \mp \varphi)} \right.$$

$$+ \tan\rho \overset{\cdot}{+} \tan(\alpha - \rho - \psi + \beta)\exp(\mp\,\psi\,\tan 2\varphi)$$

$$\left. \mp \frac{[\exp(\mp\,\psi\,\tan 2\varphi) - 1]\,(1 + \cos 2\varphi)}{\sin 2\varphi} \right\} \qquad [8.78]$$

8.6.3. Surcharge loading $(\gamma = 0,\ c = 0)$

The following rate of external work term is necessary for a backfill uniformly loaded with a surcharge q:

$$\frac{\pm qH\,V_2 \cos\rho \cos(\alpha - \rho - \psi \pm \varphi)}{\sin\alpha \cos(\alpha - \rho - \psi + \beta)} \qquad [8.79]$$

For a smooth wall $(\delta < \varphi)$

$$\begin{Bmatrix} K_{aq} \\ K_{pq} \end{Bmatrix} = \frac{\mp \sec\delta\ \cos\rho \cos(\alpha - \rho - \psi \pm \varphi)\exp(\mp\,\psi\,\tan 2\varphi)}{\left[\mp\sin\alpha + \tan\delta\,\cos\alpha - \dfrac{\tan\delta\,\cos(\alpha - \rho \pm \varphi)}{\cos(\rho \mp \varphi)}\right]\cos(\rho \mp \varphi)\cos(\alpha - \rho - \psi + \beta)}$$

$$[8.80]$$

For a rough wall $(\delta \geqslant \varphi)$

$$\begin{Bmatrix} K_{aq} \\ K_{pq} \end{Bmatrix} = \frac{\mp \sec\delta\ \cos\rho \sin(\alpha \mp \varphi)\cos(\alpha - \rho - \psi \pm \varphi)\exp(\mp\,\psi\,\tan 2\varphi)}{(\mp\sin\alpha + \tan\delta\,\cos\alpha)\sin\alpha \cos(\rho \mp 2\varphi)\cos(\alpha - \rho - \psi + \beta)} \qquad [8.81]$$

In the particular case when the angle ρ of mechanism 6 (Fig. 8.7f) is zero, the arc-sandwich mechanism becomes the arc-triangle mechanism shown in Fig. 8.7(d). The results for the coefficients of earth pressure due to weight, cohesion, and surcharge (K_γ, K_c, K_q) corresponding to this particular case can be obtained directly as the limiting case of the arc-sandwich mechanism when $\rho = 0$.

8.7. DISCUSSION OF RESULTS

8.7.1. Passive pressure

Table 8.2 shows the resultant passive pressure coefficients $K_{p\gamma}$ for three different wall inclinations $\alpha = 70°, 90°, 110°$ and a horizontal backfill ($\beta = 0$). These solutions were obtained by using the six failure mechanisms given in Fig. 8.7. Mechanisms 3 and 4 failed to yield good solutions for most cases, especially for the case of perfectly smooth walls ($\delta = 0$). Mechanism 6 was identical to the logsandwich mechanism (Mechanism 5) except that the logspiral region was replaced with the circular ($\varphi \neq 0$) shearing region discussed in section 3.4, Chapter 3. Results between the two compared favorably until high soil friction angles were encountered (Table 8.2). As expected, Mechanism 2 gives better results than Mechanism 1.

Figure 8.13(a) shows the critical lay-out for the two-triangle mechanism (Mechanism 2) while Fig. 8.13(b) shows a similar critical lay-out for the logsandwich mechanism (Mechanism 5). The effect of wall roughness on the resulting passive pressure coefficients is shown in Fig. 8.14. For a horizontal backfill ($\beta = 0$) it is seen that roughness is particularly important for walls angled into a backfill of high soil friction angle. For a sloping backfill ($\beta = 20°$) it is evident that roughness is important for all wall inclinations.

From these observations it is evident that Mechanism 2 and 5 yielded the best solutions. Although Mechanism 5 was physically more complicated, it was fully described by only two independent parameters as opposed to three for Mechanism 2. In an attempt to further reduce the number of parameters and hence simplify the minimization scheme, a study was made of the inclination of the straight line portion of the upper rigid body as it intersected the unloaded horizontal backfill surface (Fig. 8.7 with $\beta = 0$). Ideally for a Rankine failure state such lines should make angles of $\frac{1}{4}\pi - \frac{1}{2}\varphi$ with the horizontal. The results of this study are found in Table 8.3. Shown are the percent differences between the computed inclinations Ω and the theoretical Rankine inclinations: as well as the resulting differences in the coefficients $K_{p\gamma}$. It is interesting to note that for smooth walls inclined at $\alpha = 70°$ considerable deviation occurred, but that such deviations resulted in no more than 7% difference in coefficient values. This fact illustrates the apparent insensitivity of the logsandwich mechanism to changes in the angular

TABLE 8.2

Passive pressure coefficients $K_{p\gamma}(\beta = 0°)$

φ (°)	δ (°)	Mechanism (see Fig. 8.7)					
		(1)	(2)	(3)	(4)	(5)	(6)
$\alpha = 70°$							
10	0	1.36	1.36	1.52	1.47	1.36	1.36
	5	1.45	1.45	1.53	1.50	1.45	1.45
	10	1.55	1.54	1.54	1.54	1.54	1.54
20	0	1.75	1.75	2.31	2.01	1.75	1.75
	10	2.08	2.08	2.36	2.18	2.08	2.08
	20	2.49	2.44	2.46	2.47	2.44	2.47
30	0	2.27	2.27	3.86	2.75	2.28	2.30
	15	3.16	3.16	4.06	3.36	3.16	3.18
	30	4.76	4.43	4.50	4.76	4.41	4.76
40	0	3.02	3.02	7.76	3.52	3.02	3.27
	20	5.34	5.32	8.33	5.39	5.31	5.89
	40	12.80	10.00	10.10	15.50	9.88	--
$\alpha = 90°$							
10	0	1.42	1.42	1.68	1.60	1.42	1.42
	5	1.57	1.56	1.69	1.63	1.56	1.56
	10	1.73	1.68	1.71	1.68	1.68	1.67
20	0	2.04	2.04	3.07	2.60	2.04	2.04
	10	2.64	2.58	3.12	2.82	2.58	2.61
	20	3.53	3.18	3.27	3.19	3.17	3.19
30	0	3.00	3.00	6.38	4.80	3.00	3.01
	15	4.98	4.71	6.61	5.88	4.71	4.97
	30	10.10	7.24	7.37	8.31	7.10	8.31
40	0	4.60	4.60	16.10	15.40	4.60	4.67
	20	11.80	10.10	17.70	23.60	10.10	12.50
	40	92.60	22.70	21.70	67.90	20.90	--
$\alpha = 110°$							
10	0	1.76	1.74	3.10	2.06	1.74	1.74
	5	1.90	*1.83	3.12	1.97	1.96	1.96
	10	2.04	1.91	2.77	1.90	2.16	2.14
20	0	2.98	2.91	6.41	4.03	2.91	2.93
	10	3.78	3.38	6.50	3.85	3.91	3.94
	20	4.81	3.92	5.22	3.79	5.04	4.95
30	0	5.34	5.09	15.60	10.20	5.08	5.33
	15	9.22	6.99	16.10	10.10	8.93	10.20
	30	72.70	10.10	11.70	11.50	14.40	17.60
40	0	10.70	9.73	50.30	127.00	9.71	11.40
	20	89.70	17.60	53.50	141.00	25.50	69.40
	40	77.40	--	34.90	298.00	56.60	--

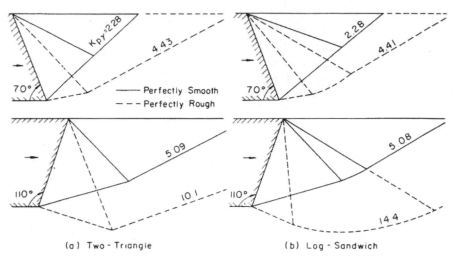

(a) Two - Triangle   (b) Log - Sandwich

Fig. 8.13. Typical passive mechanism results ($\varphi = 30°$).

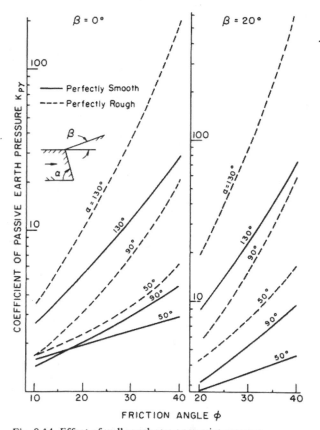

Fig. 8.14. Effect of wall roughness on passive pressure.

TABLE 8.3

Effects of Rankine constraint ($\beta = 0°$)

φ (°)	δ (°)	Passive log-sandwich (Fig. 8.9)		Active log-sandwich (Fig. 8.10)	
		Ω % diff.	$K_{p\gamma}$ % diff.	Ω % diff.	$K_{a\gamma}$ % diff.
$\alpha = 70°$					
10	0	26.30	1.47	20.00	— —
	5	— —	— —	9.20	— —
	10	1.00	— —	40.00	— —
20	0	27.40	2.24	18.40	1.85
	10	— —	— —	6.00	— —
	20	2.28	— —	3.82	— —
30	0	36.00	3.94	16.80	3.02
	15	1.66	— —	9.00	— —
	30	2.00	— —	3.50	— —
40	0	35.60	6.64	6.85	4.80
	20	1.60	— —	21.60	— —
	40	2.40	— —	6.15	— —
$\alpha = 90°$					
10	0	2.00	— —	1.40	— —
	5	2.00	— —	— —	— —
	10	1.00	— —	— —	— —
20	0	— —	— —	1.27	— —
	10	— —	— —	— —	— —
	20	2.57	— —	— —	— —
30	0	— —	— —	1.17	— —
	15	1.33	— —	2.16	— —
	30	2.00	— —	— —	— —
40	0	2.40	— —	1.07	— —
	20	4.40	— —	— —	— —
	40	7.00	— —	— —	— —
$\alpha = 110°$					
10	0	1.75	— —	— —	— —
	5	— —	— —	8.60	— —
	10	2.25	— —	4.20	— —
20	0	1.15	— —	1.82	— —
	10	4.56	— —	1.09	— —
	20	6.30	— —	7.82	— —
30	0	1.33	— —	17.50	— —
	15	5.66	— —	0.50	— —
	30	— —	— —	10.00	— —
40	0	— —	— —	4.31	— —
	20	— —	— —	14.50	— —
	40	— —	— —	0.46	— —

extent of the rigid body adjoining the wall, since the logspiral extent was virtually unaffected. It should be pointed out that the Rankine inclinations corresponded to coefficients which were greater than those obtained from the unconstrained log-sandwich mechanisms. If the mechanism is constrained such that:

$$\Omega = \tfrac{1}{2}\pi - \alpha + \rho + \psi - \varphi = \tfrac{1}{4}\pi - \tfrac{1}{2}\varphi \qquad [8.82]$$

we obtain one independent parameter:

$$\psi = \alpha - \rho + \tfrac{1}{2}\varphi - \tfrac{1}{4}\pi \qquad [8.83]$$

where ρ is the angular extent of the rigid block adjoining the wall, and ψ is the angular extent of the logspiral zone (Fig. 8.7e). Thus for horizontal backfills we can be assured that the resulting solutions will only slightly be in error.

For a backfill of varying inclination the Rankine state is defined by a line to the horizontal such that:

$$\Omega = \tfrac{1}{4}\pi - \tfrac{1}{2}\varphi + \tfrac{1}{2}\epsilon - \tfrac{1}{2}\beta \qquad [8.84]$$

where

$$\sin \epsilon = \frac{\sin \beta}{\sin \varphi} \qquad [8.85]$$

and β is the backfill inclination from the horizontal. The usefulness of [8.84] was found to be limited only to the cases of soil friction angle φ less than $30°$ and backfill angle β less than $15°$. For cases outside these limits the one-parameter solutions overestimated the corresponding two-parameter solutions by 5% to as much as 50%.

8.7.2. Active pressure

Table 8.4 shows the resultant active pressure coefficients $K_{a\gamma}$ for three different wall inclinations and a horizontal backfill. These results correspond to the six failure mechanisms given in Fig. 8.7.

Figures 8.15(a) and 8.15(b) show some critical lay-out for the two-triangle and logsandwich mechanisms, respectively. For the case of a vertical wall, $\alpha = 90°$, both two-triangle and logsandwich mechanisms yielded nearly identical results (Table 8.4). For walls angled into the backfill ($\alpha = 110°$), however, the logsandwich mechanism underestimated the upper-bound solution. Again the inclusion of a logspiral zone in place of a circular shearing zone in the logsandwich mechanism improves the solution. Table 8.3 shows that the Rankine condition at the unloaded surface was not as well obeyed as for the passive case. It should be noted, however, that the active mechanisms were much less sensitive in terms of final results, with a maximum deviation of not more than 5%. The number of independent parameters

TABLE 8.4

Active pressure coefficients $K_{a\gamma}(\beta = 0°)$

φ (°)	δ (°)	Mechanism (see Fig. 8.7)					
		(1)	(2)	(3)	(4)	(5)	(6)
$\alpha = 70°$							
10	0	0.833	0.821	0.774	0.738	0.832	0.832
	5	0.801	0.800	0.775	0.778	0.801	0.801
	10	0.786	0.787	0.786	0.826	0.787	0.786
20	0	0.648	0.616	0.576	0.447	0.647	0.647
	10	0.615	0.610	0.582	0.485	0.615	0.614
	20	0.613	0.614	0.613	0.549	0.613	0.613
30	0	0.498	0.490	0.434	0.173	0.497	0.497
	15	0.476	0.473	0.446	0.187	0.475	0.475
	30	0.501	0.501	0.501	0.230	0.501	0.501
40	0	0.375	0.320	0.328	––	0.375	0.373
	20	0.370	0.303	0.346	––	0.368	0.365
	40	0.428	0.417	0.428	––	0.428	0.418
$\alpha = 90°$							
10	0	0.704	0.704	0.622	0.572	0.704	0.704
	5	0.662	0.664	0.624	0.566	0.664	0.663
	10	0.635	0.642	0.631	0.564	0.642	0.637
20	0	0.490	0.490	0.394	0.236	0.490	0.490
	10	0.447	0.448	0.400	0.226	0.448	0.447
	20	0.427	0.434	0.420	0.222	0.434	0.427
30	0	0.333	0.333	0.250	––	0.333	0.333
	15	0.301	0.302	0.259	––	0.302	0.301
	30	0.297	0.303	0.289	––	0.302	0.297
40	0	0.217	0.215	0.155	––	0.217	0.217
	20	0.199	0.200	0.165	––	0.200	0.197
	40	0.210	0.214	0.202	––	0.214	0.210
$\alpha = 110°$							
10	0	0.644	0.649	––	0.441	0.649	0.647
	5	0.625	0.639	––	0.436	0.601	0.595
	10	0.616	0.649	––	0.435	0.569	0.561
20	0	0.380	0.387	––	0.015	0.385	0.382
	10	0.371	0.386	––	––	0.341	0.333
	20	0.378	0.417	––	––	0.319	0.307
30	0	0.212	0.218	––	––	0.216	0.212
	15	0.215	0.226	––	––	0.188	0.181
	30	0.237	0.275	––	––	0.179	0.168
40	0	0.106	0.111	––	––	0.109	0.106
	20	0.115	0.123	––	––	0.095	0.090
	40	0.146	0.180	––	––	0.095	0.077

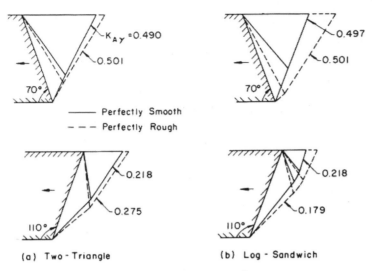

Fig. 8.15. Typical active mechanism results ($\varphi = 30°$).

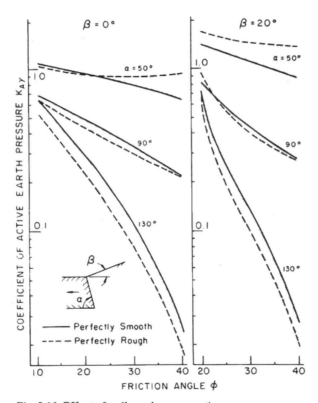

Fig. 8.16. Effect of wall roughness on active pressure.

can also be reduced to one, such that:

$$\Omega = \tfrac{1}{4}\pi + \tfrac{1}{2}\varphi - \tfrac{1}{2}\epsilon + \tfrac{1}{2}\beta \qquad [8.86]$$

where ϵ is defined by [8.85]. As in the passive case the use of [8.86] results in poor values. For sloping backfills the two-parameter mechanism must again be used. The effect of wall roughness on the active pressure coefficients is shown in Fig. 8.16. Unlike the passive case for horizontal backfills, wall roughness has a much smaller effect and is important for walls angles out of the soil ($\alpha < 90°$).

8.8. COMPARISON WITH KNOWN SOLUTIONS

8.8.1. Passive pressure

Sokolovskii's solutions

To date the best solutions have been generated by Sokolovskii (1965) using the slip-line method or method of stress characteristics. For soils with weight, solutions are obtained by an approximate numerical integration of the characteristic stress equations of the plastic equilibrium field (see for example, [6.93] and [6.95]). The question of whether such a slip-line solution is either an upper- or lower-bound solution has already been discussed in Chapter 1 and in section 6.2 of Chapter 6. It is generally agreed, however, that such solutions give good estimates of the exact values. Table 8.5 shows a comparison of limit analysis solutions for horizontal backfills with those obtained by Sokolovskii. For the cases $\alpha = 70°$ and $\alpha = 90°$ (vertical wall) good agreement exists. It is also seen that the two-triangle and logsandwich mechanisms control over the whole range of soil—wall conditions, especially for rough walls retaining soil of high friction angle ($\varphi = 30-40°$). For these cases the limit analysis method yields values which differ from the Sokolovskii solutions by no more than 15% ($\varphi = 40°$, $\delta = 40°$). For the case of the wall bearing into the backfill ($\alpha = 110°$), considerably more disagreement occurred. The maximum difference obtained was 26.6% for the same case given above. It should be noted that the Sokolovskii solutions are very limited and easily available only for horizontal backfills.

Lee and Herington's solutions

In a recent study by Lee and Herington (1972) some slip-line solutions for perfectly rough walls with sloping backfills have been formulated for both associated and non-associated flow rule materials of Coulomb type. The most important numerical results obtained by Lee and Herington are the values of the passive earth thrust coefficients K_{pc} and $K_{p\gamma}$ as a function of the internal soil friction angle φ, wall slope α, and backfill slope β. For convenience of practical applications, the

TABLE 8.5

Comparison of limit analysis with passive slip-line solutions ($\beta = 0°$)

φ (°)	δ (°)	$K_{p\gamma}$		Mechanism (Fig. 8.7)	% difference
		Sokolovskii (1965)	Limit analysis		
$\alpha = 70°$					
10	0	1.34	1.36	1,2,5,6	+ 1.49
	5	1.46	1.45	1,2,5,6	− 0.68
	10	1.53	1.54	1,2,3,4,5,6	+ 0.65
20	0	1.71	1.75	1,2,5,6	+ 2.34
	10	2.08	2.08	1,2,5,6	−
	20	2.42	2.44	2,5	+ 0.83
30	0	2.16	2.28	1,2,5	+ 7.86
	15	3.16	3.16	1,2,5	−
	30	4.30	4.41	5	+ 2.56
40	0	2.84	3.02	1,2,5	+ 6.34
	20	5.32	5.31	2,5	− 2.07
	40	9.32	9.88	5	+ 6.00
$\alpha = 90°$					
10	0	1.42	1.42	1,2,5,6	−
	5	1.56	1.56	1,2,5,6	−
	10	1.66	1.68	2,4,5,6	+ 1.20
20	0	2.04	2.04	1,2,5,6	−
	10	2.55	2.58	2,5	+ 1.18
	20	3.04	3.17	2,5	+ 4.27
30	0	3.00	3.00	1,2,5,6	−
	15	4.62	4.71	2,5	+ 1.95
	30	6.55	7.10	5	+ 8.40
40	0	4.60	4.60	1,2,5	−
	20	9.69	10.10	2,5	+ 4.23
	40	18.20	20.90	5	+14.85
$\alpha = 110°$					
10	0	1.75	1.74	2,5,6	− 0.57
	5	1.95	1.83	2	− 6.15
	10	2.10	1.90	2,4	− 9.50
20	0	2.90	2.91	2,5	+ 0.34
	10	3.80	3.38	2	−11.00
	20	4.62	3.79	4	−18.00
30	0	5.06	5.08	2,5	+ 0.40
	15	8.45	8.93	5	+ 5.65
	30	12.30	14.40	5	+17.10
40	0	9.56	9.71	5	+ 1.57
	20	22.40	25.50	5	+13.80
	40	44.70	56.60	5	+26.60

TABLE 8.6

Passive earth pressure coefficients K_{pc}, $K_{p\gamma}$ Slip-line solutions (Lee and Herington, 1972)

Backfill angle β (°)	Wall angle α (°)	Angle of internal friction φ (°)	K_{pc} vertical component plastic material	frictional material	horizontal component plastic material	frictional material	$K_{p\gamma}$ vertical component plastic material	frictional material	horizontal component plastic material	frictional material
20	90	22.5	4.15	3.85	7.85	7.75	2.30	2.15	5.60	5.60
		30	7.65	6.60	11.45	11.30	6.45	5.50	11.20	10.70
		40	20.1	14.7	22.8	21.2	29.6	19.8	35.0	31.2
	80	22.5	2.50	2.20	7.00	6.90	1.20	1.00	5.20	5.05
		30	4.05	3.80	10.05	9.90	3.60	2.80	9.60	9.40
		40	12.2	8.10	19.5	17.4	15.8	10.1	27.4	23.6
	70	22.5	1.25	1.00	6.30	6.15	0.25	0.05	4.80	4.70
		30	2.60	1.75	8.85	8.50	1.50	0.90	8.50	8.00
		40	6.80	4.00	16.1	14.3	8.00	4.20	21.7	18.7
0	90	22.5	3.15	2.90	5.10	4.90	1.40	1.35	3.45	2.35
		30	5.00	4.20	6.95	6.40	3.30	2.60	5.65	5.00
		40	11.1	7.10	12.2	9.65	12.0	6.60	14.2	10.2
		45	18.0	11.9	17.0	15.3	26.0	14.9	26.0	21.2
	80	22.5	2.05	1.80	4.65	4.45	0.65	0.60	3.00	2.95
		30	3.20	2.75	6.05	5.70	1.80	1.50	4.95	4.75
		40	6.85	4.50	10.1	9.30	6.55	4.50	11.0	9.10
		45	10.9	6.30	14.0	12.1	13.2	7.50	18.8	15.2
	70	22.5	1.15	1.00	4.10	4.05	0.15	0.05	2.75	2.65
		30	1.85	1.45	5.30	5.00	0.75	0.45	4.25	4.00
		40	4.05	2.50	8.60	7.40	3.20	2.40	8.80	7.40
		45	6.00	3.20	11.3	9.30	6.55	3.40	14.2	11.3

TABLE 8.6 (continued)

Passive earth pressure coefficients K_{pc}, $K_{p\gamma}$ Slip-line solutions (Lee and Herington, 1972)

Backfill angle β (°)	Wall angle α (°)	Angle of internal friction φ (°)	K_{pc}				$K_{p\gamma}$			
			vertical component		horizontal component		vertical component		horizontal component	
			plastic material	frictional material	plastic material	frictional material	plastic material	frictional material	plastic material	frictional material
−20	90	22.5	2.35	1.90	3.20	3.10	0.60	0.50	1.35	1.30
		30	3.40	2.95	4.10	3.85	1.30	1.00	2.25	2.15
		40	6.15	4.50	6.10	5.75	3.95	2.65	4.70	4.15
		45	9.00	5.65	8.00	7.25	7.50	4.70	7.50	6.40
	80	22.5	1.60	1.30	2.85	2.75	0.25	0.20	1.15	1.10
		30	2.30	1.75	3.50	3.40	0.70	0.55	1.90	1.70
		40	3.95	2.60	5.10	4.55	2.10	1.35	3.60	3.25
		45	5.50	3.55	6.40	5.75	3.80	2.40	5.40	4.50
	70	22.5	1.05	0.90	2.45	2.40	0.05	0.00	1.10	1.05
		30	1.45	1.25	3.00	2.95	0.30	0.25	1.55	1.65
		40	2.35	1.50	4.15	3.60	1.00	0.60	2.80	2.45
		45	3.20	1.85	5.10	4.10	1.85	1.15	4.00	3.50
−35	90	38	3.55	2.65	3.30	3.10	1.05	0.85	1.35	1.35
		40	4.00	2.90	3.50	3.25	1.30	0.95	1.50	1.45
		45	5.30	3.40	4.30	3.80	2.35	1.45	2.35	2.20
	80	38	2.35	1.70	2.65	2.45	0.55	0.35	1.05	0.85
		40	2.55	1.75	2.85	2.50	0.70	0.50	1.15	1.10
		45	3.25	2.00	3.45	2.90	1.25	0.65	1.75	1.45
	70	38	1.55		2.10		0.25		0.75	
		40	1.70		2.20		0.30		0.85	
		45	2.10		2.60		0.60		1.20	

passive thrust is resolved into the vertical and horizontal components and the passive earth thrust coefficients are defined accordingly. In Table 8.6, the components of passive thrust coefficients are tabulated for the perfectly plastic Coulomb material with the associated flow rule and the particular frictional Coulomb material which has a non-associated flow rule (section 2.6, Chapter 2). Values of φ range from a value slightly greater than the surface slope β to $45°$. Values of β are $20°$, $0°$, $-20°$ and $-35°$, the negative sign indicating an downward slope from the wall.

A study of the values in Table 8.6 shows that the differences in passive thrust components predicted for the perfectly plastic Coulomb material and the frictional Coulomb material differ by 10–20% for the horizontal components and up to 80–90% for the vertical components. The numerical values for the perfectly plastic material are seen always higher than those of the frictional material. This is expected because Lee and Herington (1972) have shown that their partial slip-line stress fields are both statically and kinematically admissible for the perfectly plastic material. It follows from the upper- and lower-bound limit theorems that such a solution gives the *correct* or *exact* solution for a perfectly plastic material with an associated flow rule. For a frictional material, Theorem X, section 2.6, Chapter 2, states that the *correct* solution for an associated flow rule material is also an upper bound of the true solution for a non-associated flow rule material with the same value of φ.

Table 8.7 shows a comparison of limit analysis solutions for horizontal backfills with those obtained by Lee and Herington. It is seen that even for relatively large values of φ ($\varphi = 40°$), the range of error is 5.7–12.4%. This range decreases with smaller φ and also decreases as α is decreased. Here errors in Table 8.7 are referred to the slip-line solutions. It should be noted that the numerical values presented in

TABLE 8.7

Comparison of limit analysis with passive slip-line solutions ($\beta = 0°$)

Wall angle α (°)	Angle of internal friction φ (°)	$K_{p\gamma}$, slip-line Lee and Herington (1972) See Table 8.6			$K_{p\gamma}$, limit analysis See Table 8.9 $\delta = \varphi$	% difference
		vertical component	horizontal component	resultant		
90	30	3.30	5.65	6.55	7.10	8.4
	40	12.00	14.20	18.60	20.90	12.4
80	30	1.80	4.95	5.26	5.45	3.6
	40	6.55	11.00	12.80	13.90	8.6
70	30	0.75	4.25	4.32	4.41	2.1
	40	3.20	8.80	9.36	9.88	5.6

Tables 8.6 and 8.7 for the slip-line solutions are represented approximately by the traditional superposition method, in which contributions to the passive earth thrust from the cohesive and self weight components are *summed*. Since soil behavior in the plastic range is nonlinear, and superposition does not hold in general, Lee and Herington introduced a correction factor for the superposition equation. This factor is a function of all of the variables of the problem, but in the cases presented in Table 8.6, it is closely equal to unity. For a vertical wall, $\alpha = 90^{\circ}$ and a flat backfill, $\beta = 0$, the range of this factor for $\varphi = \delta = 40^{\circ}$ is $1.0-1.036$. All the values given in Tables 8.6 and 8.7, however, take this factor as unity. Thus the differences between slip-line and limit analysis solutions are expected to decrease when the more accurate values of slip-line solutions are used in Table 8.7.

Remarks on the solutions

Figure 8.17 shows a comparison of some typical limit analysis results for a vertical wall with several existing solutions. As previously mentioned the power of the method of limit analysis lies in the capability of bounding the true solution. Although lower-bound solutions are much more difficult to obtain and involve the formulation of statically admissible stress fields, some limited solutions are available. As an example, for the particular case of a vertical wall with $\varphi = 40^{\circ}$, $\delta = 20^{\circ}$; a lower-bound solution of 8.97 has been obtained by Lysmer (1970). The upperbound solution for this case is 10.10. The corresponding Sokolovskii solution of 9.68 is seen to lie between these two bounds. If an upper bound can be obtained that agrees with the corresponding lower-bound solution, then of course, the exact solution will be found.

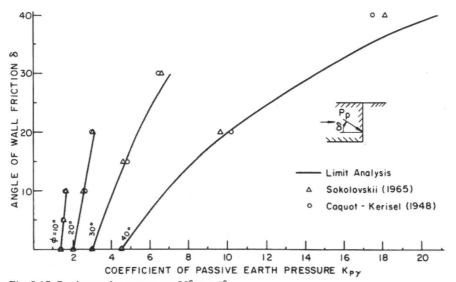

Fig. 8.17. Passive earth pressure $\alpha = 90^{\circ}$, $\beta = 0^{\circ}$.

TABLE 8.8

Comparison of limit analysis with active slip-line solutions ($\beta = 0°$)

φ (°)	δ (°)	Ka_γ		Mechanism (Fig. 8.7)	% difference
		Sokolovskii (1965)	Limit analysis		
$\alpha = 70°$					
10	0	0.826	0.833	1,5,6	+ 0.85
	5	0.794	0.801	1,2,5,6	+ 0.88
	10	0.794	0.787	1,2,3,5,6	− 0.88
20	0	0.656	0.648	1,5,6	− 1.21
	10	0.612	0.615	1,5,6	+ 0.49
	20	0.612	0.614	1,2,5,6	+ 0.33
30	0	0.521	0.498	1,5,6	− 4.41
	15	0.487	0.476	1,5,6	− 2.26
	30	0.510	0.501	1,2,3,5,6	− 1.76
40	0	0.396	0.375	1,5	− 5.30
	20	0.385	0.370	1	− 3.90
	40	0.430	0.428	1,3,5	− 0.47
$\alpha = 90°$					
10	0	0.700	0.704	1,2,5,6	+ 0.57
	5	0.670	0.664	2,5	− 0.89
	10	0.650	0.642	2,5	− 1.23
20	0	0.490	0.490	1,2,5,6	−
	10	0.450	0.448	2,5	− 0.44
	20	0.440	0.434	2,5	− 1.36
30	0	0.330	0.333	1,2,5,6	+ 0.91
	15	0.300	0.302	2,5	+ 0.67
	30	0.310	0.303	2	− 2.26
40	0	0.220	0.217	1,5,6	− 1.36
	20	0.200	0.200	2,5	
	40	0.220	0.214	2,5	− 2.73
$\alpha = 110°$					
10	0	0.665	0.649	2,5	− 2.40
	5	0.620	0.639	2	− 3.06
	10	0.596	0.649	2	+ 8.90
20	0	0.482	0.387	2	− 1.98
	10	0.356	0.386	2	+ 0.84
	20	0.344	0.417	2	+21.20
30	0	0.229	0.218	2	− 4.81
	15	0.206	0.226	2	+ 9.71
	30	0.195	0.275	2	+41.00
40	0	0.126	0.111	2	−11.90
	20	0.106	0.123	2	+16.00
	40	0.119	0.180	2	+51.30

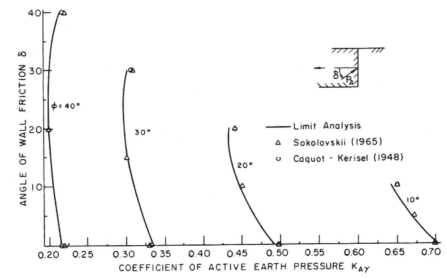

Fig. 8.18. Active earth pressure $\alpha = 90°$, $\beta = 0°$

8.8.2. Active pressure

A comparison of the active limit analysis solutions with those obtained by Sokolovskii is given in Table 8.8. In general the limit analysis results underestimate the slip-line solutions, with the greatest deviation being 6% for the cases $\alpha = 70°$ and $90°$. As in the case of passive pressures the logsandwich and two-triangle mechanisms are controlled in most cases. The exception to this, however, occurred for the case of the wall sloping into the earth backfill (i.e., $\alpha = 110°$). The two-triangle mechanism yielded results which greatly exceeded the slip-line solutions. This was particularly true for the rough wall conditions i.e., $\delta = \varphi$. For these particular cases, however, the logsandwich mechanism yielded much better agreement. The maximum deviation was reduced to 20% ($\varphi = 40°$, $\delta = 40°$). A comparison with some typical results is given in Fig. 8.18.

8.8.3. Cohesion and surcharge

In order to illustrate the case in which the effects of cohesion and surcharge are included in the analysis, two passive pressure example problems were solved using the expressions derived in the preceding sections.

The problem of a vertical wall retaining a $c - \varphi$ soil was solved (Fig. 8.19a). A comparison with the trial wedge method of limit equilibrium analysis shows excellent agreement. The problem of a wall retaining a cohesive, loaded soil is given in Fig. 8.19(b) which shows the resulting failure mechanisms for the effects of weight,

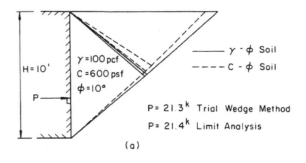

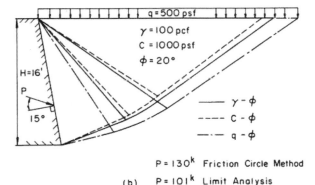

Fig. 8.19. Inclusion of cohesion and surcharge loading.

cohesion, and surcharge. The final solution was obtained by superimposing these effects. As expected, the limit analysis method yielded lower solutions for highly frictional walls. For the particular problem solved ($\varphi = 20°$, $\delta = 15°$) the limit analysis solution was over 20% lower than the one obtained by using the graphical friction circle method.

8.9. EARTH PRESSURE TABLES

In order to reduce the time required to solve the earth pressure problems of the type shown in Fig. 8.7, earth pressure tables for K_γ, K_c and K_q can be prepared for different values of φ and δ and for different values of the slope angles α and β. Elaborate earth pressure tables for the coefficients K_γ are given in Table 8.9 for the passive earth pressure case and in Table 8.10 for the active earth pressure case. The values given in these two tables are the most critical values of the logsandwich mechanisms shown in Fig. 8.7(e). In most engineering practice the backs of most walls and the surfaces of most backfills encountered are at least approximately plane, and on account of the uncertainties involved in the estimate of the values of φ and δ, the

TABLE 8.9

Passive earth pressure coefficients $K_{p\gamma}$

Angle of internal friction φ (°)	Wall friction angle δ (°)	Backfill angle β (°)	Wall angle α (°)								
			50	60	70	80	90	100	110	120	130
10	0	0	1.58	1.44	1.38	1.37	1.42	1.54	1.74	2.06	2.60
		10	1.87	1.69	1.62	1.61	1.68	1.81	2.05	2.45	3.11
	5	0	1.61	1.50	1.45	1.48	1.56	1.71	1.96	2.36	3.03
		10	2.05	1.87	1.80	1.81	1.90	2.08	2.39	2.89	3.74
	10	0	1.66	1.56	1.54	1.58	1.68	1.87	2.16	2.64	3.45
		10	2.19	2.01	1.95	1.98	2.10	2.32	2.70	3.31	4.35
15	0	0	1.75	1.62	1.57	1.59	1.70	1.91	2.24	2.78	3.70
		10	2.08	1.92	1.88	1.93	2.07	2.32	2.74	3.43	4.61
	5	0	1.78	1.68	1.67	1.74	1.89	2.15	2.57	3.24	4.40
		10	2.27	2.13	2.10	2.18	2.36	2.69	3.22	4.10	5.62
	10	0	1.84	1.77	1.79	1.89	2.08	2.40	2.90	3.72	5.13
		10	2.46	2.32	2.32	2.43	2.66	3.07	3.72	4.81	6.69
	15	0	1.91	1.87	1.91	2.04	2.27	2.64	3.23	4.20	5.87
		10	2.63	2.50	2.52	2.66	2.95	3.44	4.22	5.52	7.79
20	0	0	1.92	1.81	1.79	1.86	2.04	2.37	2.91	3.78	5.32
		10	2.29	2.17	2.18	2.30	2.56	2.98	3.68	4.85	6.91
		20	2.78	2.62	2.62	2.77	3.09	3.63	4.50	5.98	8.63
	5	0	1.98	1.90	1.92	2.04	2.30	2.72	3.39	4.49	6.45
		10	2.52	2.41	2.45	2.63	2.96	3.51	4.40	5.90	8.57
		20	3.14	2.99	3.02	3.24	3.65	4.35	5.49	7.42	10.9
	10	0	2.05	2.01	2.08	2.26	2.58	3.09	3.91	5.27	7.69
		10	2.75	2.67	2.75	2.98	3.39	4.08	5.19	7.05	10.4
		20	3.52	3.37	3.45	3.73	4.26	5.13	6.57	9.01	13.4

TABLE 8.9 (continued)

Angle of internal friction φ (°)	Wall friction angle δ (°)	Backfill angle β (°)	Wall angle α (°)								
			50	60	70	80	90	100	110	120	130
20	15	0	2.14	2.14	2.26	2.49	2.88	3.49	4.47	6.11	9.04
		10	2.99	2.93	3.05	3.34	3.85	4.68	6.02	8.29	12.4
		20	3.90	3.77	3.89	4.25	4.90	5.97	7.73	10.7	16.4
	20	0	2.26	2.29	2.44	2.71	3.17	3.89	5.04	6.95	10.4
		10	3.22	3.19	3.34	3.70	4.30	5.29	6.95	9.65	14.0
		20	4.26	4.15	4.32	4.77	5.55	6.83	8.94	12.5	18.9
25	0	0	2.14	2.05	2.06	2.18	2.46	2.98	3.81	5.23	7.80
		10	2.54	2.46	2.53	2.76	3.18	3.88	5.02	6.99	10.6
		20	3.15	3.04	3.14	3.44	4.00	4.91	6.43	9.06	13.9
	5	0	2.21	2.15	2.22	2.42	2.82	3.47	4.53	6.33	9.64
		10	2.81	2.75	2.88	3.19	3.74	4.63	6.11	8.66	13.4
		20	3.58	3.50	3.66	4.07	4.80	5.99	7.98	11.4	17.8
	10	0	2.30	2.29	2.42	2.72	3.22	4.02	5.34	7.60	11.8
		10	3.08	3.07	3.27	3.76	4.36	5.49	7.35	10.6	16.6
		20	4.04	4.00	4.24	4.78	5.70	7.23	9.75	14.1	22.4
	15	0	2.41	2.46	2.67	3.05	3.66	4.64	6.25	9.02	14.2
		10	3.39	3.43	3.69	4.20	5.05	6.44	8.74	12.7	20.2
		20	4.54	4.55	4.87	5.55	6.70	8.59	11.7	17.2	27.4
	20	0	2.56	2.67	2.94	3.40	4.13	5.31	7.23	10.6	16.8
		10	3.72	3.80	4.13	4.76	5.80	7.47	10.4	15.3	24.5
		20	5.07	5.12	5.55	6.38	7.79	10.1	14.5	21.4	34.5
	25	0	2.74	2.89	3.21	3.76	4.62	6.00	8.26	12.2	19.5
		10	4.05	4.18	4.59	5.34	6.57	8.54	12.0	17.8	29.7
		20	5.60	5.71	6.23	7.24	8.90	11.6	16.8	25.0	40.4

TABLE 8.9 (continued)

Angle of internal friction φ (°)	Wall friction angle δ (°)	Backfill angle β (°)	Wall angle α (°)								
			50	60	70	80	90	100	110	120	130
30	0	0	2.37	2.31	2.37	2.57	3.00	3.78	5.08	7.37	11.7
		10	2.82	2.79	2.95	3.34	4.01	5.12	7.00	10.3	16.8
		20	3.57	3.54	3.79	4.32	5.25	6.79	9.43	14.2	23.3
		30	4.41	4.42	4.76	5.68	6.74	8.82	12.4	18.8	31.3
	5	0	2.46	2.44	2.57	2.88	3.49	4.49	6.16	9.13	14.8
		10	3.13	3.15	3.40	3.92	4.79	6.24	8.70	13.1	21.6
		20	4.07	4.12	4.48	5.19	6.42	8.46	11.9	18.2	30.4
		30	5.19	5.26	5.76	6.79	8.39	11.2	16.0	24.5	41.3
	10	0	2.57	2.61	2.82	3.29	4.06	5.32	7.44	11.2	18.5
		10	3.47	3.55	3.91	4.58	5.70	7.56	10.7	16.4	27.4
		20	4.66	4.78	5.27	6.21	7.79	10.4	14.9	23.0	38.9
		30	6.07	6.23	6.90	8.02	10.3	14.0	20.1	31.3	53.2
	15	0	2.72	2.83	3.16	3.75	4.71	6.27	8.92	13.7	22.9
		10	3.85	4.02	4.50	5.34	6.75	9.08	13.0	20.2	34.1
		20	5.31	5.52	6.17	7.37	9.37	12.7	18.4	28.7	48.7
		30	7.05	7.32	8.21	10.3	12.6	17.2	25.0	39.2	66.0
	20	0	2.91	3.11	3.55	4.27	5.44	7.36	10.6	16.4	27.8
		10	4.29	4.54	5.15	6.20	7.94	10.8	16.1	25.2	42.9
		20	6.03	6.35	7.18	8.68	11.2	15.3	23.0	37.0	63.0
		30	8.14	8.54	9.68	12.3	15.1	20.8	30.5	49.0	82.0
	25	0	3.15	3.44	3.97	4.85	6.25	8.55	12.5	19.5	33.2
		10	4.77	5.11	5.86	7.14	9.24	12.7	19.1	30.1	51.4
		20	6.81	7.25	8.29	10.1	13.1	18.1	27.5	44.0	78.5
		30	9.32	9.87	11.3	14.4	17.9	25.0	37.6	60.0	100.
	30	0	3.42	3.77	4.41	5.45	7.10	9.80	14.4	22.7	38.8
		10	5.26	5.70	6.60	8.13	10.6	15.1	22.2	35.1	60.3
		20	7.62	8.18	9.44	11.6	15.2	21.4	32.8	54.0	94.0
		30	10.5	11.2	13.0	16.7	20.8	29.0	44.0	72.4	122.

TABLE 8.9 (continued)

Angle of internal friction φ(°)	Wall friction angle δ(°)	Backfill angle β(°)	Wall angle α(°)								
			50	60	70	80	90	100	110	120	130
35	0	0	2.67	2.64	2.76	3.07	3.69	4.87	6.92	10.7	18.3
		10	3.14	3.19	3.47	4.07	5.20	6.90	10.0	15.9	27.7
		20	4.06	4.14	4.60	5.56	7.03	9.66	14.3	23.0	40.9
		30	5.17	5.37	6.10	7.40	9.50	13.3	20.0	32.7	56.0
	5	0	2.78	2.81	3.01	3.47	4.37	5.91	8.60	13.6	23.7
		10	3.50	3.63	4.05	4.86	6.25	8.61	12.8	20.6	36.5
		20	4.66	4.88	5.53	6.60	8.79	12.3	18.6	30.4	54.5
		30	6.14	6.49	7.50	9.20	12.2	17.2	26.4	43.6	78.4
	10	0	2.92	3.02	3.34	4.01	5.19	7.17	10.7	17.2	30.4
		10	3.92	4.14	4.74	5.81	7.61	10.7	16.1	26.4	47.4
		20	5.39	5.75	6.64	8.20	10.9	15.6	23.8	39.4	71.3
		30	7.29	7.82	9.00	11.2	14.7	21.2	34.1	56.9	110.
	15	0	3.10	3.29	3.77	4.67	6.16	8.68	13.1	21.5	38.6
		10	4.40	4.76	5.55	6.93	9.24	13.2	20.2	33.5	60.6
		20	6.25	6.77	7.94	10.0	13.5	19.5	30.1	50.3	91.6
		30	8.63	9.41	11.0	13.8	18.5	27.0	43.5	73.0	138.
	20	0	3.33	3.64	4.32	5.44	7.31	10.5	16.1	26.6	48.2
		10	4.97	5.48	6.49	8.24	11.2	16.1	26.0	43.5	79.2
		20	7.23	7.96	9.48	12.0	16.5	24.1	38.7	66.0	118.
		30	10.2	11.2	13.4	17.0	23.2	34.5	55.0	96.0	175.
	25	0	3.63	4.10	4.94	6.33	8.63	12.5	19.4	32.5	59.2
		10	5.63	6.30	7.58	9.75	13.4	19.5	31.7	53.4	97.7
		20	8.35	9.32	11.2	14.2	20.0	29.4	46.8	81.0	148.
		30	11.9	13.3	16.2	20.8	29.0	43.0	70.0	122.	225.
	30	0	4.01	4.61	5.64	7.33	10.1	14.8	23.2	39.0	71.4
		10	6.36	7.21	8.79	11.4	15.8	24.2	38.0	64.3	118.
		20	9.59	10.8	13.2	16.8	23.2	35.0	57.5	98.0	188.
		30	13.9	15.7	19.0	24.5	34.8	52.5	86.0	150.	285.

TABLE 8.9 (continued)

Angle of internal friction φ (°)	Wall friction angle δ (°)	Backfill angle β (°)	Wall angle α (°)								
			50	60	70	80	90	100	110	120	130
35	35	0	4.42	5.15	6.38	8.39	11.7	17.2	27.1	45.8	84.1
		10	7.14	8.18	10.1	13.2	19.2	28.4	44.8	75.8	139.
		20	10.9	12.4	15.2	19.7	28.3	43.0	69.0	122.	225.
		30	16.0	18.1	22.0	28.5	40.0	22.6	102.	180.	350.
40	0	0	2.98	3.01	3.22	3.67	4.60	6.41	9.70	16.1	29.8
		10	3.51	3.66	4.13	5.04	6.68	9.58	14.9	25.5	48.3
		20	4.65	4.88	5.66	7.20	9.68	14.3	22.8	39.8	70.0
		30	6.11	6.59	7.70	10.0	14.0	21.0	34.2	60.5	110.
		40	7.97	8.30	9.80	12.8	19.2	30.3	52.0	91.0	162.
	5	0	3.12	3.22	3.54	4.21	5.56	7.97	12.4	21.1	39.9
		10	3.94	4.20	4.87	6.14	8.35	12.3	19.6	34.2	65.6
		20	5.38	5.84	6.94	9.0	12.4	18.7	30.5	53.9	102.
		30	7.35	8.12	9.60	12.5	18.2	28.1	46.4	82.9	150.
		40	9.89	10.5	12.6	16.5	25.0	41.1	68.0	132.	230.
	10	0	3.30	3.49	3.96	4.96	6.76	9.94	15.8	27.6	52.8
		10	4.46	4.87	5.81	7.50	10.4	15.7	25.6	45.2	87.8
		20	6.29	7.01	8.53	11.2	15.9	24.4	40.3	72.0	135.
		30	8.87	10.0	12.0	16.2	23.6	37.0	61.8	111.	210.
		40	12.3	13.5	15.3	21.8	33.0	52.5	90.0	165.	315.
	15	0	3.53	3.84	4.55	5.91	8.25	12.4	20.2	35.6	69.0
		10	5.06	5.68	6.95	9.19	13.1	20.0	33.0	59.0	115.
		20	7.42	8.44	10.5	14.0	20.3	31.5	52.6	94.7	185.
		30	10.7	12.3	15.7	20.8	31.0	48.5	81.1	147.	280.
		40	15.2	17.0	21.0	29.0	47.0	70.0	120.	225.	430.

TABLE 8.9 (continued)

Angle of internal friction φ (°)	Wall friction angle δ (°)	Backfill angle β (°)	Wall angle α (°)								
			50	60	70	80	90	100	110	120	130
40	20	0	3.82	4.30	5.31	7.06	10.1	15.4	25.5	45.5	88.9
		10	5.80	6.68	8.35	11.2	16.3	25.3	43.0	80.0	155.
		20	8.77	10.2	12.8	17.3	25.6	40.2	70.0	127.	250.
		30	12.9	15.2	19.5	26.4	41.0	63.0	106.	190.	350.
		40	18.7	21.4	27.0	37.9	60.0	94.5	164.	295.	550.
	25	0	4.21	4.92	6.23	8.45	12.3	19.1	31.8	57.3	113.
		10	6.70	7.87	10.0	13.7	20.1	31.6	56.0	102.	201.
		20	10.4	12.2	15.7	21.5	32.0	50.6	88.5	165.	300.
		30	15.6	18.5	23.8	33.0	49.0	78.0	132.	248.	450.
		40	22.8	27.0	35.0	49.0	74.0	120.	210.	375.	700.
	30	0	4.71	5.67	7.31	10.1	14.8	23.3	39.3	71.1	140.
		10	7.75	9.26	12.0	16.6	24.6	40.0	69.7	127.	251.
		20	12.2	14.6	19.0	26.5	39.5	64.0	114.	220.	400.
		30	18.7	22.5	29.0	44.0	62.0	100.	170.	315.	600.
		40	27.7	36.5	43.0	60.0	93.0	150.	260.	475.	920.
	35	0	5.33	6.52	8.54	11.9	17.8	28.2	47.7	86.6	171.
		10	8.95	10.8	14.2	19.9	30.0	50.0	88.0	160.	320.
		20	14.4	17.4	22.8	32.5	50.0	82.0	150.	290.	600.
		30	22.2	26.9	34.5	48.5	75.0	120.	210.	388.	760.
		40	32.0	38.5	51.0	72.0	108.	177.	310.	565.	1120.
	40	0	6.01	7.45	9.88	13.9	20.9	33.3	56.6	103.	204.
		10	10.2	12.6	16.6	23.4	36.0	59.4	101.	190.	365.
		20	16.6	20.3	26.8	38.5	59.5	100.	184.	360.	780.
		30	26.0	31.7	40.5	56.8	91.0	150.	265.	485.	950.
		40	36.5	44.0	59.5	82.0	125.	215.	375.	700.	1330.

TABLE 8.10

Active earth pressure coefficients $K_{a\gamma}$

Angle of internal friction φ (°)	Wall friction angle δ (°)	Backfill angle β (°)	Wall angle α (°)								
			50	60	70	80	90	100	110	120	130
10	0	0	1.11	0.943	0.832	0.756	0.704	0.669	0.650	0.641	0.641
		10	1.41	1.20	1.06	0.982	0.937	0.922	0.900	0.895	0.890
	5	0	1.09	0.917	0.801	0.720	0.664	0.626	0.601	0.586	0.577
		10	1.45	1.23	1.08	1.00	0.951	0.936	0.920	0.900	0.890
	10	0	1.07	0.911	0.787	0.702	0.642	0.600	0.570	0.549	0.533
		10	1.53	1.29	1.13	1.05	0.991	0.966	0.950	0.940	0.935
15	0	0	1.02	0.850	0.735	0.651	0.589	0.541	0.504	0.472	0.438
		10	1.27	1.04	0.893	0.782	0.701	0.643	0.595	0.555	0.516
	5	0	1.00	0.828	0.709	0.622	0.557	0.507	0.467	0.433	0.395
		10	1.28	1.04	0.885	0.764	0.679	0.612	0.560	0.516	0.442
	10	0	1.00	0.821	0.695	0.603	0.536	0.484	0.442	0.405	0.365
		10	1.32	1.07	0.889	0.758	0.663	0.591	0.536	0.489	0.473
	15	0	1.02	0.826	0.691	0.596	0.525	0.470	0.425	0.385	0.342
		10	1.38	1.11	0.903	0.760	0.657	0.581	0.522	0.471	0.420
20	0	0	0.937	0.767	0.647	0.559	0.490	0.434	0.387	0.341	0.290
		10	1.15	0.920	0.765	0.653	0.568	0.500	0.441	0.387	0.329
		20	1.44	1.17	1.01	0.901	0.822	0.781	0.759	0.749	0.732
	5	0	0.921	0.748	0.626	0.536	0.465	0.409	0.361	0.314	0.263
		10	1.14	0.915	0.754	0.634	0.546	0.474	0.414	0.360	0.301
		20	1.47	1.19	1.03	0.907	0.840	0.786	0.763	0.741	0.736
	10	0	0.924	0.742	0.614	0.520	0.448	0.391	0.342	0.295	0.243
		10	1.17	0.926	0.751	0.626	0.531	0.457	0.396	0.340	0.280
		20	1.51	1.23	1.06	0.937	0.855	0.812	0.776	0.767	0.748
	15	0	0.942	0.745	0.610	0.512	0.438	0.379	0.328	0.280	0.229
		10	1.21	0.949	0.756	0.622	0.523	0.446	0.383	0.325	0.265
		20	1.59	1.29	1.11	0.982	0.895	0.837	0.813	0.789	0.769

TABLE 8.10 (continued)

Angle of internal friction φ (°)	Wall friction angle δ (°)	Backfill angle β (°)	Wall angle α (°)								
			50	60	70	80	90	100	110	120	130
20	20	0	0.970	0.759	0.614	0.511	0.434	0.372	0.319	0.270	0.217
		10	1.29	0.984	0.771	0.626	0.521	0.441	0.375	0.315	0.253
		20	1.72	1.39	1.18	1.04	0.951	0.888	0.848	0.821	0.800
25	0	0	0.859	0.688	0.568	0.478	0.406	0.346	0.293	0.241	0.184
		10	1.03	0.814	0.661	0.549	0.462	0.389	0.327	0.267	0.203
		20	1.25	1.00	0.818	0.681	0.569	0.480	0.401	0.326	0.249
	5	0	0.848	0.674	0.552	0.459	0.387	0.327	0.275	0.223	0.168
		10	1.03	0.810	0.648	0.532	0.443	0.370	0.308	0.249	0.186
		20	1.27	1.00	0.824	0.673	0.557	0.462	0.381	0.307	0.230
	10	0	0.851	0.671	0.542	0.448	0.374	0.313	0.261	0.210	0.156
		10	1.05	0.814	0.645	0.523	0.431	0.356	0.294	0.235	0.173
		20	1.31	1.03	0.830	0.673	0.548	0.449	0.367	0.292	0.216
	15	0	0.866	0.672	0.540	0.441	0.365	0.304	0.251	0.200	0.146
		10	1.09	0.828	0.647	0.520	0.423	0.347	0.284	0.225	0.164
		20	1.37	0.107	0.853	0.678	0.545	0.441	0.357	0.282	0.206
	20	0	0.896	0.685	0.542	0.439	0.361	0.298	0.244	0.193	0.139
		10	1.14	0.856	0.658	0.521	0.420	0.342	0.277	0.217	0.156
		20	1.45	1.12	0.886	0.688	0.545	0.438	0.351	0.274	0.198
	25	0	0.925	0.725	0.552	0.443	0.361	0.296	0.240	0.187	0.134
		10	1.22	0.920	0.676	0.528	0.423	0.341	0.273	0.212	0.151
		20	1.56	1.20	0.929	0.708	0.554	0.439	0.349	0.270	0.192
30	0	0	0.787	0.617	0.497	0.406	0.333	0.272	0.218	0.165	0.108
		10	0.929	0.717	0.569	0.460	0.373	0.301	0.239	0.180	0.116
		20	1.12	0.861	0.683	0.546	0.438	0.353	0.276	0.207	0.135
		30	1.38	0.107	0.899	0.765	0.684	0.610	0.561	0.500	0.434

TABLE 8.10 (continued)

Angle of internal friction φ (°)	Wall friction angle δ (°)	Backfill angle β (°)	Wall angle α (°)								
			50	60	70	80	90	100	110	120	130
30	5	0	0.778	0.606	0.484	0.392	0.319	0.258	0.205	0.154	0.099
		10	0.932	0.715	0.559	0.446	0.359	0.287	0.226	0.168	0.108
		20	1.12	0.861	0.678	0.536	0.426	0.338	0.263	0.194	0.125
		30	1.39	1.09	0.912	0.776	0.694	0.619	0.570	0.507	0.428
	10	0	0.781	0.604	0.477	0.383	0.309	0.248	0.196	0.145	0.093
		10	0.946	0.720	0.557	0.439	0.349	0.277	0.216	0.159	0.100
		20	1.16	0.881	0.681	0.532	0.419	0.328	0.252	0.184	0.117
		30	1.43	1.14	0.934	0.795	0.712	0.634	0.570	0.506	0.426
	15	0	0.798	0.607	0.475	0.378	0.302	0.242	0.189	0.138	0.087
		10	0.972	0.728	0.558	0.437	0.343	0.270	0.209	0.152	0.095
		20	1.19	0.900	0.695	0.532	0.414	0.321	0.245	0.177	0.111
		30	1.51	1.18	0.968	0.823	0.738	0.657	0.590	0.524	0.442
	20	0	0.821	0.618	0.479	0.377	0.299	0.237	0.184	0.134	0.083
		10	1.01	0.750	0.566	0.437	0.341	0.266	0.204	0.147	0.091
		20	1.25	0.943	0.712	0.539	0.414	0.318	0.240	0.172	0.106
		30	1.59	1.24	1.01	0.885	0.773	0.688	0.618	0.549	0.448
	25	0	0.862	0.638	0.487	0.380	0.299	0.235	0.180	0.130	0.080
		10	1.08	0.785	0.581	0.442	0.342	0.265	0.201	0.143	0.087
		20	1.35	1.00	0.739	0.550	0.418	0.318	0.238	0.169	0.103
		30	1.74	1.35	1.11	0.940	0.820	0.729	0.654	0.564	0.473
	30	0	0.900	0.770	0.501	0.387	0.302	0.236	0.179	1.27	0.078
		10	1.17	0.829	0.602	0.453	0.347	0.266	0.200	0.141	0.085
		20	1.47	1.08	0.776	0.568	0.425	0.321	0.238	0.167	0.100
		30	1.88	1.46	1.19	1.01	0.882	0.783	0.701	0.604	0.489
35	0	0	0.717	0.551	0.433	0.343	0.271	0.211	0.158	0.107	0.057
		10	0.837	0.634	0.491	0.383	0.299	0.230	0.171	0.115	0.060
		20	0.986	0.741	0.572	0.443	0.342	0.261	0.191	0.128	0.066
		30	1.18	0.895	0.703	0.558	0.434	0.331	0.240	0.160	0.084

TABLE 8.10 (continued)

Angle of internal friction φ (°)	Wall friction angle δ (°)	Backfill angle β (°)	Wall angle α (°)								
			50	60	70	80	90	100	110	120	130
5	5	0	0.711	0.542	0.424	0.333	0.260	0.201	0.149	0.101	0.052
		10	0.843	0.629	0.483	0.372	0.289	0.220	0.162	0.108	0.056
		20	1.01	0.741	0.568	0.435	0.333	0.250	0.182	0.120	0.060
		30	1.20	0.904	0.708	0.557	0.426	0.320	0.230	0.151	0.078
	10	0	0.717	0.543	0.418	0.326	0.253	0.194	0.143	0.095	0.049
		10	0.849	0.635	0.480	0.368	0.282	0.213	0.155	0.103	0.052
		20	1.02	0.759	0.569	0.430	0.326	0.243	0.175	0.115	0.057
		30	1.22	0.923	0.720	0.560	0.422	0.312	0.222	0.145	0.074
	15	0	0.731	0.546	0.417	0.322	0.248	0.189	0.138	0.091	0.046
		10	0.876	0.643	0.481	0.365	0.277	0.208	0.150	0.098	0.049
		20	1.05	0.775	0.575	0.430	0.322	0.238	0.170	0.110	0.054
		30	1.27	0.975	0.753	0.567	0.421	0.308	0.216	0.140	0.070
	20	0	0.755	0.557	0.420	0.322	0.246	0.186	0.135	0.088	0.044
		10	0.915	0.664	0.488	0.367	0.275	0.205	0.147	0.095	0.047
		20	1.11	0.800	0.592	0.434	0.322	0.235	0.166	0.107	0.052
		30	1.36	1.02	0.781	0.580	0.424	0.306	0.214	0.137	0.068
	25	0	0.791	0.575	0.430	0.325	0.246	0.185	0.133	0.086	0.043
		10	0.968	0.692	0.501	0.371	0.276	0.204	0.145	0.093	0.046
		20	1.17	0.847	0.610	0.443	0.323	0.235	0.165	0.105	0.050
		30	1.44	1.08	0.819	0.598	0.429	0.307	0.213	0.134	0.066
	30	0	0.846	0.601	0.442	0.331	0.249	0.185	0.132	0.085	0.042
		10	1.04	0.730	0.519	0.379	0.280	0.205	0.144	0.092	0.044
		20	1.27	0.908	0.637	0.455	0.329	0.237	0.164	0.103	0.049
		30	1.60	1.15	0.870	0.623	0.438	0.310	0.213	0.133	0.064
	35	0	0.928	0.634	0.460	0.341	0.254	0.187	0.132	0.084	0.041
		10	1.12	0.783	0.545	0.392	0.287	0.208	0.145	0.091	0.044
		20	1.41	0.989	0.676	0.473	0.337	0.241	0.166	0.103	0.048
		30	1.75	1.29	0.951	0.656	0.457	0.318	0.216	0.134	0.064

TABLE 8.10 (continued)

Angle of internal friction φ (°)	Wall friction angle δ (°)	Backfill angle β (°)	Wall angle α (°)								
			50	60	70	80	90	100	110	120	130
40	0	0	0.649	0.491	0.374	0.287	0.217	0.160	0.111	0.065	0.024
		10	0.760	0.556	0.421	0.316	0.237	0.172	0.118	0.069	0.025
		20	0.874	0.643	0.482	0.358	0.266	0.191	0.129	0.075	0.026
		30	1.38	0.762	0.577	0.429	0.316	0.226	0.150	0.086	0.029
		40	1.22	0.920	0.751	0.614	0.511	0.443	0.364	0.277	0.160
	5	0	0.645	0.486	0.368	0.279	0.210	0.153	0.105	0.061	0.022
		10	0.759	0.553	0.415	0.310	0.230	0.166	0.112	0.064	0.023
		20	0.879	0.644	0.479	0.352	0.259	0.183	0.123	0.070	0.024
		30	1.03	0.770	0.579	0.426	0.309	0.218	0.143	0.081	0.027
		40	1.25	0.941	0.769	0.628	0.524	0.439	0.361	0.274	0.157
	10	0	0.654	0.485	0.364	0.275	0.205	0.149	0.101	0.058	0.021
		10	0.767	0.562	0.413	0.305	0.225	0.160	0.108	0.061	0.022
		20	0.894	0.651	0.480	0.349	0.254	0.178	0.118	0.067	0.023
		30	1.05	0.786	0.586	0.425	0.305	0.212	0.139	0.078	0.026
		40	1.29	0.973	0.769	0.650	0.542	0.455	0.374	0.273	0.155
	15	0	0.664	0.490	0.365	0.272	0.201	0.145	0.098	0.056	0.020
		10	0.783	0.571	0.415	0.304	0.221	0.157	0.105	0.059	0.021
		20	0.937	0.666	0.486	0.349	0.251	0.175	0.115	0.065	0.022
		30	1.07	0.811	0.590	0.426	0.304	0.209	0.135	0.075	0.024
		40	1.35	1.02	0.804	0.657	0.567	0.459	0.377	0.286	0.154
	20	0	0.690	0.503	0.367	0.273	0.200	0.143	0.096	0.054	0.019
		10	0.822	0.585	0.421	0.306	0.220	0.155	0.103	0.057	0.020
		20	0.975	0.688	0.496	0.352	0.250	0.173	0.113	0.063	0.021
		30	1.11	0.846	0.610	0.434	0.305	0.208	0.133	0.073	0.023
		40	1.43	1.08	0.849	0.693	0.578	0.485	0.399	0.290	0.155

TABLE 8.10 (continued)

Angle of internal friction φ (°)	Wall friction angle δ (°)	Backfill angle β (°)	Wall angle α (°)								
			50	60	70	80	90	100	110	120	130
40	25	0	0.717	0.518	0.376	0.276	0.200	0.143	0.095	0.053	0.018
		10	0.860	0.607	0.431	0.310	0.221	0.155	0.101	0.056	0.019
		20	1.03	0.729	0.511	0.359	0.252	0.173	0.112	0.061	0.020
		30	1.17	0.892	0.637	0.445	0.309	0.208	0.132	0.072	0.023
		40	1.53	1.15	0.908	0.741	0.617	0.518	0.409	0.297	0.156
	30	0	0.765	0.543	0.388	0.281	0.203	0.143	0.094	0.052	0.018
		10	0.927	0.643	0.447	0.317	0.224	0.156	0.101	0.055	0.019
		20	1.10	0.772	0.532	0.369	0.256	0.174	0.112	0.061	0.020
		30	1.24	0.953	0.681	0.463	0.314	0.210	0.133	0.071	0.022
		40	1.67	1.25	0.984	0.802	0.668	0.559	0.442	0.321	0.160
	35	0	0.840	0.576	0.405	0.289	0.207	0.145	0.094	0.052	0.017
		10	1.02	0.690	0.470	0.327	0.230	0.158	0.102	0.055	0.018
		20	1.22	0.840	0.562	0.383	0.263	0.177	0.113	0.060	0.020
		30	1.33	1.03	0.727	0.480	0.324	0.214	0.134	0.071	0.023
		40	1.84	1.38	1.08	0.881	0.733	0.589	0.463	0.335	0.164
	40	0	0.946	0.598	0.428	0.302	0.214	0.148	0.096	0.052	0.017
		10	1.09	0.738	0.500	0.342	0.238	0.162	0.103	0.055	0.018
		20	1.38	0.931	0.604	0.402	0.273	0.182	0.114	0.061	0.019
		30	1.50	1.17	0.792	0.512	0.338	0.221	0.136	0.072	0.023
		40	2.07	1.55	1.21	0.985	0.817	0.654	0.513	0.353	0.171

values of the earth pressure presented in Tables 8.9 and 8.10 are considered to be sufficiently accurate for practical purposes in the analysis and design of retaining walls.

8.10. SUMMARY AND CONCLUSIONS

It has been shown that the upper-bound technique of limit analysis can yield rationally founded solutions that are in good agreement with the Sokolovskii slipline results. In addition, these solutions are easily obtainable in a closed form. The formulation needed can be readily derived and has great physical appeal. An elaborate tabulation of passive pressure results for a cohesionless soil retained by a rigid wall of varying roughness δ and inclination α from the horizontal is given in Table 8.7. Table 8.8 shows the corresponding active pressure results.

The investigation of several assumed failure mechanisms has shown that significant solution improvement can especially be realized for the case of rough walls. These improvements have basically resulted from the use of a new sandwich mechanism which incorporates a logarithmic spiral shearing zone. The use of this mechanism is particularly convenient and desirable due to the fact that only two independent parameters are needed in its description. An investigation was also made to study the effects of shifting the pole of the logspiral region away from its position at the top of the wall. From the results obtained it was concluded that very little solution improvement, if any, could be expected. Such improvement is out-weighed by the complexity of the elaborate formulation and the associated optimization procedure needed.

It has also been shown that the simplicity of the upper-bound technique makes it possible to easily include the effects of cohesion and surcharge. Such problems have previously been solved using the graphical forms of limit equilibrium. The problems of retaining walls with broken backs as well as backfills with irregular slopes can also be solved when appropriate kinematically admissible failure mechanisms are constructed. With the inclusion of non-homogeneous, anisotropic, layered soils for the limit analysis of footing problems which have already been presented in the preceding chapter; and for the limit analysis of slope stability problems which are the topics of discussion in the following Chapter, the extension to earth pressure situations is also possible.

Chapter 9

STABILITY OF SLOPES

9.1. INTRODUCTION

This Chapter concerns the application of limit analysis techniques to problems involving stability of earth slopes. Since a great variety of conditions may lead to a failure of a soil slope due to its own weight, no more than a discussion of the simple slopes of stability computations will be attempted. The discussion here will be limited to rigid body *sliding* in cohesive material whose shearing resistance is determined by Coulomb's yield or failure condition. The term rigid body sliding involves a downward and outward movement of a slice of earth as shown in Fig. 9.1, and the sliding occurs along the well-defined, entire surface of contact between the slice and its base.

In order to familiarize the reader with the methods of computation, the individual problems connected with the stability of slopes will be taken up in the following sequence. First we consider the critical height of an inclined slope using a rotational failure mechanism (logarithmic spiral) passing through the toe, because it is similar to that for the stability of a vertical slope described in section 3.2, Chapter 3 (example 3.2). Then we solve in succession the following two problems: (1) the case where the logarithmic spiral failure surface may pass below the toe; and (2) the case where the soil mass may be anisotropic and non-homogeneous. Since the stress fields are not considered in these problems, solutions of the logarithmic spiral mechanism can at best give only upper bounds to the problems. However, the fact that the numerical results obtained in what follows agree with the existing results, for which satisfactory solutions already exist, would suggest that the same is also true for the more general case of the embankment stability problems described in this Chapter.

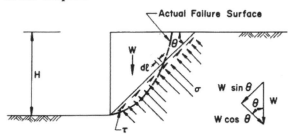

Fig. 9.1. Limit equilibrium solution for the stability of a vertical cut.

On account of the important influence of the shape of slip surface on the slope stability analysis, the method of *variational calculus* is applied in the latter part of this chapter to obtain the shape of slip surface and the corresponding normal stress distribution. It is shown that for a horizontal slope of uniform soil, the logarithmic spiral surface of angle φ is the most critical surface among all the possible slip surfaces. This result emphasizes the correctness of the logspiral failure mechanism assumed in the upper-bound method of limit analysis described in the earlier part of this chapter.

9.1.1. Limit equilibrium approach

Before we move to find the critical height of an embankment, first, let us examine the conventional *limit equilibrium approach*, developed by Fellenius (1936) and others and often referred to as the *Swedish circle method* when a circular arc is chosen as the failure surface. The comparison between limit equilibrium and plastic limit analysis can be illustrated by evaluating the stability of soil in a vertical bank. The height at which an unsupported vertical cut, as illustrated in Fig. 9.1, will collapse due to the weight of the soil will be defined as the *critical height, H_c*. The conventional analysis (limit equilibrium) will be examined first and then compared to the method of limit analysis.

It is common practice to evaluate this problem by the equilibrium method. For simplicity, the failure surface is assumed to be a plane inclined at an angle θ to the horizontal (see Fig. 9.1) and Coulomb condition of failure $\tau = c + \sigma \tan \varphi$ is applied. The distribution of σ and τ along the failure plane is unknown, but if l is the length of the shear plane:

$$\int \tau \, dl = \int c \, dl + \int \sigma \tan \varphi \, dl = cl + \tan \varphi \int \sigma \, dl \qquad [9.1]$$

Equilibrium requires that:

$$\int \sigma \, dl = W \cos \theta \qquad [9.2a]$$

$$\int \tau \, dl = W \sin \theta \qquad [9.2b]$$

Substituting [9.2] into [9.1] yields:

$$W \sin \theta = c \, \frac{H}{\sin \theta} + W \cos \theta \tan \varphi \qquad [9.3]$$

where $\quad W = \dfrac{\gamma H^2}{2 \tan \theta} \qquad [9.4]$

W defines the unit weight of the soil, γ, multiplied by the volume of the soil mass above the shear plane. If θ is minimized then:

$$\theta_c = (\tfrac{1}{4}\pi + \tfrac{1}{2}\varphi) \tag{9.5}$$

Equation [9.3] then yields the critical height:

$$H_c = \frac{4c}{\gamma} \tan (\tfrac{1}{4}\pi + \tfrac{1}{2}\varphi) \tag{9.6}$$

Assuming a curved surface instead of a plane for the sliding surface will reduce the critical height slightly. Fellenius (1927) used this condition and found the critical height to be:

$$H_c = \frac{3.85c}{\gamma} \tan (\tfrac{1}{4}\pi + \tfrac{1}{2}\varphi) \tag{9.7}$$

In this analysis, Coulomb condition of failure is only satisfied along the assumed failure plane. It is not shown, nor known, if Coulomb's condition of failure is violated at other points. A valid solution requires that equilibrium, compatibility (and boundary conditions) and the stress—strain relationship be satisfied. Here only equilibrium has been satisfied. From the limit theorems it is known that this solution is not a lower bound nor an upper bound since only equilibrium is satisfied, and not the yield criterion and its associated flow rule. The solution is merely one of many solutions that satisfy equilibrium. It is not known whether or not it is unique.

9.1.2. Limit analysis approach

The limit analysis of this problem, first performed by Drucker and Prager (1952), involves determining a lower bound on the collapse load by assuming a stress field which satisfies equilibrium and does not violate the yield criterion at any point. An upper bound is obtained by a velocity field compatible with the flow rule in which the rate of work of the external forces equals or exceeds the rate of internal energy dissipation.

Starting first with the upper bound, a mechanism (velocity field) has been selected as shown in Fig. 3.8 (example 3.2, p. 58). As the wedge, formed by the shear plane which makes an angle β with the vertical, slides downward along the discontinuity surface, there is a separation velocity, $V\sin\varphi$, from the discontinuity surface. The rate of work done by the external forces is the vertical component of the velocity multiplied by the weight of the soil wedge, while the rate of energy dissipated along the discontinuity surface is found by multiplying the area of the discontinuity surface $(H/\cos\beta)$ by c times the discontinuity in velocity across the surface $(\delta u = V\cos\varphi)$. Equating the rate of external work to the rate of internal energy dissipation and minimizing the height of the vertical cut with respect to the angle β gives [9.6]. This implies that the limit equilibrium method solution presented previously is an upper bound, unless the solution is exact.

A lower bound can be obtained by constructing a stress field composed of three regions as shown in Fig. 4.8(a) (section 4.3, p. 119). Region I, in the bank itself, is subjected to uniaxial compression, which increases with depth. Region II is under biaxial compression and region III is under hydrostatic pressure ($\sigma_x = \sigma_y$). Figure 4.8(b) shows the corresponding Mohr circles for each region. Failure occurs when the circle representing region I meets the yield curve. Therefore:

$$H_c \geqslant \frac{2c}{\gamma} \tan \left(\tfrac{1}{4}\pi + \tfrac{1}{2}\varphi\right) \tag{9.8}$$

hence, the lower-bound solution is only one half the value given by the upper-bound solution. If the average is used, $H_c = \frac{3c}{\gamma} \tan(\tfrac{1}{4}\pi + \tfrac{1}{2}\varphi)$, then the maximum error is 33%.

The upper bound may be improved by choosing for a discontinuity surface a logarithmic spiral rather than the plane used here (section 9.2). The lower bound may be improved by the choice of other stress fields, probably quite complex and which may involve tensile stresses in the soil.

9.1.3. Soils with tension cut-off

If the soil is unable to resist tension, the introduction of a tensile crack in a failure mechanism is permissible. No energy is dissipated in the formation of a simple tension crack; both normal and shear stresses are zero on the plane of separation (see the origin in Fig. 3.17).

The rotational mechanism containing a simple tension crack and a homogeneous shearing zone is shown in Fig. 3.18 (example 3.4, section 3.3, p. 71). Failure due to tipping over of the soil "slab" of thickness Δ about point A with an angular velocity ω is possible. The total rate of dissipation of internal energy for unit dimension perpendicular to the paper is given by [3.63], and the rate of external work, done by gravity, is given by [3.64]. If the rate of external work is equated to the dissipation, and Δ allowed to approach zero, it yields:

$$H_c \leqslant \frac{2c}{\gamma} \tan \left(\tfrac{1}{4}\pi + \tfrac{1}{2}\varphi\right) \tag{9.9}$$

This upper bound, which neglects the tensile stress, agrees with the lower bound for the previous case, where moderate tensile capacity was incorporated into the yield criterion. However, an examination of the stress field for this latter case reveals that no tensile stress exists. Hence, the solution is also valid for the case of a tension cut-off. Since the upper- and lower-bound solutions agree, the value of H_c is:

$$H_c = \frac{2c}{\gamma} \tan \left(\tfrac{1}{4}\pi + \tfrac{1}{2}\varphi\right) \tag{9.10}$$

This also confirms Terzaghi's solution (1943) for a tensile crack extending the full height of the bank.

9.2. LOGSPIRAL MECHANISM PASSING THROUGH THE TOE

The problem considered here is the critical height of an embankment with slope angles α and β as shown in Fig. 9.2. The limit analysis of this problem for a vertical bank ($\alpha = 0$ and $\beta = 90°$) has been discussed in the preceding section. It has been pointed out that the upper-bound solution can be improved by considering a rotational discontinuity (logarithmic spiral) instead of the translation discontinuity (plane surface) used earlier. In order to provide a more complete means of comparing the results obtained by the limit analysis approach with those obtained by the limit equilibrium approach, and also yield additional theoretical evidence as to the validity and limitations of the theory of perfect plasticity as applied to stability problems in soil mechanics, it is of value, therefore, to obtain the more elaborate upper-bound solution for the critical height of an embankment using a logarithmic spiral surface discontinuity mechanism. This is described in the present section.

9.2.1. Critical height of an embankment (Fig. 9.2)

The upper-bound theorem of limit analysis states that the embankment shown in

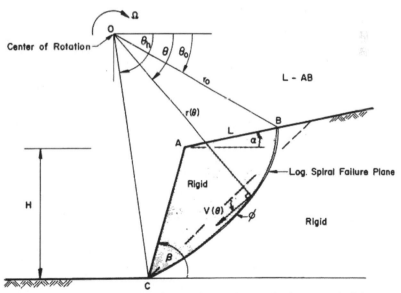

Fig. 9.2. Failure mechanism for the stability of an embankment with failure plane passing through toe.

Fig. 9.2 will collapse under its own weight if, for any assumed failure mechanism, the rate of work done by the soil weight exceeds the internal rate of dissipation. Equating external and internal energies for any such mechanism thus gives an upper bound on the critical height.

A rotational discontinuity mechanism is shown in **Fig. 9.2** in which the failure surface is assumed to pass through the toe of the slope. The possibility that the failure surface may pass below the toe for very small values of β and φ will be considered in the following section. The triangular-shaped region ABC rotates as a rigid body about the center of rotation O (as yet undefined) with the materials below the logarithmic surface BC remaining at rest. Thus, the surface BC is a surface of velocity discontinuity. The assumed mechanism can be specified completely by three variables. For the sake of convenience, we select the slope angles θ_o and θ_h of the chords OB and OC, respectively and the height H of the embankment. From the geometrical relations it may be shown that the ratios, H/r_o and L/r_o, can be expressed in terms of the angles θ_o and θ_h in the forms:

$$\frac{H}{r_o} = \frac{\sin \beta}{\sin (\beta - \alpha)} \{\sin (\theta_h + \alpha) \exp [(\theta_h - \theta_o) \tan \varphi] - \sin (\theta_o + \alpha)\} \qquad [9.11]$$

and:

$$\frac{L}{r_o} = \frac{\sin (\theta_h - \theta_o)}{\sin (\theta_h + \alpha)} - \frac{\sin (\theta_h + \beta)}{\sin (\theta_h + \alpha) \sin (\beta - \alpha)} \{\exp [(\theta_h - \theta_o) \tan \varphi]$$

$$\sin (\theta_h + \alpha) - \sin (\theta_o + \alpha)\} \qquad [9.12]$$

Rate of external work

A direct integration of the rate of external work due to the soil weight in the region ABC is very complicated. An easier alternative is first to find the rates of work $\dot{W}_1$, $\dot{W}_2$ and $\dot{W}_3$ due to the soil weight in the regions OBC, OAB, and OAC, respectively. The rate of external work for the region ABC is then found by the simple algebraic summation, $\dot{W}_1 - \dot{W}_2 - \dot{W}_3$. The steps of computation for this case of an inclined slope are essentially the same as those of a vertical slope described earlier in section 3.2, Chapter 3 (example 3.2, Figs. 3.9 and 3.10). It is found, after some simplification, that the rate at which work is done by the soil weight in the region ABC is:

$$\gamma r_o^3 \, \Omega (f_1 - f_2 - f_3) \qquad [9.13]$$

where γ is the unit weight of the soil, Ω is the angular velocity of the region ABC, the functions f_1, f_2, and f_3 are defined as:

$$f_1(\theta_h, \theta_o) = \frac{1}{3(1 + 9 \tan^2 \varphi)} \{(3 \tan \varphi \cos \theta_h + \sin \theta_h) \exp [3(\theta_h - \theta_o) \tan \varphi]$$

$$- (3 \tan \varphi \cos \theta_o + \sin \theta_o)\} \qquad\qquad [9.14]$$

$$f_2(\theta_h, \theta_o) = \frac{1}{6} \frac{L}{r_o} \left(2 \cos \theta_o - \frac{L}{r_o} \cos \alpha\right) \sin (\theta_o + \alpha) \qquad\qquad [9.15]$$

and L/r_o is a function of θ_h and θ_o ([9.12]).

$$f_3(\theta_h, \theta_o) = \frac{1}{6} \exp [(\theta_h - \theta_o) \tan \varphi] \left[\sin (\theta_h - \theta_o) - \frac{L}{r_o} \sin (\theta_h + \alpha)\right]$$

$$\left\{\cos \theta_o - \frac{L}{r_o} \cos \alpha + \cos \theta_h \exp [(\theta_h - \theta_o) \tan \varphi]\right\} \qquad\qquad [9.16]$$

Rate of internal dissipation

The internal dissipation of energy occurs along the discontinuity surface BC. The differential rate of dissipation of energy along the surface may be found by multiplying the differential area, $r d\theta / \cos \varphi$, of this surface by the cohesion c times the tangential discontinuity in velocity, $V \cos \varphi$, across the surface ([3.19]). The total internal dissipation of energy is found by integration over the whole surface ([3.38]):

$$\int_{\theta_0}^{\theta_h} c(V \cos \varphi) \frac{r d\theta}{\cos \varphi} = \frac{c r_o^2 \Omega}{2 \tan \varphi} \{\exp[2(\theta_h - \theta_o) \tan \varphi] - 1\} \qquad\qquad [9.17]$$

Critical height

Equating the external rate of work, [9.13], to the rate of internal energy dissipation, [9.17] gives:

$$H = \frac{c}{\gamma} f(\theta_h, \theta_o) \qquad\qquad [9.18]$$

where $f(\theta_h, \theta_o)$ is defined as:

$$f(\theta_h, \theta_o) = \frac{\sin \beta \{\exp[2(\theta_h - \theta_o) \tan \varphi] - 1\}}{2 \sin (\beta - \alpha) \tan \varphi (f_1 - f_2 - f_3)}$$

$$\{\sin (\theta_h + \alpha) \exp [(\theta_h - \theta_o) \tan \varphi] - \sin (\theta_o + \alpha)\} \qquad\qquad [9.19]$$

By the upper-bound theorem of limit analysis, [9.18] gives an upper bound for the critical value of the height, H_c. The function $f(\theta_h,\theta_o)$ has a minimum value when θ_h and θ_o satisfy the conditions:

$$\frac{\partial f}{\partial \theta_h} = 0 \quad \text{and} \quad \frac{\partial f}{\partial \theta_o} = 0 \qquad\qquad [9.20]$$

Solving these equations and substituting the values of θ_h and θ_o thus obtained into [9.18] yields a least upper bound for the critical height, H_c, of an inclined slope. Denoting $N_s = \min f(\theta_h,\theta_o)$, one obtains:

$$H_c \leqslant \frac{c}{\gamma} N_s \qquad\qquad [9.21]$$

The dimensionless number N_s is known as the *stability factor* of the embankment. The value of N_s is a *pure number* and depends not only on the slope angles α and β but also on the angle of internal friction, φ. The height H_c is the critical value of an inclined slope. It corresponds to the critical height of vertical slopes, determined by [3.42]. The stability factor N_s is an analogue to the earth pressure coefficients K_c, K_γ (section 8.3, Chapter 8).

9.2.2. Numerical results

A complete solution to this problem with failure plane passing through toe has been obtained by numerical methods, the numerical work being performed on a

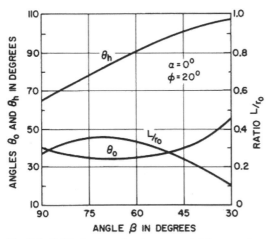

Fig. 9.3. Critical values for θ_h, θ_o and L/r_o as a function of slope angle β with failure plane passing through toe.

TABLE 9.1

Stability factor $N_s = H_c(\gamma/c)$ by limit analysis (logspiral passing through the toe, Fig. 9.2)

Friction angle ϕ (°)	Slope angle α (°)	Slope angle β (°)					
		90	75	60	45	30	15
0	0	3.83	4.57	5.25	5.86	6.51	7.35
5	0	4.19	5.14	6.17	7.33	9.17	14.80
	5	4.14	5.05	6.03	7.18	8.93	14.62
10	0	4.59	5.80	7.26	9.32	13.53	45.53
	5	4.53	5.72	7.14	9.14	13.26	45.15
	10	4.47	5.61	6.98	8.93	12.97	44.56
15	0	5.02	6.57	8.64	12.05	21.71	
	5	4.97	6.49	8.52	11.91	21.50	
	10	4.90	6.39	8.38	11.73	21.14	
	15	4.83	6.28	8.18	11.42	20.59	
20	0	5.51	7.48	10.39	16.18	41.27	
	5	5.46	7.40	10.30	16.04	41.06	
	10	5.40	7.31	10.15	15.87	40.73	
	15	5.33	7.20	9.98	15.59	40.16	
	20	5.24	7.04	9.78	15.17	39.19	
25	0	6.06	8.59	12.75	22.92	120.0	
	·5	6.01	8.52	12.65	22.78	119.8	
	10	5.96	8.41	12.54	22.60	119.5	
	15	5.89	8.30	12.40	22.37	118.7	
	20	5.81	8.16	12.17	21.98	117.4	
	25	5.71	7.97	11.80	21.35	115.5	
30	0	6.69	9.96	16.11	35.63		
	5	6.63	9.87	16.00	35.44		
	10	6.58	9.79	15.87	35.25		
	15	6.53	9.67	15.69	34.99		
	20	6.44	9.54	15.48	34.64		
	25	6.34	9.37	15.21	34.12		
	30	6.22	9.15	14.81	33.08		
35	0	7.43	11.68	20.94	65.53		
	5	7.38	11.60	20.84	65.39		
	10	7.32	11.51	20.71	65.22		
	15	7.26	11.41	20.55	65.03		
	20	2.18	11.28	20.36	64.74		
	25	7.11	11.12	20.07	64.18		
	30	6.99	10.93	19.73	63.00		
	35	6.84	10.66	19.21	60.80		
40	0	8.30	14.00	28.99	185.6		
	5	8.26	13.94	28.84	185.5		
	10	8.21	13.85	28.69	185.3		
	15	8.15	13.72	28.54	185.0		
	20	8.06	13.57	28.39	184.6		
	25	7.98	13.42	28.16	184.0		
	30	7.87	13.21	27.88	183.2		
	35	7.76	12.95	27.49	182.3		
	40	7.61	12.63	26.91	181.1		

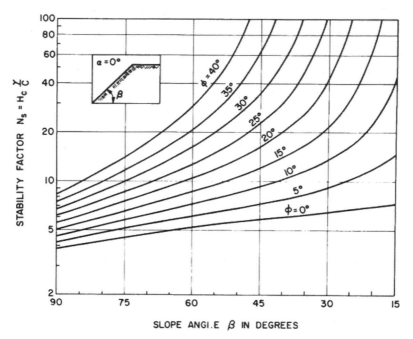

Fig. 9.4. Stability factors N_S, as a function of slope angle β with failure plane passing through toe.

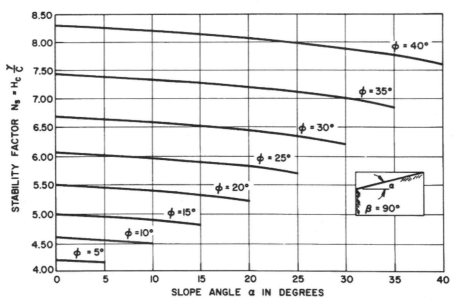

Fig. 9.5. Stability factors N_S, as a function of slope angle α with failure plane passing through toe.

CDC 6400 digital computer. The results of these computations are graphically represented in Figs. 9.3, 9.4 and 9.5 for some special cases with either $\alpha = 0$ or $\beta = 90°$, and tabulated numerically in Table 9.1 for the general case shown in Fig. 9.2.

For the case $\alpha = 0$, the critical values of θ_h, θ_o and L/r_o corresponding to [9.20] are plotted against the inclined slope angle β for $\varphi = 20°$ (Fig. 9.3). They determine the position of the center of the critical toe logspiral. The relation between the inclined slope angle β and the stability factor N_s for various values of φ is shown in Fig. 9.4. For the case $\varphi = 0$, they increase from 3.83 for $\beta = 90°$ to 7.35 for $\beta = 15°$.

It is found that the results obtained are practically identical to those results obtained by the *φ-circle method* of Taylor (1948) and by the *slice method* of Fellenius (1927). For the case $\beta = 90°$, the relation between α and N_s for various values of φ is shown in Fig. 9.5. For the case $\varphi = 30°$, the values of the stability factor N_s with respect to the slope angle α decrease from 6.69 for $\alpha = 0$ to 6.22 for $\alpha = 30°$. Table 9.1 gives the value of N_s for various combinations of the slopes α and β. Since there are no existing solutions available for the case where $\alpha \neq 0$, the present results will certainly provide useful information concerning the problem, although the answers may not be exact.

9.3. LOGSPIRAL MECHANISM PASSING BELOW THE TOE

In the preceding discussion it has been tacitly assumed that the logspiral surface of sliding passes *through* the toe. The case where the failure surface may pass *below* the toe, as for small values of friction angle φ and earth slope angle β (Fig. 9.6), was not considered. This will be described herein.

9.3.1. Critical height of an embankment (Fig. 9.6)

Rate of external work

Here, as in the preceding section, the rate of external work done by the region $ABC'CA$ can easily be obtained by first finding the rates of work $\dot{W}_1$, $\dot{W}_2$, $\dot{W}_3$, and $\dot{W}_4$ due to the soil weight in the regions $OBC'O$, $OABO$, $OAC'O$, and $ACC'A$, respectively. The rate of external work for the region $ABC'CA$ is then obtained by the simple algebraic summation, $\dot{W}_1 - \dot{W}_2 - \dot{W}_3 - \dot{W}_4$. It is found, after performing some algebraic manipulations, that the total rate of external work due to the weight of the soil in the region $ABC'CA$ is:

$$\gamma\Omega r_o^3(f_1 - f_2 - f_3 - f_4) \tag{9.22}$$

The functions $f_1(\theta_h, \theta_o)$, $f_2(\theta_h, \theta_o)$ and $f_3(\theta_h, \theta_o)$ remain identical in their form

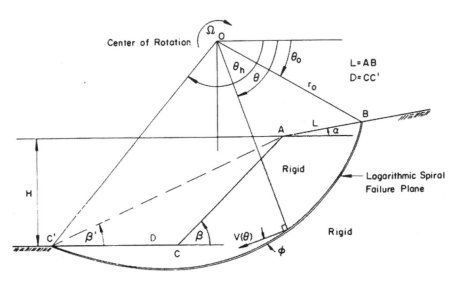

Fig. 9.6. Failure mechanism for the stability of an embankment with failure plane passing below toe.

as in the previous equations [9.14], [9.15] and [9.16], and the function $f_4(\theta_h, \theta_o)$ resulting from the region $ACC'A$ is expressed as:

$$f_4(\theta_h, \theta_o) = \left(\frac{H}{r_o}\right)^2 \frac{\sin(\beta - \beta')}{2\sin\beta\sin\beta'}\left[\cos\theta_o - \left(\frac{L}{r_o}\right)\cos\alpha - \frac{1}{3}\left(\frac{H}{r_o}\right)(\cot\beta' + \cot\beta)\right]$$

[9.23]

in which θ_o, θ_h, and β' = angular variables, specifying the assumed mechanism completely; height = H; length of $OB = r_o$; length of $AB = L$; and length of $C'C = D$ (Fig. 9.6).

Rate of internal dissipation

The internal dissipation of energy occurs along discontinuity surface BC'. The total internal dissipation of energy is identical to the expression given in [9.17].

Critical height

Equating external rate of work, [9.22], with internal rate of energy dissipation yields:

$$H = \frac{c}{\gamma} f(\theta_h, \theta_o, \beta')$$

[9.24]

in which $f(\theta_h, \theta_o, \beta')$ is now defined as:

$$f(\theta_h, \theta_o, \beta') = \frac{\sin \beta' \{\exp [2(\theta_h - \theta_o) \tan \varphi] - 1\}}{2 \sin (\beta' - \alpha) \tan \varphi (f_1 - f_2 - f_3 - f_4)}$$

$$\{\sin (\theta_h + \alpha) \exp [(\theta_h - \theta_o) \tan \varphi] - \sin (\theta_o + \alpha)\} \qquad [9.25]$$

Function $f(\theta_h, \theta_o, \beta')$ has a minimum and, thus, indicates a least upper bound when θ_h, θ_o, and β' satisfy the conditions:

$$\frac{\partial f}{\partial \theta_h} = 0; \quad \frac{\partial f}{\partial \theta_o} = 0; \quad \frac{\partial f}{\partial \beta'} = 0 \qquad [9.26]$$

with $\beta' \leqslant \beta$ (Fig. 9.6). The corresponding values for θ_h, θ_o, and β' satisfying [9.26] result in $N_s = \min.f(\theta_h, \theta_o, \beta')$. Thus, the critical height becomes [9.21] and:

$$\frac{D}{H_c} = \frac{\sin (\beta - \beta')}{\sin \beta \sin \beta'} \qquad [9.27]$$

which is the ratio between distance D and critical height H_c (Fig. 9.6).

9.3.2. Numerical results

The results for the failure plane passing below toe (Fig. 9.6) are tabulated numerically in Table 9.2 for small values of α, β, and φ. For the case of $\varphi = 5°$ and $\alpha = 0°$ the corresponding critical values of θ_h, θ_o, H/r_o, L/r_o, and D/H are plotted in Fig. 9.7. They determine the position of the center of the critical logspiral surface of sliding. Fig. 9.8 shows the transition zone where the most critical failure plane starts to pass below the toe, when $\alpha = 0$. The values of the stability factor N_s with respect to a failure along a critical logspiral surface of sliding below toe are given by the solid curves in Fig. 9.8. The results for the critical logspiral surface passing

TABLE 9.2

Stability factor $N_s = H_c(\gamma/c)$ by limit analysis for small values of α, β, and φ (Fig. 9.6)

Friction angle ϕ (°)	Slope angle α (°)	Slope angle, β (°)							
		50	45	40	35	30	25	20	15
0	0	5.52	5.53	5.53	5.53	5.53	5.53	5.53	5.53
5	0	6.92*	7.35*	7.84*	8.41*	9.13	10.02	11.46	14.38
5	5	6.76*	7.18*	7.64*	8.19*	8.83	9.65	10.99	13.71

* Failure through toe; all others failure below toe.

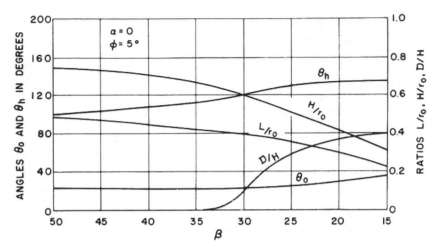

Fig. 9.7. Critical values for θ_h, θ_0, L/r_0, H/r_0, and D/H as a function of slope angle β (failure plane passing below toe).

through toe are shown by the dotted curves. For the case $\varphi = 0$, their differences are seen to increase from 0 for β near 60° to 1.82 for $\beta = 15°$.

Comparison of limit analysis results with already existing limit equilibrium solu-

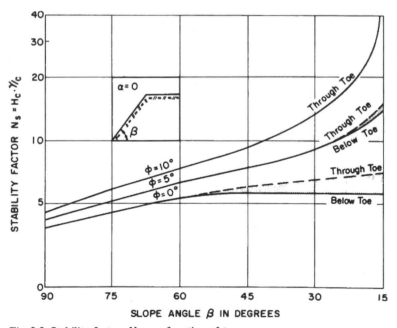

Fig. 9.8. Stability factors N_s, as a function of β.

TABLE 9.3

Comparison of stability factor $N_S = H_c(\gamma/c)$ by methods of limit equilibrium and limit analysis
$\alpha = 0$

Slope angle β (°)	Friciton angle ϕ (°)	Curved failure surface			
		Limit equilibrium			limit analysis
		slices	ϕ circle	logspiral	logspiral
90	0	3.83	3.83	3.83	3.83
	5	4.19	4.19	4.19	4.19
	15	5.02	5.02		5.02
	25	6.06	6.06	6.06	6.06
75	0	4.57	4.57	4.57	4.56
	5	5.13	5.13		5.14
	15	6.49	6.52		6.57
	25	8.48	8.54		8.58
60	0	5.24	5.24	5.24	5.25
	5	6.06	6.18	6.18	6.16
	15	8.33	8.63	8.63	8.63
	25	12.20	12.65	12.82	12.74
45	0	5.88	5.88*	5.88*	5.53*
	5	7.09	7.36		7.35
	15	11.77	12.04		12.05
	25	20.83	22.73		22.90
30	0	6.41*	6.41*	6.41*	5.53*
	5	8.77*	9.09*		9.13*
	15	20.84	21.74		21.69
	25	83.34	111.1	125.0	119.93
15	0	6.90*	6.90*	6.90*	5.53*
	5	13.89*	14.71*	14.71*	14.38*
	10		43.62		45.49

* Critical failure surface passes below toe.

tions are given in Table 9.3. It may be seen that agreement is good. Table 9.4
tabulates the value of N_s for various combinations of slopes α and β. Because there
are no existing solutions available for the case $\alpha \neq 0$, the tabulated results will be
useful in the analysis and design of such problems.

TABLE 9.4

Stability factor $N_S = H_C(\gamma/c)$ by limit analysis

Friction angle ϕ (°)	Slope angle α (°)	Slope angle β (°)						
		90	85	80	75	70	65	60
0	0	3.83	4.08	4.33	4.56	4.80	5.03	5.25
5	0	4.19	4.50	4.82	5.14	5.47	5.81	6.16
	5	4.14	4.44	4.74	5.05	5.37	5.69	6.03
10	0	4.58	4.97	5.37	5.80	6.25	6.73	7.26
	5	4.53	4.91	5.30	5.71	6.15	6.63	7.14
	10	4.47	4.83	5.21	5.61	6.03	6.48	6.99
15	0	5.02	5.50	6.01	6.57	7.18	7.85	8.63
	5	4.97	5.44	5.94	6.49	7.08	7.75	8.52
	10	4.91	5.36	5.85	6.38	6.97	7.63	8.38
	15	4.83	5.27	5.74	6.26	6.82	7.46	8.19
20	0	5.50	6.10	6.75	7.48	8.30	9.25	10.39
	5	5.46	6.04	6.68	7.40	8.21	9.16	10.28
	10	5.40	5.97	6.60	7.30	8.10	9.04	10.16
	15	5.33	5.88	6.50	7.18	7.97	8.89	9.98
	20	5.24	5.77	6.37	7.03	7.79	8.68	9.74
25	0	6.06	6.79	7.62	8.58	9.70	11.05	12.74
	5	6.01	6.73	7.56	8.50	9.61	10.96	12.64
	10	5.95	6.67	7.48	8.41	9.51	10.84	12.52
	15	5.89	6.58	7.38	8.30	9.38	10.70	12.36
	20	5.80	6.48	7.26	8.16	9.22	10.51	12.14
	25	5.70	6.35	7.10	7.97	9.00	10.26	11.84
30	0	6.69	7.61	8.67	9.94	11.48	13.44	16.04
	5	6.64	7.55	8.61	9.86	11.40	13.35	15.94
	10	6.59	7.48	8.53	9.77	11.30	13.24	15.82
	15	6.52	7.40	8.44	9.67	11.18	13.10	15.67
	20	6.44	7.31	8.32	9.54	11.03	12.93	15.47
	25	6.35	7.19	8.18	9.37	10.83	12.70	15.20
	30	6.22	7.04	7.99	9.14	10.56	12.37	14.78
35	0	7.42	8.58	9.97	11.68	13.86	16.77	20.94
	5	7.38	8.52	9.90	11.60	13.77	16.68	20.84
	10	7.32	8.46	9.82	11.51	13.68	16.58	20.73
	15	7.26	8.38	9.73	11.41	13.56	16.44	20.58
	20	7.19	8.29	9.63	11.29	13.42	16.29	20.40
	25	7.10	8.18	9.49	11,13	13.23	16.07	20.14
	30	6.99	8.04	9.33	10.93	12.99	15.78	19.78
	35	6.84	7.86	9.10	10.64	12.64	15.34	19.21
40	0	8.29	9.77	11.61	13.97	17.15	21.72	28.91
	5	8.24	9.71	11.54	13.89	17.09	21.63	28.82
	10	8.19	9.65	11.46	13.81	16.97	21.53	28.71
	15	8.13	9.57	11.38	13.71	16.86	21.40	28.57
	20	8.06	9.49	11.27	13.59	16.72	21.25	28.39
	25	7.98	9.38	11.15	13.44	16.55	21.05	28.15
	30	7.87	9.25	10.99	13.25	16.33	20.78	27.82
	35	7.74	9.09	10.78	13.00	16.02	20.39	27.32
	40	7.56	8.86	10.50	12.64	15.55	19.77	26.45

55	50	45	40	35	30	25	20	15
5.46	5.52	5.53	5.53	5.53	5.53	5.53	5.53	5.53
6.53	6.92	7.35	7.84	8.41	9.13	10.02	11.46	14.38
6.38	6.76	7.18	7.64	8.19	8.83	9.65	10.99	13.71
7.84	8.51	9.31	10.30	11.61	13.50	16.64	23.14	45.49
7.72	8.38	9.16	10.13	11.42	13.28	16.37	22.79	44.95
7.54	8.18	8.93	9.87	11.11	12.89	15.84	21.96	42.90
9.54	10.64	12.05	13.97	16.83	21.69	32.11	69.40	
9.42	10.51	11.91	13.82	16.65	21.48	31.85	69.05	
9.26	10.34	11.72	13.59	16.38	21.14	31.38	68.26	
9.04	10.09	11.42	13.23	15.92	20.49	30.25	65.17	
11.80	13.63	16.16	19.99	26.66	41.22	94.63		
11.69	13.51	16.03	19.85	26.48	41.02	94.38		
11.54	13.35	15.85	19.64	26.23	40.69	93.78		
11.35	13.12	15.58	19.32	25.82	40.09	92.90		
11.07	12.79	15.17	18.77	25.01	38.64	88.63		
14.97	18.10	22.90	31.33	50.06	119.93			
14.86	17.98	22.77	31.19	49.89	119.70			
14.73	17.83	22.60	30.99	49.63	119.35			
14.55	17.62	22.35	30.69	49.23	118.79			
14.30	17.33	21.98	30.20	48.50	117.43			
13.92	16.85	21.35	29.24	46.76	112.07			
19.71	25.41	35.54	58.27	144.20				
19.61	25.29	35.41	58.13	144.01				
19.48	25.15	35.25	57.92	143.74				
19.31	24.96	35.01	57.63	143.31				
19.08	24.68	34.67	57.16	142.54				
18.74	24.27	34.11	56.30	140.54				
18.22	23.54	33.01	54.25	134.52				
27.45	39.11	65.52	166.38					
27.34	39.00	65.39	166.22					
27.22	38.85	65.22	166.00					
27.05	38.66	64.70	165.72					
26.84	38.40	64.65	165.19					
26.53	38.02	64.12	164.30					
26.07	37.38	63.14	162.33					
25.27	36.15	60.80	154.98					
41.89	71.49	185.49						
41.78	71.37	185.35						
41.66	71.23	185.17						
41.51	71.04	184.93						
41.29	70.78	184.57						
41.00	70.41	184.04						
40.58	69.81	183.01						
39.88	68.73	180.81						
38.53	66.12	172.51						

9.4. STABILITY OF SLOPES IN ANISOTROPIC, NON-HOMOGENEOUS SOILS

In the preceding sections, we have limited our discussions to cohesive materials whose shearing resistance to sliding is determined *isotropically* and *homogeneously* by Coulomb's condition of yielding. In reality every mass of soil exhibits some anistropy with respect to shearing direction, and some non-homogeneity with respect to depth. Such variations to the orientation and depth of soil are likely to influence the critical height with respect to slope failures.

9.4.1. Statement of the problem (Fig. 9.9)

Herein, the general problem of the stability of a non-homogeneous, anisotropic slope of the type shown in Fig. 9.9 is considered. This type of slope is frequently encountered in engineering practice but design data to assess the critical height of such a slope are very scant. This lack of detailed information is largely due to the difficult procedures in analysis encountered when the conventional limit equilibrium method is used. However, as in the preceding sections on the stability of isotropic, homogeneous slopes, the upper-bound technique of limit analysis can be used to obtain the solutions in closed form for the critical height of the generalized

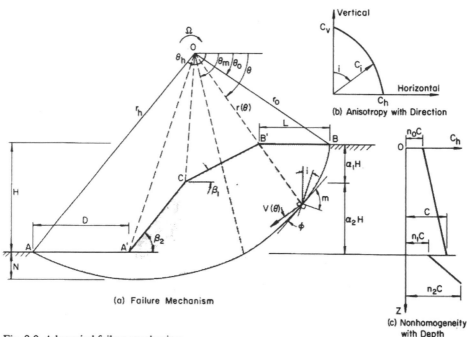

(a) Failure Mechanism

(b) Anisotropy with Direction

(c) Nonhomogeneity with Depth

Fig. 9.9. A logspiral failure mechanism.

problem. It will be seen by comparing some numerical results of the classical and limit analysis solutions, that good agreement is usually obtained.

Since the following conditions are considered in the closed form solution (1) logspiral surface of sliding, passing through and below toe; (2) non-homogeneity; (3) anisotropy; and (4) general slope, the problem considered in this section will also serve to demonstrate the usefulness and power of the upper-bound technique of limit analysis in developing new solutions.

In the following computations, we shall assume that the material of the broken inclined slope obeys the Coulomb yield condition. Since Coulomb's yield condition is described by two parameters; cohesion c and friction angle φ, we shall further assume that only the parameter c is *nonhomogeneous* and *anisotropic*. The friction angle φ is assumed to remain homogeneous and isotropic throughout the calculations, i.e., a constant value for a given type of slope. A discussion will therefore be given of the types of nonhomogeneity and anisotropy to be used in the calculations.

Nonhomogeneity

A very common case of *nonhomogeneity* of cohesion is that of a linear variation of c with respect to depth z as shown diagramatically in Fig. 9.10. In this connection, we have shown in this diagram five different types of linear variations of cohesion with respect to depth used previously by several investigators in their calculations. The variation of cohesion shown in Fig. 9.10(e) is the most general case and will therefore be adopted in the following calculations. This nonhomogeneity of cohesion with respect to depth is also sketched in Fig. 9.9(c) along with the broken inclined slope, Fig. 9.9(a). In Fig. 9.9(c), the ratios of relative cohesions at various depth are defined by n_0, n_1 and n_2.

Anisotropy

The variation of cohesion with respect to direction at a particular point has been studied by several investigators (Casagrande and Carrillo, 1954; Lo, 1965). It is found that the variation of cohesion with direction approximates to the curve

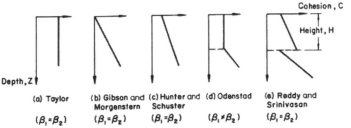

Fig. 9.10. Several types of linear variations of cohesion with depth. (a) Taylor, 1948; (b) Gibson and Morgenstern, 1962; (c) Hunter and Schuster, 1968; (d) Odenstad, 1963; (e) Reddy and Srinivasan, 1967.

shown in Fig. 9.9(b). The cohesion c_i, whose *major principal stress* direction inclines at an angle i to the vertical direction, is given by:

$$c_i = c_h + (c_v - c_h) \cos^2 i \qquad\qquad [9.28]$$

in which c_h and c_v are the cohesions whose major principal stress direction is in the horizontal and vertical directions, respectively. The cohesions c_h and c_v are called *horizontal and vertical principal cohesions*, respectively. The vertical principal cohesion, c_v, for example, can be obtained by taking vertical soil samples at any site and being tested with the major principal stress applied in the same direction.

The ratio of the principal cohesions c_h/c_v, denoted by κ, is assumed to be the same at all points in the medium (for an isotropic material, $c_i = c_h = c_v$ and $\kappa = 1.0$). The angle m as shown in Fig. 9.9(a) is the angle between the failure plane and the plane normal to the direction of the *major principal stress* at the point under consideration. This major principal stress direction at the point inclines at an angle i to the vertical direction. This angle, according to Lo's tests (1965), is found to be a constant value which is independent of the angle of rotation of the major principal stress. This particular type of anisotropy of cohesion with direction at a point as sketched in Fig. 9.9(b) is used in the following calculations.

Broken inclined slope

The design of a broken inclined slope with different slope angles β_1 and β_2 as shown in Fig. 9.9(a) is becoming more notable because the minimum volume of excavated clay is always desirable. Some slope sections of this type have already been discussed by Odenstad (1963). In the following investigations, which deal with the influence of nonhomogeneity and anistropy of cohesion on the stability of inclined slopes, it will be assumed that this inclined slope surface consists of two broken planes specified by the inclined angular parameters β_1 and β_2 and the corresponding vertical heights relative to the total height H are defined by the *depth factors* α_1 and α_2. The vertical height of the logspiral failure surface below toe is denoted by N.

9.4.2. Upper-bound solution of limit analysis

The upper-bound theorem of limit analysis states that a cut in clay shown in Fig. 9.9(a) will collapse under its own weight if, for the assumed logspiral failure mechanism which is specified completely by the three parameters θ_o, θ_h, D/r_o, the rate of external work done by the soil weight exceeds the rate of internal energy dissipation. The upper-bound values of the critical height can then be obtained by equating the external rate of work to the internal rate of energy dissipation for any such a mechanism.

Referring to Fig. 9.9(a), the region $AA'CB'BA$ rotates as a rigid body about the as yet undefined center of rotation O with the materials below the logarithmic spiral failure surface AB remaining at rest. Thus, the surface AB is a surface of velocity discontinuity.

Rate of external work

Here again, as in previous cases, the rate of external work done by the region $AA'CB'BA$ can easily be obtained from the algebraic summation of $\dot{W}_1 - \dot{W}_2 - \dot{W}_3 - \dot{W}_4 - \dot{W}_5$. The terms, $\dot{W}_1$, $\dot{W}_2$, $\dot{W}_3$, $\dot{W}_4$, and $\dot{W}_5$ represent the rates of external work done by the soil weights in the regions $OABO$, $OB'BO$, $OCB'O$, $OA'CO$, and $OAA'O$, respectively. After some simplification, the total rate of external work done by the soil weight is found to be:

$$\gamma \Omega r_o^3 (f_1 - g_2 - g_3 - g_4 - g_5) \tag{9.29}$$

in which function $f_1(\theta_h, \theta_o)$ remains identical in its form, as in the previous equation [9.14], and functions g_2, g_3, g_4 and g_5, resulting from regions $OB'BO$, $OCB'O$, $OA'CO$ and $OAA'O$, respectively, are expressed as:

$$g_2 = \frac{1}{6}\left(\frac{L}{r_o}\right) \sin\theta_o \left(2\cos\theta_o - \frac{L}{r_o}\right) \tag{9.30}$$

$$g_3 = \frac{\alpha_1}{3}\left(\frac{H}{r_o}\right)\left[\cos^2\theta_o + \frac{L}{r_o}\left(\frac{L}{r_o} - 2\cos\theta_o\right) + \sin\theta_o \cot\beta_1 \left(\cos\theta_o - \frac{L}{r_o}\right)\right.$$
$$\left. - \frac{\alpha_1}{2}\frac{H}{r_o}\cot\beta_1\left(\cos\theta_o + \frac{L}{r_o} - \sin\theta_o \cot\beta_1\right)\right] \tag{9.31}$$

$$g_4 = \frac{\alpha_2}{3}\left(\frac{H}{r_o}\right)\left\{(\cos^2\theta_h + \sin\theta_h \cos\theta_h \cot\beta_2) \exp[2(\theta_h - \theta_o)\tan\varphi]\right.$$
$$+ \left(2\frac{D}{r_o}\cos\theta_h + \frac{D}{r_o}\sin\theta_h \cot\beta_2 + \frac{\alpha_2}{2}\frac{H}{r_o}\cos\theta_h \cot\beta_2 + \frac{\alpha_2}{2}\frac{H}{r_o}\sin\theta_h \cot^2\beta_2\right)$$
$$\left. \exp[(\theta_h - \theta_o)\tan\varphi] + \left(\frac{D}{r_o}\right)^2\right\} \tag{9.32}$$

$$g_5 = \frac{1}{6}\left(\frac{D}{r_o}\right)\sin\theta_h \left\{2\cos\theta_h \exp[(\theta_h - \theta_o)\tan\varphi] + \frac{D}{r_o}\right\}\exp[(\theta_h - \theta_o)\tan\varphi] \tag{9.33}$$

From the geometrical relations, the ratios H/r_o, L/r_o, and N/r_o can be expressed as:

$$\frac{H}{r_o} = \sin\theta_h \exp[(\theta_h - \theta_o)\tan\varphi] - \sin\theta_o \tag{9.34}$$

$$\frac{L}{r_o} = \cos\theta_o - \cos\theta_h \exp[(\theta_h - \theta_o)\tan\varphi] - \frac{D}{r_o} - \left(\frac{H}{r_o}\right)(\alpha_1 \cot\beta_1 + \alpha_2 \cot\beta_2) \quad [9.35]$$

$$\frac{N}{r_o} = \cos\varphi \exp[(\tfrac{1}{2}\pi + \varphi - \theta_o)\tan\varphi] - \sin\theta_o - \frac{H}{r_o} \quad [9.36]$$

Rate of internal dissipation

The total rates of internal energy dissipation along the discontinuity logspiral failure surface AB are found by multiplying the differential area $rd\theta/\cos\varphi$ by c_i times the discontinuity in velocity, $V\cos\varphi$, across the surface and integrating over the whole surface AB. Since the layered clays possess different values of c_i, the integration is thus divided into two parts as follows:

$$\int_{\theta_o}^{\theta_h} c_i (V\cos\varphi)\frac{rd\theta}{\cos\varphi} = \int_{\theta_o}^{\theta_m} (c_i)_1 (V\cos\varphi)\frac{rd\theta}{\cos\varphi} + \int_{\theta_m}^{\theta_h} (c_i)_2 (V\cos\varphi)\frac{rd\theta}{\cos\varphi} \quad [9.37]$$

The logspiral angle (θ_m) and the anisotropic angle (i) are obtained directly from the geometric configuration shown in Fig. 9.9(a) and may be written as:

$$\sin\theta_m \exp(\theta_m \tan\varphi) = \sin\theta_h \exp(\theta_h \tan\varphi) \quad [9.38]$$

and: $i = \theta - \tfrac{1}{2}\pi - \varphi + m = \theta + \Phi$ [9.39]

in which: $\Phi = -(\tfrac{1}{2}\pi + \varphi - m)$ [9.40]

Referring to [9.28] and geometry of Fig. 9.9, $(c_i)_1$ and $(c_i)_2$ can be expressed as:

$$(c_i)_1 = c\left[n_o + \frac{(1-n_o)}{(H/r_o)}\left\{\sin\theta \exp[(\theta-\theta_o)\tan\varphi] - \sin\theta_o\right\}\right]\left[1 + \frac{(1-\kappa)}{\kappa}\cos^2 i\right]$$

[9.41a]

$$(c_i)_2 = c\left[n_1 + \frac{(n_2-n_1)}{(N/r_0)}\left\{\sin\theta \exp[(\theta-\theta_o)\tan\varphi] - \sin\theta_m \exp[(\theta_m-\theta_o)\tan\varphi]\right\}\right]$$
$$\left[1 + \frac{(1-\kappa)}{\kappa}\cos^2 i\right]$$

[941.b]

Substituting [9.41] into [9.37] and, after integration and simplification, [9.37] reduces to:

$$c\Omega r_o^2(q_1 + q_2 + q_3) \quad [9.42]$$

The functions q_1, q_2 and q_3 appearing in the above expression are defined as:

$$q_1 = \left\{ \frac{n_o}{\exp(2\theta_o \tan \varphi)} \left[\psi(\theta) + \frac{(1-\kappa)}{\kappa} \lambda(\theta) \right] \right\}_{\theta_o}^{\theta_m}$$

$$+ \left\{ \frac{n_1}{\exp(2\theta_o \tan \varphi)} \left[\psi(\theta) + \frac{(1-\kappa)}{\kappa} \lambda(\theta) \right] \right\}_{\theta_m}^{\theta_h} \qquad [9.43]$$

$$q_2 = \frac{(1-n_o)}{(H/r_o) \exp(3\theta_o \tan \varphi)} \left\{ \xi(\theta) - \psi(\theta) \sin \theta_o \exp(\theta_o \tan \varphi) + \frac{(1-\kappa)}{\kappa} \right. \qquad [9.44]$$

$$\left. [\rho(\theta) - \lambda(\theta) \sin \theta_o \exp(\theta_o \tan \varphi)] \right\}_{\theta_o}^{\theta_m}$$

$$q_3 = \frac{(n_2 - n_1)}{(N/r_o) \exp(3\theta_o \tan \varphi)} \left\{ \xi(\theta) - \psi(\theta) \sin \theta_m \exp(\theta_m \tan \varphi) + \frac{(1-\kappa)}{\kappa} \right.$$

$$\left. [\rho(\theta) - \lambda(\theta) \sin \theta_m \exp(\theta_m \tan \varphi)] \right\}_{\theta_m}^{\theta_h} \qquad [9.45]$$

in which:

$$\xi(\theta) = \frac{(3 \tan \varphi \sin \theta - \cos \theta) \exp(3\theta \tan \varphi)}{1 + 9 \tan^2 \varphi} \qquad [9.46]$$

$$\psi(\theta) = \frac{\exp(2\theta \tan \varphi)}{2 \tan \varphi} \qquad [9.47]$$

$$\rho(\theta) = \frac{\exp(3\theta \tan \varphi)}{2} \left\{ \cos 2\Phi \left[\frac{(\cos \theta - 3 \tan \varphi \sin \theta)}{2(1 + 9 \tan^2 \varphi)} + \frac{(\tan \varphi \sin 3\theta - \cos 3\theta)}{6(1 + \tan^2 \varphi)} \right] \right.$$

$$- \sin 2\Phi \left[\frac{(\sin \theta + 3 \tan \varphi \cos \theta)}{2(1 + 9 \tan^2 \varphi)} - \frac{(\tan \varphi \cos 3\theta + \sin 3\theta)}{6(1 + \tan^2 \varphi)} \right]$$

$$\left. + \left[\frac{(3 \tan \varphi \sin \theta - \cos \theta)}{1 + 9 \tan^2 \varphi} \right] \right\} \qquad [9.48]$$

$$\lambda(\theta) = \frac{\exp(2\theta \tan \varphi)}{2} \left\{ \cos 2\Phi \left[\frac{\tan \varphi \cos 2\theta + \sin 2\theta}{2(1 + \tan^2 \varphi)} \right] - \sin 2\Phi \right.$$

$$\left. \left[\frac{\tan \varphi \sin 2\theta - \cos 2\theta}{2(1 + \tan^2 \varphi)} \right] \right\} + \frac{\exp(2\theta \tan \varphi)}{4 \tan \varphi} \qquad [9.49]$$

Equating the total rates of external work, [9.29], to the total rates of internal energy dissipation, [9.42], one obtains:

$$H = \frac{c}{\gamma} f\left(\theta_o, \theta_h, \frac{D}{r_o}\right) \tag{9.50}$$

in which c denotes the horizontal principal cohesion at the level of the toe (Fig. 9.9c) and function $f(\theta_o, \theta_h, D/r_o)$ is now defined as:

$$f\left(\theta_o, \theta_h, \frac{D}{r_o}\right) = \left(\frac{H}{r_o}\right) \frac{(q_1 + q_2 + q_3)}{(f_1 - g_2 - g_3 - g_4 - g_5)} \tag{9.51}$$

TABLE 9.5

Comparison of stability factor, $N_s = H_c(\gamma/c_v)$ for anisotropic but homogeneous slopes ($\varphi = 0$, $m = 55°$)

Slope angle β (°)	Anisotropy factor κ	Curved failure surface	
		limit equilibrium	limit analysis
		ϕ-circle (Lo, 1965)	logspiral
90	1.0	3.83	3.83
	0.9	—	3.81
	0.8	—	3.79
	0.7	—	3.78
	0.6	—	3.76
	0.5	—	3.74
70	1.0	4.79	4.79
	0.9	4.72	4.72
	0.8	4.65	4.65
	0.7	4.58	4.58
	0.6	4.49	4.49
	0.5	4.41	4.41
50	1.0	5.68	5.68
	0.9	5.54	5.54
	0.8	5.35	5.38
	0.7	5.19	5.23
	0.6	5.09	5.09
	0.5	4.85	4.95
30	1.0	—	7.45
	0.9	—	7.20
	0.8	—	6.95
	0.7	—	6.70
	0.6	—	6.45
	0.5	—	6.19

The function $f(\theta_o, \theta_h, D/r_o)$ has a minimum and, thus, indicates a least upper bound when θ_o, θ_h, and D/r_o satisfy the conditions:

$$\frac{\partial f}{\partial \theta_o} = 0; \quad \frac{\partial f}{\partial \theta_h} = 0; \quad \frac{\partial f}{\partial (D/r_o)} = 0 \tag{9.52}$$

The corresponding values for θ_o, θ_h, and D/r_o satisfying [9.52] result in $N_s = \min f(\theta_o, \theta_h, D/r_o)$. Thus, the critical height can be obtained from [9.21].

TABLE 9.6

Comparison of stability factor, $N_s = z(\gamma/c_v)$ for anisotropic soil with cohesion increasing linearly with depth, Fig. 9.10(b) ($\varphi = 0$, $m = 55°$)

Slope angle β ($°$)	Anisotropy factor κ	Curved failure surface	
		limit equilibrium	limit analysis
		ϕ-circle (Lo, 1965)	logspiral
90	1.0	2.00	2.00
	0.9	2.00	2.00
	0.8	2.00	2.00
	0.7	2.00	2.00
	0.6	2.00	2.00
	0.5	2.00	2.00
70	1.0	2.77	2.77
	0.9	2.73	2.73
	0.8	2.69	2.69
	0.7	2.65	2.65
	0.6	2.61	2.61
	0.5	2.50	2.52
50	1.0	3.78	3.78
	0.9	3.66	3.66
	0.8	3.56	3.56
	0.7	3.45	3.45
	0.6	3.31	3.31
	0.5	3.17	3.20
30	1.0	5.50	5.50
	0.9	–	5.22
	0.8	5.00	5.00
	0.7	–	4.69
	0.6	–	4.41
	0.5	4.18	4.16

9.4.3. Numerical results

The complete numerical results of the stability number are obtained by the CDC 6400 digital computer. The optimization technique reported by Powell (1964) is used to minimize the function of [9.51] without calculating the derivatives. The results are then compared in Tables 9.5 and 9.6 with the existing limit equilibrium solutions.

Comparison with Lo's results

For the cases of isotropic and homogeneous slopes with constant value of cohesion (Fig. 9.10a), the values for the stability factor N_s are found to be identical to those presented in the preceding sections (Table 9.4). Table 9.5 shows a comparison of stability numbers obtained from the limit equilibrium and limit analysis for anisotropic but homogeneous slopes. The only existing solutions available for comparison were given by Lo (1965). Herein, as in Lo's results, the value for the angle m is taken to be $55°$ and the value of friction angle, φ is put equal to zero so that the logspiral failure surface becomes a circular arc. It can be seen from the table that both results are in excellent agreement.

For the particular case of an anisotropic slope with cohesion increasing linearly with depth as shown in Fig. 9.10(b), one of the coefficients n_0 shown in Fig. 9.9(c) is zero, and the other two coefficients n_1 and n_2 have the values 1 and $(1 + N/H)$, respectively. Using [9.34] and [9.38], it follows from [9.42] that the total rate of internal energy dissipation along the logspiral surface of sliding reduces to:

$$
\frac{c\Omega r_o^2}{\left(\dfrac{H}{r_o}\right)\exp(3\theta_o \tan\varphi)} \left\{ \xi(\theta) - \psi(\theta)\sin\theta_o \, \exp(\theta_o \tan\varphi) + \frac{(1-\kappa)}{\kappa} \right.
$$

$$
\left. [\rho(\theta) - \lambda(\theta)\sin\theta_o \, \exp(\theta_o \tan\varphi)] \right\}_{\theta_o}^{\theta_h}
\qquad\qquad [9.53]
$$

and the function $f(\theta_o, \theta_h, D/r_o)$ defined in [9.51] has the form:

$$
f\left(\theta_o, \theta_h, \frac{D}{r_o}\right) =
$$

$$
\frac{\left\{ \xi(\theta) - \psi(\theta)\sin\theta_o \, \exp(\theta_o \tan\varphi) + \dfrac{1-\kappa}{\kappa} \, [\rho(\theta) - \lambda(\theta)\sin\theta_o \, \exp(\theta_o \tan\varphi)] \right\}_{\theta_o}^{\theta_h}}{(f_1 - g_2 - g_3 - g_4 - g_5)\exp(3\theta_o \tan\varphi)}
$$

$$
\qquad\qquad [9.54]
$$

It is worth pointing out that for this particular case the ratio c_h/z is equal to c/H, Fig. 9.9(c). Expression [9.21] for a least upper bound for the critical height H_c, of an inclined slope may be rewritten as:

$$N_s \geqslant \frac{\gamma H_c}{c} = \frac{\gamma z}{c_h} \qquad [9.55]$$

in which $N_s = \min.f(\theta_o, \theta_h, D/r_o)$. The stability factor N_s can now be compared in Table 9.6 with those obtained previously by Lo (1965) using the limit equilibrium method. Good agreement is again observed.

Stability factor for anisotropic and/or nonhomogeneous soils (Tables 9.7 and 9.8)

Function [9.51] or [9.54] is now used to generate the values of stability factor N_s for the friction angle φ ranging from 0 to 40° and anisotropy factor κ varying from 0.5 to 1. The values of this stability factor are given in Table 9.7 for the case of an anisotropic but homogeneous slope and in Table 9.8 for the case of an anisotropic slope with cohesion increasing linearly with depth, Fig. 9.10(b). The value for the angle m shown in Fig. 9.9(a) is taken to be $(\frac{1}{4}\pi + \frac{1}{2}\varphi)$.

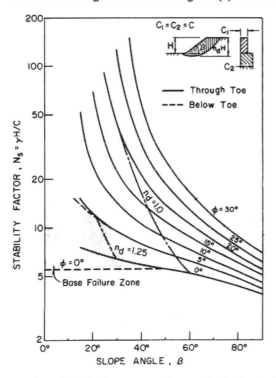

Fig. 9.11. Stability factor versus slope angles for isotropic-homogeneous soils.

TABLE 9.7

Stability factor $N_s = H_c(\gamma/c_v)$ by limit analysis for an anisotropic but homogeneous slope

Slope angle β (°)	Anisotropy factor κ	Stability number N_s				
		friction angle (°) φ = 0 m = 45°	friction angle (°) φ = 10 m = 50°	friction angle (°) φ = 20 m = 55°	friction angle (°) φ = 30 m = 60°	friction angle (°) φ = 40 m = 65°
90	1.0	3.83	4.58	5.50	6.78	8.52
	0.9	3.82	4.57	5.49	6.75	8.49
	0.8	3.81	4.56	5.48	6.73	8.46
	0.7	3.79	4.54	5.47	6.70	8.42
	0.6	3.78	4.53	5.45	6.67	8.39
	0.5	3.76	4.51	5.44	6.65	8.39
70	1.0	4.79	6.24	8.29	11.48	17.22
	0.9	4.72	6.20	8.24	11.42	17.13
	0.8	4.65	6.15	8.18	11.35	17.04
	0.7	4.58	6.09	8.12	11.28	16.94
	0.6	4.49	6.03	8.06	11.21	16.85
	0.5	4.41	5.97	7.99	11.14	16.75
50	1.0	5.68	8.51	13.64	25.74	—
	0.9	5.58	8.43	13.44	25.40	—
	0.8	5.47	8.29	13.24	25.08	—
	0.7	5.37	8.15	13.04	24.75	—
	0.6	5.27	8.02	12.83	24.43	—
	0.5	5.16	7.86	12.63	24.11	—
30	1.0	7.45	26.74	—	—	—
	0.9	7.28	26.10	—	—	—
	0.8	7.12	25.45	—	—	—
	0.7	6.96	24.80	—	—	—
	0.6	6.79	24.16	—	—	—
	0.5	6.63	23.51	—	—	—

TABLE 9.8

Stability factor $N_s = z(\gamma/c_v)$ by limit analysis for an anisotropic slope with cohesion increasing linearly with depth, Fig. 9.10(b)

Slope angle β (°)	Anisotropy factor κ	Stability number N_s				
		friction angle (°) φ = 0 m = 45°	friction angle (°) φ = 10 m = 50°	friction angle (°) φ = 20 m = 55°	friction angle (°) φ = 30 m = 60°	friction angle (°) φ = 40 m = 65°
90	1.0	2.00	2.42	2.90	3.75	4.66
	0.9	2.00	2.40	2.87	3.74	4.66
	0.8	2.00	2.38	2.85	3.73	4.65
	0.7	2.00	2.38	2.85	3.72	4.64
	0.6	2.00	2.38	2.85	3.72	4.64
	0.5	2.00	2.38	2.85	3.71	4.63
70	1.0	2.83	3.68	4.74	6.73	9.81
	0.9	2.77	3.54	4.68	6.63	9.76
	0.8	2.74	3.53	4.65	6.43	9.71
	0.7	2.73	3.51	4.63	6.40	9.66
	0.6	2.71	3.49	4.61	6.36	9.60
	0.5	2.69	3.47	4.58	6.33	9.55
50	1.0	3.94	5.44	8.62	15.50	—
	0.9	3.85	5.35	8.45	15.23	—
	0.8	3.76	5.26	8.28	14.96	—
	0.7	3.61	5.16	8.10	14.69	—
	0.6	3.52	5.06	7.93	14.42	—
	0.5	3.45	4.95	7.76	14.09	—
30	1.0	5.47	19.33	—	—	—
	0.9	5.31	18.72	—	—	—
	0.8	5.14	18.11	—	—	—
	0.7	4.98	17.50	—	—	—
	0.6	4.82	16.89	—	—	—
	0.5	4.66	16.28	—	—	—

Stability factor for two-layer soils (Figs. 9.12–9.15)

A very common case of nonhomogeneity is that of a two-layer soil as shown in the inset of Fig. 9.11. The deepest point of the logspiral surface of sliding is located at a depth $n_d H$ from the top of the slope. It is obvious that when the *depth factor* denoted by n_d is greater than unity, a portion of the slip surface will lie within the soil stratum below the toe of the slope. If this lower portion of the stratum is weaker or stronger than the one above the toe level, the stability factor N_s will definitely be altered.

For an isotropic, homogeneous slope, calculations show that for slope angles of more than 60°, the depth factor is equal to unity. The surface of logspiral sliding rises from the toe of the slope toward the top of it. On the other hand, if $\beta < 60°$, the deepest part of the surface of sliding may be located beneath the level of the toe of the slope depending on the values of φ. For all slopes with angle β larger than 53°, the surface of sliding is always passing through the toe of the slope. If β is less than 53°, the surface of sliding may pass through the toe or may pass below it. For all values of φ greater than approximately 5°, the surfaces of sliding are always passing through the toe of the slope. This is demonstrated graphically in Fig. 9.11. In this figure the values of stability factor N_s are plotted against slope angles β for

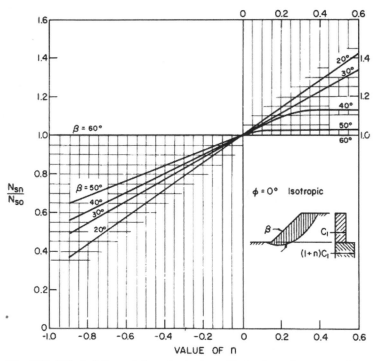

Fig. 9.12. Effect of the variation of cohesion with depth on the stability factor for isotropic soil with $\varphi = 0°$.

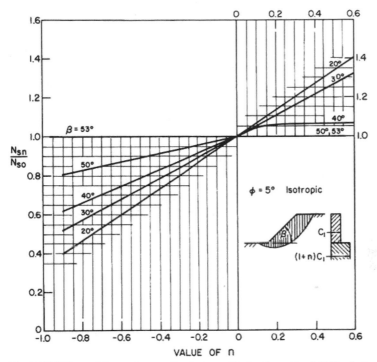

Fig. 9.13. Effect of the variation of cohesion with depth on the stability factor for isotropic soil with $\varphi = 5°$.

several values of φ. The depth factor n_d for the particular values $n_d = 1.0$ and $n_d = 1.25$ are also marked by the dash-dotted lines in the figure. Slope failures controlled by the surfaces of sliding passing below the toe of the slope are represented in Fig. 9.11 by the shaded area which starts at $\beta = 0°$ and ends at $\beta = 53°$.

The values of the stability factor N_s obtained for the two-layer soil shown in the inset of Fig. 9.11 are given in Figs. 9.12–9.15 for four values of $\varphi = 0°, 5°, 10°$ and $15°$. At a given value of the slope angle β, the stability factor N_s increases with increasing values of the relative cohesion $n = (c_2 - c_1)/c_1$. If N_{sn} is the value of N_s for given values of β, φ and n and N_{s0} is the corresponding value for $n = 0$ which corresponds to the homogeneous case, the ratio N_{sn}/N_{s0} represents the effect of a two-layer soil. Its value indicates the influence of the relative cohesion between the two layers separated at the toe level for given values of β, φ and n.

All the curves in Figs. 9.12–9.15 corresponding to different values of φ are similar in shape to that for the case $\varphi = 0$. The following points in connection with these curves shown in Figs. 9.12–9.15 are worth noting:

(1) When the slope angle β is greater than $60°$ or the friction angle φ is greater than $20°$ approximately, the depth factor n_d is always unity and the ratio N_{sn}/N_{s0} is also unity.

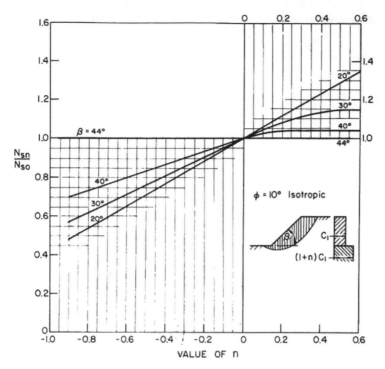

Fig. 9.14. Effect of the variation of cohesion with depth on the stability factor for isotropic soil with $\varphi = 10°$.

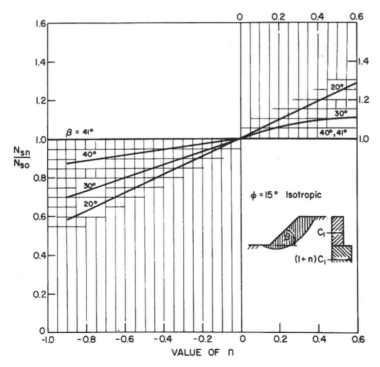

Fig. 9.15. Effect of the variation of cohesion with depth on the stability factor for isotropic soil with $\varphi = 15°$.

(2) When the relative cohesion n is negative, the lower layer is weaker than the upper one. The value of the ratio N_{sn}/N_{s0} for any given values of β and φ increases almost linearly as n increases until it reaches the limiting value of the homogeneous case $n = 0$. At this special or limiting point, the ratio N_{sn}/N_{s0} has the value unity.

(3) When the relative cohesion n is positive, the lower layer has a greater strength than the upper one. For any given values of β and φ, the ratio N_{sn}/N_{s0} raises as the relative strength of the lower layer rises, but at the same time a smaller proportion of the total length of the logspiral surface cuts through the lower stronger layer. At a limit point, the logspiral slip surface becomes tangential to the upper surface of the lower layer, and after this, any further increase in the strength of the lower layer will not influence the value of the ratio N_{sn}/N_{s0}, as the slip surface is now lying entirely inside the upper layer. This is represented by the rapid change of some curves in the figures to the horizontal line at a certain value of n. After this point, the value of N_{sn}/N_{s0} for any particular values of β and φ is unaltered as n is increased.

In order to reduce the time required to solve the stability problems of a two-layer soil of the type shown in the inset of Fig. 9.11, a tabulation of the values of the stability factor N_{sn} for various cohesion ratios $c_2/c_1 = n + 1$, slope angle β and

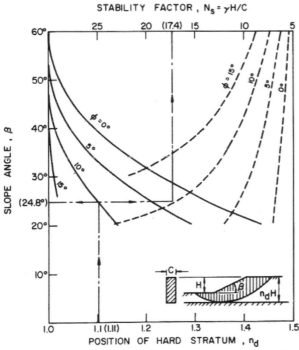

STABILITY FACTOR, $N_s = \gamma H/C$

SLOPE ANGLE, β

POSITION OF HARD STRATUM, n_d

Fig. 9.16. Relationships between depth factor, slope angle, and stability factor for homogeneous and isotropic soil.

TABLE 9.9

Stability factor, $N_{sn} = \gamma H_c/c_1$ for a two-layer soil

Slope angle $\beta°$	Friction angle $\varphi°$	Cohesion ratio, $c_2/c_1 = n+1$										
		0.1	0.2	0.4	0.6	0.8	1.0	1.1	1.2	1.3	1.4	1.5
20	0	2.61	3.10	4.08	5.06	6.04	7.02	7.51	8.00	8.49	8.98	9.46
	5	4.63	5.48	7.05	8.63	10.20	11.72	12.51	13.33	14.04	14.92	15.70
	10	11.13	12.48	15.26	17.96	20.55	23.12	24.38	25.74	27.38	28.63	30.26
30	0	3.13	3.51	4.26	5.03	5.76	6.48	6.89	7.24	7.63	7.99	8.35
	5	4.69	5.16	6.22	7.20	8.15	9.13	9.61	10.06	10.53	10.93	11.21
	10	7.54	8.25	9.65	11.02	12.30	13.50	14.23	14.71	15.10	15.26	15.44
	15	15.22	15.85	17.44	18.90	20.38	21.69	22.22	22.74	23.25	23.65	24.04
	20	41.21	41.21	41.21	41.21	41.21	41.21	41.21	41.21	41.21	41.21	41.21
40	0	3.40	3.71	4.30	4.89	5.47	6.06	6.34	6.60	6.83	6.87	6.87
	5	4.79	5.15	5.84	6.52	7.21	7.83	8.11	8.34	8.41	8.41	8.41
	10	6.96	7.41	8.25	8.97	9.71	10.29	10.56	10.68	10.68	10.68	10.68
	15	12.00	12.38	12.88	13.23	13.60	13.97	13.97	13.97	13.97	13.97	13.97
	20	19.99	19.99	19.99	19.99	19.99	19.99	19.99	19.99	19.99	19.99	19.99
	25	31.33	31.33	31.33	31.33	31.33	31.33	31.33	31.33	31.33	31.33	31.33
	30	58.27	58.27	58.27	58.27	58.27	58.27	58.27	58.27	58.27	58.27	58.27
50	0	3.70	3.93	4.37	4.82	5.27	5.67	5.82	5.83	5.83	5.83	5.83
	5	5.50	5.78	6.10	6.17	6.62	6.92	6.97	6.97	6.97	6.97	6.97
	10	8.51	8.51	8.51	8.51	8.51	8.51	8.51	8.51	8.51	8.51	8.51
	15	10.64	10.64	10.64	10.64	10.64	10.64	10.64	10.64	10.64	10.64	10.64
	20	13.62	13.62	13.62	13.62	13.62	13.62	13.62	13.62	13.62	13.62	13.62
	25	18.09	18.09	18.09	18.09	18.09	18.09	18.09	18.09	18.09	18.09	18.09
	30	25.14	25.14	25.14	25.14	25.14	25.14	25.14	25.14	25.14	25.14	25.14

TABLE 9.10

Depth factor, n_d for a two-layer soil

Slope angle $\beta°$	Friction angle $\varphi°$	Cohesion ratio, $c_2/c_1 = n + 1$										
		0.1	0.2	0.4	0.6	0.8	1.0	1.1	1.2	1.3	1.4	1.5
20	0	1.45	1.45	1.45	1.45	1.45	1.44	1.44	1.44	1.44	1.44	1.43
	5	1.36	1.36	1.36	1.36	1.36	1.33	1.33	1.33	1.33	1.32	1.32
	10	1.16	1.16	1.15	1.15	1.15	1.14	1.13	1.13	1.12	1.12	1.11
30	0	1.21	1.20	1.20	1.20	1.20	1.23	1.19	1.19	1.19	1.18	1.17
	.5	1.15	1.15	1.15	1.15	1.15	1.14	1.13	1.11	1.11	1.08	1.07
	10	1.10	1.10	1.10	1.09	1.08	1.06	1.06	1.05	1.03	1.02	1.0
	15	1.03	1.03	1.03	1.03	1.02	1.01	1.01	1.01	1.01	1.01	1.01
	20	1.0	1.0	1.0	1.0	1.0	1.0	1.0	1.0	1.0	1.0	1.0
40	0	1.13	1.13	1.12	1.12	1.12	1.12	1.08	1.07	1.05	1.0	1.0
	5	1.08	1.08	1.08	1.08	1.08	1.04	1.03	1.02	1.0	1.0	1.0
	10	1.04	1.04	1.04	1.04	1.03	1.01	1.01	1.0	1.0	1.0	1.0
	15	1.01	1.01	1.01	1.01	1.01	1.0	1.0	1.0	1.0	1.0	1.0
	20	1.0	1.0	1.0	1.0	1.0	1.0	1.0	1.0	1.0	1.0	1.0
	25	1.0	1.0	1.0	1.0	1.0	1.0	1.0	1.0	1.0	1.0	1.0
	30	1.0	1.0	1.0	1.0	1.0	1.0	1.0	1.0	1.0	1.0	1.0
50	0	1.07	1.07	1.07	1.06	1.06	1.03	1.02	1.0	1.0	1.0	1.0
	5	1.03	1.01	1.01	1.01	1.01	1.01	1.0	1.0	1.0	1.0	1.0
	10	1.0	1.0	1.0	1.0	1.0	1.0	1.0	1.0	1.0	1.0	1.0
	15	1.0	1.0	1.0	1.0	1.0	1.0	1.0	1.0	1.0	1.0	1.0
	20	1.0	1.0	1.0	1.0	1.0	1.0	1.0	1.0	1.0	1.0	1.0
	25	1.0	1.0	1.0	1.0	1.0	1.0	1.0	1.0	1.0	1.0	1.0
	30	1.0	1.0	1.0	1.0	1.0	1.0	1.0	1.0	1.0	1.0	1.0

friction angle φ is given in Table 9.9. Table 9.10 shows the corresponding values of the depth factor n_d.

Slope on a very firm stratum (Fig. 9.16)

Investigation is also made on the effect of the presence of a very firm stratum on the value of the stability factor. In the inset of Fig. 9.16, a mass of soil is shown resting at some finite depth on a very firm stratum. As an example of application of the preceding results, consider the case that we know the friction angle, $\varphi = 10°$ and depth factor $n_d = 1.11$ for the mass of soil, and we want to determine the slope which should be given to the sides of a cut and the corresponding value of the stability factor.

Figure 9.16 gives graphically the most critical slope with $\beta = 24.8°$ and the corresponding value of the stability factor $N_s = \gamma H/c = 17.4$ or $H = 17.4 \ c/\gamma$. Therefore the slope angle $\beta = 24.8°$ is *admissible* provided that the height of the cut H does not exceed the value of $17.4 \ c/\gamma$.

TABLE 9.11

Stability factor $N_s = H_c(\gamma/c)$ by limit analysis for a general slope with two different slope-angles for isotropic-homogeneous soils ($\varphi = 0$), Fig. 9.9(a)

Depth factor		Slope angle β_1	Slope angle $\beta_2(°)$						
α_1	α_2	(°)	90	80	70	60	50	40	30
$\frac{2}{3}$	$\frac{1}{3}$	90	3.83	4.06	4.30	4.58	4.92	5.38	6.06
		80	4.11	4.33	4.56	4.82	5.13	5.55	6.17
		70	4.39	4.59	4.80	5.04	5.32	5.73	6.30
		60	4.67	4.85	5.04	5.25	5.51	5.92	6.72
		50	4.97	5.12	5.28	5.46	5.72	6.17	6.99
		40	5.28	5.39	5.53	5.73	6.03	6.49	7.43
		30	5.61	5.74	5.92	6.15	6.48	6.99	7.89
$\frac{1}{2}$	$\frac{1}{2}$	90	3.83	3.96	4.10	4.26	4.47	4.76	5.23
		80	4.20	4.33	4.46	4.61	4.81	5.08	5.52
		70	4.57	4.68	4.80	4.94	5.12	5.37	5.78
		60	4.92	5.02	5.12	5.25	5.41	5.63	6.04
		50	5.27	5.35	5.43	5.54	5.69	5.93	6.37
		40	5.60	5.67	5.76	5.89	6.06	6.35	6.81
		30	6.05	6.15	6.27	6.41	6.62	6.92	7.45
$\frac{1}{3}$	$\frac{2}{3}$	90	3.83	3.89	3.95	4.02	4.11	4.24	4.45
		80	4.27	4.33	4.39	4.45	4.54	4.67	4.87
		70	4.69	4.74	4.80	4.86	4.95	5.07	5.26
		60	5.10	5.14	5.19	5.25	5.32	5.43	5.61
		50	5.47	5.51	5.55	5.60	5.67	5.77	5.95
		40	5.85	5.89	5.93	5.99	6.07	6.19	6.39
		30	6.37	6.42	6.48	6.55	6.65	6.79	7.02

Stability factor for broken inclined slope (Table 9.11)

The design of a broken inclined slope with two different slope angles β_1 and β_2 and the corresponding depth factors α_1 and α_2 as shown in Fig. 9.9(a) is frequently encountered in engineering practice and design data to assess the critical height of such a slope are often needed. Function [9.51] can be used to generate the values of stability factor N_s for the problem. The values of this stability factor as given in Table 9.11 are for the case of an isotropic-homogeneous slope with the slope angles β_1 and β_2 varying from $30°$ to $90°$ and depth factors α_1 and α_2 ranging from $1/3$ to $2/3$. The value for the internal friction angle φ of the soils is taken to be $0°$.

9.5. SHAPE OF CRITICAL SLIP SURFACE AND ITS ASSOCIATED NORMAL STRESS DISTRIBUTION

In the preceding computations of the critical height of inclined slopes, we have assumed that the shape of surface of sliding is a logspiral surface. This is justified because within the framework of limit analysis the plane surface and the logarithmic spiral surface of angle φ, $[r_o \exp(\theta \tan \varphi)]$ are the only two surfaces of discontinuity which permit rigid body motions relative to a fixed surface. In the following discussion, we will show that this assumption is also justified from the viewpoint of *variational calculus*. The logspiral mechanism utilized in the upper-bound method of limit analysis is therefore also appropriate in the framework of limit equilibrium methods. The numerical results reported in the preceding sections thus provide the best possible solution to the problem.

Before we move to find the shape of critical slip surface and its associated normal stress distribution, first, let us examine the shapes and assumptions used in conventional limit equilibrium methods. For a nearly vertical slope, the analysis based on the assumption that failure occurs on a plane through the toe of the slope leads to generally acceptable accuracies. The method of analysis based on this plane assumption is called *Culmann method*. However, the Culmann method does not give satisfactory accuracy for flat slopes.

We can improve the accuracy of the Culmann method by considering a curved surface of sliding. The most common shapes used to facilitate such calculations are that of a circular arc and a logarithmic spiral. Taylor (1948) showed that calculations made with a logarithmic spiral of angle φ and a circular arc as failure surfaces give very close results. It seems to appear that the use of the logarithmic spiral is in some ways more inconvenient than the use of the circle as the surface of sliding in the limit equilibrium approach. However, there is a very important advantage of using the logarithmic spiral of *angle φ*. The main advantage of such spiral $[r_o \exp(\theta \tan \varphi)]$ is that all intergranular forces acting on the spiral slip surface are directed toward the center of the spiral. Because of this condition, the equation for equilibrium of moment with respect to the sliding mass as a whole involves no term which depends on the pattern of normal stress distribution along the spiral surface and thus offers no difficulty in the stability analysis. In other words, the analysis is *statically determinate* in the sense that the computation of the critical height can be carried out without an assumption relative to the normal stress distribution.

More generally, when the angle of a spiral curve is not equal to the friction angle φ of the soil, $[r_o \exp(\theta \tan \psi)$ where $\psi \neq \varphi]$, the stability analysis by limit equilibrium methods requires an assumption concerning the pattern of normal stress distribution along the surface of sliding. For example, when the angle ψ is zero, then the surface is a circular arc. Fellenius (1927), in his *method of slices*, proposed that the lateral forces be assumed to be equal on the two vertical sides of each slice. This assumption, in fact, establishes a pattern of normal stress distribution along the arc of sliding. Taylor (1948), in the solution based on the so-called *friction-circle method* (or sometimes known as φ-circle method), assumes a stress distribution having zero values at the ends of the arc and sinusoidal stress variation in between. For the analysis of a general spiral surface ($\psi \neq \varphi$), Spencer (1969) assumes that the interslice forces are parallel which is another type of normal stress distribution assumption along the spiral surface.

Fortunately, the various normal stress assumptions for the circular arc slip surface or the general spiral surface lead only to minor differences in the stability solutions. These results indicate that the stability analysis is not very sensitive to the various shapes of the spiral surface for which the circular arc is a special case ($\psi = 0$).

Suppose the question is asked what is the most critical shape of surface of sliding in the slope stability analysis. Since we know that the shape of slip surface and the normal stress distribution are interrelated, any conventional methods of analysis will not be able to answer the above-mentioned question properly. Both the shape of slip surface and the normal stress distribution should be considered as the independent parameters in such analysis. Hence the proper method of analysis to answer this question is to consider all the possible shapes of the slip surfaces as well as all the possible distributions of normal stress on the slip surface. Since the *shape function* and *distribution function* of the problem must be chosen in such a way that the sum of all boundary forces along the slip surface must equal the total weight of the sliding mass in the vertical direction and must equal zero in the horizontal direction. In addition, the moment equilibrium condition must also be satisfied. There is of course no reason to suppose that these conditions will lead to a *unique* solution which gives the most *critical* stability condition to the problem. Clearly, this is an optimization problem and can therefore be approached by applying the method of calculus of variations. This is described in the following computations.

9.5.1. *Some physical facts and their significance*

A typical slope of homogeneous soil under a uniform surcharge load, q is shown in Fig. 9.17. The slope remains stable as long as the stress developed within the soil mass does not exceed soil strength. Instability is initiated as the applied load q reaches its critical value and the collapse of the slope may be described by the rigid body slide of soil mass along one of many "potential" surfaces, S_n as shown in Fig. 9.17. At the incipient of collapse, the conditions of static equilibrium of the sliding mass:

$$\Sigma H = 0, \quad \Sigma V = 0, \quad \Sigma M = 0 \qquad\qquad [9.56]$$

as well as the yield or failure criterion must be satisfied everywhere along the surface of sliding. The most critical of all these potential surfaces of sliding is theoretically the one which allows minimum applied load. In absence of surcharge load ($q = 0$), the gravitational weight of the soil mass acts solely as the external load applied on the slope.

As an example, consider a uniform slope of Fig. 9.18. The positions and values of stability factors, $N_s = H_c \gamma / c$ for several critical slip surfaces (plane, circular and logspiral) have been given by Taylor (1948). It is possible, then, to sketch in one figure the three types of slip surfaces and compare the volume of the sliding mass for each surface. This is illustrated in Fig. 9.19 for slopes having base angles of $\beta = 90°$ and $\beta = 70°$. The results show clearly that the most critical shape is the logspiral surface which also corresponds to the minimum weight W of the sliding mass. It can, therefore, be concluded that, of all potential slip surfaces, the one which allows the minimum weight W of the sliding mass gives the most critical situation. This condition will be used as the criterion of optimization in the following mathematical formulation.

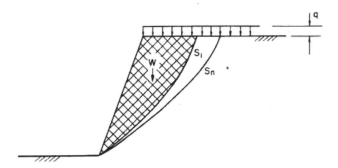

Fig. 9.17. Slope with potential slip surface.

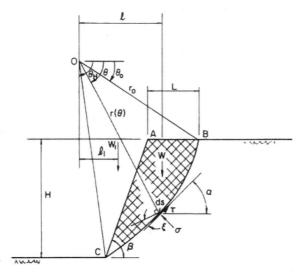

Fig. 9.18. Slope of uniform soil.

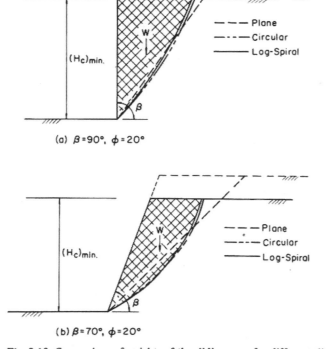

(a) $\beta = 90°$, $\phi = 20°$

(b) $\beta = 70°$, $\phi = 20°$

Fig. 9.19. Comparison of weights of the sliding mass for different slip surfaces.

9.5.2. Mathematical formulation of the problem

As stated earlier, in absence of load q the weight of the sliding mass W is the only applied load on slope and may be defined by a functional:

$$W = \int_{\theta_o}^{\theta_h} P_w \, d\theta \qquad\qquad [9.57]$$

where: $P_w = \dfrac{\gamma r^2}{2} - W_1/(\theta_h - \theta_o)$ [9.58]

in which W_1 is weight of the imaginary area $OBAC$ as shown in Fig. 9.18 and $r(\theta)$ is an unknown function defining the shape of the slip surface.

Referring to [9.56] and Fig. 9.18, the three equilibrium equations can be written as follows.

Σ horizontal forces = 0 gives:

$$\int_s [\tau \cos\alpha - \sigma \sin\alpha] \, ds = 0 \qquad\qquad [9.59]$$

Σ vertical forces = 0 gives:

$$\int_s [\tau \sin\alpha + \sigma \cos\alpha] \, ds - W = 0 \qquad\qquad [9.60]$$

Σ moment about the center of rotation = 0 gives:

$$\int_s [\sigma r \sin\xi - \tau r \cos\xi] \, ds + Wl = 0 \qquad\qquad [9.61]$$

where: $\alpha = \hat{\pi} - \theta - \arctan\left(\dfrac{r}{r'}\right)$ [9.62a]

$\xi = \frac{1}{2}\pi - \arctan\left(\dfrac{r}{r'}\right)$ [9.62b]

$ds = dr/\sin\xi$ [9.62c]

in which $r' = dr/d\theta$. The tangential shear stress, τ and normal stress, σ are related through the Coulomb failure or yield criterion, $\tau = c + \sigma\tan\varphi$. Using the Coulomb criterion, [9.59], [9.60] and [9.61] become:

$$\int_{\theta_o}^{\theta_h} P_1 \, d\theta = 0, \qquad \int_{\theta_o}^{\theta_h} P_2 \, d\theta = 0, \qquad \int_{\theta_o}^{\theta_h} P_3 \, d\theta = 0 \qquad\qquad [9.63a,b,c]$$

in which:

$$P_1 = (-\sigma)\,[(r\cos\theta)'\tan\varphi + (r\sin\theta)'] - c(r\cos\theta)' \qquad\qquad [9.64]$$

$$P_2 = \sigma[(r\cos\theta)' - (r\sin\theta)'\tan\varphi] - c(r\sin\theta)' + \frac{\gamma}{2}r^2 - \frac{W_1}{\theta_h - \theta_o} \qquad\qquad [9.65]$$

$$P_3 = \sigma(rr' - r^2\tan\varphi) - cr^2 + \frac{\gamma}{3}r^3\cos\theta - \frac{W_1 l_1}{\theta_h - \theta_o} \qquad\qquad [9.66]$$

where $r(\theta)$ and $\sigma(\theta)$ are as yet two unknown functions and the prime indicates differentiation of a function with respect to variable θ. The problem of finding the critical slip surface and its associated normal stress distribution on the surface may now be stated as follows: Given the slope shown in Fig. 9.18, determine the shape function $r(\theta)$ and stress function $\sigma(\theta)$ so as to minimize the weight functional, W of [9.57] subjected to the constraint conditions of equations (9.63a,b,c). With Lagrange's multiplier denoted by λ_1, λ_2 and λ_3, one can write:

$$I = P_w + \lambda_1 P_1 + \lambda_2 P_2 + \lambda_3 P_3 \qquad\qquad [9.67]$$

Since all integrands in P_w, P_1, P_2 and P_3 involve only $r(\theta)$, $\sigma(\theta)$ and the first derivative of $r(\theta)$, the Euler differential equation will be first order, and can be represented by:

$$\frac{d}{d\theta}\left[\frac{\partial I}{\partial\sigma'(\theta)}\right] - \frac{\partial I}{\partial\sigma(\theta)} = 0 \qquad\qquad [9.68a]$$

and:
$$\frac{d}{d\theta}\left[\frac{\partial I}{\partial r'(\theta)}\right] - \frac{\partial I}{\partial r(\theta)} = 0 \qquad\qquad [9.68b]$$

in which $\sigma' = d\sigma/d\theta$. After substitution, integration and simplification of [9.68], it follows that the two unknown functions $r(\theta)$ and $\sigma(\theta)$ must satisfy the following first-order differential equations:

$$r'[\lambda_2(\cos\theta - \tan\varphi\sin\theta) - \lambda_1(\tan\varphi\cos\theta + \sin\theta)]$$

$$+ r[\lambda_1(\tan\varphi\sin\theta - \cos\theta) - \lambda_2(\sin\theta + \tan\varphi\cos\theta)] + (rr' - r^2\tan\varphi)\lambda_3 = 0$$

[9.69]

independently of the normal stress distribution $o(\theta)$, and:

$$o'[\lambda_2(\cos\theta - \tan\varphi\sin\theta) - \lambda_1(\tan\varphi\cos\theta + \sin\theta) + \lambda_3 r] + o(2r\lambda_3\tan\varphi)$$

$$- \gamma r(1 + \lambda_2) + \lambda_3(2cr - \gamma r^2\cos\theta) = 0 \qquad [9.70]$$

The shape of the most critical slip surface can therefore be obtained by first solving [9.69] for $r(\theta)$. Once the funtion $r(\theta)$ is determined, [9.70] can then be used for the determination of $o(\theta)$ which describes the corresponding normal stress distribution along the critical slip surface obtained earlier.

9.5.3. Shape of slip surface

For convenience of solution, [9.69] is now transformed from polar to Cartesian coordinates (Fig. 9.20):

$$\lambda_1 y' - \lambda_2 - \lambda_3(yy' + x) + \tan\varphi[\lambda_1 + \lambda_2 y' - \lambda_3(y - xy')] = 0 \qquad [9.71]$$

where $x = r\cos\theta$ and $y = r\sin\theta$.

Equation [9.71] can also be written in the form:

$$y'\left(y - \frac{\lambda_1}{\lambda_3}\right) + \left(x + \frac{\lambda_2}{\lambda_3}\right) + \tan\varphi\left[\left(y - \frac{\lambda_1}{\lambda_3}\right) - y'\left(x + \frac{\lambda_2}{\lambda_3}\right)\right] = 0 \qquad [9.72]$$

Let:

$$X = x + \frac{\lambda_2}{\lambda_3}, \quad Y = y - \frac{\lambda_1}{\lambda_3} \qquad [9.73]$$

Equation [9.72] now becomes:

$$- Y'Y - X + (- Y + Y'X)\tan\varphi = 0 \qquad [9.74]$$

By substitution into [9.74] the following terms:

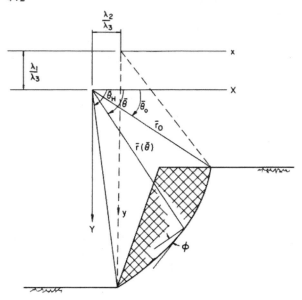

Fig. 9.20. Transformation of coordinates.

$$X = \bar{r} \cos \bar{\theta}, \qquad Y = \bar{r} \sin \bar{\theta} \tag{9.75a}$$

$$Y' = \frac{\bar{r} \cos \bar{\theta} + \bar{r}' \sin \bar{\theta}}{\bar{r}' \cos \bar{\theta} - \bar{r} \sin \bar{\theta}} \tag{9.75b}$$

the complicate form of [9.69] now reduces to the simple form:

$$\bar{r}^2 \tan \varphi - \bar{r}\,\bar{r}' = 0 \tag{9.76}$$

from which: $\bar{r}(\bar{\theta}) = \bar{r}_o \exp[(\bar{\theta} - \bar{\theta}_o) \tan \varphi]$ \hfill [9.77]

is the general solution. Equation [9.77] obviously represents the simplest form of logspiral surface of angle φ having $\bar{r}_o$ as an arbitrary constant.

9.5.4. Normal stress distribution

Rewriting [9.70] with respect to the new coordinates, one obtains:

$$\sigma' + 2\sigma \tan \varphi + 2c - \gamma\bar{r} \cos \bar{\theta} - \frac{\gamma}{\lambda_3} = 0 \tag{9.78}$$

Equation [9.78] is a linear, first-order differential equation from which there exists an exact solution of the form:

$$\sigma(\bar{\theta}) = \frac{\gamma}{2\lambda_3 \tan\varphi} - \frac{c}{\tan\varphi} + \frac{\gamma\bar{r}_o \exp[(\bar{\theta} - \bar{\theta}_o)\tan\varphi](3\tan\varphi\cos\bar{\theta} + \sin\bar{\theta})}{1 + 9\tan^2\varphi}$$

$$+ D\exp(-2\bar{\theta}\tan\varphi) \tag{9.79}$$

in which the Lagrange multiplier, λ_3 and the integration constant, $\dot{D}$ are as yet to be determined.

Since the moment equation [9.63c] is independent of $\sigma(\bar{\theta})$ for a logspiral slip surface of angle φ, the two remaining force equations [9.63a,b] may therefore be satisfied by the proper choices of λ_3 and D. Substitute $\bar{r}(\bar{\theta})$ of [9.77] and $\sigma(\bar{\theta})$ of [9.79] into [9.63a,b] and solve for λ_3 and D, the final form of the non-dimensionalized $\sigma(\bar{\theta})$ can be expressed by:

$$\frac{\sigma(\bar{\theta})}{\gamma H} = A_1 + \left[\frac{3\tan\varphi}{(H/\bar{r}_o)(1 + 9\tan^2\varphi)\exp(\bar{\theta}_o\tan\varphi)}\right]\left[\cos\bar{\theta}\exp(\bar{\theta}\tan\varphi)\right.$$

$$\left. + \frac{\sin\bar{\theta}\exp(\bar{\theta}\tan\varphi)}{3\tan\varphi}\right] + A_2\exp(-2\bar{\theta}\tan\varphi) \tag{9.80}$$

wherein A_1 and A_2 are two pure numbers. For a given slope angle β and internal friction angle φ, the values of A_1 and A_2 depend only on the angles θ_o and θ_h whose value determines the position of the center of the *toe spiral*. The constants A_1 and A_2 have the following forms:

$$A_1 = \frac{1}{\dfrac{u_3(\theta)|_{\theta_o}^{\theta_h}}{u_3(-\theta)|_{\theta_o}^{\theta_h}} + \dfrac{u_1(\theta)|_{\theta_o}^{\theta_h}}{u_1(-\theta)|_{\theta_o}^{\theta_h}}} \left\{ \frac{[-3(1+\tan^2\varphi)u_4(2\theta) - 3\tan\varphi\,u_1(2\theta) - u_3(2\theta)]|_{\theta_o}^{\theta_h}}{4(H/r_o)(1+9\tan^2\varphi)u_4(\theta_o)u_1(-\theta)|_{\theta_o}^{\theta_h}} \right.$$

$$+ \frac{(\cos\varphi + 17\tan\varphi)u_4(2\theta)|_{\theta_o}^{\theta_h} + u_1(2\theta)|_{\theta_o}^{\theta_h} - 3\tan\varphi\,u_3(2\theta)|_{\theta_o}^{\theta_h}}{4(H/r_o)(1+9\tan^2\varphi)u_4(\theta_o)u_3(-\theta)|_{\theta_o}^{\theta_h}}$$

$$- \frac{[\tan\varphi\,u_1(\theta) + u_3(\theta)]|_{\theta_o}^{\theta_h}}{f(1+\tan^2\varphi)u_1(-\theta)|_{\theta_o}^{\theta_h}} - \frac{[u_1(\theta) + \tan\varphi\,u_3(\theta)]|_{\theta_o}^{\theta_h}}{f(1+\tan^2\varphi)u_3(-\theta)|_{\theta_o}^{\theta_h}}$$

$$\left. - \frac{(L/r_o)\sin\theta_o\,u_4(\theta_o) - [\sin\theta_h\cos\theta_o + (L/r_o)\sin\theta_h - \sin\theta_o\cos\theta_h]u_4(\theta_h)}{2(H/r_o)u_3(-\theta)|_{\theta_o}^{\theta_h}} \right\}$$

$$\tag{9.81}$$

$$A_2 = \frac{[\tan\varphi\, u_1(\theta) - u_3(\theta)]\big|_{\theta_o}^{\theta h}}{f(1 + \tan^2\varphi)\, u_1(-\theta)\big|_{\theta_o}^{\theta h}} - \frac{u_1(\theta)\big|_{\theta_o}^{\theta h}}{u_1(-\theta)\big|_{\theta_o}^{\theta h}} (A_1)$$

$$- \frac{[3(1 + \tan^2\varphi)\, u_4(2\theta) + 3\tan\varphi\, u_1(2\theta) + u_3(2\theta)]\big|_{\theta_o}^{\theta h}}{4(H/r_o)(1 + 9\tan^2\varphi)\, u_4(\theta_o)\, u_1(-\theta)\big|_{\theta_o}^{\theta h}} \qquad [9.82]$$

in which the functions $u(\theta)$ are defined as:

$$u_1(\theta) = (\tan\varphi \cos\theta + \sin\theta)\exp(\theta \tan\varphi) \qquad [9.83a]$$

$$u_2(\theta) = (\tan\varphi \sin\theta + \cos\theta)\exp(\theta \tan\varphi) \qquad [9.83b]$$

$$u_3(\theta) = (\tan\varphi \sin\theta - \cos\theta)\exp(\theta \tan\varphi) \qquad [9.83c]$$

$$u_4(\theta) = \exp(\theta \tan\varphi) \qquad [9.83d]$$

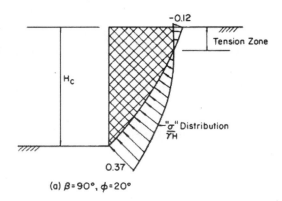

(a) $\beta = 90°$, $\phi = 20°$

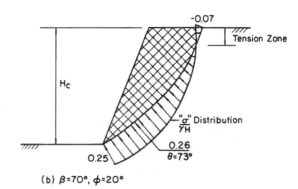

(b) $\beta = 70°$, $\phi = 20°$

Fig. 9.21. Normal stress distribution.

and the ratios H/r_o, L/r_o and the functions f_1, f_2, f_3 and f appearing in the above expressions correspond to the special case of the same form of equations given in section 9.2, when $\alpha = 0$.

As an illustration, Fig. 9.21 shows the normal stress distribution obtained from [9.80], for slopes with base angles of $\beta = 90°$ and $\beta = 70°$ and soil friction angle $\varphi = 20°$. The distributions correspond to the *critical height* of an inclined slope, determined by [9.21]. The values of θ_o and θ_h correspond to the critical angles of the toe spiral, determined by conditions [9.20]. The values of the normal stress along the critical toe spiral are shown by the arrows in the diagram in Fig. 9.21. They increase from a maximum compressive stress near the toe to a maximum tensile stress at the top of the slope. The portion of the distributions marked 'tension zone' in the diagram indicates the possible development of a tension crack in a slope.

9.6. SUMMARY AND CONCLUSION

Slope stability solutions based on the upper-bound technique of limit analysis are presented in this chapter in terms of stability factor for isotropic and homogeneous as well as anisotropic, non-homogeneous slopes. The general formulation of such problems is straight forward and simple and the numerical results for all cases agree well with the existing limit equilibrium solutions. The rotational failure mechanism (logarithmic spiral) utilized in all the upper-bound solutions is shown by variational calculus also to be the most appropriate surface of sliding in the framework of limit equilibrium methods. It can be concluded, therefore, that the upper-bound technique of limit analysis can be applied to predict the critical height of slopes. In many cases it may be much more convenient to use the upper-bound method than the existing limit equilibrium methods and it also places the matter of stability analysis on a much more logical ground.

Chapter 10

BEARING CAPACITY OF CONCRETE BLOCKS OR ROCK

10.1. INTRODUCTION

The problem considered in this Chapter is that of a rigid punch bearing on a flat surface of a large concrete block or rock resting on a smooth or rough rigid base, Fig. 10.1. It is assumed that the force P on the punch is normally and centrally loaded and is increased until penetration occurs as a result of plastic flow and eventually fracture in the material. There are many practical applications of this problem, and we list three important ones below:

(1) The design of foundations whereby the column load is transmitted to the soil mass through a footing.

(2) The design of the end bearing of a post-tensioned concrete beam.

(3) The solution forms the basis of the theory of the indirect split cylinder test (also known as Brazilian test) whereby the tensile strength of a brittle material may be determined by splitting tests on cylinders or cubes of such materials.

A complete *elastic* solution to the problem under conditions of plane strain has been obtained by Goodier (1932). The same problem under the conditions of axial symmetry has been obtained by Sundara and Yogananda (1966). The three-dimen-

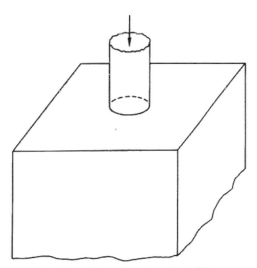

Fig. 10.1. Three-dimensional punch problem.

sional general punch problem (Fig. 10.1) is complicated and probably not possible
for an elastic–plastic and fracture analysis, since such an analysis requires the basic
knowledge of the stress–strain relations for concrete in the elastic as well as the
inelastic range and a fracture criterion. No such general relations and criterions have
been determined as yet for inelastic concrete (Mroz, 1972) though an encouraging
start has been made recently (Chen and Chen, 1973) but much more remains to be
done.

Although rock and concrete are materials of very limited deformability under
tension in our ordinary experience, there are theoretical and experimental indica-
tions that the load carrying capacity of blocks subjected to load over part of their
surface which produces a triaxial stress state in the material can be computed on
the basis of the upper- and lower-bound theorems of limit analysis. In the following
sections, a simplified material model that can be used in evaluation of plastic limit
loads in concrete blocks or rock is first postulated and a discussion is given, therefore, of the significance of the limit analysis in terms of the real behavior of
concrete or rock and its idealizations. Using this material model, the upper- and
lower-bound techniques of limit analysis are then employed in what follows to
obtain the bearing pressure which can be applied by strip loading or circular and
square punches.

The first problem to be considered under this approach is the incipient collapse
of *two* dimensional blocks compressed by a strip loading or rigid punch (Fig. 10.2).

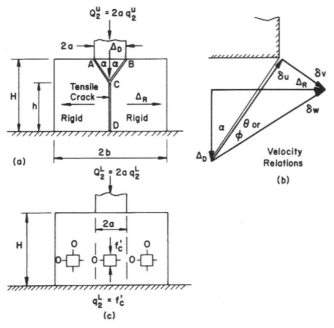

Fig. 10.2. Bearing capacity under strip loading.

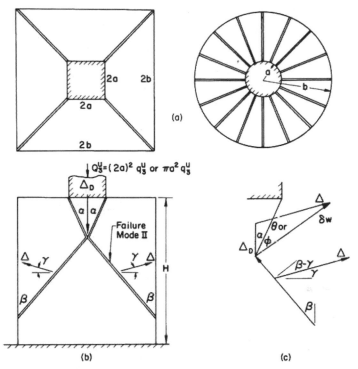

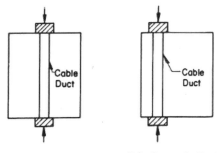

Fig. 10.3. Bearing capacity in three dimensions.

The second is to extend this result to obtain the solution of a circular cylinder or a square block compressed by a force applied *centrally* through a circular or square punch (Fig. 10.3) (Item 1 as listed above). Finally, the solutions to the problem of a circular cylindrical concrete block or a square prismatic concrete block with a longitudinal cable duct loaded by two circular or two square punches applied on the cable duct at both ends (Fig. 10.4) are presented (Item 2 as listed above). As a .

(a) Concentric Duct   (b) Eccentric Duct

Fig. 10.4. Simplified problem of the end bearing in a post-tensioned concrete beam.

result of these studies, a new indirect tension test (known as the *double-punch test*) for the tensile strength of rock-like materials is developed (Item 3 as listed above). The double-punch method will be presented in the following Chapter.

10.2. A SIMPLIFIED MATERIAL MODEL

10.2.1. Material behavior

Under ordinary experience, rock and concrete are brittle in tension and limitedly deformable in compression. A typical uniaxial stress−strain curve for concrete is shown in Fig. 10.5(a). It is seen that during tension the material behaves in a brittle manner and the maximum stress f_t' is about 8−12 times *less* than the maximum stress f_c' in compression. Similar proportions occur between corresponding strains ϵ_t and ϵ_c: whereas ϵ_c takes on values of about 0.2%, ϵ_t does not usually exceed 0.01%. The crushing of concrete is seen always followed by an unstable portion of the compressive stress−strain curve. Thus, it may come as a surprise to many of us

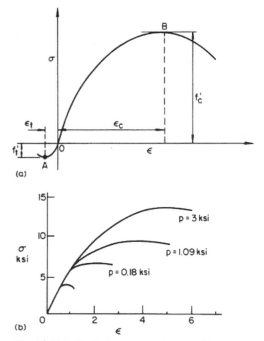

(a)

(b)

Fig. 10.5(a). Typical stress−strain curve for concrete under uniaxial tension and compression.

Fig. 10.5(b). Typical stress−strain curve for concrete under compression and lateral pressure. (1 ksi = 6.89 MN/m^2.)

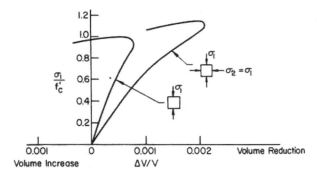

Fig. 10.6. Typical stress−strain curve for concrete volume change under biaxial compression.

to learn that under certain conditions rock and concrete too can be very ductile.

Ordinary experience simply is not adequate to predict the behavior of rocks or concrete blocks subjected to load over part of their surface. Von Karman (1911) and Böker (1915) demonstrated forcefully, if qualitatively, that rocks or concrete subjected to sufficient confining pressure will flow. Much more sophisticated work has been done since the early work of Von Karman and Böker, and the change from brittle to ductile behavior has been clearly demonstrated to be a function of confining pressure and temperature (see for example, Donath, 1970). The stress−strain curves of Fig. 10.5(b) illustrate the change in behavior of a concrete as it deforms under increasingly higher confining pressure. Hydrostatic pressure is seen largely to increase both maximum stress and maximum strain during compression, (Richart et al., 1928), and the unstable portion gradually vanishes for increasing pressures.

Large dilatational effects accompany uniaxial or multiaxial compressive strains in advanced stages of deformation; and Fig. 10.6 illustrates this effect for the case of biaxial compression of rectangular disks (Kupfer et al., 1969). The effect of dilatancy typical for numerous rocks and concrete can be attributed to progressive growth of internal cracks and creation of new cracks within the aggregate. Similar effects for triaxial compression of rock are reported by Swanson and Brown (1972).

An up-to-date book containing both a survey of the field and a rather extensive list of references is given in the Proceedings of an International Conference on the Structure of Concrete and Its Behavior Under Load edited by Brooks and Newman (1965).

10.2.2. Idealizations

If the tensile strength of concrete or rock is assumed to be zero for dimensions of engineering interest, and not just a small fraction of the compressive strength, then the limit theorems of perfect plasticity will hold rigorously for this idealization. They can be applied easily when a tensile crack is introduced in the failure mechanism. The early application of the limit theorems approach to the analysis and design of Voussoir arches is a good example. Kooharian (1952) made the extreme but not unreasonable assumption that concrete is unable to take tension and behaves as a *rigid*, infinitely strong material in compression. This idealization was selected because it was simple and provided a first approximation to the behavior of a real Voussoir arch. Similarly, the complete ignoring of the ability to carry tensile stress was found to predict the critical height of a free standing vertical bank of soil (example 3.4, section 3.3, p. 71) and to give a good first approximation to the load carrying capacity of a prestressed concrete beam (Drucker, 1961b).

If concrete and rock are assumed to have zero tensile strength and a finite compressive strength, they may almost be considered real materials. The assumption of infinite ductility in tension at zero strength is rigorous and conservative for the application of limit analysis but the assumption of infinite ductility in compression at selected compressive strength is more questionable. Brittleness in compression and a falling stress—strain curve after a maximum strength is usually observed, as indicated in Fig. 10.7. This is not compatible with the limit theorems. If the compressive strain of concrete is small and is not repeated often, then the deformability of concrete in compression prior to an appreciable fall-off of stress may be sufficient to permit the consideration of limit theorems with the concrete idealized as *perfectly plastic* at a yield stress in compression that approximates the ultimate strength f_c'.

However, with the safe assumption that concrete or rock is unable to resist any tension, the computed bearing pressure of concrete under a loaded area is shown in section 10.3 to be not greater than the unconfined compressive strength of the prism of the material directly below the load. It is a known fact that, on the contrary, the bearing pressure is raised substantially up to some limit by an increase of the ratio of the unloaded to the loaded area. Therefore, the tensile strength of rock or concrete must be taken into account in order to achieve correlation be-

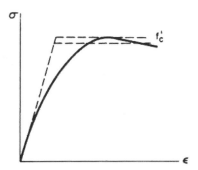

Fig. 10.7. Idealized stress—strain curve for concrete.

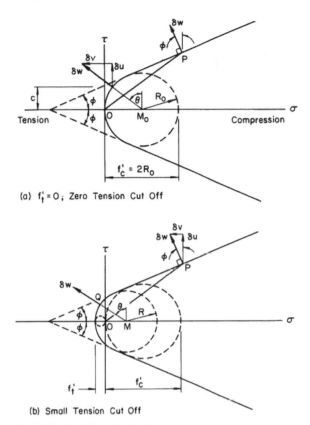

(a) $f'_t = 0$; Zero Tension Cut Off

(b) Small Tension Cut Off

Fig. 10.8. Modified Coulomb criterion.

tween prediction and this experimental fact. Fortunately, in bearing capacity problems, there appears to be fair experimental justification of the assumption that sufficient *local* tensile deformability does exist to permit the application of limit theorems to concrete idealized as perfectly plastic with a Coulomb failure surface in compression and a small but non-zero tension cut-off, Fig. 10.8(b). The permissibility of using the Coulomb failure surface as a yield surface is more questionable. An upper bound obtained with this assumption is an upper bound for a frictional material, but lower bounds are of less definite meaning (Theorem X, section 2.6, p. 44).

10.2.3. Some physical facts

Tensile strains of the order of four times those obtained in a simple tension have been reported in a concrete cylinder loaded by a concentric punch (Campbell, 1957). Experiment shows that the average strains near ultimate failure are greater in the splitting test than in the flexure and direct tension tests. Kaplan (1963) has observed that on the average, the tensile strains are about 30% greater than in the flexure tests and twice as large as the average strain in simple tension tests. It should be noted here that the average strain referred to is the average strain within the length of the strain gage. As concrete is a heterogeneous material, the actual local stress and local strain are much greater than those measured because of high stress concentration around the aggregate. Dantu (1958) has shown experimentally that aggregate inclusions cause internal strain concentration to be in the order of five times that of the average strain. These experimental results indicate that the local tensile strain reached in concrete prior to complete rupture may be in the order of ten times greater than that obtained in a simple tension test.

Even the more moderate point of view as reported by Winter (1960), by Robinson (1967), and many others which indicates completely brittle, but initially stable fracturing of concrete in tension does not rule out the application of the assumption of perfectly plastic failure in tension. It is only necessary for the tensile strength to be maintained for the required range of tensile strain in order for the limit theorems to apply.

It is worth mentioning that moderate cyclic moisture changes in concrete must induce a marked plastic or stable inelastic deformation of the mortar. A tensile strain of the order 0.0006 inch/inch or more in the mortar has been reported (Bloem and Stanton, 1952). Furthermore, it is important to note that a substantial increase in the tensile strength and thus tensile strain of concrete may be achieved with short lengths of wire in random orientation, but nearly uniform spacing throughout the concrete (Romualdi and Mandel, 1964). This will provide even better correlation between the theoretical predictions of limit analysis and test results than was obtained by a rigid punch on a plain concrete block. These test results on strength and deformation of concrete blocks will be discussed further in the latter part of this Chapter.

10.2.4. Scope

In the following work, deformability in tension and in compression of the material is supposed sufficient to permit the application of limit analysis to the bearing capacity of concrete or rock. The analysis assumes a modified Coulomb yield criterion with a small tension cut-off, Fig. 10.8(b). Theoretical estimations are found to be in good agreement with test results.

In section 10.3, bearing capacity is computed with the safe assumption that the concrete or rock is unable to take any tension. The lower and upper bounds obtained with the limit theorems indicate that the bearing capacity then is equal to the unconfined compressive strength of the material. In section 10.4, a modified criterion with a non-zero but small tension cut-off is postulated, the expressions for energy dissipation are written in a suitable form for calculations; and the evidence supporting such a postulate is discussed.

In sections 10.5 and 10.6, upper and lower bounds are obtained for the two-dimensional case of a strip loading and checked against published test results. In section 10.7, the upper-bound technique of limit analysis is employed to obtain the limiting loads which can be applied by circular and square punches. The upper-bound results are found to be in good agreement with published test results. The lower-bound theorem of limit analysis is used in section 10.8 to demonstrate that the upper-bound results are adequate in accuracy for many engineering purposes. Section 10.9 treats the effect of friction between the block and the base on the bearing capacity of the block. Finally, the theoretical and experimental analyses of the generalized bearing capacity problem, as shown in Fig. 10.4 are presented.

The theoretical part of the analysis includes upper-bound calculations of the bearing pressure for short as well as long concrete blocks. The experimental part of analysis includes: (1) the elastic–plastic and the incipient plastic flow strain field during indentation of a circular punch on a circular concrete block; (2) the bearing capacity of the concrete blocks with various punch eccentricity ratios. Since calculations of the upper-bound solutions are sometimes laborious, simple approximate evaluations of the general bearing capacity problem for practical purposes are suggested.

10.3. A MODIFIED COULOMB STRESS CRITERION WITH ZERO TENSION CUT-OFF (FIG. 10.8a)

10.3.1. Dissipation of energy

In this section, the material is assumed unable to take any tension and to obey a Coulomb yield criterion (straight line Mohr envelope of failure) in compression. This criterion of yielding is represented in Fig. 10.8(a) where the requirement of zero tension is met by the circle termination as shown (Chen and Drucker, 1969). The Coulomb envelope intersects the horizontal axis at an angle φ and the vertical axis at a distance c from the origin. Only rigid body discontinuous velocity solutions will be considered in this discussion. The rate of dissipation of energy per unit area of discontinuity surface has the form: ·

$$D_A = \tau \delta u - \sigma \delta v = \delta u (\tau - \sigma \tan \theta) \qquad [10.1]$$

in which the jump in tangential velocity across the discontinuity is denoted by δu, the separation velocity by $\delta v = \delta u \tan \theta$, and the relative velocity vector δw is at an angle of $\theta \geqslant \varphi$ to the surface. It can be seen that [10.1] is a simple modification of [3.18] by substituting θ for φ in [3.18]. Since the modified Coulomb yield criterion, Fig. 10.8(a), must be satisfied at the discontinuous surface, $\tau = R_0 \cos \theta$ and $\sigma = R_0 - R_0 \sin \theta$, it follows from [10.1] that:

$$D_A = c\delta u \, \frac{\tan (45 + \frac{1}{2}\varphi)}{\tan (45 + \frac{1}{2}\theta)} \qquad [10.2]$$

10.3.2. Two-dimensional punch (Fig. 10.2)

Figure 10.2(a) shows a two-dimensional velocity pattern consisting of a rigid wedge region ABC of angle 2α and a simple tension crack, CD, perpendicular to the smooth (frictionless) base. The wedge moves downward as a rigid body and displaces the surrounding material sideways. The relative velocity vector δw at each point along the lines of discontinuity, AC, BC is inclined at an angle φ to these lines. The compatible velocity relations are drawn in Fig. 10.2(b). No energy is dissipated in the formation of a simple tension crack; both normal and shear stress are zero on the plane of separation (see the origin in Fig. 10.8a). Equating external rate of work to the dissipation given by [10.2], gives an upper bound on the bearing capacity:

$$Q_2^u \, \delta w \cos (\varphi + \alpha) = \frac{c \, \delta w \cos \varphi \, 2a}{\sin \alpha} \qquad [10.3]$$

or:

$$q_2^u = \frac{Q_2^u}{2a} = \frac{c \cos \varphi}{\cos (\varphi + \alpha) \sin \alpha} \qquad [10.3b]$$

in which q_2^u denotes an upper bound on the average bearing pressure in this two-dimensional problem.

Minimization of the right-hand side of [10.3b] gives:

$$\alpha = \tfrac{1}{4}\pi - \tfrac{1}{2}\varphi \qquad [10.4]$$

and: $q_2 \leqslant q_2^u = 2c \tan (\tfrac{1}{4}\pi + \tfrac{1}{2}\varphi) = f_c' \qquad [10.5]$

in which f_c' denotes the unconfined compressive strength of a prism (or cylinder) of the material. The corresponding Mohr's circle for this stress state is the dashed circle in Fig. 10.8(a) ($f_c' = 2R_0$).

A lower bound is obtained by observing that the elementary discontinuous stress distribution shown in Fig. 10.2(c) is in equilibrium and satisfies the boundary condition on stress. Therefore, from the lower-bound theorem:

$$q_2 \geqslant q_2^l = 2c \tan (\tfrac{1}{4}\pi + \tfrac{1}{2}\varphi) = f_c' \qquad [10.6]$$

where q_2^l denotes a lower bound on the average bearing pressure.

Since the upper and lower bounds coincide in this case, the unconfined compressive strength f_c' is the correct value of the bearing pressure in this problem with this idealized material.

10.3.3. Three-dimensional punch

The same result can readily be extended to any loaded area in a three-dimensional problem when the material is unable to take any tension whatsoever. This extension follows directly from the absence of energy dissipation on the surface of a simple tension crack in the upper-bound calculations and the admissibility of the uniaxial compression field used to obtain the lower bound for the two-dimensional case. Fig. 10.3 shows the three-dimensional velocity patterns for circular and square punches. They consist of simple tension cracks and cone-shape or pyramid-shape rupture surfaces directly beneath the loaded areas. A straight-forward computation then shows that the bearing pressure is f_c' for any loaded area when the assumption of zero tensile strength is made.

10.4. A MODIFIED COULOMB CRITERION WITH A SMALL BUT NOT ZERO TENSION CUT-OFF (FIG. 10.8b)

Many experimental studies on rock and concrete have led to the conclusion that a Coulomb criterion is suitable for fracture. If taken too literally, this criterion predicts that the largest principal tensile stress at fracture increases with increasing mean stress. This contradicts the experimental evidence (Nadai, 1950) that for practical purposes the tensile strength is not affected by moderate normal stress components in directions perpendicular to the direction of tensile. The Coulomb criterion must be modified to take tensile strength and local deformability into consideration in bearing capacity problems. Figure 10.8(b) shows the modified criterion with a circle termination. Mohr's circles for unconfined compression and for simple tension also are drawn in the figure. The circles intersect the horizontal axis at distance f_c' and f_t' respectively from the origin.

The expression for the rate of dissipation of energy can be derived in a straightforward manner on the not-so-justifiable assumption of perfect plasticity. Within this concept, if the velocity coordinates are superimposed on the stress coordinates as in Fig. 10.8(b), the vector representing the tangential slip and normal separation velocity components δu, δv across the failure surface is normal to the yield envelope (this is the normality condition as described in section 2.3, Chapter 2). The rate of dissipation of energy per unit area D_A may be interpreted as the dot product of a stress vector (σ, τ) with a velocity vector $(\delta v, \delta u)$. The stress vector is shown on Fig. 10.8(b) as OP or OQ and the velocity vector is shown as δw. By considering the stress vector OQ to be the sum of vectors OM and MQ, the result takes the form:

$$D_A = R\delta w - (R - f_t') \, \delta v \tag{10.7}$$

$$\text{or: } D_A = \delta w \left(f_c' \frac{1 - \sin\theta}{2} + f_t' \frac{\sin\theta - \sin\varphi}{1 - \sin\varphi} \right) \tag{10.8}$$

$$\text{where: } \tan\theta = \frac{\delta v}{\delta u} \geqslant \tan\varphi \tag{10.9}$$

$$R = \tfrac{1}{2} f_c' - f_t' \frac{\sin\varphi}{1 - \sin\varphi} \tag{10.10}$$

Equation [10.8] can also be derived directly from [10.1] by noting that τ and σ must satisfy the relations (see Fig. 10.8b):

$$\tau = R\cos\theta, \, \sigma = [R\sin\theta - (R - f_t')] \tag{10.11}$$

For the particular cases of *simple tensile separation* and *simple sliding* for which $\theta = \tfrac{1}{2}\pi$ and $\theta = \varphi$ respectively, the expression [10.8] takes the simple forms:

$$D_A = f_t' \, \delta v \text{ for } \theta = \tfrac{1}{2}\pi \tag{10.12}$$

$$D_A = f_c' \frac{1 - \sin\varphi}{2} \delta w \text{ for } \theta = \varphi \tag{10.13}$$

10.5. BEARING CAPACITY UNDER A STRIP LOADING – UPPER BOUND

In the following work, the upper-bound technique of limit analysis is applied to obtain the load carrying capacity under a strip loading using a modified Coulomb criterion of yielding with a small tension cut-off.

The velocity boundary conditions for this problem are that the constant area of loading and the base of the block must remain plane because both the punch and the base are assumed to be rigid. When a velocity field which satisfies these conditions has been found, the lowest value of the average pressure over the punch, which is computed by equating the external rate of work to the dissipation in [10.8], is an upper bound for the collapse pressure.

10.5.1. Short blocks (Fig. 10.2a)

We return now to Fig. 10.2 with its strip loading Q_2 applied on a width $2a$, to a block of thickness H and width $2b$. A compatible velocity field with the parameters α and θ also are shown in the figure, from which the rate of internal dissipation of energy along the surfaces of discontinuity can be calculated easily. The upper bound now is found to be a function of the tensile strength f_t' and the compressive strength f_c':

$$q_2^u = \frac{Q_2^u}{2a} = \frac{f_c'\left(\frac{1-\sin\theta}{2}\right) + f_t'\left[\left(\frac{\sin\theta - \sin\varphi}{1-\sin\varphi}\right) + \sin(\alpha+\theta)\left(\frac{H}{a}\sin\alpha - \cos\alpha\right)\right]}{\sin\alpha \cos(\alpha+\theta)}$$

[10.14]

The upper bound has a minimum value when $\theta = \varphi$ and α satisfies the condition:

$$\cot\alpha = \tan\varphi + \sec\varphi\left[1 + \frac{\dfrac{H}{a}\cos\varphi}{\left(\dfrac{f_c'}{f_t'}\right)\left(\dfrac{1-\sin\varphi}{2}\right) - \sin\varphi}\right]^{1/2}$$

[10.15]

while [10.14] reduces to:

$$q_2^u = f_t'\left[\frac{H}{a}\tan(2\alpha+\varphi) - 1\right]$$

[10.16]

The technique in obtaining the minimum value of [10.14] has been described in detail in section 3.5, Chapter 3. The upper bounds, given by the minimum values of α in [10.15], are plotted in Fig. 10.9 for values of $f_c' = 5f_t'$, $10f_t'$ and $\varphi = 0°$, $20°$ (solid curves).

It is of interest to note that for the particular material for which $f_t' = 0$, [10.14] reduces to [10.3b] with a proper substitution of φ by θ in [10.3b] and the expression has a minimum value with respect to θ for a given α when:

$$\theta = \tfrac{1}{2}\pi - 2\alpha \geqslant \varphi$$

[10.17a]

or: $\alpha = \tfrac{1}{4}\pi - \tfrac{1}{2}\theta$

[10.17b]

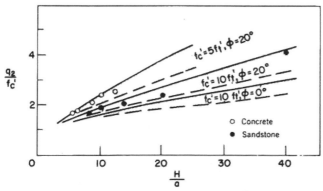

Fig. 10.9. Comparison of upper and lower bounds with test data for strip loading on concrete (Graf, 1934) and sandstone (Bach and Baumann, 1924).

Again, the unconfined compression strength f_c' is obtained for the average bearing pressure.

There is no slip between the concrete and the punch in the assumed velocity field. Consequently, the upper bound is applicable to either a smooth or a rough punch. The base, however, is assumed smooth. When the angle α is arbitrarily chosen equal to the value $\frac{1}{4}\pi - \frac{1}{2}\varphi$ and when $\theta = \varphi$, the bearing pressure q_2^u of [10.14] reduces to the simple form:

$$\frac{q_2^u}{f_c} = 1 + \tan\left(\tfrac{1}{4}\pi + \tfrac{1}{2}\varphi\right)\frac{H}{a}\frac{f_t'}{f_c} - \tan^2\left(\tfrac{1}{4}\pi + \tfrac{1}{2}\varphi\right)\frac{f_t'}{f_c} \qquad [10.18]$$

This relationship indicates that the bearing capacity under a strip loading is approximately proportional to the ratio of block thickness to loading width for a given material. This conclusion is identical with that of Meyerhof (1953).

10.5.2. Long blocks (Fig. 10.3b)

The rigid block sliding and splitting pattern of Fig. 10.2 is suitable for a short block (say roughly $H < 2b$). For the case of a long block, the velocity field may be modified to the type as shown in Fig. 10.3(b) (failure mode II). Instead of a simple tensile "crack" along the centerline of the block, a plane sliding with an angle β to the vertical is taken. The downward movement Δ_D of the strip loading is accomodated by the sideway movements Δ of the two surrounding rigid blocks, inclining at an angle γ to the horizontal. The relative velocity vector δw at each point along the wedge surface directly beneath the punch is inclined at an angle $\theta \geqslant \varphi$ to these surfaces.

The plane sliding involves shearing and separation so that expression [10.8] can be used to calculate the rate of dissipation of energy per unit area. The total dissipation of energy in the block can then be found by adding the rates of dissipation at various discontinuity surfaces. Equating the rate at which work is done by the force on the punch to the total rate of energy dissipation in the block, it is found that an upper bound on the average bearing capacity of the strip loading is:

$$q_2^u = \frac{Q_2^u}{2a} = \frac{\cos\gamma}{\sin\alpha\cos(\alpha + \gamma + \theta)}\left[f_c'\left(\frac{1 - \sin\theta}{2}\right) + f_t'\left(\frac{\sin\theta - \sin\varphi}{1 - \sin\varphi}\right)\right]$$

$$+ \left(\frac{b}{a}\right)\frac{\sin(\alpha + \theta)}{\sin\beta\cos(\alpha + \gamma + \theta)}\left[f_c'\left(\frac{1 - \cos(\beta - \gamma)}{2}\right) + f_t'\left(\frac{\cos(\beta - \gamma) - \sin\varphi}{1 - \sin\varphi}\right)\right]$$

$$[10.19]$$

It is of interest to note that [10.14] may be interpreted as the limiting case of [10.19] when $\beta \to 0$, $\gamma \to 0$, and $b/\sin\beta \to (H - a/\tan\alpha)$.

The best upper bound in [10.19] is found by minimizing the function q_2^u with respect to the variables α, β, γ for a given value θ. The expression $\partial q_2^u/\partial\alpha = 0$ gives:

$$\frac{\cos(2\alpha + \gamma + \theta)}{\sin^2\alpha} = \frac{\left(\dfrac{b}{a}\right)\left[\left(\dfrac{f_c'}{f_t'}\right)\left(\dfrac{1 - \cos(\beta - \gamma)}{2}\right) + \left(\dfrac{\cos(\beta - \gamma) - \sin\varphi}{1 - \sin\varphi}\right)\right]}{\sin\beta\left[\left(\dfrac{f_c'}{f_t'}\right)\left(\dfrac{1 - \sin\theta}{2}\right) + \left(\dfrac{\sin\theta - \sin\varphi}{1 - \sin\varphi}\right)\right]} \qquad [10.20]$$

The expression $\partial q_2^u/\partial\beta = 0$ gives:

$$\frac{\cos\gamma}{\cos\beta} = 1 + \frac{2}{\left(\dfrac{f_c'}{f_t'}\right) - \left(\dfrac{2}{1 - \sin\varphi}\right)} \qquad [10.21]$$

independently of the value of α. The expression $\partial q_2^u/\partial\gamma = 0$ gives:

$$\left[\frac{1}{2}\left(\frac{f_c'}{f_t'}\right) - \left(\frac{1}{1 - \sin\varphi}\right)\right]\sin(\alpha + \beta + \theta) - \left[\frac{1}{2}\left(\frac{f_c'}{f_t'}\right) - \left(\frac{\sin\varphi}{1 - \sin\varphi}\right)\right]\sin(\alpha + \gamma + \theta)$$

$$= \left(\frac{a}{b}\right)\frac{\sin\beta}{\sin\alpha}\left[\left(\frac{f_c'}{f_t'}\right)\left(\frac{1 - \sin\theta}{2}\right) + \left(\frac{\sin\theta - \sin\varphi}{1 - \sin\varphi}\right)\right] \qquad [10.22]$$

with the condition that $-\gamma \leqslant \dfrac{\pi}{2} - \beta - \varphi$.

The minimum values of the function q_2^u were obtained approximately by a process of trial and error for certain values of f_c'/f_t', b/a, φ, and θ. It was found that the best upper-bound value was obtained when taking $\theta = \varphi$ for various given values of f_c'/f_t', b/a and φ.

For the typical set of values: $f_c' = 10 f_t'$, $b/a = 10$ and $\theta = \varphi = 0°$, the upper bound has a minimum value near the point $\alpha = 21°$, $\beta = 37.9°$, $\gamma = 10°$, where it has the value $2.33\, f_c'$; while for $f_c' = 10 f_t'$, $b/a = 10$ and $\theta = \varphi = 20°$, the corresponding values are $\alpha = 15.6°$, $\beta = 38.9°$, $\gamma = -0.4°$ and $q_2^u = 3.53\, f_c'$. The values of the upper bound obtained were found to be higher than those obtained by the splitting mode of Fig. 10.2 when Fig. 10.9 was interpreted as q_2/f_c' vs. $2b/a$ (solid lines only).

There is no relative motion across the punch base or the block base so that no distinction need be made between rough and smooth punches as well as rough and smooth block bases.

10.5.3. Semi-infinite blocks

For a very large value of b/a, the bearing capacity for a strip loading tends to a limiting value, the bearing capacity of a strip loading on the surface of a semi-infinite solid as given by [3.81]. With the value $\varphi = 35°$, the limiting pressure is found to be 12 times the cylinder strength (q_2/f_c'), which checks remarkably well with the results of loading tests (11.9 times the cylinder strength; Pohle (1951)). Here the compressive cylinder strength was estimated to be 0.8 times the cube strength).

10.6. BEARING CAPACITY UNDER A STRIP LOADING – LOWER BOUND

10.6.1. Triangular stress field (Fig. 10.10)

To obtain a lower bound for the average bearing pressure, the discontinuous triangular stress pattern described in section 4.7, Chapter 4 is found to be especially useful. The triangular stress pattern is shown again here in Fig. 10.10 where the pressures Q, q and P used previously in Fig. 4.28(a) are now denoted by q, r and P in Fig. 10.10(a) respectively for the consistency of the notations used in this Chap-

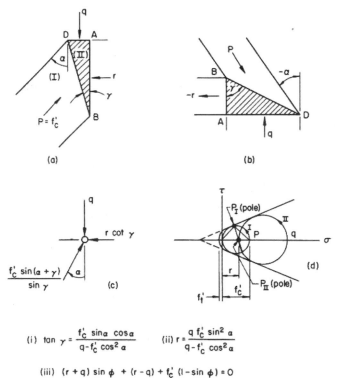

(i) $\tan \gamma = \dfrac{f_c' \sin\alpha \cos\alpha}{q - f_c' \cos^2 \alpha}$ (ii) $r = \dfrac{q f_c' \sin^2 \alpha}{q - f_c' \cos^2 \alpha}$

(iii) $(r + q) \sin \phi + (r - q) + f_c' (1 - \sin \phi) = 0$

Fig. 10.10. Triangular stress field used as a basic element to build up a lower-bound stress solution.

ter. The pressures q, r, P and the inclination angle α are taken to be positive as shown in Fig. 10.10(a). Fig. 10.10(b) shows the stress field when the inclination angle α of P and the stress r are taken to be negative (this field has been rotated 180° clockwise in comparison with Fig. 10.10a). The overall equilibrium of forces acting on the triangular element ABD in the vertical and horizontal directions is shown in Fig. 10.10(c) and the corresponding two equilibrium equations are labeled as (i), (ii) in the figure. The Mohr circles for the triangular region ABD and the region marked I with P taken to be f_c' are shown in Fig. 10.10(d). The modified Coulomb yield criterion is also shown and labeled as (iii).

Using this triangular stress pattern as the basic element, a statically admissible stress field for the bearing pressure can be built up. An extension into three dimensions of the two-dimensional stress field will be discussed in section 10.8.

10.6.2. Stress fields for the case $\varphi = 0$ (Fig. 10.12)

The geometrical configuration of the rectangular block, the lack of friction between the block and the base, and the small tensile strength of concrete or rock suggest that the rectangular block behaves as a truss to carry the load as shown in

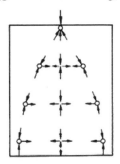

(a) Truss Action

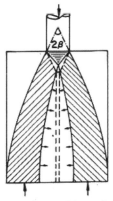

(b) Arch - Shape Stress Pattern

Fig. 10.11. A pin-connected truss is imagined to carry the load inside the body.

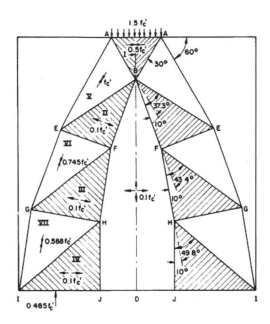

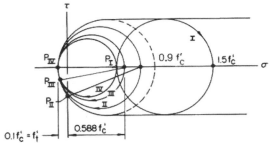

Fig. 10.12(a). Statically admissible stress field for the bearing pressure, $\varphi = 0°, f'_c = 10 f'_t$, $H/2a = 5.1., b/a = 4.9$.

Fig. 10.12(b). System of Mohr's circles.

Fig. 10.11(a). This concept of the appropriate choice of a pin-connected truss to support loads for obtaining lower bounds on plastic limit loads has been discussed in details in Chapter 4. In the following work a combined numerical treatment and graphical construction are presented. As a first step, a truncated arch angle 2β is chosen to start the construction (Fig. 10.11b).

Fig. 10.12(a) shows the stress field when the angle of the truncated arch is taken to be $60°$ and φ is taken to be zero. The Mohr's circles for the regions marked I, II, III, IV in the right half leg of Fig. 10.12(a) are shown in Fig. 10.12(b). The points labeled P_I, P_{II}, P_{III}, and P_{IV} are the poles of the corresponding Mohr circles. The stresses acting in the various regions are denoted by arrows and shown in Fig. 10.12(a). The field is built up by the basic triangular field in the following way. A uniaxial compression of amount f_c' in the region ABE produces a vertical compression $1.5\,f_c'$ and a horizontal compression $0.5\,f_c'$ in the region AAB, AB being the line of stress discontinuity (this region is stressed to the point of yielding). The uniaxial compression f_c' is supported and carried down by the five triangles marked II, III, IV, VI, and VII. The three shaded triangles II, III, IV are under a biaxial state of tension–compression with a tension of amount $f_t' = 0.1\,f_c'$ being taken. The angle α (Fig. 10.10b) is taken to be -10 degrees for these three regions. The regions marked VI, VII are under uniaxial compression. The region $JHFBFHJ$ directly below the line of indentation AA is in a state of hydrostatic tension of amount $f_t' = 0.1\,f_c'$. The graphical construction, starting from the top triangle AAB down to the bottom triangle HIJ, gives the corresponding rectangular block dimensions, that is, $H/2a = 5.10$, $b/a = 4.90$.

10.6.3. Stress fields for the case $\varphi > 0$ (Fig. 10.13)

More generally, when the angle of the arch is 2β and φ is not equal to zero, the pressure on AA has the value $c \cot \varphi\ [(\exp 2\beta \tan \varphi)\ \tan^2(\frac{1}{2}\pi + \frac{1}{2}\varphi) - 1]$ as given by [4.63]. Fig. 10.13(a) shows a statically admissible stress field with $\beta = 30^\circ$ and $\varphi = 20^\circ$. Lines AB and AM are two lines of stress discontinuity to approximate the continuous changing of stresses in the Prandtl field (logarithmic spiral zone) as already described in section 4.8, Chapter 4. The triangular region BMM vertically below AA is subjected to hydrostatic pressure $0.245\,f_c'$ where the line BM is parallel to the major principal direction of the stress in the region ABM. The hydrostatic pressure $0.245\,f_c'$ is then carried down to the base through the rectangular region $MNNM$ in a state of biaxial tension–compression which does not violate the yield condition. The rest of the constructions are obvious and similar to that of Fig. 10.12(a). The Mohr circles for regions I_1, I_2 and V in the right half leg of Fig. 10.13(a) are shown in Fig. 10.13(b). Line $P_{I_2}-1$ is parallel to line AM in Fig. 10.13(a) and $P_{I_2}-P_{I_1}-2$ is parallel to AB. The stress point 1 gives the stress acting on the line of discontinuity AM of regions I_2 and V, while the stress point 2 gives the stress acting on line AB of regions I_1 and I_2.

10.6.4. Numerical results

The value of q_2^l/f_c' is plotted against H/a in Fig. 10.9 for various arch angle 2β, and jointed by a smooth curve (denoted by dashed lines). Using $f_t' = 0.1\,f_c'$ and taking $\varphi = 20^\circ$ for sandstone, and $f_t' = 0.2\,f_c'$ for concrete, the above analysis is found to be in good agreement with the results of footing tests (Bach and Baumann, 1924; Graf, 1934). Experimental results are denoted by small circles in the figure.

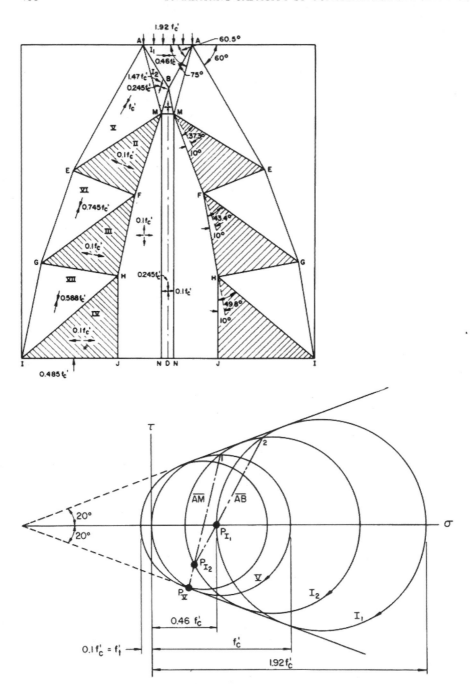

Fig. 10.13(a). Statically admissible stress field for the bearing pressure, $\varphi = 20°, f'_c = 10\,f_t$, $H/2a = 6.4$, $b/a = 6.0$.

Fig. 10.13(b). Mohr's circles for regions I_1, I_2, and V in (a).

It is of interest to note that in the process of constructing the various statically admissible stress fields to obtain the lower bound curve for various $H/2a$ ratios, we observe that the $H/2a$ ratio is always equal to or a little bit greater than the b/a ratio. It may be concluded, therefore, that for a block with $2b \geqslant H$, the bearing capacity depends mainly on the splitting depth H (Fig. 10.2a) as well as the tensile strength f_t', while the angle of internal friction φ can only effect the shearing strength adjacent to the punch locally. The shearing failure of the block along an inclined surface is not likely to govern for the case when the tensile strength of the material is low and $2b \geqslant H$.

10.7. THREE-DIMENSIONAL SQUARE AND CIRCULAR PUNCHES – UPPER BOUND

10.7.1. Short blocks

Considering first the splitting mode of failure, a simple discontinuous velocity field is shown diagrammatically in Fig. 10.2(a) (section view) and Fig. 10.3(a) (plan view). They consist of simple tension cracks and cone-shape or pyramid-shape rupture surfaces directly beneath the punches, and it can be seen that this velocity field is a direct extension into three dimensions of the two-dimensional velocity field in Fig. 10.2. The relative velocity between the cone-shape or pyramid-shape surfaces and the surrounding material is taken to be at an angle φ.

It is a simple matter to calculate the areas of the surfaces of discontinuity. The rate of dissipation of energy then is found by multiplying the area of each discontinuity surface by f_t' times the separation velocity δv across the surface for a *simple tensile crack* (expression [10.12]) or f_c' $(1 - \sin \varphi)/2$ times the relative velocity δw across the surface for the Coulomb *shearing type of discontinuity* (expression [10.13]). It is found by equating the rate at which work is done by the force on the punch to the rate of internal dissipation of energy, that the value of the upper bound on the average bearing pressure is:

$$q_3^u = \frac{1 - \sin \varphi}{\sin \alpha \cos (\alpha + \varphi)} \frac{f_c'}{2} + \tan (\alpha + \varphi) \left(\frac{2bH}{a^2} - \cot \alpha \right) f_t' \qquad [10.23]$$

This expression is an upper bound for a square punch on a square block or a circular punch on a circular cylinder. The upper bound has a minimum value with respect to α when:

$$\cot \alpha = \tan \varphi + \sec \varphi \left[1 + \frac{\dfrac{2bH}{a^2} \cos \varphi}{\left(\dfrac{f_c'}{f_t'} \right) \left(\dfrac{1 - \sin \varphi}{2} \right) - \sin \varphi} \right]^{1/2} \qquad [10.24]$$

and expression [10.23] can be reduced to:

$$q_3^u = f_t' \left[\frac{2bH}{a^2} \tan(2\alpha + \varphi) - 1 \right]$$ [10.25]

It can be seen that expression [10.24] or [10.25] is a simple modification of the two-dimensional expression [10.15] or [10.16] in which the ratio H/a is simply substituted by the ratio $2bH/a^2$ for the three-dimensional case.

10.7.2. Long blocks

The velocity field of Fig. 10.3 is an extension into three dimensions of the two-dimensional velocity field discussed in section 10.5 which can be used to give reasonable upper bounds for the indentation pressure for large values of $H/2b$. Taking $\theta = \varphi$, the values of the upper bound on the average bearing pressure are found to be:

$$q_3^u = \frac{\cos\gamma(1 - \sin\varphi)}{\sin\alpha\cos(\alpha + \gamma + \varphi)} \frac{f_c'}{2}$$

$$+ \left(\frac{b}{a}\right)^2 \frac{\sin(\alpha + \varphi)}{\sin\beta\cos(\alpha + \gamma + \varphi)} \left\{ f_c' \left[\frac{1 - \cos(\beta - \gamma)}{2} \right] + f_t' \left[\frac{\cos(\beta - \gamma) - \sin\varphi}{1 - \sin\varphi} \right] \right\}$$

$$+ \frac{\cos\gamma\sin(\alpha + \varphi)}{\cos(\alpha + \gamma + \varphi)} f_t' \left[\left(\frac{b}{a}\right)^2 \cot\beta + \left(2\frac{b}{a} - 1\right)\cot\alpha \right]$$ [10.26]

This is an upper-bound solution for a square punch on a square block or a circular punch on a circular cylinder. The last term in the right-hand side of expression [10.26] is the contribution due to the simple tension cracks which do not appear in the corresponding two-dimensional calculations ([10.19]). Minimization of the expression with respect to variables α, β, and γ follows. The equation $\partial q_3^u/\partial\alpha = 0$ gives:

$$\left[\frac{1 - \sin\varphi}{2} \frac{f_c'}{f_t'} \right] \frac{\cos(2\alpha + \gamma + \varphi)}{\sin^2\alpha} + \frac{\left(2\frac{b}{a} - 1\right)}{\sin^2\alpha} \left[\sin(\alpha + \varphi)\cos(\alpha + \gamma + \varphi) - \frac{\cos\gamma\sin2\alpha}{2} \right]$$

$$= \frac{\left(\frac{b}{a}\right)^2}{\sin\beta} \left\{ \left[\frac{1 - \cos(\beta - \gamma)}{2} \right] \frac{f_c'}{f_t'} + \left[\frac{\cos(\beta - \gamma) - \sin\varphi}{1 - \sin\varphi} \right] \right\} + \left(\frac{b}{a}\right)^2 \cos\gamma\cot\beta$$ [10.27]

The equation $\partial q_3^u/\partial\beta = 0$ gives:

$$\frac{\cos\gamma}{\cos\beta} = 1 + \frac{4}{\left(\frac{f_c'}{f_t'}\right) - \left(\frac{2}{1 - \sin\varphi}\right) - 2}$$ [10.28]

independently of the value of α, and the equation $\partial q_3^u/\partial\gamma = 0$ gives:

$$\left[\frac{1}{2}\left(\frac{f_c'}{f_t'}\right) - \left(\frac{1}{1-\sin\varphi}\right)\right]\sin(\alpha+\beta+\varphi) - \left[\frac{1}{2}\left(\frac{f_c'}{f_t'}\right) - \left(\frac{\sin\varphi}{1-\sin\varphi}\right)\right]\sin(\alpha+\gamma+\varphi)$$

$$= \left(\frac{a}{b}\right)^2\frac{\sin\beta}{\sin\alpha}\left[\left(\frac{f_c'}{f_t'}\right)\left(\frac{1-\sin\varphi}{2}\right) + \left(2\frac{b}{a}-1\right)\sin(\alpha+\varphi)\cos\alpha\right] + \cos\beta\sin(\alpha+\varphi)$$

[10.29]

with the condition that $-\gamma \leqslant \dfrac{\pi}{2} - \beta - \varphi$.

The minimum values of the average bearing pressure q_3^u were obtained approximately by a process of trial and error for certain values of f_c'/f_t', φ, and b/a. For the typical set of values: $f_c' = 10\,f_t'$ and $b/a = 2$. The q_3^u has the minimum value $2.30\,f_c'$ or $2.94\,f_c'$ when α, β, and γ are approximately $26.5°$, $55.2°$, and $-16.8°$, for $\varphi = 0°$ or $22.6°$, $57.4°$, and $-12.6°$ for $\varphi = 20°$, respectively.

Using $f_c' = 10\,f_t'$ and $\varphi = 20°$, the above analysis is found to give good upper-bound values to the loading tests on mortar and concrete cylinders loaded axially by circular steel punches over a part of the area (Campbell, 1957; Meyerhof, 1953). However, the test points in Fig. 10.14 indicate that when the ratio b/a is over 5 approximately ($H/2a$ over 5.5 approximately), the local deformability of concrete in tension is not sufficient to permit the application of limit analysis. Rupture of the mortar or concrete penetrates progressively downwards from the tip of the cone- or pyramid-shape formed directly under the punch.

10.7.3. Semi-infinite blocks

When the concrete block is reinforced to prevent splitting, the bearing capacity

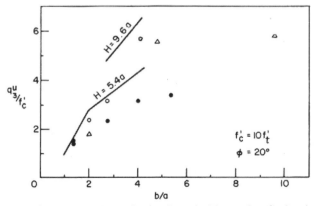

Fig. 10.14. Comparison of upper bound with test data for bearing capacity of circular cylinders and square blocks loaded by a concentric circular punch.

becomes equal to that of a circular punch on the surface of a semi-infinite solid. Some check of this limiting value for a circular punch can be obtained from the test results of loading a short high-tensile steel cylinder on the surface of relatively large concrete blocks reinforced to prevent splitting. The average bearing capacity was found to be 24 times the cylinder strength (q_3/f_c'), which corresponds to a theoretical angle of internal friction of about $35°$ according to the slip-line analysis (Table 7.1, Chapter 7). This checks remarkably well with the experimental estimation by Meyerhof (1953).

10.8. THREE-DIMENSIONAL SQUARE AND CIRCULAR PUNCHES – LOWER BOUNDS

In this section, lower bounds are obtained for the value of the average indentation pressure over an octagonal area of contact. It is thought that the results will be a good approximation to the circular punch problem. (The circle circumscribes the octagonal area of contact.) Furthermore, the stress fields used for the octagonal area can be adjusted in a simple manner in order to give lower bounds for the indentation pressure under a square or rectangular punch.

Physically, it is clear that the bearing capacity of the three-dimensional punch problem is always larger than that of two-dimensional strip loading on account of the hoop stresses and the greater area to distribute the stresses produced adjacent to the cone- or wedge-shape directly beneath the punch.

10.8.1. Stress fields for the case $\varphi = 0$ (Figs. 10.15 and 10.16)

A three-dimensional stress field

The extension of the two-dimensional stress field of Fig. 10.12(a) to a three-dimensional octagonal punch is shown in Fig. 10.15(a). Since the stress field is symmetrical about the four vertical planes passing through the opposite corners of the octagon, we need only to consider that part of the stress field which supports the pressure on one-eighth of the punch area. The stress field is also symmetrical about the vertical planes passing through the mid-points of opposite sides of the octahedral.

In Fig. 10.15(a), OA_1A_2 is one-eighth of the area of contact between the punch and concrete, A_1A_2 being one side of the octagon. The region AAB in Fig. 10.12(a) becomes the cone-shape volume A_1A_2AB. The triangular faces of the volume A_1A_2AB are all inclined at an angle of $60°$ to the horizontal. Only one of the eight "legs" is shown in Fig. 10.15(a). The bearing pressure $1.5 f_c'$ on the octagonal area is supported by the uniaxial compression of amount f_c' in the eight *legs*, with an all-around horizontal compression of amount $0.5 f_c'$ being produced in the volume A_1A_2AB. The *legs* with uniaxial compression f_c' are then carried down and spread

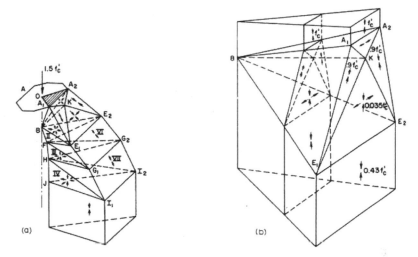

Fig. 10.15(a). Stress field for three-dimensional octahedral punch.

Fig. 10.15(b). Details for region $BA_1A_2E_2E_1$ in (a).

out through the truncated wedge stress field shown in Fig. 10.15(b). For clarity, the two uniaxial compression prisms, f_c' and $0.43 f_c'$ which are shown in Fig. 10.15(b), are not shown in the perspective in Fig. 10.15(a) since only the volume $BA_1A_2E_2E_1$ is used in the construction. Figure 10.16(a) shows the plane section through the side A_1A_2 of the octagonal area inclined at an angle $60°$ to the horizontal. Sections by planes parallel to the plane $A_1A_2E_2E_1$ are similar in shape to the section shown in Fig. 10.16(a), but are of varying size.

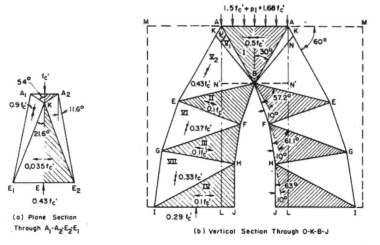

(a) Plane Section Through A_1-A_2-E_2-E_1

(b) Vertical Section Through O-K-B-J

Fig. 10.16. Plan and vertical sections in Fig. 10.15(a) for $H/2a = 2.75$, $b/a = 3.35$, $f_c' = 10 f_t'$, $\varphi = 0°$ and $q_3^l = 1.68 f_c'$.

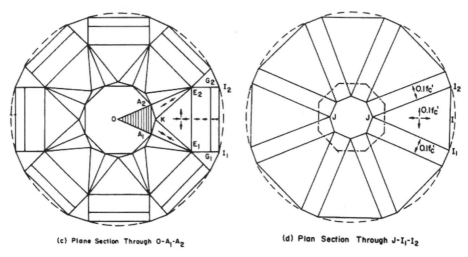

(c) Plane Section Through O-A₁-A₂   (d) Plan Section Through J-I₁-I₂

Fig. 10.16. Continued.

The uniaxial compression of amount $0.43 f_c'$ (Fig. 10.15b) is supported and carried down by the five quadrifrontal-shape volumes marked II, III, IV, VI, VII (Fig. 10.15a). The volumes, II, III, IV are under a triaxial state of tension–tension–compression with a tension of amount $f_t' = f_c'/10$ being taken. Fig. 10.10(b) shows a vertical section through the line of symmetry of one of these three quadrifrontal-shape volumes. The angle α is again taken to be $-10°$ for these three volumes. The two volumnes marked VI, VII are under uniaxial compression. The uniaxial compression $0.43 f_c'$ in Fig. 10.16(a) is obtained by trial and adjustment such that the plane BE_1E_2 (Fig. 10.15a) of stress discontinuity between the truncated wedge volume $BA_1A_2E_2E_1$ and the quadrifrontal volume BFE_1E_2 marked II is consistent. A uniaxial compression of amount $0.9 f_c'$ is taken for the truncated wedge stress field (Fig. 10.16a) for the purpose which will be obvious and justified in a later improvement of the lower bound. Figure 10.16(b) is then the vertical cross-section through the mid-points of the opposite sides of the octagonal. Figures 10.16(c) and 10.16(d) show the plan sections through AA and JJ, respectively. The volume whose vertical section is $BFHJJHFB$ (Fig. 10.16) directly below the octagonal area of contact is stress-free.

The graphical construction, starting from the top triangle AAB down to the bottom triangle HIJ, (Fig. 10.16b) then gives the corresponding concrete block dimensions, that is, $H/2a = 2.75$; $b/a = 3.35$.

An improved stress field

To improve the lower bound, a vertical compression of amount p_i is superimposed in the volume vertically below the octagonal area of contact, increasing the pressure on the area to $1.5 f_c' + p_i$, and a horizontal compression of amount p_i',

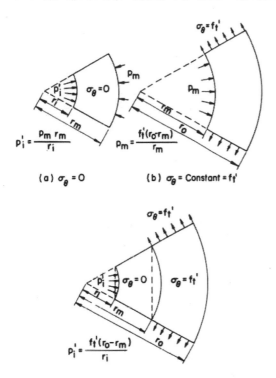

(a) $\sigma_\theta = 0$ (b) σ_θ = Constant = f_t'

(c) Combination of (a) and (b)

Fig. 10.17. Stress fields for the axial symmetrical plane stress problem.

varying from top surface AA down to the level $N'N'$ (Fig. 10.16b) is superimposed in the cone AAB circumscribing the volume A_1A_2AB (Fig. 10.15a). The horizontal pressure p_i' acting on the surface of the cone AAB is spread out and transmitted by the hoop-tension stress field (Fig. 10.16b, denoted by dashed lines) where the cone surface AAB and the octagonal surface $ALLA$ are the surfaces of stress discontinuity.

Figure 10.17(c) shows a portion of the plan view of the hoop-tension stress field by the plane passing through the cone AAB and parallel to the octagonal area of contact. The radii r_i and r_m are the horizontal distance measured from the center of the block to the edges BA_1 (or BA_2) and A_1E_1 or (A_2E_2) (Fig. 10.15a), respectively. The radius r_0 is equal to the block width b, that is $r_0 = b = 3.35a$ in this particular case.

The hoop-tension stress field is obtained as follows: The equation of equilibrium for an axial symmetrical plane stress problem is:

$$\frac{\partial \sigma_r}{\partial r} + \frac{\sigma_r - \sigma_\theta}{r} = 0 \qquad [10.30]$$

Two particular solutions which satisfy the prescribed boundary conditions of Fig. 10.17(a and b) for the above equation are obtained easily. In the first case (Fig. 10.17a) σ_θ is taken to be zero and the solution is $p_i' = (p_m r_m)/r_i$, where p_i' and p_m are the inner and outer pressure acting on the thick wall cylinder, respectively. In the second case σ_θ is taken to be constant of amount f_t'. The solution is $p_m = f_t'$ $(r_0 - r_m)/r_m$, where p_m now denotes the inner pressure of the thick wall tube as shown in Fig. 10.17(b). The hoop-tension stress field of Fig. 10.17(c) then follows from simple combination of the above two particular fields.

The pressure p_i' which is carried out through the hoop-tension stress field must be so chosen that the yield condition is nowhere violated in the resulting stress field. p_i' must not be greater than $0.21 f_c'$ in order that the yield condition be not violated at the corner points of the octagonal area (i.e., points A in Fig. 10.16b) since $p_i' = f_t' (r_0 - r_m)/r_i = 0.1 f_c' (3.35a - 1.1a)/1.1a = 0.21 f_c'$.

Since $p_i' \leqslant 0.21 f_c'$, it is found that N' (Fig. 10.16b) are the critical points which decide which values of p_i' and p_i. The value which gives a maximum value of p_i is $0.18 f_c'$ ($p_i' = 0.18 f_c' = p_i$ at N') so that the field is statically admissible since the modified Coulomb yield criterion is nowhere violated in the resulting field. Hence, by the Lower-Bound Theorem of limit analysis, $1.5 f_c' + 0.18 f_c' = 1.68 f_c'$ is a lower bound for collapse value of the average bearing pressure.

With $H/2a = 2.75$, $b/a = 3.35$, $\varphi = 0°$ and $f_c' = 10 f_t'$, a corresponding upper-bound value can be computed. It is found that the upper-bound value is $2.80 f_c'$. Although this differs considerably from the lower bound, $1.68 f_c'$, the value $2.24 f_c'$ cannot differ from the true limit load by more than about 25%.

10.8.2. Stress fields for the case $\varphi > 0$ (Figs. 10.18 and 10.19)

Fig. 10.18 shows a similar extension of Fig. 10.15 and 10.16 when the angle of friction φ is taken to be $20°$. It is found, after the graphical construction, that the stress field will be statically admissible if the average pressure acting on the octagon is taken to be $2.40 f_c'$, with $H/2a = 3.25$ and $b/a = 3.75$.

In order to increase the lower bound, the pressure in the region ABK (or $A_1 A_2 K$, Fig. 10.19) is increased from the value f_c' to $1.1 f_c'$ with an adjustment of the truncated wedge region $A_1 A_2 E_2 E_1$ to insure that the modified Coulomb yield criterion will nowhere be violated. Fig. 10.19 shows this improved field accompanied by the modified region $A_1 A_2 E_2 E_1$. The construction is self-explanatory in the figure.

With $H/2a = 3.30$, $b/a = 3.80$, $\varphi = 20°$, and $f_c' = 10 f_t'$, the corresponding upper-bound value is found to be $4.70 f_c'$ (splitting mode governs). The upper and lower bounds, $4.70 f_c'$ and $2.50 f_c'$, respectively, determine the pressure with a maximum possible error of about 30% if the value $3.60 f_c'$ is taken for the limit load.

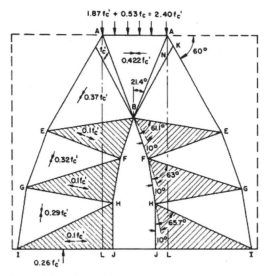

Fig. 10.18. Stress field for three-dimensional octahedral punch, $H/2a$ = 3.25, b/a = 3.75, f_c' = 10 f_t', φ = 20° and q_3' = 2.40 f_c'.

10.8.3. Remarks on the stress fields

The stress fields so constructed have always resulted in a uniformly distributed pressure over the concrete, and so are not distinguished from a uniformly loaded

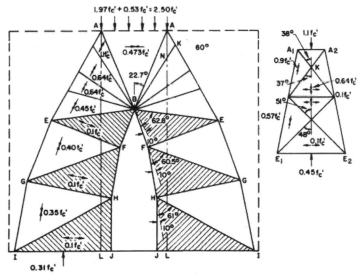

Fig. 10.19. Improved stress field of Fig. 10.18, $H/2a$ = 3.3, b/a = 3.8, f_c' = 10f_t', φ = 20° and q_3' = 2.50 f_c'.

area. No account is taken of the effects due to roughness of a punch and the rigidity of a punch which is assumed to be able to take any kind of pressure distribution over the concrete. Physically, it is clear that the bearing pressure is greater for a rigid punch than for a uniformly loaded area.

Shield (1955b, 1955c, 1960) has shown that the average pressure in the indentation by a perfectly rough circular flat-ended rigid punch of a semi-infinite solid of an elastic–perfectly plastic material, for which $\varphi = 0$, is about 6% (3.03 f_c' vs. 2.85 f_c') greater than that of a smooth punch, and about 18% (3.03 f_c' vs. 2.57 f_c') larger than the pressure over a uniformly loaded area. The pressure distribution over a punch is not uniform; and it varies from 2.50 f_c' at the edge of a punch to 3.60 f_c' at the center for a smooth punch and from 2.80 f_c' at the edge to 3.20 f_c' at the center for a rough punch (see Fig. 7.22).

More generally, when φ is not equal to zero, Cox et al. (1961) have shown numerically that the average indentation pressure for a smooth rigid punch is about 35% (7.00 f_c' vs. 5.20 f_c') greater than for a uniformly loaded area and the pressure distribution varies from 5.30 f_c' at the edge to 11.80 f_c' at the center. Here φ is taken to be $20°$ (see Fig. 7.24).

Hence, a somewhat more involved stress field in which account has to be taken of the effects due to the roughness and the non-uniform pressure distribution over the punch, could further push the lower-bound value up to the vicinity of the upper-bound solution; since the upper-bound solution is applicable to either a smooth or a rough punch. This would be expected because it has been shown to be possible for the case of semi-infinite solid as discussed above. It seems unlikely, for there are no new physical complications, that a similar increase cannot be found for the case of finite block. This suggests the conclusion that the upper-bound solution obtained in section 10.7 is much closer to the correct value than the lower-bound value.

10.9. FRICTION EFFECTS ON THE BEARING CAPACITY OF BLOCKS

In the analysis presented of the splitting failure mode, there is slip between the block and the base. Therefore, the rate of dissipation of energy due to friction on this discontinuity surface should be taken into account in the computation of the bearing capacity of the block. The rate of dissipation of energy due to friction may be computed by multiplying the discontinuity in velocity $\Delta_D \tan(\alpha + \varphi)$, across the surface by μ (friction coefficient between the block and the base) times the normal force Q_3^μ acting on this surface (Fig. 10.2b). The total rate of dissipation of energy is then obtained by adding the additional dissipation to the previous calculated dissipation when the block base is smooth. Equating external rate of work to the total dissipation then shows that the bearing capacity of the block is increased by a factor $1/[1 - \mu \tan(\varphi + \alpha)]$, over that for the smooth base.

TABLE 10.1

Values of upper bound and loading tests of Campbell (1957) and Meyerhof (1953)

$\dfrac{H}{2a}$	$\dfrac{b}{a}$	$[1 - \mu \tan (\alpha + \phi)]^{-1}$	Bearing capacity, q_3^u/f_c'		
			smooth	rough	test
2.20	4.10	1.15	3.79	4.36	4.38
1.20	4.80	1.17	2.88	3.37	2.97
2.40	4.80	1.13	3.87	4.36	4.09

Using $f_c' = 10f_t'$, $\varphi = 20°$ and $\mu = 0.2$, which are representative of practice, the values of the upper bound q_3^u obtained for a smooth as well as a rough block base and the average values of the loading tests of a circular steel punch on concrete blocks (Campbell, 1957; and Meyerhof, 1953) are shown in Table 10.1. The base friction is seen to increase the bearing capacity of a concrete block by about 15%, for blocks with relatively low ratios of $H/2a$.

10.10. CONCRETE BLOCKS WITH A CONCENTRIC CABLE DUCT (FIG. 10.4a)

In the following work, only the upper-bound theorem of limit analysis is employed to obtain the bearing pressure which can be applied by circular and square punches to the blocks shown in Fig. 10.4. The problem to be considered in this section is the incipient collapse of a circular cylinder or a square block with a co-axial cable duct compressed by two forces applied centrally through two circular or square punches (Fig. 10.4a). The problem to be considered in the following two sections is to extend this result to obtain the solution of a block with an eccentrically located cable duct and eccentrically loaded by two rigid punches (Fig. 10.4b). Finally, experimental results for the bearing capacity of such concrete blocks are discussed as is the limited applicability of so drastic an idealization of the real behavior of a material as brittle as concrete.

Since the lower-bound theorem of limit analysis is not considered in the present analysis, the solutions so obtained can at best give only upper bounds to the problem. However, the fact that the upper-bound solutions obtained in the previous sections are found to be very close to the correct values, suggests that the same is also true for the more general case of the bearing capacity problem described in the following sections.

10.10.1. Short circular and square blocks (Fig. 10.20)

Fig. 10.20 shows a failure mechanism consisting of simple tension cracks and

MECHANISM I

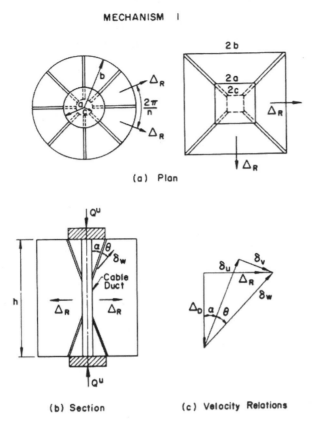

(a) Plan

(b) Section

(c) Velocity Relations

Fig. 10.20. Short circular and square blocks with a concentric cable duct.

truncated cone or truncated pryamid rupture surfaces directly beneath the punches. The two truncated cones or pryamids of angle 2α move toward each other as rigid bodies, and displace the surrounding material horizontally sideways. The relative velocity vector δw at every point on the truncated cone or pryamid surfaces is inclined at an angle θ to the surfaces. The compatible velocity relations are also shown in Fig. 10.20 from which the rate of internal dissipation of energy on the surfaces of discontinuity can be calculated easily. Since the sliding surfaces of the truncated cones or pryamid involve shearing and separation, the rate of dissipation of energy is found by multiplying the area of each truncated cone or pryamid surface by the dissipation function, D_A, as given by [10.8]. To this internal rate of dissipation has to be added the rate of dissipation obtained as the product of the area of the discontinuity surfaces for simple tensile cracks multiplied by f_t' times the separation velocity δv ([10.12]). It is found, by equating the rate at which work is done by the forces on the punches to the rate of the total internal dissipation of energy, that the value of the upper bound on the average bearing pressure over the net bearing area is:

$$q^u = \frac{\left(\dfrac{1-\sin\theta}{2}\right)f'_c + \left(\dfrac{\sin\theta-\sin\varphi}{1-\sin\varphi}\right)f'_t}{\sin\alpha\cos(\alpha+\theta)} + \frac{\tan(\alpha+\theta)\left[\dfrac{h}{a}\left(\dfrac{b}{a}-\dfrac{c}{a}\right) - \left(1-\dfrac{c}{a}\right)^2\cot\alpha\right]f'_t}{1-\left(\dfrac{c}{a}\right)^2}$$

[10.31]

The upper bound has a minimum value when $\theta=\varphi$ and α satisfies the condition:

$$\cot\alpha = \tan\varphi + \sec\varphi\left\{1 + \frac{\dfrac{h}{a}\left(\dfrac{b}{a}-\dfrac{c}{a}\right)\cos\varphi}{\left[1-\left(\dfrac{c}{a}\right)^2\right]\left(\dfrac{1-\sin\varphi}{2}\right)\left(\dfrac{f'_c}{f'_t}\right) - \left(1-\dfrac{c}{a}\right)^2\sin\varphi}\right\}^{1/2}$$

[10.32]

valid for:

$$\left(1-\frac{c}{a}\right)\cot\alpha \leqslant \frac{h}{2a}$$

[10.33]

and [10.31] is reduced to:

$$q^u = \frac{\left[\dfrac{h}{a}\left(\dfrac{b}{a}-\dfrac{c}{a}\right)\tan(2\alpha+\varphi) - \left(1-\dfrac{c}{a}\right)^2\right]f'_t}{1-\left(\dfrac{c}{a}\right)^2}$$

[10.34]

For the special case for which $c/a = 0$, [10.31] to [10.34] reduce to [10.23] to [10.25] obtained previously in section 10.7.

The value of q^u/f'_c is plotted against $h/2a$ in Fig. 10.21 for a punch for which $b/a = 4$ and $c/a = 0$. Experimental results are denoted by small circles in the figure. The theoretical curve computed for $\varphi = 20°$ and $f'_c = 12\,f'_t$ is found to be in good agreement with the results of tests. Details of the tests will be discussed later.

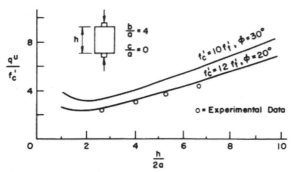

Fig. 10.21. Comparison of upper bounds with test data for two circular punches on a short circular block.

MECHANISM 2

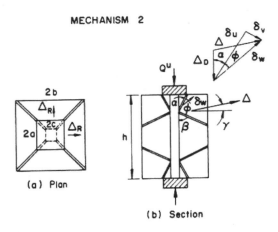

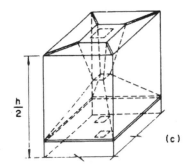

Fig. 10.22. Long circular and square blocks with a concentric cable duct.

10.10.2. Long circular and square blocks (Fig. 10.22)

Mechanism 1 for a short block may be modified to Mechanism 2 for a long block as shown in Fig. 10.22. Instead of simple tensile cracks along the total height of the block, eight planes of sliding all inclined at an angle of β to the vertical are assumed. The two punches move towards each other with a relative velocity, $2\Delta_D$, and are accomodated by the sideway movements, Δ, of the eight surrounding rigid blocks, inclined at an angle of γ to the horizontal. These planes of sliding involve shearing and separation so that [10.8] may be used to compute the rate of dissipation of energy per unit area. The total dissipation of energy in the block can then be found by adding to this rate the rates of dissipation at other discontinuity surfaces, i.e., those due to simple tensile cracks plus truncated cone or pyramid rupture surfaces. Equating the rate at which work is done by the force on the punches to the total rate of energy dissipation in the block, it is found that an upper bound on the average bearing capacity of the punch loading is:

$$q^u = \frac{\left(1 - \frac{c}{a}\right)^2}{1 - \left(\frac{c}{a}\right)^2} \frac{(1 - \sin\varphi)\cos\gamma}{\sin\alpha\cos(\alpha + \gamma + \varphi)} \frac{f'_c}{2}$$

$$+ \frac{\left(\frac{b}{a} - \frac{c}{a}\right)^2}{1 - \left(\frac{c}{a}\right)^2} \frac{\sin(\alpha + \varphi)}{\sin\beta\cos(\alpha + \gamma + \varphi)} \left[f'_c \frac{1 - \cos(\beta - \gamma)}{2} + f'_t \frac{\cos(\beta - \gamma) - \sin\varphi}{1 - \sin\varphi} \right]$$

$$+ \frac{\cos\gamma\sin(\alpha + \varphi) f'_t f_1(\alpha, \beta)}{\left[1 - \left(\frac{c}{a}\right)^2\right]\cos(\alpha + \gamma + \varphi)} \qquad\qquad [10.35]$$

where:

$$f_1(\alpha, \beta) = \left(\frac{b}{a} - \frac{c}{a}\right)^2 \cot\beta + \left(2\frac{b}{a} - \frac{c}{a} - 1\right)\left(1 - \frac{c}{a}\right)\cot\alpha \qquad [10.36]$$

This is an upper-bound solution for a square punch on a square block or a circular punch on a circular cylinder. For the special case of no cable duct, $c/a = 0$, [10.35] reduces to [10.26] obtained previously in section 10.7. The upper-bound solution has a minimum value when α, β, and γ satisfy the conditions:

$$\frac{\partial q^u}{\partial\alpha} = 0, \quad \frac{\partial q^u}{\partial\beta} = 0, \quad \text{and} \quad \frac{\partial q^u}{\partial\gamma} = 0 \qquad\qquad [10.37]$$

TABLE 10.2

Mechanism 2 bearing capacity of circular and square long blocks with a concentric cable duct, $\varphi = 20°$ (Fig. 10.22)

f'_c/f'_t	b/a	c/a	Angle (°)			q^u/f'_c	Minimum $h/(2a)$
			α	β	γ		
10	2	0	22.7	57.3	−12.7	2.9	3.6
		0.6	16.0	57.3	−12.7	1.5	4.8
	4	0	14.7	57.3	−12.7	7.4	6.4
		0.6	9.0	57.3	−12.7	5.9	8.8
14	2	0	22.8	49.6	−20.4	2.6	4.0
		0.6	15.4	49.6	−20.4	1.4	5.4
	4	0	13.9	49.6	−20.4	6.9	7.4
		0.6	8.1	49.6	−20.3	5.6	10.4

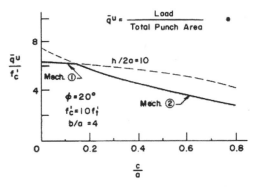

Fig. 10.23. Concrete blocks with a concentric cable duct.

Solving these equations and substituting the values of α, β, and γ thus obtained in [10.35], yields a least upper-bound solution for the bearing capacity problem. The results of these computations are presented in Table 10.2 for $\varphi = 20°$ and $f_c'/f_t' = 10$ and 14, respectively. The critical values of β and γ are seen to be insensitive to the dimension ratios b/a and c/a but depend mainly on the concrete strength ratio f_c'/f_t'.

The upper bounds, given by Mechanism 1 and Mechanism 2, are plotted in Fig. 10.23 for values of $f_c' = 10 f_t'$, $b/a = 4$, $h/2a = 10$, and $\varphi = 20°$. (Note: $\bar{q}$ = load/ total punch area). It can be seen that the concrete bearing strength is relatively insensitive to c/a ratios of the cable duct when these are very small, but the strength is considerably reduced when the c/a ratios are large. When $h/2a \geqslant 12$, Mechanism 2 almost always governs. The concrete blocks with $h/2a \geqslant 12$ can therefore be considered as long blocks.

10.11. CONCRETE BLOCKS WITH AN ECCENTRIC CABLE DUCT – SMALL ECCENTRICITY (FIG. 10.4b)

10.11.1. Short circular and square blocks (Fig. 10.24)

Mechanism 1 in Fig. 10.20 can be modified to provide an upper bound for the collapse pressure for the case of a block with an eccentrically located cable duct and eccentrically loaded by two rigid punches. Only the plan view of the modified mechanisms is shown in Fig. 10.24. For a circular block, it is convenient to approximate the circular punch by a regular polygon of $n = 8, 16, 32, ...$ Mechanism 3, shown in Fig. 10.24, is for $n = 8$.

Following the same procedure described for Mechanism 1, the bearing capacity

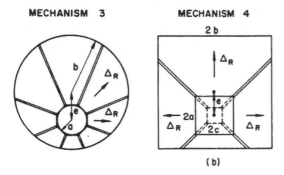

Fig. 10.24. Short concrete blocks with an eccentric cable duct — small eccentricity.

for a circular block is found to be:

$$q^u = \frac{n}{\pi} \frac{(1 - \sin \varphi) \tan \frac{\pi}{n}}{\sin \alpha \cos (\alpha + \varphi)} \left(\frac{f_c'}{2}\right) + \frac{4 \tan (\alpha + \varphi) \tan \frac{\pi}{n}}{\pi \left[1 - \left(\frac{c}{a}\right)^2\right]}$$

$$\left\{\frac{b}{a} \frac{h}{a} \cos \frac{\pi}{n} \sum_{i=1}^{\frac{1}{4}n} \left[1 - \left(\frac{e}{b}\right)^2 \sin^2 \frac{(2i - 1) \pi}{n}\right]^{1/2} - \frac{c}{a} \frac{h}{a} - \frac{n}{4} \left(1 - \frac{c}{a}\right)^2 \cot \alpha\right\} f_t' \qquad [10.38]$$

the value of q^u is minimum when:

$$\cot \alpha =$$

$$\tan \varphi + \sec \varphi \left\{1 + \frac{\frac{h}{a} \frac{b}{a} \cos \varphi \cos \frac{\pi}{n} \sum_{i=1}^{\frac{1}{4}n} \left[1 - \left(\frac{e}{b}\right)^2 \sin^2 \frac{(2i - 1) \pi}{n}\right]^{1/2} - \frac{c}{a} \frac{h}{a} \cos \varphi}{\frac{n}{8} \frac{f_c'}{f_t'} (1 - \sin \varphi) \left[1 - \left(\frac{c}{a}\right)^2\right] - \frac{n}{4} \left(1 - \frac{c}{a}\right)^2 \sin \varphi}\right\}^{1/2} \qquad [10.39]$$

and [10.38] reduces to:

$$q^u = \left(\frac{4}{\pi}\right) \frac{\tan \frac{\pi}{n} \tan (2\alpha + \varphi)}{1 - \left(\frac{c}{a}\right)^2} \left\{\frac{b}{a} \frac{h}{a} \cos \frac{\pi}{n} \sum_{i=1}^{\frac{1}{4}n} \left[1 - \left(\frac{e}{b}\right)^2 \sin^2 \frac{(2i - 1) \pi}{n}\right]^{1/2}\right.$$

$$\left. - \frac{c}{a} \frac{h}{a} - \frac{n}{4} \left(1 - \frac{c}{a}\right)^2 \cot (2\alpha + \varphi)\right\} f_t' \qquad [10.40]$$

valid for:

$$\cot \alpha \leqslant \frac{h}{2a} \qquad [10.41]$$

By comparing the values of the average bearing pressure computed from [10.40] for $n = 8$ and $n = 64$, respectively, it is found that the bearing pressure is not sensitive to the value of n, and thus $n = 8$ may be considered as a suitable value in [10.40] for all practical purposes.

Similarly, the bearing capacity for a square punch on a square block is found to be (Mechanism 4, Fig. 10.24):

$$q^u = \frac{1 - \sin\varphi}{\sin\alpha\cos(\alpha + \varphi)}\frac{f'_c}{2} + \frac{\frac{h}{a}\left[2\left(\frac{b}{a} - \frac{c}{a}\right) - \frac{e}{b}\frac{b}{a}\right]\tan(\alpha + \varphi)}{1 - \left(\frac{c}{a}\right)^2}\frac{f'_t}{2}$$

$$- \frac{\left(1 - \frac{c}{a}\right)^2}{1 - \left(\frac{c}{a}\right)^2}\cot\alpha\tan(\alpha + \varphi)f'_t \qquad\qquad [10.42]$$

The value of q^u is minimum when:

$$\cot\alpha = \tan\varphi + \sec\varphi\left\{1 + \frac{\frac{1}{2}\frac{h}{a}\left[2\left(\frac{b}{a} - \frac{c}{a}\right) - \frac{e}{b}\frac{b}{a}\right]\cos\varphi}{\left[1 - \left(\frac{c}{a}\right)^2\right]\frac{1 - \sin\varphi}{2}\frac{f'_c}{f'_t} - \left(1 - \frac{c}{a}\right)^2\sin\varphi}\right\}^{1/2} \qquad [10.43]$$

and [10.42] reduces to:

$$q^u = \frac{f'_t}{2}\frac{\frac{h}{a}\left[2\left(\frac{b}{a} - \frac{c}{a}\right) - \frac{e}{b}\frac{b}{a}\right]\tan(2\alpha + \varphi) - 2\left(1 - \frac{c}{a}\right)^2}{1 - \left(\frac{c}{a}\right)^2} \qquad [10.44]$$

valid for:

$$\left(1 - \frac{c}{a}\right)\cot\alpha \leqslant \frac{h}{2a} \qquad\qquad [10.45]$$

10.11.2. Long square blocks (Fig. 10.25)

A direct extension of the eccentrically loaded situation of Mechanism 2 (Fig. 10.22) for a square block is shown in Fig. 10.25 (vertical section only). Mechanism 2' is evident in Fig. 10.25. It is found that the equation for computing the bearing capacity pressure remains identical in its form as in the previous solution ([10.35]), but the function $f_1(\alpha,\beta)$ defined by [10.36] must be substituted by:

MECHANISM 2'

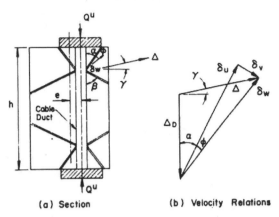

(a) Section   (b) Velocity Relations

Fig. 10.25. Long concrete blocks with an eccentric cable duct – small eccentricity.

$$f_2(\alpha, \beta) = \left[\frac{b}{a} \left(2 - \frac{e}{b} \right) - \frac{c}{a} - 1 \right] \left(1 - \frac{c}{a} \right) \cot \alpha$$

$$+ \frac{1}{2} \left\{ \left(\frac{b}{a} - \frac{c}{a} \right)^2 + \left[\frac{b}{a} \left(1 - \frac{e}{b} \right) - \frac{c}{a} \right]^2 \right\} \cot \beta \qquad [10.46]$$

10.12. CONCRETE BLOCKS WITH AN ECCENTRIC CABLE DUCT – LARGE ECCENTRIC-
ITY (FIG. 10.4b)

Mechanisms 5 through 8 are shown in Figs. 10.26 and 10.27 and need no detailed explanation. The procedure in obtaining the bearing capacity equations for
various mechanisms is identical to the previous cases, and, hence, only the final
results are recorded here for the sake of brevity.

10.12.1. Short circular blocks (Mechanism 5, Fig. 10.26a)

$$q^u = \frac{f'_c}{\pi} \frac{(1 - \sin \varphi) K}{\left[1 - \left(\frac{c}{a} \right)^2 \right] \sin \alpha \cos (\alpha + \varphi)}$$

$$+ \frac{f_t}{\pi} \frac{\left\{ \frac{b}{a} + \frac{b}{a} \left[2 - \left(\frac{e}{b} \right)^2 \right]^{1/2} - 3 \frac{c}{a} \right\} \frac{h}{a} - 2 \left[\left(1 - \frac{c}{a} \right)^2 + \frac{b}{a} \left(1 - \frac{e}{b} \right) - \frac{c}{a} \right] \cot \alpha}{\left[1 - \left(\frac{c}{a} \right)^2 \right] \cot (\alpha + \varphi)}$$

$$[10.47]$$

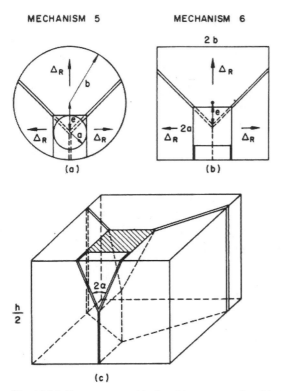

(c)

Fig. 10.26. Short concrete blocks with an eccentric cable duct — large eccentricity.

in which:

$$K = 1 - 2\left(\frac{c}{a}\right)^2 + \left[\left(\frac{b}{a}\right)^2 - 1\right]^{1/2} - \frac{e}{b}\frac{b}{a} + \frac{1}{2}\left(\frac{b}{a}\right)^2$$

$$\left\{\frac{\pi}{2} - \sin^{-1}\left[1 - \left(\frac{a}{b}\right)^2\right]^{1/2} - \left[1 - \left(\frac{a}{b}\right)^2\right]^{1/2}\cos\sin^{-1}\left[1 - \left(\frac{a}{b}\right)^2\right]^{1/2}\right\}$$

[10.48]

The value of q^u is minimum when:

$$\cot \alpha = \tan \varphi + \sec \varphi$$

$$\left\{1 + \frac{\dfrac{h}{a}\dfrac{b}{a} + \dfrac{h}{a}\dfrac{b}{a}\left[2 - \left(\dfrac{e}{b}\right)^2\right]^{1/2} - 3\dfrac{h}{a}\dfrac{c}{a}}{K(1 - \sin\varphi)\sec\varphi\left(\dfrac{f_c'}{f_t'}\right)^i - 2\left[\left(1 - \dfrac{c}{a}\right)^2 + \dfrac{b}{a}\left(1 - \dfrac{e}{b}\right) - \dfrac{c}{a}\right]\tan\varphi}\right\}^{1/2}$$

[10.49]

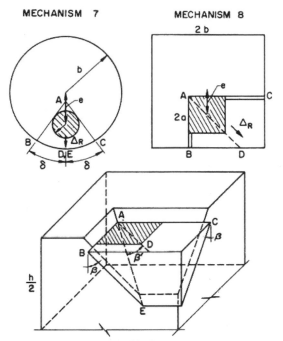

Fig. 10.27. Long concrete blocks with an eccentric cable duct – large eccentricity.

and [10.47] can be reduced to:

$$q^u = \frac{f_t'}{\pi} \frac{\left\{ \frac{b}{a} + \frac{b}{a} \left[2 - \left(\frac{e}{b} \right)^2 \right]^{1/2} - 3 \frac{c}{a} \right\} \frac{h}{a} \tan(2\alpha + \varphi) - \left[\left(1 - \frac{c}{a} \right)^2 + \frac{b}{a} \left(1 - \frac{e}{b} \right) - \frac{c}{a} \right]}{1 - \left(\frac{c}{a} \right)^2}$$

[10.50]

valid for $\cot \alpha \leqslant h/2a$ ([10.41]).

10.12.2. Short square blocks (Mechanism 6, Fig. 10.26b)

$$q^u = \frac{f_c'}{4} \frac{\left[\frac{b}{a} \left(1 - \frac{e}{b} \right) - 2 \left(\frac{c}{a} \right)^2 + 1 \right] (1 - \sin \varphi)}{\left[1 - \left(\frac{c}{a} \right)^2 \right] \sin \alpha \cos (\alpha + \varphi)}$$

$$+ \frac{f_t'}{2} \frac{\frac{3}{2} \frac{h}{a} \left(\frac{b}{a} - \frac{c}{a} \right) - \frac{1}{2} \frac{h}{a} \frac{b}{a} \frac{e}{b} - \left[\left(1 - \frac{c}{a} \right)^2 + \frac{b}{a} \left(1 - \frac{e}{b} \right) - \frac{c}{a} \right] \cot \alpha}{\left[1 - \left(\frac{c}{a} \right)^2 \right] \cot (\alpha + \varphi)}$$

[10.51]

The value of q^u is minimum when:

$$\cot \alpha = \tan \varphi + \sec \varphi$$

$$\left\{ 1 + \frac{\dfrac{h}{2a}\left[3\left(\dfrac{b}{a} - \dfrac{c}{a}\right) - \dfrac{e}{b}\dfrac{b}{a}\right]\cos\varphi}{\left[\dfrac{b}{a}\left(1 - \dfrac{e}{b}\right) - 2\left(\dfrac{c}{a}\right)^2 + 1\right]\dfrac{(1 - \sin\varphi)}{2}\left(\dfrac{f_c'}{f_t'}\right) - \left[\left(1 - \dfrac{c}{a}\right)^2 + \dfrac{b}{a}\left(1 - \dfrac{e}{b}\right) - \dfrac{c}{a}\right]\sin\varphi} \right\}^{1/2}$$

$$[10.52]$$

and [10.51] can be reduced to:

$$q^u = \frac{f_t'}{2}\frac{\left[3\left(\dfrac{b}{a} - \dfrac{c}{a}\right) - \dfrac{e}{b}\dfrac{b}{a}\right]\dfrac{h}{2a}\tan(2\alpha + \varphi) - \left[\left(1 - \dfrac{c}{a}\right)^2 + \dfrac{b}{a}\left(1 - \dfrac{e}{b}\right) - \dfrac{c}{a}\right]}{1 - \left(\dfrac{c}{a}\right)^2} \qquad [10.53]$$

valid for $\cot \alpha \leqslant h/2a$ ([10.41]).

10.12.3. Long circular blocks (Mechanism 7, Fig. 10.27)

The velocity vector, δw, of the volume $ABDCE$ is inclined at an angle φ to the two sliding surfaces ABE and ACE, respectively (or at an angle ψ to the diagonal line AE). The line AE makes an angle β' to the horizontal.

$$q^u = \frac{f_c'}{2\pi}\frac{(1 - \sin\varphi)\left[1 + \csc^2\delta \tan^2\beta'\right]^{1/2}}{\sin(\beta' - \psi)}\left\{\cot\delta + \left(\frac{b}{a}\right)^2\left[\delta + \sin^{-1}\left(\frac{a}{b} - \frac{e}{b}\sin\delta\right)\right]\right.$$

$$\left. + \left(1 - \frac{e}{b}\frac{b}{a}\sin\delta\right)\left[\left(\frac{b}{a}\right)^2 - \left(1 - \frac{e}{b}\frac{b}{a}\sin\delta\right)^2\right]^{1/2} - \frac{e}{b}\frac{b}{a}\left(2 - \frac{e}{b}\frac{b}{a}\sin\delta\right)\cos\delta\right\}$$

$$[10.54]$$

valid for $\cot \alpha \leqslant h/2a$ ([10.41]).

in which the angle ψ must satisfy the geometric condition:

$$\sin \varphi = [\tan\beta' - \tan(\beta' - \psi)]\cos(\beta' - \psi)[\sin\tan^{-1}(\sin\delta \cot\beta')] \qquad [10.55]$$

The upper-bound solution has a minimum value when β' and δ satisfy the conditions:

$$\partial q^u/\partial\beta' = 0 \quad \text{and} \quad \partial q^u/\partial\delta = 0 \qquad [10.56]$$

Solving these equations and substituting the values of β' and δ thus obtained into [10.54] yields a least upper-bound solution. Thus, for a punch for which $b/a = 4$ and $\varphi = 20°$, for example, the upper bound has the minimum value $2.86\, f'_c$ or $2.24 f'_c$ when β', δ, and ψ are approximately $60°$, $60°$, and $22.4°$ for $e/b = 2/3$ or $60°$, $55°$, and $23°$ for $e/b = 3/4$, respectively.

10.12.4. Long square blocks (Mechanism 8, Fig. 10.27)

$$q^u = \frac{f'_c}{8} \frac{\left(1 + \dfrac{b}{a}\right)\left[1 + \dfrac{b}{a}\left(1 - \dfrac{e}{b}\right)\right](1 - \sin\varphi)(2 + \tan^2\beta)}{(1 + \cos^2\beta - 2\sin^2\varphi)^{1/2}\tan\beta - 2\sin\varphi} \qquad [10.57]$$

in this case the angle ψ must satisfy the geometric condition:

$$\sin\psi = \frac{\sqrt{2}\sin\varphi}{\cos\beta\,(2 + \tan^2\beta)^{1/2}} \qquad [10.58]$$

The upper bound has a minimum value when it satisfies the condition $\partial q^u/\partial\beta = 0$. Thus, for example, for a punch for which $b/a = 4$ and $\varphi = 20°$, the

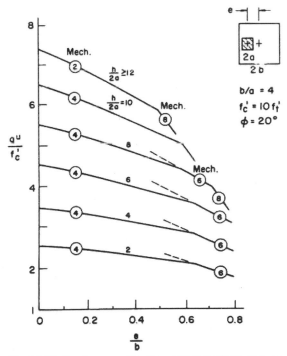

Fig. 10.28. Bearing capacity of an eccentrically loaded square block.

upper bound has the minimum value near the point $\beta = 60°$ and $\psi = 25.6°$. The value is $4.52 f_c'$ for $e/b = 2/3$ and $3.85 f_c'$ for $e/b = 3/4$.

Figure 10.28 shows the values of the q^u/f_c' ratio for square blocks with various eccentricity ratios e/b. The results of calculations made with various mechanisms are calculated for concrete with $f_c' = 10 f_t'$ and $\varphi = 20°$ and for a punch with $b/a = 4$ and $c/a = 0$.

Tresca's yield criterion for metals may be considered as a special concrete for which $\varphi = 0°$. Equations [10.54] and [10.57] then reduce to the upper-bound solutions [7.20] and [7.18] respectively, which were obtained previously in section 7.5.2, Chapter 7 for the plastic indentation of finite metal blocks by flat punches.

10.13. EXPERIMENTAL STUDY OF THE STRAIN FIELD

The experimentally determined strain field for the indentation of a circular

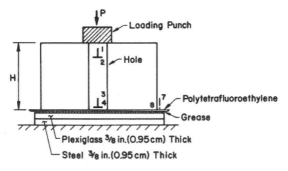

(a) Plastic Base Arrangement

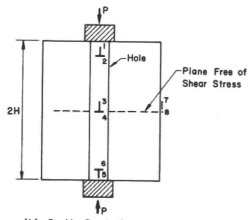

(b) Double Punch Arrangement

Fig. 10.29. Cylinder loaded by a double punch.

punch on a circular block having various block base conditions will be described herein to ascertain whether the strain field developed in the concrete block at the instant of collapse is sufficient to allow complete plasticity.

Mortar and concrete cylinders at different sizes (6 inches diameter by 12 inches height, and 6 inches diameter by 6 inches height) were used in a research program at Lehigh University for the bearing capacity studies of blocks. For some 6 X 6 inches cylinders, a thin sheet of teflon was inserted between the cylinder and the supporting base so as to minimize friction effects (Fig. 10.29a). To obtain a smooth condition at the base, the 6 X 12 inches cylinders were tested by a double punch (short steel cylinders, 2 inches in diameter) in which the load was applied to a cylinder at two opposite faces (Fig. 10.29b). This condition ensured zero shear stress over the mid-height section of the cylinder.

Each specimen contained a center hole 5/8 inch in diameter along the axis of the cylinder. Strain gauges were attached to a typical cylinder at the positions indicated in Fig. 10.29. The gauges inside the center hole were mounted to the specimen by applying internal pressure to a rubber hose.

10.13.1. Results for plain concrete blocks (Fig. 10.30)

The vertical and horizontal strain distributions along the axis of the cylinder are shown in Fig. 10.30 for various base friction conditions. These curves differ in the numerical values, but show the same general distribution of strain with depth in the cylinder. A marked increase in vertical compressive strain of the order of at least twice that obtained in a simple compression test $(300 \cdot 10^{-5}$ inch/inch) near the region directly below the punch. The horizontal tensile strains are seen to be distributed rather uniformly along the axis of a cylinder, but reverse to a compressive strain near the bottom, for blocks with teflon and steel bases.

The important point to be noted from the strain distribution curves is that the average tensile strains along the axis of the cylinder near ultimate failure in the punch tests are greater than in the flexure and direct tension tests. On the average, the tensile strains for smooth punch tests (teflon base or double punch) are about four times those in the flexure tests, and five times as large as the average strain in simple tension tests. Although the average tensile strains in rough punch tests (steel base) are much smaller than those in smooth cases, the average values reached in the specimen are still as large as the average strain in flexure tests.

The strain readings for the gauges on the surface of a block (gauge 8 in Fig. 10.29a) indicate that an average horizontal tensile strain of up to 80% or more of the average tensile strain in flexure tests is reached in the cylinder just prior to collapse. This suggests that the tensile plastification tends to relieve the more highly stressed parts in the center portion of the specimen, and to throw the tensile stress onto those parts near the surface of the specimen where the stress is lower.

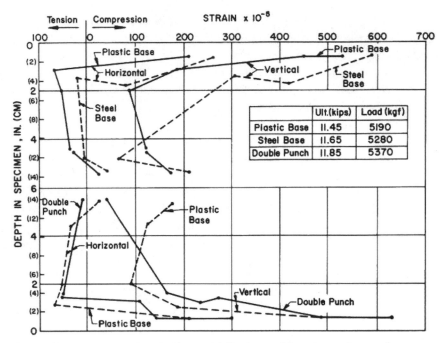

Fig. 10.30. Horizontal and vertical strain distributions at failure in punch tests. Specimen tested at about 34-days age. (1 kip = 4.45 kN.)

For the case of a double punch, the vertical strain distribution on the middle horizontal section of the specimen is found to be almost uniform (gauge 7 in Fig. 10.29b). However, the vertical strain distribution over the teflon or steel base is not uniform, and it varies from about zero at the edge of a cylinder to a compressive strain of order $200 \cdot 10^{-5}$ inch/inch or more at the center. This suggests that the load carrying capacity for a double punch may be higher than those of similar specimens with steel or teflon supporting bases. Indeed, this is found to be the case, and is in agreement with most of the results observed (Hyland and Chen, 1970).

10.13.2. Results for random wire reinforced mortar blocks (Fig. 10.31)

In order to increase the tensile strength and ductility of concrete, 2% in volume of 1-inch long wire was added into and mixed randomly with the concrete. Fig. 10.31 represents the measured strain distributions on specimens of random wire reinforced mortar under double-punch loading condition (Fig. 10.29b). Details of the results are reported elsewhere (Chen and Carson, 1974). The plot is of the tensile (horizontal) and compressive (vertical) strains in the principle directions versus the distance from the nearest punch when the loads are increased continuously until failure. Fig. 10.31 shows that for the random wire reinforced material

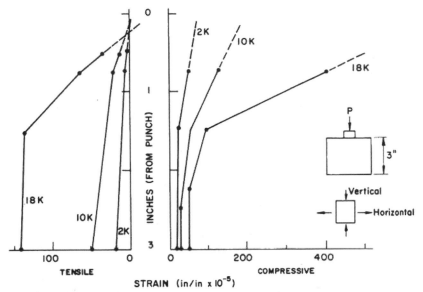

Fig. 10.31. Strain distribution for reinforced mortar (6-inch cylinders). (1 kip = 4.45 kN, 1 inch = 25.4 mm.)

the order of the maximum tensile strain at the load $P = 18$ kips is approximately $140 \cdot 10^{-5}$ inch/inch. This value is approximately 14 times the usual values measured for specimens tested in direct tension, and roughly twice those values measured in indirect tensile testing.

To construct the stress distribution directly from the measured strain field is not possible at the present stage of knowledge, since the essential to this construction is the knowledge of concrete behavior in a combined state of stress. Unfortunately, as already pointed out earlier, the description of the stress–strain relationship for concrete to date is restricted only to the most elementary aspects of simple tests. No general relations have been determined as yet. For the present purpose, the tensile strength of concrete may be assumed to be unaffected by moderate normal stress components in directions perpendicular to the direction of tension. Experimental evidence has been reported in support of this assumption (Nadai, 1950).

As a basis for constructing the associated stress field, the material is said to have yielded when the vertical strains and horizontal strains in the mortar reach values of $300 \cdot 10^{-5}$ inch/inch and $10 \cdot 10^{-5}$ inch/inch, respectively. These are about the average maximum strains in simple compression tests and in simple tension tests. It is emphasized, however, that this is an arbitrary criterion; nevertheless, these readings represent the critical stages in the loading history of the concrete in such tests.

The strain distribution diagrams thus suggest that a high compressive stress more comparable with a triaxial test is developed in the region directly beneath the punch. The plane passing through the axis of the cylinder exhibits an almost uniform tensile stress over that plane for the double-punch test condition and supports the earlier assumption that the concrete can strain sufficiently to develop complete plasticity throughout the material so that the limit analysis technique can be applied.

10.14. COMPARISON OF TEST RESULTS WITH CALCULATED STRENGTHS

10.14.1. Circular punches on circular blocks (Table 10.3)

Test results and theoretical upper-bound solutions are summarized in Table 10.3. Details of the results are reported elsewhere (Chen and Covarrubias, 1971). The critical mechanism which gives the lowest value of the bearing capacity pressure among the mechanisms considered, is listed in the column "Mechanism". The ratios of calculated to measured strengths in Table 10.3 show that the upper-bound solutions predict the test results remarkably well for the case $e/b = 0$ (typical failures are shown in Fig. 11.19, Chapter 11), and reasonably well for $e/b = 1/3$. The ratios range from 0.98 to 1.03 for $e/b = 0$ and 1.07 to 1.30 for $e/b = 1/3$. For the large ratio of $e/b = 2/3$, the upper-bound solutions may over-estimate the bearing capacity by as much as 54%. This difference may be explained by the fact that local plastic flow of concrete for the case of large ratios of e/b is considerably less than that of small ratios of e/b, because hydrostatic pressure (or lateral confinement of the material due to the hoop stress) around the punch cannot be induced high enough to permit the application of limit analysis.

10.14.2. Square punches on square blocks

The test results reported by Hawkins (1968) and those calculated from the present upper-bound solutions are compared in Table 10.4 for specimens with $h/2a \leqslant 8$ and $b/a \leqslant 4$. The direct tensile strength of the concrete used in the theoretical calculations is estimated to be $f_c'/12$. The ratios of calculated to measured strengths listed in the column "Theory/test" show that the upper-bound limit analysis for a square punch on a square block predicts the results reasonably well. Although all the tests correspond to the extreme eccentricity ratio $e/b = 1 - a/b$, yet the calculated to measured strengths are in error only for the range from 1.19 to 1.53. Moreover, it is reported in the preceding section that the load carrying capacity for a double punch specimen (Fig. 10.29b) is higher than that of a similar specimen which is supported directly on the steel bed of a testing machine, as was the case in Hawkins' tests. This would suggest that a better correlation between the theoretical predictions of upper-bound limit analysis and the results of double punch tests may be expected.

TABLE 10.3

Bearing capacity of circular punches on circular blocks — double punch test*

Test number	f'_c in kips per square inch (MN/m^2)	f_t in kips per square inch (MN/m^2)	f'_c/f_t	$h/(2a)$	e/b	q_3/f'_c tests	theory**	mechanism	approximation**	Theory test	Approx. test
1	6.37 (43.92)	0.56 (3.86)	11.4	2.66	0	2.53	2.49	1 or 3	2.49	0.98	0.98
2					1/3	2.23	2.38	3	2.18	1.07	0.98
3					2/3	1.75	2.18	3	1.58	1.25	0.90
4	6.45 (44.47)	0.57 (3.93)	11.3	4.00	0	3.10	3.19	1 or 3	3.19	1.03	1.03
5					1/3	2.53	3.09	3	2.70	1.22	1.07
6					2/3	1.85	2.85	3	1.79	1.54	0.97
7	6.39 (44.06)	0.55 (3.79)	11.5	5.33	0	3.74	3.82	1 or 3	3.82	1.02	1.02
8					1/3	2.93	3.70	3	3.08	1.26	1.05
9					2/3	1.93	2.86	7	2.06	1.48	1.06
10	6.32 (43.57)	0.54 (3.72)	11.7	6.66	0	4.35	4.40	1 or 3	4.40	1.01	1.01
11					1/3	3.29	4.27	3	3.51	1.30	1.07
12					2/3	2.13	2.86	7	2.30	1.34	1.08

* $b/a = 4$ for all specimens.

** $f'_c = 12 f_t$, $\phi = 20°$.

TABLE 10.4

Bearing capacity of square punches on square blocks — Hawkins' tests* (1968)

Test number	2a, in inches	b/a	h/(2a)	e/b	q_3/f_c' tests	theory**	mecha-nism	Theory test
1	1.72	3.50	7.00	0.714	1.99	2.98	6	1.50
2	2.42	2.48	4.96	0.597	1.75	2.08	6	1.19
3	3.00	2.00	4.00	0.500	1.40	1.91	4	1.36
4	1.72	3.50	7.00	0.714	1.95	2.98	6	1.53
5	2.00	3.00	6.00	0.667	1.77	2.45	6	1.38
6	3.00	2.00	4.00	0.500	1.52	1.91	4	1.26
7	2.00	3.00	6.00	0.667	1.91	2.45	6	1.28
8	3.00	2.00	4.00	0.500	1.71	1.91	4	1.12
9	2.00	3.00	6.00	0.667	1.68	2.45	6	1.46
10	3.00	2.00	4.00	0.500	1.38	1.91	4	1.39

* All specimens are 6-inches cubes (1 inch = 25.4 mm).
** $f_c' = 12 f_t$; $\phi = 20°$

10.14.3. Punches on blocks with eccentricity and cable duct

Experimental verifications of the upper-bound limit analysis solutions are de-scribed above for the special case of concrete blocks with punch eccentricity but without cable duct. Experimental data are also available for concrete blocks with both punch eccentricity and cable duct. Test results for specimens with variations of duct diameters and eccentricities with both plain and random wire reinforced concrete blocks are reported (Chen, 1973). All specimens were tested by double punch set-up (Fig. 10.29b). Before the specimens with ducts were tested, snug fitted steel rods were inserted into each end. The length of the rods was such that the two rods did not make contact with each other, therefore, they could not carry any of the applied compressive load. The test results were compared with the theoretical linear elastic analysis and perfectly plastic limit analysis.

The results obtained indicate that when the concrete in bearing is confined as in the *concentrically* loaded blocks, the upper-bound limit analysis solutions control; while in the case where there is little confinement, as in the *eccentrically* loaded blocks with large eccentricities, the results are found to be bounded by the upper-bound limit analysis solutions and the linear elastic fracture solutions.

As expected, the greater ductility of the random wire reinforced concrete allows greater plastic flow to occur and, hence, has resulted closer to the upper-bound solution than those obtained for plain concrete. Furthermore, square specimens are found to have relatively lower strength than the corresponding circular specimens.

This may be explained by the high stress concentrations near the corners of the square punch and also in the corners of the square ducts. These stress concentrations may induce early premature cracks and hence lower load carrying capacities.

10.15. APPROXIMATE SOLUTION

10.15.1. Approximate upper bound solutions

Hawkins (1968) had developed an approximate expression for bearing capacity using grossly simplified plasticity concepts and failure models fashioned according to failure modes observed in tests. The expression for concentrically loaded specimens has the form:

$$\frac{q_3}{f_c'} = 1 + \frac{M}{\sqrt{f_c'}} (\sqrt{R} - 1) \qquad\qquad [10.59]$$

where M = a constant dependent on the characteristics of the concrete mix and equal to 50 for design purposes, and R = unloaded/load area.

It was concluded that the ratio, q_3/f_c' was independent of the depth of the block provided the formation of the failure cone was not restricted by the proximity of the base of the specimen. This restriction is equivalent to requiring that the ratio of the height of the block to the diameter or side length of the punch should be greater than about 2.0.

Values of q_3/f_c' calculated from [10.59] are compared with the test values reported by Hyland and Chen (1970) for their 6-inch high specimens in Table 10.5. The value of M in [10.59] is taken as 50. The ratios of the average test strengths for three different types of base to the calculated strengths are listed in the last column. It is apparent that [10.59] gives a reasonable though nonconservative prediction of the measured strengths.

Since the test results become increasingly sensitive to friction effects at the base for small height to punch diameter ratios $(H/2a)$, [10.59] should be limited accordingly. Based on the test results of Hyland and Chen (1970), for height to punch diameter ratios $(H/2a)$ less than 2.5, values of q_3/f_c' calculated from [10.59] should be decreased linearly to unity at an $H/2a$ ratio of zero.

It is interesting to note that an approximate equation similar to that of [10.59] for bearing capacity of concentrically loaded blocks may be derived from the perfect plasticity equations developed herein. Considering the splitting mode of failure, for example, the value of the upper bound on the average bearing pressure is given by [10.23] in section 10.7. If the cone angle α in [10.23] is arbitrarily chosen equal to the value $\frac{1}{4}\pi - \frac{1}{2}\varphi$, where φ is the angle of internal friction of concrete, the bearing pressure, q_3^u, [10.23], reduces to:

TABLE 10.5

Comparison of measured and calculated strengths

Set	Square root of block area divided by punch area $\sqrt{R}$	Calculated bearing pressure q_3 divided by f_c' for			(q_3/f_c') test*
		steel base	plastic base	double punch	(q_3/f_c') calc
1	4.00	3.10	3.26	3.26	1.01
4	3.00	2.40	2.51	2.53	0.80
7	4.38	3.29	3.24	3.34	0.92
10	3.14	2.48	2.46	2.45	0.88
11	3.14	2.52	2.49	2.59	0.73
14	4.00	3.35	3.33	2.87	0.99
17	3.00	2.25	2.55	2.29	0.95
20	4.38	3.07	–	3.06	1.02
23	3.14	2.37	–	2.32	0.78

* Average for all 3 base conditions.

$$\frac{q_3^u}{f_c'} = 1 + \tan\left(\tfrac{1}{4}\pi + \tfrac{1}{2}\varphi\right)\left[\frac{2bH}{a^2} - \tan\left(\tfrac{1}{4}\pi + \tfrac{1}{2}\varphi\right)\right]\frac{f_t'}{f_c'} \qquad [10.60]$$

Using the empirical formula $f_t' = m\sqrt{f_c'}$, where m is a constant dependent on the characteristics of the concrete, the bearing capacity of [10.60] becomes:

$$\frac{q_3^u}{f_c'} = 1 + 2m \tan\left(\tfrac{1}{4}\pi + \tfrac{1}{2}\varphi\right)\left[\frac{bH}{a^2} - \frac{1}{2}\tan\left(\tfrac{1}{4}\pi + \tfrac{1}{2}\varphi\right)\right]\frac{1}{\sqrt{f_c'}} \qquad [10.61]$$

Since the present analysis assumes an arbitrary choice of the cone angle α, the value of the upper bound on the average bearing pressure will be too high and, thus, m and φ may be treated as the arbitrary curve-fitting material parameters. For the limiting values of the ratios $H/2a = 2.5$ or $b/a = 4$, the bearing pressure in [10.61] will be independent of the depth of the block or the width of the block, respectively. Therefore, for any ratio $b/a > 4$ or $H/2a > 2.5$ the limiting value $b = 4a$ or $H = 5a$ should be used in [10.61] for the computation. For example, for the dimensions of a block with $H/2a > 2.5$, and assuming $m = 2.5$, $\varphi = 20°$, [10.61] gives:

$$\frac{q_3^u}{f_c'} = 1 + \frac{36}{\sqrt{f_c'}}\left[\sqrt{R} - 0.15\right] \qquad [10.62]$$

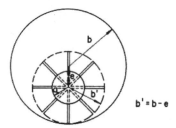

b' = b - e

Fig. 10.32. Approximate solution to an eccentrically loaded block.

Using the values of R listed in the second column of Table 10.5, values of q_3^u/f_c' calculated from [10.62] are found to be very close to the values listed in Table 10.5, calculated from Hawkins' approximation [10.59].

10.15.2. Axially loaded block approximation (Fig. 10.32)

The bearing capacity of a block with an eccentric punch load (Fig. 10.4b) may be estimated directly from the solution of the axially loaded situation (Fig. 10.4a) by assuming that the rigid punch load acts only across an effective block width $2b'$, and this width forms a concentric block with the punch. Thus, for example, for the case of a circular punch on a circular cylinder, this effective cylinder width may be taken to be $b' = b - e$ as shown by the dashed circle in Fig. 10.32. The material outside the radius b' is assumed to have no effect on the bearing capacity of the cylinder. These approximate solutions are summarized in Table 10.3. The observed bearing capacity is seen to be in good agreement with the approximate theoretical estimates.

10.16. SUMMARY AND CONCLUSIONS

This Chapter deals with the bearing capacity of concrete blocks or rock. The limit theorems of perfect plasticity are applied herein to obtain bearing capacity in two dimensions (strip loading or strip rigid punches) and in three dimensions (circular and square rigid punches). The approach is based on the assumption that sufficient local deformability of concrete in tension and in compression does exist to permit the application of limit analysis to concrete idealized as a perfectly plastic material. A modified Coulomb failure surface in compression and a small but non-zero tension cut-off is utilized. The predicted bearing capacity of concrete blocks is found to be in good agreement with published test results. However, when the ratio of the unloaded area to the loaded area is great (approximately 25:1) crack propagation does enter for large-size blocks. An appropriate fracture mechanics for concrete then is needed. Nevertheless, the solutions obtained herein still provide a reasonable upper bound for fractured concrete.

The assumption of concrete to be a perfectly plastic body when determining the bearing capacity of concrete blocks is also examined experimentally in this Chapter. Available data show that the concrete can be strained sufficiently to develop almost complete plasticity throughout the material during indentation of a circular punch on a circular concrete block. The adoption of the limit analysis approach to concrete media appears therefore to be reasonably justified.

The implications of this basic assumption for concrete are far reaching and, when applied, often provide good predictions of bearing capacity. For example, the behavior of the bearing capacity of a concrete block is closely related to the behavior of an indirect tensile test (splitting), in which compressive load is applied to a cylinder along two opposite generators. The relevant formula for computing the tensile strength of the indirect tensile test has been analyzed by the theory of elasticity. A plasticity treatment of this problem will be given in the following Chapter. It is found that the result obtained by limit analysis is identical to that derived from elasticity theory. This does give some indication of the role that plasticity plays in concrete.

The new approach has proved to be very fruitful for bearing capacity problems in concrete, for one arrives thereby at mathematical formulations which not only permit problems to be solved in a relatively simple mathematical form but also give promise of providing very satisfactory agreement with observations. It seems clear, at this stage, that more problems of theoretical significance and practical importance must be investigated so that the implications of plasticity to this class of material may be better understood. Thus, the present theory should be considered as only a first step in this particular application of limit analysis, and it should be extended to practical problems to be useful.

Chapter 11

DOUBLE-PUNCH TEST FOR TENSILE STRENGTH OF CONCRETE, ROCK AND SOILS

11.1. INTRODUCTION

Measurement of the tensile strength of concrete, rock and soils has always been important because the majority of cracks occurring in concrete structures, highway pavements and earthfill dams can be attributed to some form of tensile stress. There are now a variety of techniques which give some measure of the tensile strength of these materials. These include the *direct pull test*, the *flexure test*, the *ring test*, and the *split cylinder test*.

In the *direct pull tests* on briquetts and bobbins, the specially shaped specimens are loaded through gripping devices. The drawbacks of direct pull tests include the difficulty in eliminating eccentricity of the line of action of the load and the development of stress concentrations near the gripping devices. In *flexure tests*, the concrete or soil beams of square sections are subjected to single point or two point loading. The theoretical maximum tensile stress in the bottom fiber, known as the *modulus of rupture*, is calculated by assuming a linear distribution of flexural stress across the section at failure (linear elastic analysis). Although the flexural tests are easier, the tensile strength determined in this way is found to be significantly higher than the direct pull tensile strength.

11.1.1. Splitting tests (Fig. 11.1)

In the *splitting tests* on cylinders (also known as *Brazilian test*), a concrete or rock cylinder is laid horizontally between the loading platens of the testing machine and compressed along two opposite generators until the specimen splits across the vertical diametric plane as indicated in Fig. 11.1(a). In countries where the compressive strength of concrete is determined from cubes rather than from cylinders, the tensile strength has been obtained using a split cube or a cube specimen tested diagonally (Fig. 11.1b,d). A formula for computing the tensile strength of splitting tensile tests can be obtained from the theory of *linear elasticity*. The elastic stress distribution in splitting tests will be presented in the following section. The advantages of the splitting tests are that, besides being simple and easy, they enable similar specimens, and the same testing machine, to be used for both tensile and compressive strength tests. In addition, splitting tests on cylinders give more consis-

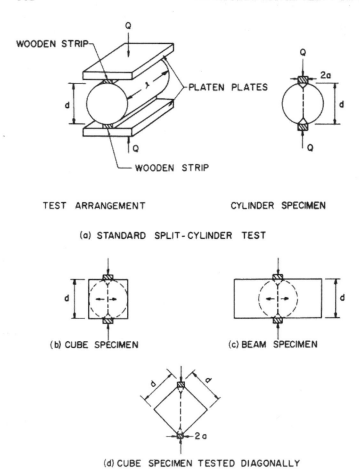

Fig. 11.1. Four possible splitting test arrangements

tent results with the measured strengths being between those of the other two tests (direct pull and flexure tests).

11.1.2. Double punch test (Fig. 11.2)

More recently, a plasticity solution of this problem has been reported (Chen, 1970b). The result derived from the theory of *perfect plasticity* using the limit analysis technique is found to be identical to that derived from the theory of linear elasticity. The success in applying the theory of perfect plasticity to the problem of the split tensile test suggests an alternative new testing technique for the determination of the tensile strength of concrete.

The *double punch test* will be described in section 11.6–11.9 for concrete, rock and soils. In this test, a concrete cylinder is placed vertically between the loading

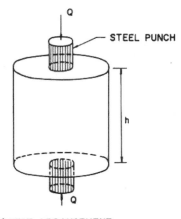

(a) TEST ARRANGEMENT

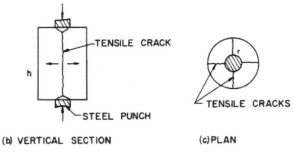

(b) VERTICAL SECTION   (c) PLAN

Fig. 11.2. Apparatus for double-punch tests.

platens of the test machine and compressed by two steel punches located concentrically on the top and bottom surfaces of the cylinder (Fig. 11.2). It is observed that, although the specimen splits across the vertical diametric plane in a manner exactly similar to that observed in a split tensile test; the necessary test arrangement in obtaining the tensile strength of these materials may be reduced.

11.2. ELASTIC STRESS DISTRIBUTION IN SPLITTING TESTS

11.2.1. Exact solution

It will be shown herein that two equal and opposite line forces Q acting along a diametral plane AB (Fig. 11.3b) will give rise to an *almost* uniform tensile stress over that plane. The solution to this problem can be obtained most conveniently by the method of superposition of the following two basic elastic stress distributions.

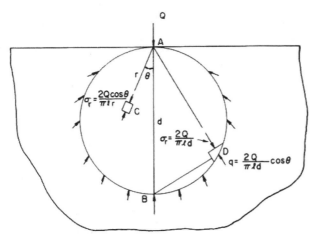

(a) STRESS IN A PLATE

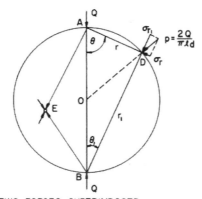

(b) TWO FORCES SUPERIMPOSED

Fig. 11.3. Stress in a circular disk.

First basic elastic stress distribution

The first basic elastic stress distribution is that due to a concentrated force Q (Fig. 11.3a) acting on the edge of a plate of thickness l bounded by one straight edge but otherwise unlimited in extent (Fig. 11.3a). This solution is well known (Boussinesq, 1892) and called the *simple radial distribution* in the two-dimensional theory of elasticity. Any element C at a distance r from the point of application of the load is subjected to a simple compression:

$$\sigma_r = \frac{2Q}{\pi l}\frac{\cos\theta}{r} \tag{11.1}$$

in the radial direction. Now taking a circle of diameter d which is tangent to the edge at point A as shown in Fig. 11.3(a), we have, for any point D of the circle, $d\cos\theta = r$. Hence from [11.1]:

$$\sigma_r = \frac{2Q}{\pi l d} \tag{11.2}$$

i.e., the stress is the same at all points on the circle, except the point A, the point of application of the load. Considering the equilibrium of the triangular element at point D, we have:

$$q = \frac{2Q}{\pi l d} \cos \theta \tag{11.3}$$

Let the circular area be removed from the plate and a similar system inverted may be superimposed (Fig. 11.3b). We now have a disc subjected to two opposite forces Q acting along a diameter, and two sets of stresses, acting on the circumference, of magnitude:

$$\frac{2Q}{\pi l d} \frac{r}{d} \text{ along } AD \quad \text{and} \quad \frac{2Q}{\pi l d} \frac{r_1}{d} \text{ along } BD$$

These two external stresses are proportional to AD and BD and may therefore be represented by these lines in a parallelogram of stresses. The resultant is clearly $OD = p = 2Q/\pi l d$, which is constant and passes through the center point O in Fig. 11.3(b). Thus the two concentrated forces superimposed together are equivalent to a uniform radial compression of magnitude $2Q/\pi l d$. Any element within the disc, for instance point E in Fig. 11.3(b), is now subjected to two radial stresses.

Second basic elastic stress distribution

The second basic stress distribution is that a circular disc subjected to a uniform tension $p = 2Q/\pi l d$ round the edge. This condition gives rise to a uniform biaxial tension $p = 2Q/\pi l d$ in any direction within the disc.

Superposition of the two stress distributions

Let us now superimpose this second stress distribution to the first stress distribution shown in Fig. 11.3(b). The resultant distributed loads round the edge of the disc now vanish, and we are left with the conditions of the problem; i.e., a disc subjected to two opposite forces acting along a diameter (Fig. 11.1a). The exact stresses at any point in the disc can thus be calculated readily. In particular, on the vertical diameter, $\theta = \theta_1 = 0$, we have the vertical stress component (compressive):

$$\sigma_r|_{\theta=\theta_1=0} = \frac{2Q}{\pi l} \frac{1}{r} + \frac{2Q}{\pi l} \frac{1}{d-r} - \frac{2Q}{\pi l d} = \frac{2Q}{\pi l d} \left(\frac{d}{r} + \frac{d}{d-r} - 1 \right) \tag{11.4}$$

and the horizontal stress component (tensile):

$$f_t' = \sigma_\theta|_{\theta=\theta_1=0} = -\frac{2Q}{\pi l d} \tag{11.5}$$

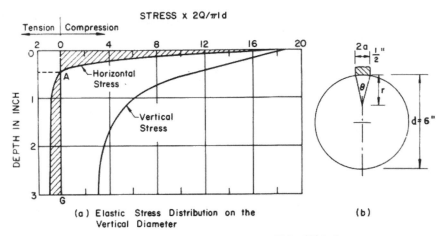

(a) Elastic Stress Distribution on the
Vertical Diameter

(b)

Fig. 11.4. Stress distributions in cylinder loaded over a width of ½ inch.

By considering a cylinder of concrete or rock as a number of such discs, we see that a uniform tensile stress is developed over the vertical diametral plane.

Remarks on the solution (Fig. 11.4)

The calculations outlined above form an exact solution for the ideal case considered. However, to prevent premature failure near the loading points it is usual in practice to insert packings of width $\frac{1}{12}d$ between the specimen and the platen plates. Under these conditions, the tensile stress on the central about three-fourths of the diameter remains at f_t' but changes rapidly to a maximum compressive stress of about 18 times this value under the packings, as shown in Fig. 11.4(a). If the load is assumed to be uniformly distributed over the width $2a$ (say $\frac{1}{2}$ inch), it can be shown that if $2a < \frac{1}{10}d$, the stresses on the vertical diameter may be adequately approximated by:

$$\text{vertical stress } \sigma_r = \frac{2Q}{\pi l d} \left[\frac{d}{4a} \left(\theta + \sin \theta \right) + \frac{d}{d-r} - 1 \right] \qquad [11.6]$$

$$\text{horizontal stress } \sigma_\theta = \frac{-2Q}{\pi l d} \left[1 - \frac{d}{4a} \left(\theta - \sin \theta \right) \right] \qquad [11.7]$$

Term θ is the angle subtended by the loaded area at the point considered (Fig. 11.4b), and tensile stress is taken as negative. The stress distribution along the vertical diameter, calculated for $2a = \frac{1}{12}d$, is shown in Fig. 11.4(a). Therefore, the material in zones immediately beneath the punch is in a state of biaxial compression for plane stress conditions or triaxial compression for plane strain conditions.

TABLE 11.1

Strength and coefficients of variations of specimens by various testing methods (Wright, 1955)

Method	Average strength psi (kg/cm²)	Coefficient of variation (%)
Direct pull	275 (19.5)	7
Modulus of rupture	605 (42.5)	6
splitting-cylinders	405 (28.5)	5
Compressive-cubes	5980 (421.0)	3.5

11.2.2. Finite element solutions

Rigorous theoretical solutions to the critical stress distributions developed in splitting tests on specimens of various shapes as shown in Fig. 11.1 (b, c, d) have been reported recently by Davies and Bose (1968) using the *stiffness method of finite element analysis*. It is found that specimens (a), (b), and (c) in Fig. 11.1 have similar distribution patterns of tensile stress. Equation [11.5] may be considered valid for the specimens (a), (b), and (c) in Fig. 11.1 under point or distributed loading. The maximum splitting stress for a cube specimen tested diagonally is found to be 0.77 of that predicted by [11.5].

11.2.3. Remarks on the splitting tests

An appraisal of the various testing techniques on the splitting concrete cylinder test has been given by Wright (1955) and splitting tests on cube specimens have been described by Nilssons (1961). The most attractive part of the splitting tests is that they enable similar specimens, and the same testing machine, to be used for both tensile and compressive strength tests. In addition, the splitting tests on cylinders give more consistent results with the measured strengths being between those of the other two tests. The typical results in Table 11.1 indicate their relative accuracies on control tests.

11.3. LIMIT ANALYSIS OF SPLITTING TENSILE TESTS

The distributions of stress and hence the relevant formulas for computing the tensile strength of various splitting tests have been analyzed in the preceding section by the theory of elasticity. A plasticity treatment of this problem is given below. Three types of analysis are considered in what follows for the condition of plane strain: (1) limit analysis; (2) slip-line field; and (3) finite element analysis. Each type of analysis corresponds to a somewhat different stress–strain relationship for concrete or rock. Nevertheless, it is demonstrated that analyses (2) and (3) presented in section 11.4 and section 11.5 give distribution patterns of tensile stress similar to that of elastic stress distribution. In particular, the relevant formulas for computing the tensile strength of various splitting tests obtained by the three plasticity analyses are found to be similar with that of elasticity analysis. Equation [11.5] may therefore be considered valid for the three specimens (cylinder, cube and beam) under point or distributed line loading. The maximum splitting stress for a cube specimen tested diagonally is about 0.75 of that predicted by [11.5].

11.3.1. Limit analysis – Upper bound (Fig. 11.5b)

The limit analysis is based on the fundamental assumption that the local tensile

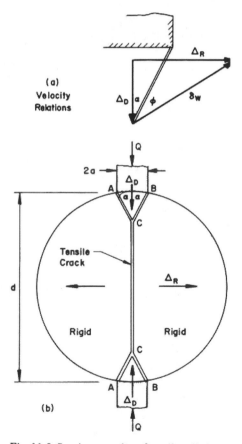

Fig. 11.5. Bearing capacity of a split-cylinder test.

strain of concrete is sufficient to permit the application of limit analysis. In addition, it is assumed, as in Chapter 10, that the concrete or rock may be idealized as perfectly plastic with the modified Coulomb failure surface as the yield surface in compression, and a small but non-zero tension cut-off (Fig. 10.8b). In the following we shall consider first the cylinder specimen compressed by two rigid punches (Fig. 11.5b).

The Upper-Bound Theorem of limit analysis states that the concrete or rock cylinder will collapse, if, for any assumed failure mechanism, the rate of work done by the applied loads exceeds the internal rate of dissipation. Equating external and internal energies for any such mechanism thus gives an upper bound on the collapse load.

Figure 11.5(b) shows a failure mechanism consisting of two rigid wedge regions ABC, and a simple tension crack CC connecting these two wedges. The wedges move toward each other as a rigid body, and displace the surrounding material horizontally sideways. The relative velocity vector δw at each point along the lines of discontinuity AC and BC is inclined at an angle φ to these lines. The compatible velocity relations are shown in Fig. 11.5(a). The rate of dissipation of energy along the wedge surfaces may be found by multiplying the area of these discontinuous surfaces by f_c' $\frac{1}{2}(1 - \sin \varphi)$ times the discontinuity in velocity δw across the surfaces ([10.13]). Similarily, the rate of dissipation of energy along the separation surface CC is found by multiplying the area of separation by f_t' times the relative separation velocity $2\Delta_R$ across the separation surface ([10.12]). Equating the external rate of work to the total rate of internal dissipation yields:

$$Q^u = \left(\frac{a}{\sin \alpha}\right) \left[\frac{f_c' l(1 - \sin \varphi)}{\cos (\alpha + \varphi)} - 2f_t' l \cos \alpha \tan (\alpha + \varphi)\right] + f_t' l d \tan (\alpha + \varphi) \qquad [11.8]$$

in which d is the diameter and l is the length of the cylinder.

The upper bound has a minimum value when α satisfies the condition $\partial Q^u / \partial \alpha = 0$, which is:

$$\cot \alpha = \tan \varphi + \sec \varphi \left\{1 + \frac{\dfrac{d}{2a} \cos \varphi}{\left(\dfrac{f_c'}{f_t'}\right) \left[\dfrac{1 - \sin \varphi}{2}\right] - \sin \varphi}\right\}^{1/2} \qquad [11.9]$$

Equation [11.8] is in fact a simple modification of the two-dimensional solution obtained in [10.14]. For the dimensions used in the standarized splitting tension test on cylinder (ASTM-C649-62T): $2a = \frac{1}{2}$ inch (say)* and $d = 6$ inches, and the

---

* ASTM specificies that the width of the plywood strip placed between the punch and the concrete cylinder is 1 inch, however, the load is actually distributed through the plywood to the concrete cylinder over a band of appreciably less width (say $\frac{1}{2}$ inch)

average values for concrete: $f_c'/f_t' = 10$ and $\varphi = 30°$, the upper bound has a minimum value at the point $\alpha = 16.1°$, and:

$$Q \leqslant Q^u = 1.83\ ldf_t' \qquad\qquad\qquad [11.10]$$

Therefore: $f_t' \geqslant 0.548\ \dfrac{Q}{ld}$ $\qquad\qquad\qquad\qquad\qquad$ [11.11]

11.3.2. Limit analysis – Lower bound (Fig. 11.4a)

The Lower-Bound Theorem of limit analysis states that if an equilibrium distribution of stress can be found in the concrete cylinder which nowhere exceeds the modified Coulomb yield criterion in Fig. 10.8(b) then the loads imposed can be carried without collapse, or still just be at the point of collapse. Clearly, any stress distribution obtained through the theory of elasticity will give a safe or lower bound on the collapse load, provided that the chosen stress fields nowhere violate the yield criterion.

Stress distribution in a plane disc (plane stress) subjected to loads perpendicular to the axis of a disc (Fig. 11.5b) has been discussed thoroughly in the preceding section. The stress distribution for the disc will be a statically admissible stress field if the magnitude of the maximum shearing stress on any section through the concrete in the disc is not greater than an amount which depends linearly upon the hydrostatic pressure (Fig. 10.8b). If the load is assumed to be uniformly distributed over the width $2a$ (say 1/2 inch), it is demonstrated that if $2a < \frac{1}{10}d$, the stresses on the vertical diameter may be adequately approximated by [11.6] and [11.7]. The stress distribution along the vertical diameter, calculated for $2a = \frac{1}{12}d$, is shown in Fig. 11.4(a).

It can be shown that the critical points which decide the maximum value of Q are those points along the vertical diameter jointing the applied forces. The vertical stress at point A (Fig. 11.4(a), $r = 0.45$ inch) must not be greater than f_c' in order that the yield condition be not violated, since the horizontal stress is zero at this point. Since the cylinder is long, it approximates closely to plane strain condition. The condition above point A is thus comparable with a triaxial test. Yielding does not, therefore, occur here. The points vertically below A ($r > 0.45$ inch) are under a biaxial state of compression–tension. with a tension of amount near $2Q/\pi ld$. It is found that the critical point along the vertical diameter plane, which first reaches the yield condition, is the point $r = 0.5$ inch [$\sigma_r = -10.45(2Q/\pi ld)$, $\sigma_\theta = 0.236(2Q/\pi ld)$]. The modified Coulomb yield condition can be written (Fig. 10.8b):

$$\sigma_r = \sigma_\theta \tan^2\left(\tfrac{1}{4}\pi + \tfrac{1}{2}\varphi\right) - f_c' \qquad\qquad [11.12]$$

For $\varphi = 30°$ and $f_c' = 10\,f_t'$, [11.12] reduces to:

$$\sigma_r = 3\sigma_\theta - 10\,f_t' \qquad\qquad\qquad\qquad [11.13]$$

Substituting the values σ_r and σ_θ at $r = 0.5$ in. into [11.13], a lower bound on the collapse load of the splitting tensile test is thus obtained:

$$Q \geqslant Q^l = 1.41 \, ldf'_t \qquad\qquad [11.14]$$

Therefore: $f'_t \leqslant 0.71 \dfrac{Q}{ld} \qquad\qquad [11.15]$

Therefore, the stress field of Fig. 11.4(a) and the velocity field of Fig. 11.5(b) show that the tensile strength of concrete for the indirect tensile test lies within $\pm 13\%$ of the value $0.629 \, Q/ld \approx 2Q/\pi ld$. It is interesting to note that the average value of the upper- and lower-bound solutions given previously, is identical to that derived from elasticity theory.

11.3.3. Observed mode of failure (Fig. 11.6)

As pointed out earlier in Chapter 10, the ductility of both the mortar and the concrete materials increases with the increase of percentage of short fiber wire reinforcement. Chen and Carson (1971) observed that during splitting cylinder test, the increase in the tensile ductility actually becomes visible in the form of gross

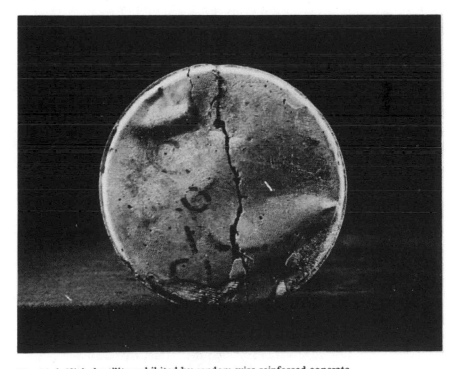

Fig. 11.6. High ductility exhibited by random wire reinforced concrete.

deformations of the planar surfaces under strip loading. Fig. 11.6 shows the gross deformation of an originally planar surface in concrete which is reinforced to 2.00% by volume with 1 inch (2.54 cm) long fibers.

In the test, the type of failure also varied with the increase in the percentage of wire reinforcement. The unreinforced specimens fail in a manner closer to a planer mode, characteristic of brittle materials. As the percentage reinforcement increased, the mode of failure progresses toward the conical failure predicted by limit analysis which is characteristic of more ductile materials (see Figs. 11.5(b) and 11.6).

11.3.4. Splitting tests on cube and beam specimens (Figs. 11.1b,c,d)

The upper and lower bounds obtained in [11.11] and [11.15] are also applicable for splitting tests on concrete specimens of other shapes such as, for example, the cases shown in Fig. 11.1(b) (cube specimen) and Fig. 11.1(c) (beam specimen). This follows directly from the facts that the assumed failure mechanism in the upper-bound calculations for a cylinder specimen is also applicable to a cube specimen or a beam specimen (see dashed lines in Figs. 11.1b and 11.1c) and the admissibility of the elastic stress field used to obtain the lower bound for the cylinder specimen or directly from the theorem that the addition to a body of weightless material cannot result in a lower collapse load (Theorem IV, section 2.5, p. 38). This then indicates that the formula $f'_t = 2Q/\pi l d$ may be considered valid for the splitting tests on cube specimen or beam specimen. This conclusion is identical with that of Davies and Bose (1968) using a finite element method with linear elastic idealization for concrete.

The extension of the preceding information to obtain a relevant formula for computing the tensile strength of a cube specimen tested diagonally is evident. An identical failure mechanism of the problem is shown in Fig. 11.1(d). For the dimensions: $2a = \frac{1}{2}$ inch, as an example, and $d = 6$ inches, the upper-bound equation [11.8] has a minimum value at the point $\alpha = 14.6°$, and:

$$Q \leqslant Q^u = 2.16 \, ldf'_t \qquad\qquad [11.16]$$

(Note: The appropriate value d in [11.8] and [11.9] is $\sqrt{2}\, d - 2a = 8$ inches.)

Therefore: $f'_t \geqslant 0.463 \dfrac{Q}{ld}$ \qquad\qquad [11.17]

The elastic solution of Davies and Bose (1968) will give a safe or lower bound on the collapse load, provided that the stress field nowhere violates the modified Coulomb yield condition. (The solution is valid for the case of point load, however, as pointed out by Davies and Bose, the pattern of stress distribution for the point load and distributed load is similar except near the loading zones. Thus, it will not effect the analyses herein.) It is found that the critical point, which decides the maximum value of Q, is the point at the center of the diagonal cube specimen. It can then be shown that when the maximum tensile stress at this point $[\sigma_\theta = 0.77 (2Q/\pi ld)]$ reaches f_t', a maximum lower-bound load of the diagonal cube specimen is obtained:

$$Q \geqslant Q^l = 2.04 \, ldf_t' \qquad\qquad\qquad [11.18]$$

Therefore: $f_t' \leqslant 0.49 \dfrac{Q}{ld}$ $\qquad\qquad\qquad\qquad\qquad$ [11.19]

Thus, the tensile strength of a diagonal cube specimen lies within ± 3% of the value $0.476 \, Q/ld = 0.75 \, (2Q/\pi ld)$. It is interesting to note that the diagonal cube specimen has been used in the Soviet Union and the relevant formula for computing the tensile strength is assumed to be about 80% of the value $2Q/\pi ld$. This agrees quite well with the present limit analysis.

11.4. PLASTIC STRESS DISTRIBUTION IN SPLITTING TESTS BY SLIP-LINE METHOD

11.4.1. Material model (Fig. 11.7)

Similar to that for the modified Coulomb yield condition used in the limit analysis, we assume herein that in the compressive domain the limiting maximum

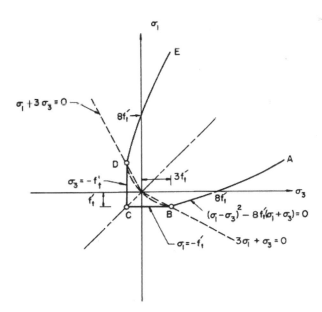

Fig. 11.7. A yield condition.

shear stress $s = \frac{1}{2}(\sigma_1 - \sigma_3)$ on any physical plane at a point is a function of the mean normal stress $p = \frac{1}{2}(\sigma_1 + \sigma_3)$, thus:

$$s = f(p) \tag{11.20}$$

where $\sigma_1 > \sigma_2 > \sigma_3$. In our discussion, we shall consider a particular form of [11.20], namely:

$$f_1 = (\sigma_1 - \sigma_3)^2 - 8f_t'(\sigma_1 + \sigma_3) = 0 \tag{11.21}$$

valid for: $3\sigma_1 + \sigma_3 \geqslant 0, \; \sigma_1 + 3\sigma_3 \geqslant 0$ \hfill [11.22]

In the tensile domain the yield condition is governed by the maximum principal tensile stress, thus, for this particular case, we have:

$$f_2 = \sigma_1 + f_t' = 0, \quad f_3 = \sigma_3 + f_t' = 0 \tag{11.23}$$

valid for: $3\sigma_1 + \sigma_3 \leqslant 0, \; \sigma_1 + 3\sigma_3 \leqslant 0$ \hfill [11.24]

Equations [11.21] to [11.24] are represented in the principal stress space by the curve marked *EDCBA* shown in Fig. 11.7. When the power 2 appearing in the term $(\sigma_1 - \sigma_3)$ of [11.21] is taken to be 1, [11.21] to [11.24] reduce to the modified Coulomb yield condition with small tension cut-off shown in Fig. 10.8b. Such a yield condition has already been applied in the preceding section and also in Chapter 10 for some bearing capacity problems in concrete or rock. General discussion of the yield or fracture criterion applicable for brittle materials can be found, for instance, in Paul (1962), Izbicki (1972) and Mroz (1972).

11.4.2. Slip-line solutions (Fig. 11.8)

Consider now a block of thickness d indented by two rigid flat punches of width $2a$, Fig. 11.8(a). It can be expected that the plastic domain spreads in the vicinity of the punch axis while more distant zones are rigid and essentially unstressed. For the case of large $d/2a$ ratio, a combined solution for which the tensile yield stress ([11.23]), is reached near the axis of symmetry LL while the non-linear shear yield condition [11.21] occurs near loaded edges. The net of slip-line is shown in Fig. 11.8(a) and Fig. 11.8(b) shows the set of stress characteristics or slip-line field along with the stress distribution along the line OG. The details of the stress solutions are discussed in Izbicki's papers (1972). It is seen that the plastic zone reduces to a line FG when the limiting tension condition is fulfilled.

When the stress distribution along OG is compared with that obtained by linear elastic solution (Fig. 11.4a), it is clear that the two material models provide qualitatively similar distributions: the stress along OG is tensile near the block symmetry axis and changes its sign below the punch. However, the linear elasticity analysis

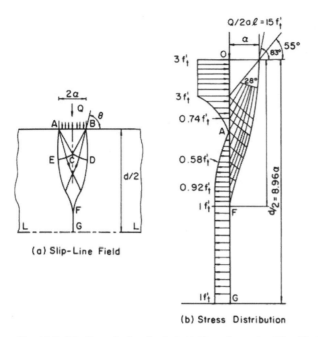

(a) Slip-Line Field

(b) Stress Distribution

Fig. 11.8. Slip-line solution for indentation of punch with a block.

and limit analysis suggest a linear relationship between the limit punch pressure $Q/2al\,f_t'$ and the ratio $d/2a$ ([11.5], [11.11] and [11.15]), whereas the slip-line solution predicts a non-linear dependence. Elastic and upper- and lower-bound limit analysis solutions along with the upper and lower bounds on the limit load of slip-line solutions in terms of $d/2a$ are shown in Fig. 11.9.

Since it is possible to associate velocity fields with the slip-line stress fields shown in Fig. 11.8(b) (Izbicki, 1972), the solutions obtained by slip-line construction are therefore *incomplete* but always give an upper bound to the limit load. It remains to be shown that the plastic stress field shown in Fig. 11.8(b) can also be extended into the rigid region in a satisfactory manner (statically admissible) so that these slip-line solutions are *complete* and thus give the correct limit load. By applying the general method of Bishop (1953) and extending the stress field such as the one shown in Fig. 11.8(b) into the rigid zone, Izbicki (1972) obtains various minimal shapes of block corresponding to different $d/2a$ ratios for which the solution of Fig. 11.8(b) is valid. This then gives the lower-bound solutions to the slip-line fields (see Fig. 11.9).

Using the plastic stress field shown in Fig. 11.8(b), the maximum tensile stress f_t' which occurs at the middle central portion of the specimen is found to be 1.065 of the tensile value calculated by [11.5] ($Q/2al = 15\,f_t'$ and $d/2a = 8.96$) and therefore agrees quite well with the formula [11.5] based on the linear theory of elasticity.

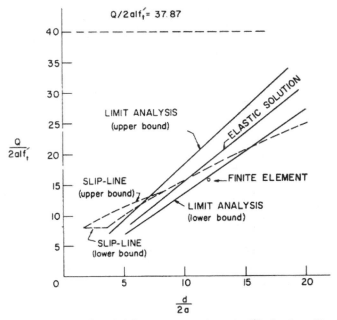

Fig. 11.9. Dependence of the pressure on the ratio $d/2a$ for the split-cylinder test.

11.5. PLASTIC STRESS DISTRIBUTION IN SPLITTING TESTS BY FINITE ELEMENT METHOD

In the preceding sections, concrete or rock has been idealized as a linear elastic–fracture material (section 11.2) or as an elastic-perfectly plastic material (sections 11.3 and 11.4). The elastic solution, limit analysis solution and slip-line solution can all be obtained either in closed form, or by means of relatively simple graphical or numerical computer methods. However, the local cracking of the concrete at an early stage of loading causes stress redistribution in the block, so that the stress distributions presented in the preceding sections may be questioned. Furthermore, nonlinear stress–strain behavior of concrete or rock before cracking or crushing should also be considered for a realistic assessment of the stress distribution in the blocks.

Herein, the concrete or rock is considered to be a linear elastic, plastic strain-hardening and fracture material. The nonlinear effects caused by the plasticity of compression concrete and cracking of the tension concrete are considered in the analysis of splitting cylinder test under plane strain condition. An incremental finite element method is used for the nonlinear analysis of the problem. The aim of this section is to help clarify the above-mentioned questions by presenting the results of the more *refined* theory. Detailed discussion of this model in reference to yielding and deformation of concrete material will be presented in Chapter 12.

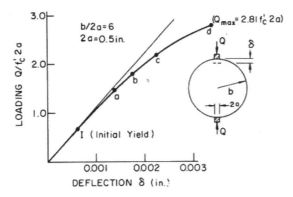

Fig. 11.10. Load-deflection curve of the concrete cylinder subjected to rigid punches under plane strain condition.

In the following, the plastic stress distributions in the splitting cylinder test are presented. The solution is obtained by the stiffness method of finite element analysis using the stress—strain relations to be described in Chapter 12.

Fig. 11.10 shows a typical load-deflection curve of the problem under the condition of plane strain. Nonlinear behavior of the curve is seen to occur at a very early stage of loading $[Q/(2af_c') = 0.752]$. Yielding starts beneath the corner of the punch and the spreading of the plastic zone is found to be limited to the area around the punch until a crack is developed at the corner of the punch. After the initiation of the first crack, the region along the vertical diametric plane becomes plastic very quickly and the plastic zone keeps on spreading as the load increases. When the load is about 78% of the ultimate load (point c on the curve), the crack beneath the corner of the punch starts to propagate while the plastic zone keeps on spreading. Finally, when most of the elements underneath the punch corner have been fractured, the cylinder collapses. The maximum load obtained is $2.81\ (2af_c')$.

The redistributions of the vertical and horizontal stresses along the vertical diametric plane AA' containing the loads, and the vertical plane BB' containing the edges of the punches are shown in Fig. 11.11. The various stages of loading are denoted by the letters I, a, b, c and d in the figure. The corresponding loading points are also marked on the load-deflection curve shown in Fig. 11.10.

It can be seen that there is practically no difference in the distributions of the horizontal and vertical stresses across the vertical diametric plane for the splitting cylinder test (Fig. 11.11a) as compared with that obtained by linear elastic solution (Fig. 11.4a) and slip-line solution (Fig. 11.8b). The tensile stress is almost uniform near the center two-third of the cylinder and the maximum tensile stress occurs at about $\frac{1}{2}b$ from the loading points and drops slightly at the midpoint. The maximum tensile strength is found to be 1.07 of the f_t' value calculated by [11.5] and therefore agrees quite well with the elastic and limit analysis and slip-line solutions.

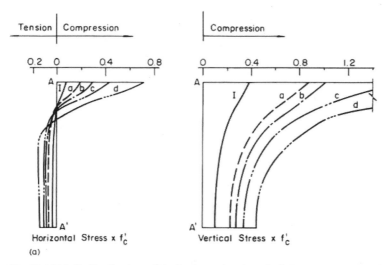

Fig. 11.11(a). Redistributions of the horizontal and vertical stress components along two vertical planes. (a) vertical diametric plane containing the loads.

The corresponding stresses along the vertical plane BB' passing through the edges of the punches are seen in Fig. 11.11(b). The similarity of the curves for the planes BB' and AA' are evident except for the discrepancies adjacent to the loading points. This is probably due to the development of cracks in concrete beneath the edges of the punch. Nevertheless, the average value of the tensile stress is very close to that by [11.5] and it would be reasonable to conclude that this expression is not significantly affected by the early development of concrete cracks in such a test.

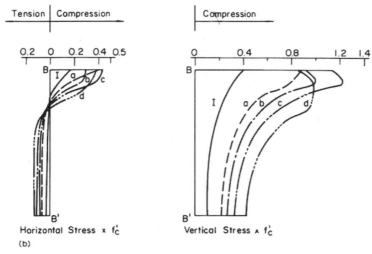

Fig. 11.11(b). Vertical plane containing the punch edges.

As a further example, a cube specimen having a block width to punch width ratio equal to 6 is also investigated using the refined theory. The maximum tensile stress across the vertical axis occurs at about center half of the cube and the average value is approximately 0.90 of the f_t' value predicted by [11.5]. The vertical and horizontal stress distributions on the vertical axis of the cube are found to be very similar to that of the cylinder except for the discrepancies adjacent to the loading points. From a practical viewpoint the precise values of the stresses in these zones are not significant. Therefore, [11.5] should give a good estimate of the maximum tensile stress.

11.6. PLASTIC STRESS DISTRIBUTION IN DOUBLE-PUNCH TEST AND LIMIT ANALYSIS SOLUTION

A new indirect tensile test known as *double-punch test* has been introduced recently by Chen (1970c), in which a compressive load is applied to a cylinder or cube along two opposite faces (Figs. 11.12 and 11.13). This condition sets up an almost uniform tensile stress over the vertical planes containing the applied load, and the specimen splits across these planes similar to the splitting tests (Fig. 11.14). Here, as in splitting tests, the test is carried out in a compression testing machine, and thus it enables similar specimens, and the same testing machine to be used for both tensile and compressive strength tests.

In the following, we shall first develop a formula for computing the tensile strength in the new test using the limit analysis techniques. It is shown that the formula for computing the tensile strength of concrete or rock is very simple. A rigorous theoretical solution for the plastic stress distribution of this problem is

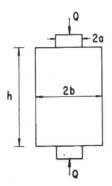

DOUBLE PUNCH TEST

Fig. 11.12. Specimen configuration.

Fig. 11.13. Test arrangement.

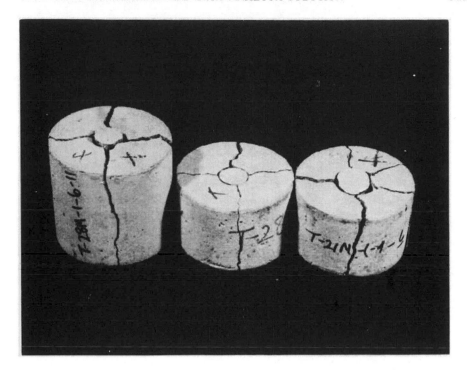

Fig. 11.14. Failure modes.

then presented. Experimental investigations of the double-punch test for concrete, rock and soil materials will be discussed in the sections that follow.

11.6.1. Limit analysis solution

Since the behavior of a concrete or rock block during a bearing capacity test is closely related to the behavior of a double-punch test, the relevant formula of the double-punch test can therefore be obtained directly from a simple modification of results reported in Chapter 10, [10.23] to [10.25], by substituting h for $2H$ in these equations. For ease of reference, however, we shall rederive these equations again in this section.

Upper-bound computation (Fig. 11.15)

Fig. 11.15 shows diagrammatically an ideal failure mechanism for a double-punch test on a cylinder specimen. It consists of many simple cracks along the radial direction and two cone-shape rupture surfaces directly beneath the punches. The cone shapes move toward each other as a rigid body and displace the surrounding material horizontally sideways. The relative velocity vector δw at each point

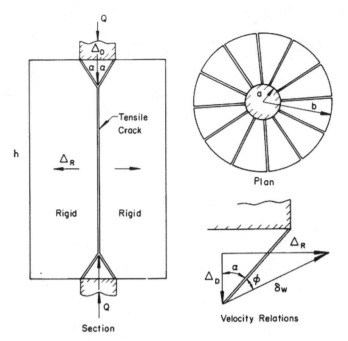

Fig. 11.15. Bearing capacity of double-punch test.

along the cone surface is inclined at an angle φ to the surface. The compatible velocity relation is also shown in Fig. 11.15. It is a simple matter to calculate the areas of the surfaces of discontinuity. The rate of dissipation of energy then is found by multiplying the area of each discontinuity surface by f_t' times the separation velocity $2\Delta_R$ across the surface for a simple tensile crack ([10.12]) or f_c' $\frac{1}{2}(1 - \sin \varphi)$ times the relative velocity δw across the cone-shape rupture surface for a simple shearing ([10.13]). Equating the external rate of work to the total rate of internal dissipation yields the value of the upper bound on the applied load Q:

$$\frac{Q^u}{\pi a^2} = \frac{1 - \sin \varphi}{\sin \alpha \cos (\alpha + \varphi)} \frac{f_c'}{2} + \tan (\alpha + \varphi) \left(\frac{bh}{a^2} - \cot \alpha\right) f_t' \qquad [11.25]$$

in which α is the yet unknown angle of the cone, a is the radius of the punch and b and h are the specimen dimensions (Fig. 11.15).

The upper bound has a minimum value when α satisfies the condition $\partial Q^u / \partial \alpha = 0$, which is:

$$\cot \alpha = \tan \varphi + \sec \varphi \left[1 + \frac{\dfrac{bh}{a^2} \cos \varphi}{\dfrac{f_c'}{f_t'} \left(\dfrac{1 - \sin \varphi}{2}\right) - \sin \varphi}\right]^{1/2} \qquad [11.26]$$

Equation [11.26] is valid for $\alpha \geqslant \tan^{-1}(2a/h)$ Now [11.25] can be reduced to:

$$\frac{Q^u}{\pi a^2} = f_t'\left[\frac{bh}{a^2}\tan(2\alpha + \varphi) - 1\right]$$ [11.27]

Using typical values of $f_c' = 10\ f_t'$ and $\varphi = 30°$ and assuming $2a = 1.5$ inches (3.80 cm), $2b = 6$ inches (15.30 cm), and $h = 8$ inches (20.40 cm), the upper bound has a minimum value at the point $\alpha = 10°$ and [11.27] gives:

$$Q \leqslant Q^u = \pi(1.19\ bh - a^2)f_t'$$ [11.28]

It is found that the value of the coefficient 1.19 appearing in [11.28] is not too sensitive to the dimensions used in a double-punch test. For a typical 6-inches diameter cylinder, for example, assuming $2a = 1.0$ inches (2.55 cm) or 1.5 inches (3.80 cm), the value of the coefficient varies from 1.20 to 0.97 or 1.32 to 1.11 for h varies from 4 inches (10.20 cm) to 8 inches (20.40 cm) or 6 inches (15.30 cm) to 10 inches (25.50 cm), respectively.

Formula determining tensile strength

As concluded in Chapter 10, the upper-bound solution so obtained is close to the correct value. It seems therefore reasonable to take [11.28]:

$$f_t' = \frac{Q}{\pi(1.20\ bh - a^2)}$$ [11.29]

as a working formula for computing the tensile strength in a double-punch test.

It is important to note that earlier bearing capacity tests indicate that when the ratio b/a or $h/2a$ is greater than 5 approximately, the local deformability of concrete in tension is not sufficient to permit the application of limit analysis. Rupture of the mortar or concrete penetrates progressively downwards from the tip of the cone shape formed directly under the punch and crack propagation dominates. In such circumstances, the applied load Q becomes equal to that of a double-punch test with the ratio $b/a = 5$ or $h/2a = 5$. Therefore, for any ratio $b/a > 5$ or $h/2a > 5$, the limiting value $b = 5a$ or $h = 10a$ should be used in [11.29] for the computation of the tensile strength in a double-punch test. For example, for the dimensions used in a cylinder compression test: $2b = 6$ inches (15.30 cm) and $h = 12$ inches (30.60 cm), assuming the same punch diameter $2a = 1.5$ inches (3.80 cm), the appropriate value for h in [11.29] is 7.5 inches (19.20 cm) instead of the value 12 inches (30.60 cm).

Equation [11.29] may be considered also valid for the case of a circular double punch on a square block specimen. However, the restrictions on the limiting value of the ratio $b/a = 5$ (specimen width/punch diameter) or $h/2a = 5$ should be taken into account in a similar manner.

The following example shows a typical double-punch test for a cylinder speci-

TABLE 11.2

Tensile strength computed from double-punch test

Make	Cylinder height h inches (cm)	Ultimate load Q kip (kg)	f'_t psi (kg/cm^2)	$\dfrac{f'_c}{f'_t}$
Mortar	6 (15.30)	26.5 (12.0)	400 (28.0)	11.0
	6 (15.30)	26.4 (11.9)	400 (28.0)	10.5
	4 (10.20)	20.6 (9.3)	474 (33.2)	9.4
	4 (10.20)	20.1 (9.1)	463 (32.5)	9.1
Concrete	6 (15.30)	32.2 (14.6)	487 (34.1)	13.2
	6 (15.30)	29.8 (13.5)	452 (31.7)	12.3
	4 (10.20)	27.0 (12.3)	620 (43.4)	10.4
	4 (10.20)	25.3 (11.5)	582 (40.8)	9.6

Punch diameter = $2a$ = 1.5 inches (3.80 cm). Cylinder width = $2b$ = 6 inches (15.30 cm).

men: $Q = 26,500$ lb. (12 kg), $2a = 1.5$ inches (3.80 cm), $2b = 6$ inches (15.30 cm), and $h = 6$ inches (15.30 cm):

$$f'_t = \frac{26,500}{\pi[1.20 \times 3 \times 6 - (0.75)^2]} = 402 \text{ psi} \left(\approx \frac{1}{11} f'_c\right) \qquad [11.30]$$

Table 11.2 shows the tensile strength computed from the results of a number of double-punch tests reported by Hyland and Chen (1970).

Since the size of the punch is small compared to that of the specimen, the second term in the parentheses on the right hand-side of [11.29] is small compared to the first term. For instance, in the example chosen, $a^2 = 0.56$, which is small compared to $1.2\ bh = 21.6$. Hence, [11.29] could be further simplified as:

$$f'_t = \frac{Q}{3.67\ bh} \qquad [11.31]$$

11.6.2. Plastic stress distributions by finite element method

To see the stress distributions along the vertical and horizontal center planes of the double-punch test, the analytical results using the elastic, strain-hardening and fracture model described in the preceding section is given below.

The contact surface between the punch and the block is assumed to be perfectly rough. The geometry used in the numerical analysis is $h/2b = 1$, $b/a = 4$ and $2a = 1.5$ inches (3.80 cm) (see Fig. 11.12).

Fig. 11.16 shows a typical load-deflection curve of the problem. Nonlinear behavior is seen to occur at a very early stage of loading ($Q/\pi a^2 = 0.61\ f'_c$). The finite

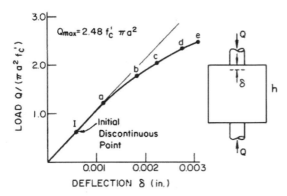

Fig. 11.16. Typical load-deflection curve for a concrete cylinder subjected to double punches.

element subdivision for a quarter of the cylinder is shown in Fig. 11.17. Plastic yielding is seen to start near the edge of the punch. The spreading of the plastic zone is seen to form a cone-shaped zone underneath the punch (Fig. 11.17a) until a crack is developed at edge B (Fig. 11.17b). Here, the Arabic or Roman numbers indicated in the elements show the order in which the elements become plastic or

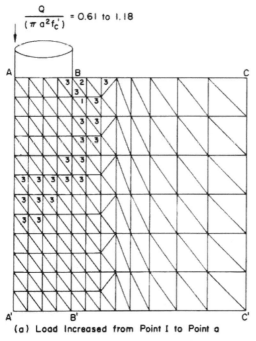

(a) Load Increased from Point I to Point a

Fig. 11.17. Spreading of elastic–plastic fracture zones to a concrete cylinder subjected to double punches.

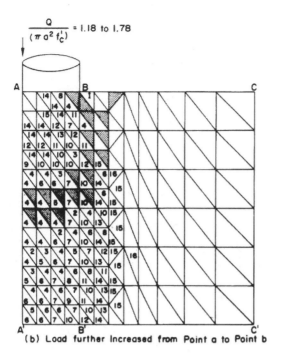

(b) Load further Increased from Point a to Point b

Fig. 11.17. Continued.

fracture during the loading interval in question, and the dark and shaded areas indicate the regions already yielded and fractured previous to that loading interval.

After the initiation of the first crack, the plastic zone spreads extremely fast; for load increases of less than 5% of the ultimate load, the region near the geometrical center of the cylinder becomes fully plastic (Fig. 11.17b). When the loading is about 80% of the total load (point c on the load-deflection curve, Fig. 11.16), the crack beneath the edge of the punch starts to propagate while the plastic zone keeps on spreading. Finally, when most of the elements underneath the punch edge have fractured, the cylinder collapses and the maximum load obtained is $2.48 f_c' \pi a^2$ which agrees rather well with the experimental values $2.64 f_c' \pi a^2$.

The redistribution of the vertical contact stress between the punch and the cylinder and the vertical stress on the horizontal middle plane of the cylinder are shown in Fig. 11.18. At the initial yield load (point marked I in Fig. 11.16), the ratio of the vertical stress at the punch edge to that at the punch center is about 2.7. As the load increases, this ratio decreases and becomes 1.6 just before collapse.

The contact plane AB between the punch and the cylinder and the horizontal middle plane $A'C'$ are labelled in Fig. 11.17 along with the finite element sub-division. The stress distributions along the vertical axis AA' and BB' on the vertical plane are given in Figs. 11.19 and 11.20, respectively. The similarity of the curves

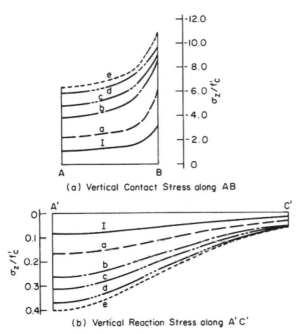

(a) Vertical Contact Stress along AB

(b) Vertical Reaction Stress along A'C'

Fig. 11.18. Redistribution of contact stresses along *AB* and *A'C'* of a concrete cylinder subjected to double punches.

along the axis *AA'* and *BB'* are evident except for the differences adjacent to the punch. This is probably caused by the cracking of the material near the punch edge *B*. The maximum tensile stress distribution across the center vertical planes occurs along the axis *AA'* (Fig. 11.19). The maximum tensile stress on this axis occurs at

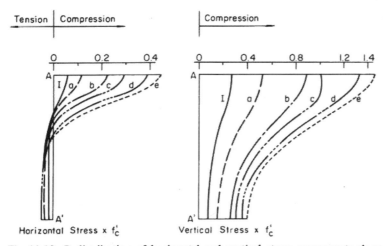

Fig. 11.19. Redistribution of horizontal and vertical stress components along *AA'* for a concrete cylinder subjected to double punches.

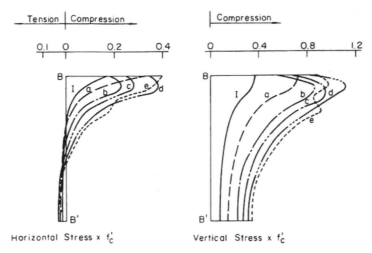

Fig. 11.20. Redistribution of the horizontal and vertical stress components along *BB'* for a concrete cylinder subjected to double punches.

the center point and is about 0.75 of the f_t' value calculated by [11.29]. From a practical viewpoint, the precise maximum values of the tensile stresses in any indirect tensile test are not essential. Therefore, it is considered that the limit analysis solution ([11.29]), is useful and convenient in predicting the comparative orders of magnitude of the tensile strength of materials developed under the double-punch test.

11.7. EXPERIMENTAL RESULTS OF DOUBLE-PUNCH TEST FOR CONCRETE MATERIALS

In the following, the double-punch test for tensile strength of concrete materials is examined experimentally, and the effect of several variables upon the observed strength and uniformity of the test results are discussed. Once these effects have been evaluated, a standard testing procedure for obtaining uniform results can be considered. In addition, comparisons of tensile strength of concrete determined from double-punch tests and split-cylinder tests are also presented.

Before we move on to discuss the experimental results we note that an ideal failure mode for a double-punch test on a cylinder specimen consists of several tensile crack planes radiating from a central axis and two cone-shaped rupture regions sheared in the compression zones directly beneath the punches (Fig. 11.14). The cone shapes approach each other and are shown in the preceding section to produce over the diametral planes involved, an almost uniform tensile stress given by [11.29]. Equation [11.29] gives an average tensile stress which exists over all of the cracked diametral planes (Fig. 11.14) and, thus, a large, more effective sample area than that of the split cylinder test is obtained. Many cracks also indicate more even stress distribution in the test specimen. Where the specimen's top and bottom surfaces are very rough or not parallel to each other, the specimen may fail in only two cracks, and usually at a significantly lower load. Most specimens fail in three or four cracks.

In the split cylinder test the plane of failure of the specimen is predetermined. That is, it will crack vertically whether that plane happens to be the strongest or weakest area of the specimen. In contrast to this, the double-punch test does not predetermine the failure plane and so will fail in the *weakest* planes. This explains the consistently *lower* strengths obtained for the double-punch test than that of the split-cylinder test.

11.7.1. *Experimental results*

In the following we summarize the results of experimental investigations on double-punch tests. Details of the experiments are given elsewhere (Chen and Trumbauer, 1972; Chen and Colgrove, 1974).

Effect of dimensions
The effect of changing the ratio of specimen surface area to the loaded area was investigated by varying the punch diameter from 1.0 (2.55 cm) to 1.5 (3.80 cm) to 2.0 (5.10 cm) inches, while keeping the specimen diameter constant at 6 inches (15.20 cm). By varying the cylinder height it was possible to determine the effect of height, diameter, and loaded area versus tensile strength.

The relationship between the standard split-cylinder tensile strength and double-punch tensile strength ratio with various specimen heights is shown in Fig. 11.21. The results included in Fig. 11.21 show that the double-punch tensile strength is uniform for all specimen heights with a 1.50-inch (3.82-cm) diameter punch except for the 4-inch (10.20-cm) cylinder height. This agreement is shown in Table 11.3.

Table 11.4, a comparison between the tensile strength of the 6-inch-high cylinders and the 6-inch cubes is shown. Note that the values of f_c'/f_t' in Table 11.4, are almost identical. The values used for f_c' are those of the standard 6×12-inch (30.4-cm) compression cylinder and the 6-inch compression cube. The values of f_t' are double-punch tensile strengths calculated from [11.29].

Effect of concrete mix
A comparison between the double-punch and split-cylinder test at different concrete mix proportions was made by varying the water cement ratio as given in Table 11.5.

As can be seen, the double-punch tensile strengths closely parallel the strengths

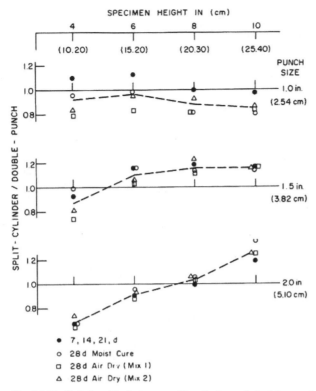

Fig. 11.21. Relationship between split cylinder and double-punch tensile strength.

given by the split-cylinder tests with significantly lower variability in the upper water cement ratios.

Effect of curing and aging

The test was undertaken to determine if specimens tested by the double-punch method reflected the same strength changes throughout their curing period as those tested by the split-cylinder test method. Both regular and lightweight concrete was

TABLE 11.3

Double-punch tensile strength in psi (kg/cm^2) with 1.5-inch (3.82 cm) punch

Curing condition (days)	Specimen height, inch(cm)			
	10(25.40)	8(20.30)	6(15.20)	4(10.20)
14-day moise cure	432 (30.3)	424 (30.0)	434 (30.5)	538 (37.8)
28-day moise cure	484 (34.0)	484 (34.0)	474 (33.3)	624 (43.8)
28-day air dry mix (1)	450 (31.5)	480 (33.7)	508 (35.7)	710 (49.8)
28-day air dry mix (2)	501 (35.2)	472 (33.1)	550 (38.6)	723 (50.8)

TABLE 11.4

Double-punch test for 6-inch high cylinders and cubes

Sample	Curing condition (days)	Punch diameter inch (cm)	Average ultimate load,* Q kip (kg)	Double-punch, f_t psi(kg/cm²)	Simple compression, f'_c psi(kg/cm²)	$\dfrac{f'_c}{f'_t}$
6-inch cylinder	28 moist	1.0 (2.54)	26.70 (12.1)	575 (40.3)	6510 (457)	11.3
		1.5 (3.82)	31.32 (14.2)	473 (33.2)		13.7
		2.0 (5.10)	38.15 (17.3)	590 (41.4)		11.0
	28 air dry	1.0 (2.54)	31.82 (14.5)	685 (48.1)	6960 (488)	10.2
		1.5 (3.82)	33.72 (15.3)	510 (35.8)		13.6
		2.0 (5.10)	38.94 (17.6)	602 (42.2)		11.5
6-inch cube	28 moist	1.0 (2.54)	28.96 (13.1)	624 (43.8)	7190** (505)	11.5
		1.5 (3.82)	34.18 (15.5)	549 (38.6)		13.1
		2.0 (5.10)	39.62 (18.0)	613 (43.0)		11.7
	28 air dry	1.0 (2.54)	28.70 (13.0)	619 (43.4)	7190** (505)	11.6
		1.5 (3.82)	33.46 (15.2)	507 (35.6)		14.1
		2.0 (5.10)	40.74 (18.5)	631 (44.3)		11.4

* Average of three tests with natural surface.
** Compression cube.

TABLE 11.5

Regular concrete mix with various water cement ratios

Water cement ratio	Simple compression, psi (MN/m²)	Split cylinder, psi (MN/m²)	Coefficient of variation (%)	Double punch psi (MN/m²)	Coefficient of variation (%)
0.4	5396 (27.21)	505 (3.48)	2.16	376 (2.60)	4.89
0.5	4907 (33.83)	506 (3.49)	3.17	394 (2.71)	6.96
0.6	4176 (28.79)	461 (3.18)	9.76	373 (2.57)	1.79
0.7	3634 (25.06)	398 (2.74)	9.68	354 (2.44)	3.76

studied in this test. Figure (11.22) shows the parallel correlation between the two tests for both types of concrete. This therefore indicates the sensitivity of both methods to record the strength changes with time.

Effect of wooden bearing discs

Plywood discs, $\frac{1}{8}$ inch thick and with diameters corresponding to those of the metal punches, were used to determine the effects of surface roughness between the punch and the specimen. The ratio between the double-punch tensile strength with

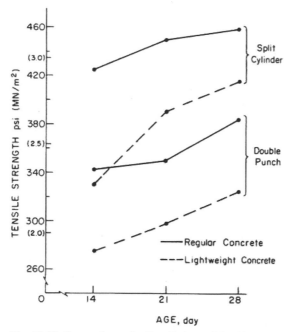

Fig. 11.22. Comparison of split cylinder and double-punch tests throughout curing period.

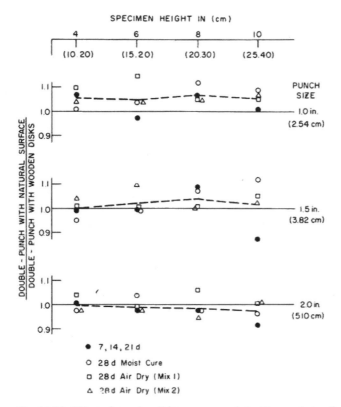

Fig. 11.23. Effect of wooden disks on measured double-punch tensile strength.

the punch bearing directly on the surface and the double-punch tensile strength with wooden discs between the punch and the specimen is shown in Fig. 11.23. It can be seen that the difference is not significant; if the surfaces of the specimen are troweled smooth during casting, no wooden discs are necessary.

Effects of stressing rate

The influence of the rate of stressing on the double-punch test was determined using lightweight and regular concrete specimens stressed at ranges of from 100 psi/min (0.69 MN/m^2/min) to 1000 psi/min (6.89 MN/m^2/min). The results are shown in Table 11.6 and from that it is suggested to a stressing rate of 100–200 psi/min (0.69–1.38 MN/m^2/min).

Effect of molds

In split-cylinder testing cylinders cast in cardboard molds give specimens with lower strengths and higher variability than specimens cast in steel molds.

Regular and lightweight concrete specimens were cast in both cardboard and

TABLE 11.6

Effect of stressing rate

Rate, psi/min. (MN/m² /min.)	Regular concrete		Lightweight concrete	
	strength,* psi (MN/m²)	coefficient of variation, %	strength,* psi (MN/m²)	coefficient of variation,%
100 (0.69)	379 (2.62)	2.74	264 (1.82)	1.59
200 (1.38)	390 (2.69)	4.14	261 (1.80)	2.18
300 (2.07)	364 (2.51)	7.23	276 (1.90)	6.46
500 (3.45)	368 (2.54)	4.18	287 (1.98)	4.75
1000 (6.89)	362 (2.50)	14.00	267 (1.84)	3.72

* Average of six test results at 28 days.

steel cylinder molds. Standard 12-inch (30.50-cm) cardboard molds were cut to 6 inches (15.20 cm) heights and false bottoms were made for the steel molds. Cube specimens were also cast in both plywood and steel molds.

The results in Table 11.7 show the double-punch test consistently reflects greater strengths and lower coefficients of variation in the case of steel molds. This therefore indicates the sensitivity of the double-punch method to record these changes.

Effect of testing machine

Testing-machine conditions may significantly affect the measured strength of concrete. Care must be taken to accurately align the punches and specimen in the testing machine. Each of the testing machines used was fitted with a spherical bearing block on the upper platen. Tests were made on the type of lubricant used on the upper platen. With a poor lubricant, the platen is able to move initially but then breaks down under load and becomes effectively fixed. With a high-pressure lubricant the spherical bearing is able to adjust throughout the loading.

TABLE 11.7

Effect of mold type on specimen strength

		Cylinder		Cube	
		cardboard	steel	plywood	steel
Regular	Strength, psi (MN/m²)	333 (2.30)	364 (2.51)	335 (2.31)	354 (2.44)
	coefficient of variation	5.98	2.07	1.53	1.32
Light-weight	strength, psi (MN/m²)	261 (1.80)	264 (1.82)	271 (1.87)	274 (1.89)
	coefficient of variation	2.18	2.64	2.99	1.78

TABLE 11.8

Effect of two types of lubricant on measured strengths

	Strength, psi (MN/m²)	Coefficient of variation (%)
High-pressure lubricant	361 (2.49)	1.69
Poor lubricant	329 (2.27)	11.3

In this test a low-grade all-purpose grease was compared to a high-pressure graphite lubricant. Again, as with the mold test, the double-punch test was sensitive to this condition and able to accurately reflect the changes. In the case of the high-pressure graphite lubricant the strength was significantly higher due to the more evenly distributed load and the coefficient of variation was sharply lower than with the poor lubricant (Table 11.8).

It was also decided to investigate the effect, if any, of the size of testing machine used. The results are given in Table 11.9. Three machines, a 300 kip (1334 kN) Baldwin hydraulic machine, a 120 kip (534 kN) Tinius-Olsen mechanical machine, and a 60 kip (267 kN) Baldwin hydraulic machine were used for this test. The measured double-punch tensile strength of concrete is seen to be insensitive to the size of testing machine.

Effect of sawed specimens

Full size 6 × 12 inch (15.30 × 30.50 cm) cylinders were sawed in half to make two 6-inch cylinders useable for double-punch testing. Top and bottom halves were tested to determine any differences in double-punch strength. The tests were undertaken at different laboratories, Fritz Engineering Laboratory, Lehigh University, and at the Lehigh Portland Cement Company (LPC). As can be seen by the data in Table 11.10, the strengths of the two halves are not the same, and the bottom specimens are significantly stronger than that of the top specimen.

TABLE 11.9

Effect of three types of testing machine on the measured strength

	60 kip	120 kip	300 kip
Strength, psi (MN/m²)	361 (2.49)	362 (2.50)	358 (2.47)
Coefficient of variation, %	1.49	1.61	3.10

TABLE 11.10

Effect of sawed specimens

	Fritz lab.	LPC lab.
Strength, top half, psi (MN/m^2)	333 (2.30)	324 (2.24)
Strength, bottom half, psi (MN/m^2)	374 (2.58)	366 (2.52)

11.7.2. Double-punch versus split cylinder testing procedure

The double-punch test is considered to have three major advantages over the popular split-cylinder test: (1) the relative simplicity of performing the test; (2) the fact that a lower capacity testing machine can be used; and (3) a *truer* tensile strength.

In the split-cylinder test (see Fig. 11.1a) it is necessary to lay the specimen lengthwise between the platens of the testing machine, being careful to keep the specimen properly centered. Wooden strips must be placed on the top and bottom contact surfaces of the cylinder, and sometimes metal plates are placed over the strips. The head of the machine must then be lowered until contact is made with the specimen being careful that the specimen remains centered. Upon failure, the specimen frequently falls after it has split and breaks into pieces. A special device must be used in order to keep the specimen intact.

In the double-punch test the punches are centered on the top and bottom surfaces with the templates, proper alignment is carefully insured, an elastic rope is tied around the perimeter of the cylinder, then the head of the machine is lowered, and loaded to failure. The elastic rope will hold the specimen together after failure.

The split-cylinder test requires a load at failure of approximately 50–70 kips (222–311 kN), while the double-punch test requires only 30–40 kips (133–178 kN). Since a smaller machine is required for the double-punch test, it is practical to perform the test in the field on small portable testing machines, or in laboratories which do not have large machines. For example, the double-punch test could be tied in with routine CBR or compaction tests for soils, hence, no additional equipment is needed for the determination of the tensile strength of soils (or stabilized materials). Similar examples may be found in the testing of rock.

The double-punch test also works satisfactorily under the above-mentioned procedure for a cube specimen and is considered much easier than performing a diagonal split-cube test.

The double-punch test gives an average tensile strength which exists over all of the failure planes, and a "truer" strength than the split cylinder test because of the *weak link theory*.

11.7.3. Recommended procedure

The testing procedure recommended to minimize variability is based on a 6-inches high by 6 inches diameter cylinder with two 1.5-inch diameter punches. Use of wooden discs between the punches and the surfaces of the specimens is not necessary, providing the surfaces are reasonably smooth.

For those countries in which cubes are used for testing, a 6-inch cube with 1.5-inch-diameter punches is recommended. Specimens should be loaded at a rate of 100–200 psi per minute and a high-pressure lubricant should be used on the spherical bearing block for lower testing variability during the double-punch test.

11.8. EXPERIMENTAL RESULTS OF DOUBLE-PUNCH TEST FOR ROCKS

This section presents some experimental results of the tensile strength of rocks as determined by the double-punch test method. Comparisons of tensile strength determined from double-punch and split-cylinder tests are also presented.

Before we move on to discuss the experimental results we note the fact that regardless of the strength parameters sought or the manner of testing; the significant strength of rock is often difficult to determine due to the inherited irregularities in the rock. If tests are made on specimens cut from cores, the selection of the specimen can greatly influence the result. Recognizing this problem associated with rock testing, some rock strength tests using the double-punch, split-cylinder and compression methods are discussed in what follows. The main purpose here is to compare the results of the double-punch test to those of the split-cylinder test. The inclusion of the compression test here is for additional rock classification purposes.

Test specimens can be prepared from available rock core samples. Typical core samples including depths to the core recovery, and types of rock are given in Table 11.11. From the rock descriptions given in the table, the strength is expected to range from medium to high. The specimen length/diameter ratios used for the test results listed in Table 11.11 are 1 for split-cylinder and double-punch tests and 2 for simple compression test. The punch diameter/specimen diameter ratio used for the double-punch test results is approximately $\frac{1}{4}$, following the conclusions reached in section 11.7.

11.8.1. Test results

Table 11.11 shows the test results and computed tensile and compressive strengths. Details of the experiments are given elsewhere (Dismuke et al., 1972). The split-cylinder tensile strength is calculated from [11.5] and the double-punch tensile strength is calculated from [11.29].

TABLE 11.11

Compilation of some test results

Core No.	Rock type	Depth (ft)	Dia (inch)	Unit weight		Double-punch f_t		Split-cylinder f_t		Compressive strength f_c'		$\dfrac{f_c'}{f_t}$*
				psf	(g/cm³)	psi	(kg/cm²)	psi	(kg/cm²)	psi	(kg/cm²)	
1	Epidote	202	1.644	213.7	(3.419)	1643	(116)	1793	(126)	13,113	(922)	8.0
2	Syenite	644	1.274	170.5	(2.728)	1947	(137)	2184	(154)	10,824	(761)	5.6
4	Epidote	853	1.269	200.5	(3.208)	1783	(125)	2575	(181)	33,228	(2336)	18.6
5	Syenite-epidote-Grenville	310	1.269	198.6	(3.178)	2136	(150)	2730	(192)	33,045	(2323)	15.5
7	Altered dolomite	302	1.626	200.2	(3.203)	2176	(153)	3188	(224)	8,185	(575)	3.8
9	Epidote magnetite	550	1.269	210.2	(3.363)	1657	(117)	1877	(132)	20,646	(1451)	12.5
11	Altered grenville	214	1.269	176.2	(2.835)	2626	(185)	2368	(166)	13,102	(921)	5.0
12	Altered Grenville-syenite	913	1.269	165.5	(2.648)	1214	(85)	1198	(84)	7,577	(533)	6.2
13	Syenite	452	1.270	166.1	(2.658)	1051	(74)	1147	(81)	8,603	(605)	8.2
15	Epidote	330	1.267	212.7	(3.403)	2410	(169)	2835	(199)	10,451	(735)	4.3
16	Syenite	222	1.269	170.4	(2.726)	2625	(185)	2700	(190)	13,339	(938)	5.1
17	Epidote Grenville	210	1.267	200.4	(3.206)	2442	(172)	2985	(210)	9,849	(692)	4.0
24	Diabase-feldspar	1673	1.634	188.3	(3.013)	2960	(208)	2735	(192)	17,024	(1197)	5.8
27	Gneiss	900	1.645	175.0	(2.800)	2940	(207)	2976	(209)	16,141	(1135)	5.5
33	Quartz conglomerate	1290	1.640	159.6	(2.554)	1213	(85)	1416	(100)	4,733	(333)	3.9
35	Gneiss	1430	1.630	182.7	(2.923)	2233	(157)	2404	(169)	16,659	(1171)	7.5
36	Metamorphic feldspar and magnesite	1496	1.660	173.7	(2.779)	1476	(104)	1534	(108)	8,545	(601)	5.8
38	Diabase	1190	1.650	186.0	(2.976)	1346	(95)	1700	(120)	13,811	(971)	10.3

* f_t' = double-punch tensile strength.

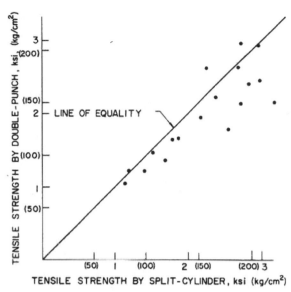

Fig. 11.24. Tensile strength determined by split-cylinder and double-punch methods.

Fig. 11.24 shows the comparison between the tensile strength as determined by the double-punch and split-cylinder test methods. The compressive and tensile strength are plotted with respect to the unit weight in Fig. 11.25.

The tensile strength determined by the double-punch test, as shown in Fig. 11.24, is lower than that determined from the split-cylinder tensile test. This is partly due to the fact that the presence of weak planes or discontinuous planes in a rock specimen causes the radial breaks in the double-punch test to follow the weaker planes. In contrast to the above, the strength obtained from the split-cylinder test is influenced — to a degree — by orientation of the specimen between the platens. If a weak plane in the specimen is placed parallel to the platens a much higher strength may be indicated than if it is rotated 90°. In tests Nos. 1, 7, and 17 the specimens broke along joints in both the double-punch and compression tests.

It is of interest to note that the split-cylinder test specimens usually developed a wedge at the line of load application. Approximately one-half of the double-punch specimens listed in Table 11.11 did not develop cones under both punches. This may have been due to irregularities in the specimens.

As seen in Fig. 11.25, the compressive strength and — to a lesser extent — tensile strength of the specimens generally increased with unit weight, which is to be expected.

11.8.2. Summary and conclusions

The double-punch test method for rocks appears to give lower values than the

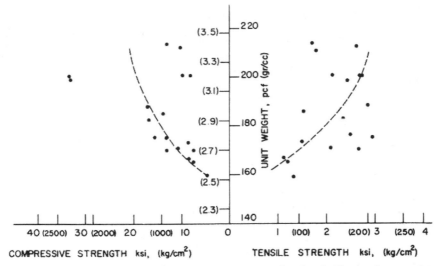

Fig. 11.25. Unit weight vs. compressive and tensile strengths of rock. The tensile strength is the average of double-punch and split-cylinder tests.

split-cylinder test. This occurs, primarily, because in the double-punch test the failure can occur on any one of the infinitely many radial planes in contrast to the split-cylinder test in which the plane of failure is predetermined. Therefore, the double-punch test is probably better for most applications than the split-cylinder test. The split-cylinder test for rock specimen is easier to shape and the test is easier to perform than the double-punch test.

The test results discussed herein amplify the fact that core recovery location and selection of test samples from the core will largely determine the strength values. Only experienced personnel should select core samples as the existence (orientation and strength) of weak planes in the core vitally affects the results.

11.9. EXPERIMENTAL RESULTS OF DOUBLE-PUNCH TEST FOR SOILS

In the following we shall present some typical experimental results on the tensile strength of cohesive soils as determined by the double-punch test method. Since the Proctor mold and CBR mold are readily available in soil laboratories or field construction, it is convenient to prepare the soil samples directly either from the Proctor mold (4 × 4.6 inches) or the CBR mold (6 × 6 inches) using 1 inch (2.54 cm) or 1.33 inch (3.38 cm) (CBR piston) punch. It should be noted that the formula [11.29] derived previously for computing the tensile strength of concrete or rock is not directly applicable here for soils. A simple modification of this formula is needed. It is found (Fang and Chen, 1971) that the value of the coefficient 1.20 appeared in [11.29] should be modified to the value of 1.0 for the case of soils.

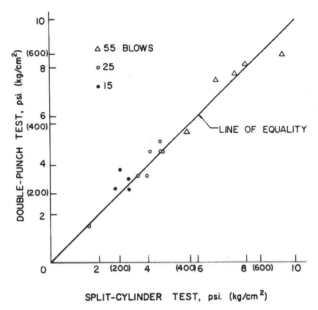

Fig. 11.26. Comparisons of tensile strength of soil determined by double-punch and split-cylinder tests.

Fig. 11.26 shows the comparisons of the tensile strength as determined by double-punch and split-cylinder tests with different values of compaction effort. Good agreement between these two tests is observed. Since the double-punch test can be directly tied in with routine CBR or compaction tests, no additional equipment is needed. It would be reasonable to conclude that the double-punch test is valid and convenient for both laboratory and field construction control of soils.

Chapter 12

SOIL PLASTICITY – THEORY AND APPLICATION

12.1. INTRODUCTION

Soil plasticity along with all other branches of mechanics of solids requires the consideration of geometry or compatibility and of equilibrium or dynamics, and of the relation between stress and strain. Equilibrium equations are determined directly by summation of forces (see [2.23]); compatibility equations ensure that strain or strain-rate components are consistent with a displacement or velocity field from which they must be derivable (see [2.24]). Compatibility and equilibrium equations are therefore independent of material properties and hence valid for elastic as well as elastic–plastic problems. The differentiating feature is the relation between stress and strain. The extreme difficulty in obtaining an exact plastic solution even with the aid of digital computers is due mainly to the fact that the stress–strain relationship in the plastic range is far more complicated than Hooke's law for linearly elastic materials.

Plastic action is load path-dependent and almost always requires step-by-step calculations that follow the history of loading. They are further complicated by the fact that the elastic–plastic boundary is changing with continued loading and the stress–strain relationships for loading and unloading are different. Even without this complication, there are few solutions available for problems that consider nonlinear elasticity.

It is apparent that an exact elastic–plastic solution of a practical soil mechanics problem is unlikely. Drastic simplifications and idealizations are essential for a reasonable approximate solution. The geometry or compatibility, the stress–strain relations and the equations of equilibrium must all be idealized to accomplish a solution.

For example, the material may be idealized as *perfectly plastic*. This ignores work-hardening or work-softening as symbolized by the two straight dashed lines in Fig. 2.1. As indicated previously in Chapter 2, perfect plasticity is an appropriate idealization for stability problems in soil mechanics. This idealization contains the essential features of certain soil behavior: the tangent modulus when loading in the plastic range is small compared with the elastic modulus, and the unloading response is elastic. Further, the strain level of interest in a problem determines the choice of flow stress. Thus, in a sense, this perfectly plastic flow stress represents somewhat an averaging work-hardening or work-softening of the material over the field of flow.

Rather recently Calladine (1963), Schofield, and Roscoe and his students (1958–63) have found that "wet" clays can be described remarkably well by a simple *isotropic work-hardening* idealization. More recently we find that a combination of isotropic hardening and kinematic hardening model can properly describe the behavior of rock-like materials such as concrete. In this Chapter this simple idealizations such as perfect plasticity, isotropic hardening, and kinematic hardening models for soil and rock-like materials will be presented.

The difficulty of an exact analysis of even a perfectly plastic material has led to the subdivision of the original continuous media into an assemblage of discrete elements. This discretization of continuous media by a finite number of elements is known as the *finite element* method. The finite element discretization leads to the approximate formulation of equilibrium equations in terms of generalized stresses and strain rates, such as nodal forces and nodal displacement rates. The compatibility conditions are satisfied by a proper *choice* of a displacement function for the element. As a consequence of such an approximation, equilibrium condition is satisfied only for the discretized body, that is, overall nodal forces for every element are in equilibrium but the continuum equations of equilibrium are in general not satisfied by a finite element.

In this chapter, we discuss two elastic–plastic soil models – the extended Von Mises perfectly plastic model of Drucker and Prager (1952) (section 12.2) and a strain hardening model of Roscoe and Burland (1968) (section 12.3). In addition we present an elastic–plastic strain hardening-fracture model for describing the behavior of the rock-like material such as concrete (section 12.4). Incremental constitutive relationships are developed for all these models. Incremental equilibrium equations in terms of an individual finite element are then formulated in section 12.5 and subsequently used to develop the instantaneous stiffness of a continuous medium in a finite element formulation. Incremental integration techniques are reviewed in section 12.6 and some details of two integration schemes are also described. Two example problems and numerical experiments are presented in sections 12.7 and 12.8. A summary and some conclusions are given in section 12.9.

12.2. EXTENDED VON MISES PERFECTLY PLASTIC MODEL FOR SOIL

12.2.1. General

The mechanical behavior of all materials is complex and must be drastically idealized in order to make mathematical analysis tractable. For example, metal behavior has been extensively investigated, yet in most analytical work metal is idealized as being a perfectly plastic, isotropic hardening or kinematic hardening material. In certain instances all three idealizations can be shown to fall short of real metal behavior. The proper idealization is, of course, problem-dependent. For instance, most engineers would feel justified in using an elastic model for initial settlement analysis of a footing when the working load was far below the maximum bearing capacity of the footing.

Soil is less amenable to simple modeling than is metal. Unlike metal, soil behavior is affected by hydrostatic pressure, and the tensile and compressive behavior of soils differ. Here we treat soil as a plastic material and assume that plasticity theory applies. We discuss in this and following sections two elastic—plastic soil models — an elastic—perfectly plastic model and an elastic—plastic strain hardening model. We view both models as useful computational tools not as highly accurate predictors of detailed soil stress—strain behavior.

12.2.2. Typical soil stress—strain behavior

Some representative stress—strain curves for soil are shown in Fig. 12.1. For the moment we think in terms of a strain-controlled triaxial test, and except where noted stress means *effective stress.*

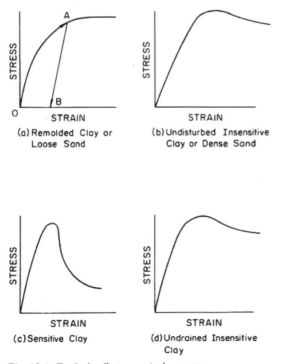

Fig. 12.1. Typical soil stress—strain curves.

The stress-strain behavior of loose sand or remolded clay is characterized by a highly nonlinear response curve which rises to a maximum and remains there as straining is continued. The behavior of undisturbed insensitive clay is characterized by an initial linear portion and peak stress followed by softening to a residual stress. Sensitive clay behaves similarly except that the difference between the peak stress and the residual stress is substantial. Finally, undrained total stress behavior of clay is characterized by an initial linear portion and peak stress with perhaps some strain softening.

In the most fundamental sense, soil is a plastic material rather than an elastic material. For example, considering Fig. 12.1(a), if we strain the soil to point A and then reverse the strain direction such that complete unloading takes place, we find that we are left with a residual strain, OB. A nonlinear elastic material would unload along loading path OA, and it is in this sense that soil is plastic rather than elastic.

It is of course not necessarily the case that soil stress–strain behavior can be successfully modeled using the classical theory of plasticity. It is in fact possible that nonlinear elastic models may be more suitable for most loading histories. Duncan and Chang (1970) have used a nonlinear elastic model (with linear unloading) to successfully predict the response of a sand in the triaxial test when a fairly complex stress history was prescribed.

Here we choose to use elastic–plastic models to describe soil stress–strain behavior. We discuss first the extended Von Mises (or Drucker-Prager) perfectly plastic model followed by a Cambridge strain hardening model.

12.2.3. The extended Von Mises yield criterion

The stress–strain curves shown in Figs. 12.1(a, b, and d) can all be approximated to some degree by a linear elastic–perfectly plastic model. It is unlikely however that a perfectly plastic idealization would be a useful model for sensitive clays. The complete description of an elastic–perfectly plastic model entails appropriate elastic constants, a yield function and a flow rule.

There exist a number of failure criteria which reflect a fundamental feature of soil behavior, that is, soil failure, unlike metal yield, is in some way a function of the hydrostatic stress component. The Coulomb criterion (see Bishop, 1966) is certainly the best known of these criteria. Shield (1955a) presented a pictorial representation of the Coulomb criterion in three-dimensional principal stress space (see Fig. 2.3) and also discussed the criterion in the context of perfect plasticity and the associated flow rule. Drucker and Prager (1952) discussed an extension of the well known Von Mises yield condition which included the hydrostatic component of the stress tensor ([2.7] and Fig. 2.5), and subsequently Drucker (1953) presented the so-called extended Tresca yield condition (see Fig. 2.5).

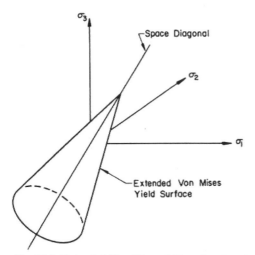

Fig. 12.2. Extended Von Mises yield surface in principal stress space.

The extended Von Mises yield function, as viewed in three-dimensional principal stress space, is shown in Fig. 12.2. The space diagonal is a line defined by $\sigma_1 = \sigma_2 = \sigma_3$ where σ_1, σ_2 and σ_3 are the principal stresses. Any plane perpendicular to the space diagonal is referred to as an octahedronal plane. The extended Von Mises yield condition is a cone with the space diagonal as its axis. The extended Tresca criterion is a pyramid with a regular hexagonal base and the space diagonal as its axes, while the Coulomb criterion is a pyramid with an irregular hexagonal base and the space diagonal as its axis. The intersections of the π-plane $\sigma_1 + \sigma_2 + \sigma_3 = 0$ with these yield surfaces are shown in Fig. 2.5.

Bishop (1966) has attempted to correlate all three criteria with experimental data and has concluded that the Coulomb criterion best predicts soil failure. Roscoe et al. (1963) content that the available experimental data (particularly triaxial extension tests) are not sufficiently reliable to allow one of the criteria to be favored over the others. They thus recommend the extended Von Mises criterion because of its simplicity. Furthermore, for the plane strain case it can be shown that in the limit state (where elastic strains are identically zero) both the extended Von Mises and the extended Tresca criteria reduce to a Coulomb type expression (Drucker, 1953; Drucker and Prager, 1952). This implies that we can adjust the constants of the extended Von Mises and extended Tresca criteria such that all three criteria will give identical limit loads. We note however that the three soil models will give different predictions for soil response below the limit load. In the spirit of the Cambridge soil models (Schofield and Wroth, 1968) and with the above discussion in mind, we utilized therefore in Chapter 5 the Drucker-Prager condition (extended Von Mises) for an analytical study of the static response of a homogeneous clay stratum to footing loads.

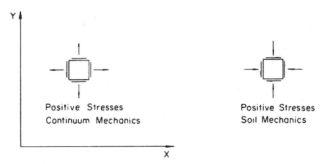

Fig. 12.3. Stress sign conventions.

Sign convention for stress

The typical continuum mechanics sign convention (tensile stresses positive) and the usual soil mechanics sign convention (compressive stresses positive) are shown in Fig. 12.3. In this Chapter the continuum mechanics sign convention is utilized while the stresses reported in other chapters follow the soil mechanics convention.

Yield function

Referring the components of stress and strain to any Cartesian coordinate system with axes x, y and z, the extended Von Mises yield criterion may be written as:

$$\alpha p + J_2^{1/2} = k \tag{12.1}$$

where p is the hydrostatic component of the stress tensor:

$$p = (\sigma_x + \sigma_y + \sigma_z)/3 \tag{12.2}$$

and J_2 is the second invariant of the deviatoric stress tensor:

$$J_2 = \tfrac{1}{6}[(\sigma_x - \sigma_y)^2 + (\sigma_y - \sigma_z)^2 + (\sigma_z - \sigma_x)^2] + \tau_{xy}^2 + \tau_{yz}^2 + \tau_{zx}^2 \tag{12.3}$$

where σ_x, σ_y, σ_z, τ_{xy}, τ_{yz}, τ_{zx} are Cartesian stress components at a point in the soil, and α and k are material constants. If α is zero, [12.1] reduces to the Von Mises yield condition. Referring again to the principal stress space shown in Fig. 12.2, $\sqrt{3}k$ corresponds to the radius of the cone at $p = 0$. We will find later that α is related to plastic volumetric strain.

In order that the extended Von Mises and Coulomb criteria give identical limit loads, α and k must be defined as follows (Drucker and Prager, 1952):

$$\alpha = \frac{3 \tan \varphi}{\sqrt{(9 + 12 \tan^2 \varphi)}} \tag{12.4}$$

$$k = \frac{3c}{\sqrt{(9 + 12 \tan^2 \varphi)}} \tag{12.5}$$

where c is the cohesion and φ is the friction angle of the soil.

12.2.4. Drucker's postulate

In the preceding sections we have repeatedly used the terms "work-hardening" and "work-softening" media by which we have meant the ascending and descending branches respectively of the soil stress–strain curve. In Fig. 12.1 the typical soil stress–strain curves (σ–ϵ curves) are symbolized by the simple triaxial tests.

In the case of the ascending branches of the σ,ϵ curves, an additional loading $\Delta\sigma > 0$ give rise to an additional strain $\Delta\epsilon > 0$, with the product $\Delta\sigma\,\Delta\epsilon > 0$. The additional stress $\Delta\sigma$ does positive work. Material of this kind is called *stable*. Work-hardening material is obviously a stable material.

In the case the stress–strain curve has a descending branch, where the strain increases with decreasing stress. On this segment the additional stress does negative work, i.e., $\Delta\sigma\,\Delta\epsilon < 0$. This material is called *unstable*. Work-softening material is an unstable material.

Drucker (1951) gives strong physical arguments for deducing the stress/strain rate relations directly from the yield surface for a stable material. Shield (1953) has investigated soil mechanics problems with Coulomb yield surface using such a theory. The results germain to this discussion are usually summarized as "*Drucker's Postulate*".

The formulation which results from the above considerations is most easily visualized in stress σ_{ij} space and strain rate $\dot{\epsilon}_{ij}$ space. In σ_{ij} space, the yield function determines a surface in this space. If the $\dot{\epsilon}_{ij}$ coordinate directions are taken to coincide with the σ_{ij} coordinate directions and $\dot{\epsilon}_{ij}$ is taken as a free vector, then "Drucker's Postulate" for a stable material leads to normality of $\dot{\epsilon}_{ij}$ to the yield surface $f(\sigma_{ij}) = 0$. This result then specifies the stress/strain rate relation since the length $\dot{\epsilon}_{ij}$ is indeterminant in a perfectly plastic or nonwork-hardening theory.

In the following sections such a formulation is introduced and developed for the case of three-dimensional flow.

12.2.5. Three-dimensional stress–strain relations

In order to determine the elastic–plastic, incremental stress–strain relations, we start with the normality or associated flow rule mentioned above and write, in indicial notation:

$$\dot{e}^p_{ij} = \lambda \frac{\partial f}{\partial \sigma_{ij}} \qquad [12.6]$$

where: $f = \alpha p + J_2^{1/2} - k$ \qquad\qquad [12.7]

and $\dot{\epsilon}^p_{ij}$ is the infinitesimal strain tensor with superscript p denoting plastic strain and the super dot denoting strain rate. The Cartesian stress tensor is denoted by σ_{ij}, and λ is an arbitrary non-negative number which is greater than zero for plastic loading $[f(\sigma_{ij}) = 0]$ and equal to zero for plastic unloading or if the stress state lies within the yield surface $[f(\sigma_{ij}) < 0]$. In general if the current stress state is known and the stress rate tensor is prescribed, the strain rate tensor is not uniquely determined since the plastic strains can only be defined to within the indeterminate parameter λ. Conversely if the strain rate is prescribed, the stress rate is uniquely determined. Since a displacement formulation is to be utilized here, we wish to develop an expression relating stress rate as a function of strain rate. Thus λ will be determined as a function of the strain rate tensor.

The elastic rate relationship between the stress and strain is:

$$\dot{\sigma}_{ij} = \frac{E}{1+\nu} \left(\dot{\epsilon}^e_{ij} + \frac{\nu}{1-2\nu} \dot{\epsilon}^e_{kk} \, \delta_{ij} \right) \tag{12.8}$$

where E is Young's modulus, ν is Poisson's ratio, δ_{ij} is the Kronecker delta, $\dot{\epsilon}^e_{ij}$ is the elastic component of the strain tensor and we sum over repeated indices (see section 2.3, Chapter 2 for summation convention). Noting that total strain rate $\dot{\epsilon}_{ij}$ is the sum of elastic and plastic strain rates, we can relate stress rate to total strain rate, as follows:

$$\dot{\sigma}_{ij} = \frac{E}{1+\nu} \left[\left(\dot{\epsilon}_{ij} - \lambda \frac{\partial f}{\partial \sigma_{ij}} \right) + \frac{\nu}{1-2\nu} \left(\dot{\epsilon}_{kk} - \lambda \frac{\partial f}{\partial \sigma_{kk}} \right) \delta_{ij} \right] \tag{12.9}$$

The stress rate–strain rate equation is fully defined once λ is known. To determine λ we note that during plastic loading, the stresses must lie on the yield surface, $f(\sigma_{ij}) = 0$, and:

$$df = \frac{\partial f}{\partial \sigma_{ij}} \, \dot{\sigma}_{ij} = 0 \tag{12.10}$$

that is, the stress rate vector must be tangent to the yield surface.

Equations [12.9] and [12.10] permit the determination of λ. To compute λ we start with some preliminaries as follows. From [12.7]:

$$\frac{\partial f}{\partial \sigma_{ij}} = \alpha \frac{\partial p}{\partial \sigma_{ij}} + \tfrac{1}{2} J_2^{-1/2} \frac{\partial J_2}{\partial \sigma_{ij}} \tag{12.11}$$

where: $\dfrac{\partial p}{\partial \sigma_{ij}} = \tfrac{1}{3} \delta_{ij}$ (note, $p = \tfrac{1}{3} \sigma_{kk}$) $\tag{12.12}$

$$\frac{\partial J_2}{\partial \sigma_{ij}} = s_{ij} \quad \text{(note, } J_2 = \tfrac{1}{2} s_{ij} \, s_{ji} \text{)} \tag{12.13}$$

and $s_{ij} = \sigma_{ij} - p\delta_{ij}$ is the deviatoric stress tensor. Hence we can rewrite [12.11] as:

$$\frac{\partial f}{\partial \sigma_{ij}} = \tfrac{1}{3}\alpha\,\delta_{ij} + \tfrac{1}{2}\,J_2^{-1/2}\,s_{ij} \qquad [12.14]$$

From [12.14] we can write:

$$\frac{\partial f}{\partial \sigma_{kk}} = \alpha \qquad [12.15]$$

We note from [12.15] that:

$$\dot{\epsilon}_{kk}^p = \lambda\frac{\partial f}{\partial \sigma_{kk}} = \lambda\alpha \qquad [12.16]$$

Thus for α other than zero, plastic volume change is nonzero. Finally, using [12.9], [12.14] and [12.15] we can rewrite [12.10] as:

$$df = 0 \Rightarrow \{\tfrac{1}{3}\alpha\,\delta_{ij} + \tfrac{1}{2}J_2^{-1/2}\,s_{ij}\}$$

$$\left\{\frac{E}{1+\nu}\left[\left(\dot{\epsilon}_{ij} - \frac{\lambda\,\alpha\,\delta_{ij}}{3} - \frac{\lambda J_2^{-1/2}}{2}\,s_{ij}\right) + \frac{\nu}{1-2\nu}\,(\dot{\epsilon}_{kk} - \lambda\,\alpha)\,\delta_{ij}\right]\right\} \qquad [12.17]$$

This equation can be solved for λ, and after some simplification we obtain:

$$\lambda = \frac{GJ_2^{-1/2}\,s_{pq}\,\dot{\epsilon}_{pq} + B\dot{\epsilon}_{kk}}{G + \alpha B} \qquad [12.18]$$

where G is the elastic shear modulus:

$$G = \frac{E}{2(1+\nu)} \qquad [12.19]$$

and: $B = \dfrac{2\alpha G}{3}\left(\dfrac{1+\nu}{1-2\nu}\right) \qquad [12.20]$

To obtain the desired stress rate—strain rate relationship, [12.18] is substituted into [12.9] to give:

$$\dot{\sigma}_{ij} = D_{ijpq}\,\dot{\epsilon}_{pq} \qquad [12.21]$$

where:

$$D_{ijpq} = \frac{E}{1+\nu}\left[\delta_{ip}\,\delta_{jq} + \frac{\nu}{1-2\nu}\,\delta_{ij}\,\delta_{pq}\right]$$

$$-\frac{[B\delta_{ij} + GJ_2^{-1/2}\,s_{ij}]}{G + \alpha B}\,[GJ_2^{-1/2}\,s_{pq} + B\delta_{pq}] \qquad [12.22]$$

The matrix D_{ijpq} is referred to here as the elastic–plastic constitutive matrix.

12.2.6. Plane strain constitutive matrix

For the plane strain case ($\gamma_{yz} = \gamma_{xz} = \epsilon_z = 0$) we can write, in matrix form:

$$\begin{Bmatrix} \dot{\sigma}_x \\ \dot{\sigma}_y \\ \dot{\tau}_{xy} \\ \dot{\sigma}_z \end{Bmatrix} = D \begin{Bmatrix} \dot{\epsilon}_x \\ \dot{\epsilon}_y \\ \dot{\gamma}_{xy} \end{Bmatrix} \qquad [12.23]$$

where the z axis is normal to the plane, $\dot{\gamma}_{xy}$ is the so-called engineering shearing strain rate:

$$\dot{\gamma}_{xy} = 2\dot{\epsilon}_{xy} \qquad [12.24]$$

and:

$$D = \frac{E}{(1+\nu)(1-2\nu)} \begin{bmatrix} 1-\nu & \nu & 0 \\ \nu & 1-\nu & 0 \\ 0 & 0 & (1-2\nu)/2 \\ \nu & \nu & 0 \end{bmatrix}$$

$$-\frac{1}{G+\alpha B} \begin{bmatrix} H_1^2 & H_1 H_2 & H_1 H_3 \\ H_2 H_1 & H_2^2 & H_2 H_3 \\ H_3 H_1 & H_3 H_2 & H_3^2 \\ H_4 H_1 & H_4 H_2 & H_4 H_3 \end{bmatrix} \qquad [12.25]$$

and:

$$H_1 = B + GJ_2^{-1/2} s_x \qquad [12.26a]$$

$$H_2 = B + GJ_2^{-1/2} s_y \qquad [12.26b]$$

$$H_3 = GJ_2^{-1/2} \tau_{xy} \qquad [12.26c]$$

$$H_4 = B + GJ_2^{-1/2} s_z \qquad [12.26d]$$

12.2.7. Concluding remarks

What we have done to this point is to develop a set of incremental stress–strain equations for soil using an elastic–perfectly plastic model. To the extent that soil failure can be predicted with the conventional material parameters c and φ, this model can capture soil failure (at least for plane strain). However this drastic idealization can not capture some important characteristics of soil behavior. For example, it has been often noted that the amount of dilation at failure predicted by the perfectly plastic model is considerably in excess of that observed experimentally (Drucker, 1961a). According to Drucker (1961a, 1966a) the failure surface for soil may not be the limit of a sequence of yield surfaces as is normally considered to be the case for metals. Accordingly, in the next section we discuss a strain hardening model which for some loading histories may more closely predict soil behavior than can a perfectly plastic model.

12.3. AN ELASTIC–PLASTIC STRAIN HARDENING MODEL FOR SOIL

12.3.1. A brief historical account

Considering again Fig. 12.1(a), we note that long before the maximum stress has been reached, some irreversible straining has occurred as evidenced by the fact that reloading from point A leaves a residual strain. In the context of the theory of plasticity this soil might be referred to as a strain hardening material since the onset of plastic yielding is not synonymous with the maximum stress. A few researchers have investigated the possibility of modeling soil as a strain hardening material, and in particular this has been one of the major thrusts of the soil mechanics group at Cambridge University for the past twenty years (Roscoe, 1970).

Apparently Drucker et al. (1957) were the first to suggest that soil might be modeled as an elastic–plastic strain hardening material. They proposed that successive yield functions might resemble extended Von Mises cones with convex end caps. As the soil strain hardens both the cone and cap expand. Drucker (1961a) again discussed this concept in a later paper in which he suggested that the failure surface may not be a yield surface. This point is further emphasized by Drucker (1966a) who noted that successive loading surfaces or yield surfaces do not approach the failure surface.

In 1958 Roscoe et al. published a paper which contained the basis for a number of subsequent strain hardening models for soil. The paper was concerned primarily with the behavior of soil in the triaxial test and contained the so-called "state boundary surface" (called a yield surface in the 1958 paper) postulate and the "critical state line" postulate. These concepts were utilized by Roscoe and Poorooshasb (1963) to develop a stress–strain theory for clay which was not, however, based upon the theory of plasticity. Calladine (1963) suggested an alternate interpretation of this theory using concepts from strain hardening plasticity. Subsequently Roscoe et al. (1963a) utilized the strain hardening theory of plasticity to formulate a complete stress–strain model for normally consolidated or lightly over-consolidated clay in the triaxial test. This model has since become known as the Cam-clay model (Schofield and Wroth, 1968).

Burland (1965) suggested a modified version of the Cam-clay model and this model was subsequently extended to a general three-dimensional stress state by Roscoe and Burland (1968). It is the modified Cam-clay model that we are con-cerned with here. We will see later that for certain stress histories modified Cam-clay strain softens rather than strain hardens.

12.3.2. Modified Cam-clay

Modified Cam-clay is an isotropic, nonlinear elastic–plastic strain hardening ma-terial. Only volumetric strain is assumed to be partially recoverable, that is, elastic distortional strain (shearing strain) is assumed to be identically zero. Elastic volu-metric strain is nonlinearly dependent on hydrostatic stress and is independent of deviatoric stresses.

In order to introduce the reader gradually to the idea of soil as a plastic strain hardening material, we consider first the response of soil to pure hydrostatic stress. Typical response for a real soil is shown in Fig. 12.4, where void ratio e_v is plotted versus the natural logarithm of the negative of the hydrostatic stress (p is of course negative here). If the current pressure, denoted by point A, is the greatest the soil has experienced, then upon application of increased pressure the soil will load along line AB. If the pressure is then decreased, the soil will unload along curve BC and upon further application of pressure will reload along curve CD. If we continue to apply pressure, the response curve tends to approach asymptotically to the line ABE (*virgin isotropic consolidation line*).

An idealized version of this response is pictured in Fig. 12.5. The virgin isotropic consolidation line is assumed to be linear. The rebound and reloading curves are

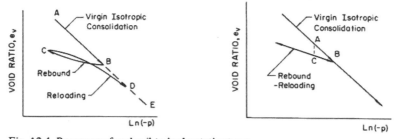

Fig. 12.4. Response of real soil to hydrostatic stress.

Fig. 12.5. Response of idealized soil to hydrostatic stress.

assumed to be identical and linear, and all rebound–reloading curves are parellel. Thus the equation for the virgin isotropic consolidation line is:

$$e_v = e_{v1} - \Lambda \ln (-p) \qquad [12.27]$$

where e_v is soil void ratio, the natural logarithm (logarithm to the base e) is denoted by ln, and e_{v1} and Λ are material constants. A generic rebound–reloading curve is defined by:

$$e_v = e_{v2} - \eta \ln (-p) \qquad [12.28]$$

where η is a material constant, and e_{v2} defines a particular rebound–reloading curve.

Referring still to Fig. 12.5, consider an infinitesimal increment of loading from A to B followed by unloading from B to C. From [12.27], the void ratio change from A to B is:

$$\dot{e}_v = - \Lambda \frac{\dot{p}}{p} \qquad [12.29]$$

where p is the current hydrostatic stress and $\dot{p}$ is the increment of hydrostatic stress (stress rate), and, from [12.28], the void ratio recovered when unloading from B to C is:

$$\dot{e}_v = -\eta \frac{\dot{p}}{p} \qquad [12.30]$$

Now: $\dot{\epsilon}_{ii} = \dfrac{\dot{e}_v}{1 + e_v} \qquad [12.31]$

where: $\dot{\epsilon}_{ii} = \dot{\epsilon}_{11} + \dot{\epsilon}_{22} + \dot{\epsilon}_{33} \qquad [12.32]$

and in the context of the theory of plasticity, the recoverable or elastic component of the volumetric strain is:

$$\dot{\epsilon}_{ii}^e = - \frac{\eta \dot{p}}{(1 + e_v) p} \qquad [12.33]$$

while, from [12.29] and [12.30], the plastic or irrecoverable component of the volumetric strain is:

$$\dot{\epsilon}_{ii}^p = - \frac{(\Lambda - \eta)\dot{p}}{(1 + e_v) p} \qquad [12.34]$$

since: $\dot{\epsilon}_{ii}^p = \dot{\epsilon}_{ii} - \dot{\epsilon}_{ii}^e \qquad [12.35]$

12.3.3. The state boundary surface and the critical state line

Although the Cambridge models were originally formulated in order to describe the behavior of soil in the triaxial test, we are concerned here with general states of stress and will hence use general stress invariants rather than those peculiar to the triaxial test. It is postulated that there exists a unique "state boundary surface" in a three-dimensional space of hydrostatic stress p, J_2, and void ratio e_v. A point in this space is referred to as a state point, and the state boundary surface is said to delimit admissible and inadmissible state points.

A portion of the state boundary surface is shown in Fig. 12.6. State points below the state boundary are admissible, while those points above the surface are inadmissible. A continuous sequence of state points is referred to as a state path. From the point of view of the theory of plasticity, state paths which lie below the state boundary surface are associated with elastic behavior, while those which lie on the state boundary surface are associated with strain hardening.

It is further postulated that there exists on the state boundary surface a "critical state line" where unlimited distortional strain may occur with no corresponding change in the stress state or the void ratio.

Consider now a soil specimen stressed uniformly (for example in the triaxial test). We load the soil until failure and plot the results in state space. Referring to Fig. 12.6, curve $HIJK$ corresponds to the state path of the soil. We assume that the initial state point (denoted by point H) lies below the state boundary surface. Thus the initial portion of the state path, HI, corresponds to elastic behavior. At point I the state path intersects the state boundary surface and at point K failure occurs. State path IJK lies on the state boundary surface and is thus associated with strain hardening, while point K lies on the critical state line.

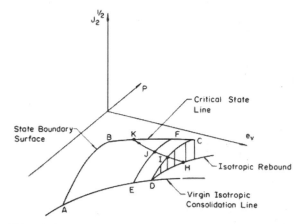

Fig. 12.6. Part of "state boundary surface".

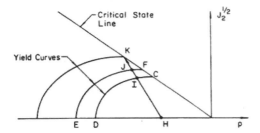

Fig. 12.7. Stress path.

As the soil strain hardens the stress state passes through a sequence of yield surfaces until failure is reached at the critical state line. These yield curves can be uniquely represented in $p - J_2$ space. For example the initial yield curve is curve *CID* on the state boundary surface and is also denoted by curve *CID* in Fig. 12.7 where the stress path is also shown. Curve *EJF*, shown in Figs. 12.6 and 12.7 corresponds to a subsequent yield curve.

The reader should recall that we have assumed that elastic distortional strain is identically zero and that elastic volumetric strain is independent of deviatoric stresses. Thus if we apply a deviatoric stress increment to the soil sample in its initial state (point *H* in Fig. 12.6), the void ratio remains unchanged and the state path corresponds to a vertical line. If we apply a hydrostatic stress increment the state path must be defined by [12.28]. We say then that state path *HI* lies in a so-called vertical "elastic wall". The elastic wall intersects the $J_2 = 0$ plane along a rebound–reloading curve and intersects the state boundary surface along a yield curve. Each yield curve is thus associated with a particular isotropic rebound–reloading curve.

12.3.4. Modified Cam-clay yield surface

A modified Cam-clay yield surface and the projection of the critical state line in $p - J_2$ space are shown in Fig. 12.8. The yield curve is elliptical and is defined by:

$$f = p^2 - p_0 p + \frac{J_2}{M^2} = 0 \qquad\qquad [12.36]$$

where M is a material constant and p_0 is a strain hardening parameter. The critical state line intersects the ellipse at its maximum point and is defined by:

$$J_2^{1/2} = - Mp \qquad\qquad [12.37]$$

that is, the critical stress state is defined by an extended Von Mises-type expression.

The stress-strain model will be completely defined once we have specified the relationship between the strain hardening parameter and the strains. Consider now

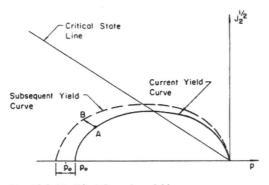

Fig. 12.8. Modified Cam-clay yield curves.

an infinitesimal stress increment denoted by line AB in Fig. 12.8. Point A lies on the current yield curve while point B lies on the subsequent yield curve. Associated with the current yield curve is an elastic wall and an isotropic rebound—reloading curve, and, of course, there is also an elastic wall and a rebound—reloading curve associated with the subsequent yield curve.

Referring to Figs. 12.6 and 12.7, the current value of the strain hardening parameter is defined by the intersection of the current isotropic rebound—reloading curve and the isotropic virgin consolidation curve. Both the current and subsequent isotropic rebound—reloading curves are shown in Fig. 12.9 along with the projection of the incremental state path on to $e_v, \ln(-p)$ space. If we now allow the soil sample to unload, the unloading state path lies in an elastic wall and its projection is denoted by BC in Fig. 12.9. Recalling [12.34] it is clear that we can relate the plastic volumetric strain rate $\dot{e}^p_{ii}$ and the change in the strain hardening parameter $\dot{p}_0$ as follows:

$$\dot{e}^p_{ii} = -\frac{(\Lambda - \eta)\,\dot{p}_0}{(1 + e_v)\,p_0} \qquad\qquad [12.38]$$

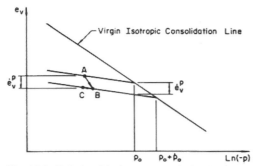

Fig. 12.9. Relationship between strain hardening parameter increment and soil ratio increment.

$$\text{or: } \dot{p}_0 = - \left(\frac{1 + e_v}{\Lambda - \eta} \right) p_0 \, \dot{e}^p_{ii} \qquad [12.39]$$

12.3.5. Behavior of modified Cam-clay in the triaxial test

We consider now a hypothetical drained triaxial test on modified Cam-clay. The test sample is first subjected to an all around confining pressure sufficient to cause plastic yielding. Thus after applying the pressure the value of the strain hardening parameter, p_0, is identical to the applied pressure. The pressure is then reduced to some value denoted by p_1. The specimen is subsequently strained axially while the confining pressure is held constant.

The stress path is denoted by line $ABCD$ in Fig. 12.10. The specimen yields at point B and strain hardening begins. At points B, C and D we have superimposed on the stress space a plastic strain rate vector. The horizontal component of the strain rate vector corresponds to volumetric strain while the vertical component corresponds to distortional strain. We use the associated flow rule here and hence the plastic strain rate vector is normal to the yield surface.

At point B the plastic component of the volumetric strain is decreasing, and thus the load increases as we continue to strain the body. We see from [12.39] that a decrease in the plastic volumetric strain is associated with an expanding yield surface. As we continue to strain the body further, the sample volume continues to decrease so that strain hardening continues. However, as the critical state line is approached the rate of the plastic volumetric strain decreases until at the critical state the plastic volumetric strain is identically zero, as indicated by the vertical strain rate vector at point D. Hence as we continue to strain the sample the vertical load remains constant and we have thus reached the failure condition. If we were to plot the axial stress versus axial strain, the curve would resemble that shown in Fig. 12.1(a).

We should emphasize that the function which defines failure is not a yield curve

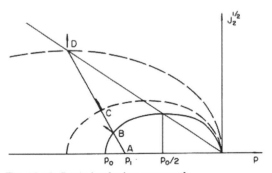

Fig. 12.10. Strain hardening stress path.

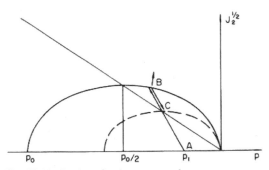

Fig. 12.11. Strain softening stress path.

nor is it the limit of a sequence of yield curves. In addition, although the extended Von Mises model and the modified Cam-clay model incorporate similar failure functions, the former predicts dilation at failure while the latter predicts zero dilation at failure.

Although the modified Cam-clay model was developed in order to predict the strain hardening behavior of clay, the model will in fact predict strain softening for certain stress histories. We consider now the same type of test as described previously in which an initial hydrostatic stress (p_0) is applied and subsequently reduced to a value denoted by p_1. In this case p_1 is considerably less than p_0 as indicated in Fig. 12.11.

Referring to Fig. 12.11, as the test specimen is strain axially, plastic yielding first occurs at point B. However here the plastic volumetric strain is positive and the specimen begins to strain soften with the axial load reducing. From [12.39] we see that an increase in the plastic volumetric strain is associated with a contracting yield surface. The specimen continues to strain soften until point C is reached where the plastic volumetric strain rate is identically zero. Continued axial straining produces no change in the axial load.

The peak stress is no longer defined by an extended Von Mises expression but is rather dependent on the maximum past hydrostatic stress. It is the residual stress that is defined by the extended Von Mises expression. The test specimen dilates at the peak stress whereas at the residual stress no dilation occurs.

We might refer to the first specimen as lightly overconsolidated and the second specimen as heavily overconsolidated. The lightly overconsolidated specimen strain hardened when sheared while the heavily overconsolidated specimen strain softened when sheared.

12.3.6. Incremental stress–strain equations suitable for numerical analysis

Although modified Cam-clay is suitable for predicting the response of soil in the triaxial test, it is not suitable for general stress analysis. Since the model is rigid-plastic with respect to shearing deformation, the function we require in a displacement formulation, stress as a function of strain, is singular. The possibility of strain softening is also a problem. Any numerical approach which requires inversion of the tangent stiffness may break down in the presence of strain softening. In addition the solution of boundary value problems involving strain softening materials is not yet well defined. In general it can not be shown that such solutions are unique. We thus present a variant of the modified Cam-clay model which is suitable for a tangent stiffness formulation.

Considering first elastic response, from [12.33], we have:

$$\dot{\epsilon}_{ii}^e = -\frac{\eta}{1+e_v}\frac{\dot{p}}{p} \tag{12.40}$$

If elastic distortional strain is identically zero, the elastic strain rate–stress rate equation is:

$$\dot{\epsilon}_{ij}^e = -\frac{\eta}{9(1+e_v)p}\,\dot{\sigma}_{kk}\,\delta_{ij} \tag{12.41}$$

In order to invert [12.41] we introduce some distortional flexibility as follows:

$$\dot{\epsilon}_{ij}^e = -\left[\frac{\eta}{9(1+e_v)p}+\frac{1}{3\beta}\right]\dot{\sigma}_{kk}\,\delta_{ij}+\frac{1}{\beta}\dot{\sigma}_{ij} \tag{12.42}$$

where β is the instantaneous elastic shear modulus of the soil. Equation [12.40] is still valid. Finally, [12.42] can be inverted to give:

$$\dot{\sigma}_{ij} = -\left[\frac{p(1+e_v)}{\eta}+\frac{\beta}{3}\right]\dot{\epsilon}_{kk}^e\,\delta_{ij}+\beta\,\dot{\epsilon}_{ij}^e \tag{12.43}$$

If we desire to keep the computational model as close as possible to modified Cam-clay, the shear modulus, β, can be made quite large, perhaps one hundred times the plastic bulk modulus.

In order to determine the elastic–plastic, incremental stress–strain relations, we again start with the associated flow rule after Drucker and write:

$$\dot{\epsilon}_{ij}^p = \lambda\frac{\partial f}{\partial \sigma_{ij}} \tag{12.44}$$

where: $f(\sigma_{ij}, p_0) = p^2 - p_0 p + \dfrac{J_2}{M^2}$ \hfill [12.45]

Noting that: $\dot{\epsilon}_{ij}^e = \dot{\epsilon}_{ij} - \dot{\epsilon}_{ij}^p$ \hfill [12.46]

we rewrite [12.43] as:

$$\dot{\sigma}_{ij} = -\left[\frac{p(1+e_v)}{\eta}+\frac{\beta}{3}\right]\left(\dot{\epsilon}_{kk}-\lambda\frac{\partial f}{\partial \sigma_{kk}}\right)\delta_{ij}+\beta\left(\dot{\epsilon}_{ij}-\lambda\frac{\partial f}{\partial \sigma_{ij}}\right) \tag{12.47}$$

The problem again is to determine λ, and we note that during plastic loading the stress state and the strain hardening parameter change such that the new stress state lies on the subsequent yield surface defined by the new value of the strain hardening parameter. At the beginning of the increment:

$$f(\sigma_{ij}, p_0) = 0 \tag{12.48}$$

and at the end of the increment:

$$f(\sigma_{ij} + \dot{\sigma}_{ij}, p_0 + \dot{p}_0) = 0 \tag{12.49}$$

Thus: $df = \dfrac{\partial f}{\partial \sigma_{ij}}\, \dot{\sigma}_{ij} + \dfrac{\partial f}{\partial p_0}\, \dot{p}_0 = 0 \tag{12.50}$

Equations [12.47] and [12.50] permit the determination of λ.

Now, from [12.45]:

$$\frac{\partial f}{\partial \sigma_{ij}} = \tfrac{1}{3}\,(2p - p_0)\,\delta_{ij} + \frac{s_{ij}}{M^2} \tag{12.51}$$

and: $\dfrac{\partial f}{\partial p_0} = - p \tag{12.52}$

Also, repeating [12.39]:

$$\dot{p}_0 = - \left(\frac{1 + e_v}{\Lambda - \eta}\right) p_0\, \dot{\epsilon}^p_{kk} = - \left(\frac{1 + e_v}{\Lambda - \eta}\right) p_0\, \lambda\, \frac{\partial f}{\partial \sigma_{kk}} \tag{12.53}$$

Utilizing [12.51], [12.52] and [12.53] and substituting [12.47] into [12.50] gives:

$$df = \left\{ \frac{(2p - p_0)}{3}\,\delta_{ij} + \frac{s_{ij}}{M^2} \right\} \left\{ \beta \left[\dot{\epsilon}_{ij} - \lambda \left(\frac{(2p - p_0)}{3}\,\delta_{ij} + \frac{s_{ij}}{M^2} \right) \right] \right.$$

$$\left. - L\,[\dot{\epsilon}_{kk} - \lambda(2p - p_0)]\,\delta_{ij} \right\} - p\,R\,\lambda\,(2p - p_0) = 0 \tag{12.54}$$

where: $L = p\,\dfrac{(1 + e_v)}{\eta} + \tfrac{1}{3}\beta \tag{12.55}$

and: $R = - p_0 \left(\dfrac{1 + e_v}{\Lambda - \eta}\right) \tag{12.56}$

Simplifying [12.54] and solving for λ gives:

$$\lambda = \frac{\dfrac{\beta s_{ij}}{M^2}\,\dot{\epsilon}_{ij} + H\,\dot{\epsilon}_{kk}}{(2p - p_0)\,H + \dfrac{2\beta J_2}{M^4} + p\,(2p - p_0)\,R} \qquad [12.57]$$

where: $H = (2p - p_0)\,(\tfrac{1}{3}\beta - L)$ $\qquad$ [12.58]

Finally, substituting [12.57] into [12.47] yields the incremental elastic–plastic stress–strain relationship:

$$\dot{\sigma}_{ij} = \beta\,\dot{\epsilon}_{ij} - L\,\dot{\epsilon}_{mm}\,\delta_{ij} - \frac{1}{\psi}\left(H\,\delta_{ij} + \frac{\beta}{M^2}\,s_{ij}\right)\left(\frac{\beta}{M^2}\,s_{kl}\,\dot{\epsilon}_{kl} + H\,\dot{\epsilon}_{mm}\right) \qquad [12.59]$$

where: $\psi = (2p - p_0)\,H + \dfrac{2\beta J_2}{M^4} + p\,(2p - p_0)\,R$ $\qquad$ [12.60]

Equation [12.59] is applicable in both the strain hardening and strain softening regions and is, of course, applicable at the critical state. If $p \leqslant p_0/2$ the model is either strain hardening or at the critical state and [12.59] is suitable for the numerical formulation used here. If $p > p_0/2$ the model is strain softening and hence would not be suitable for application here.

In order to get around this problem we could introduce a perfectly plastic idealization in the strain softening region which would be compatible with the modified Cam-clay model. For instance we might use the critical state line as a perfectly plastic yield surface. A simpler approach would utilize the current value of the modified Cam-clay yield surface as a perfectly plastic yield surface. The stress–strain equations would still be defined by [12.59], however, ψ would now be defined by:

$$\psi = (2p - p_0)\,H + \frac{2\beta J_2}{M^4} \qquad [12.61]$$

If we utilized [12.59] for the incremental stress–strain equation and define ψ by [12.60] for $p \leqslant p_0/2$ and by [12.61] for $p > p_0/2$, we have then a complete stress–strain model suitable for use in a tangent stiffness approach. This model has one potential drawback, that is, if the hydrostatic stress component is zero, the incremental stress–strain equations are singular. This may be troublesome for some boundary value problems. To avoid this problem, Zienkiewicz and Naylor (1971b) have suggested using a model which is linear in the elastic region.

12.3.7. Boundary value problems and the Cambridge soil models

Smith (1970) has used the so-called Cam-clay model to analyze the plane strain, drained behavior of a pressurized thick cylinder of clay. Smith and Kay (1971) analyzed the same problem using modified Cam-clay as well as Cam-clay. In both papers elastic strains were assumed to be identically zero. Zienkiewicz and Naylor (1971a) have analyzed the drained behavior of modified Cam-clay in the triaxial test. Some elastic distortional flexibility was introduced into the model.

Zienkiewicz and Naylor (1971b) have considered a clay layer consolidating under a footing load. The soil skeleton was modeled by a variant of modified Cam-clay. Elastic behavior was assumed to be linear and in the strain hardening region a modified Cam-clay yield curve was used. In the strain softening region a softening Coulomb-type expression with a nonassociated flow rule was utilized.

12.4. AN ELASTIC–PLASTIC STRAIN HARDENING-FRACTURE MODEL FOR CONCRETE

In this section, a general three-dimensional plasticity theory is presented for describing the inelastic behavior of homogeneous, isotropic rock-like material such as concrete. Normality of the strain rate vector to the loading surface is incorporated into the stress–strain rate relationship used in the theory. The concrete material is assumed to be a linear elastic–plastic strain hardening-fracture material with a dependence of the yield, and fracture strengths on the hydrostatic component of stress. A particular type of initial yield surface, subsequent loading surface and fracture surface is assumed in order to obtain an explicit stress–strain relation for concrete. This particular type of surfaces used is representative of the behavior of concrete.

12.4.1. Typical concrete stress–strain behavior

Fig. 12.12 shows a typical stress–strain curve for concrete under uniaxial compression. The non-linear behavior is readily apparent. For 40–60% of the ultimate stress, the departure from linearity is slight (curve marked OA). Also in this range, the deformations appear to be completely recoverable. Thus, in the conventional sense, point A on the curve represents the elastic limit. With increase in load beyond A, internal disruption occurs in the material starting with bond microcracking and this disruption is irrecoverable in a manner similar to that of plastic deformation of metals. Point A is therefore termed elastic or *initial discontinuous point*. Unloading beyond this point will follow the straight line CFD and permanent deformation OD occurs after complete unloading (Fig. 12.12). When the material is

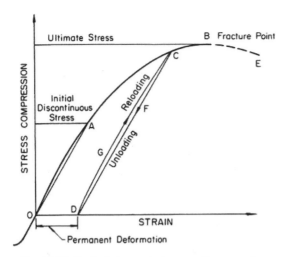

Fig. 12.12. Typical stress–strain curve for concrete.

reloaded from point D, the path follows the slightly curved line DGC. This small deviation is usually ignored in the analysis and we shall assume here that the reloading path follows DFC. With further loading, significant small cracks in mortar develop and these cracks spread and coalesce. At or near the ultimate load, the specimen fractures into a large number of separate pieces. Although qualitatively shown, descenting or softening path BE seems to be largely influenced by the characteristics of the testing machine and must be ignored in the context of current theories of plasticity which do not account for strain softening materials. Herein, the ultimate stress point B is taken as the fracture point.

Referring now to the two-dimensional picture of Fig. 12.13, the *fracture or failure curve* is the limiting curve in the biaxial principal stress σ_1, σ_2 space, since the material ruptures on reaching this curve and cannot resist further load. The stresses drop abruptly to zero. The *initial discontinuous curve* is analogous to that of initial yield curve for metals, and for stress states within the curve, deformations are elastic and reversible. As pointed out earlier, the major point of difference between metals and concrete or rock is with regard to permanent deformations. Such deformation in metals is associated with shear slip or plastic yielding. But, in concrete, it is due to the incipience of micro-cracks. The term *discontinuous curve* instead of *initial yield curve* appears to be more appropriate and will therefore be used herein when referring to concrete or rock. More generally, the term *discontinuous surface* and *fracture or failure surface* will be used to emphasize the fact that three or more components of stress may be independent variables. Two-dimensional pictures only are drawn, however.

Increasing the stress beyond the initial yield surface, into the strain hardening range, produces both plastic and elastic action in case of metals. At each stage of

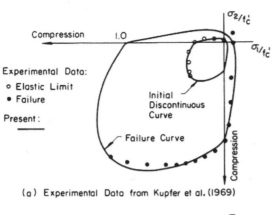

(a) Experimental Data from Kupfer et al. (1969)

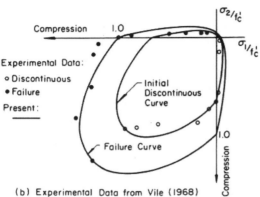

(b) Experimental Data from Vile (1968)

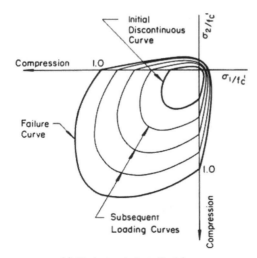

(c) Strain-hardening Model

Fig. 12.13. Initial discontinuous, subsequent loading, and failure curves in biaxial principal stress space.

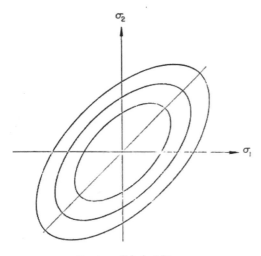

σ_1 , σ_2 Principal Stresses

Fig. 12.14. Isotropic hardening model.

plastic deformation, a new yield surface, called a *loading surface*, is established. Now, if the state of stress is changed so that the stress point representing it moves about inside the new yield surface, no plastic deformation will take place. Analogous remarks apply for concrete also. Three such subsequent loading surfaces are shown in Fig. 12.13(c).

12.4.2. Loading function

In the theory of plasticity, there are two widely used strain hardening models. The first one is the isotropic hardening model which assumes the material to have gradually and uniformly expanding initial yield (or state) surfaces as shown in Fig. 12.14. The second is the kinematic model which assumes that the initial yield surface will not expand but only translate as a rigid body (see Fig. 12.15). A proper combination of these two models leads to a model of the type shown in Fig. 12.16. The modification on this combined model with a proper consideration of the differences in properties in tension and compression leads to the model of the type shown in Fig. 12.13(c). The essential change from Fig. 12.16 to Fig. 12.13(c) is the difference in shape of the loading surface in the area where stresses are tensile. Once the loading surface is defined, incremental plastic stress–strain relations based on the *normality condition* are applicable to such a material model and the corresponding constitutive equations may then be derived.

Now, let us formulate the above-mentioned discontinuous surfaces mathematically. In our discussion, we shall assume that the shapes of the subsequent loading surfaces are similar in form to those of the initial discontinuous surface and the

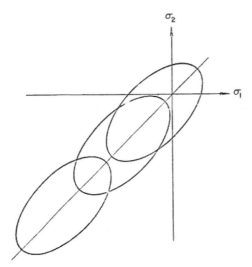

Fig. 12.15. Kinematic hardening model.

failure surface. Accordingly, the kinematics of the loading surfaces adopted herein is that the current loading surface translates along the $\sigma_1 = \sigma_2 = \sigma_3$-axis and simultaneously expands uniformly from the initial discontinuous surface to the failure surface. The current loading surface is established solely by the highest value of the stress invariants p and J_2 reached during the prior loading history. A vertical cross-section of these loading surfaces generated by this kinematic model on the plane $\sigma_3 = 0$ in the principal stress, $\sigma_1, \sigma_2, \sigma_3$ space is shown schematically in Fig. 12.13(c) and 12.16.

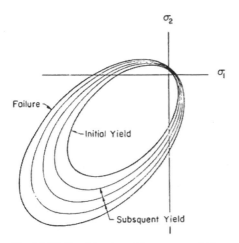

Fig. 12.16. Combination of isotropic and kinematic hardening models.

Herein, we shall assume that the *fracture surface* has the form:
in the compression domain:

$$f_u(\sigma_{ij}) = J_2 + \tfrac{1}{3} A_u\, p = \tau_u^2 \qquad [12.62]$$

in the tension and tension-compression domains:

$$f_u(\sigma_{ij}) = J_2 - \tfrac{1}{6} p^2 + \tfrac{1}{3} A_u\, p = \tau_u^2 \qquad [12.63]$$

Similarly, the *initial discontinuous surface* takes the same form as that of the fracture surface:
in the compression domain:

$$f_0(\sigma_{ij}) = J_2 + \tfrac{1}{3} A_0\, p = \tau_0^2 \qquad [12.64]$$

in the tension and tension–compression domains:

$$f_0(\sigma_{ij}) = J_2 - \tfrac{1}{6} p^2 + \tfrac{1}{3} A_0\, p = \tau_0^2 \qquad [12.65]$$

where A_0, τ_0, A_u and τ_u are material constants. The values of these constants can be determined from the uniaxial or biaxial test. This is demonstrated in Fig. 12.13(a and b).

The subsequent loading surfaces are bounded by the initial discontinuous surface and the failure surface and their shapes are assumed to be similar in form to those of the initial discontinuous surface and the failure surface. Noting that the kinematics of the loading surfaces adopted here is that the loading surface translates along the $\sigma_1 = \sigma_2 = \sigma_3$-axis and expands isotropically simultaneously, we therefore assume here that the *loading surface* has the form:
in the compression domain:

$$f(\sigma_{ij}) = \frac{J_2 + \tfrac{1}{3} A\, p}{1 - \tfrac{1}{3} B p} = \tau^2 \qquad [12.66]$$

in the tension and tension-compression domains:

$$f(\sigma_{ij}) = \frac{J_2 - \tfrac{1}{6} p^2 + \tfrac{1}{3} A\, p}{1 - \tfrac{1}{3} B p} = \tau^2 \qquad [12.67]$$

In the two extreme cases, when the value of τ^2 approaches τ_0^2 or τ_u^2, the value $B\tau^2 + A$ must approach A_0 or A_u, so that the loading surface approaches the initial discontinuous surface or failure surface respectively. Using these conditions, the constants A and B can be expressed as:

$$A = \frac{A_0\, \tau_u^2 - A_u\, \tau_0^2}{\tau_u^2 - \tau_0^2} \qquad [12.68]$$

$$B = \frac{A_u - A_0}{\tau_u^2 - \tau_0^2}$$

[12.69]

In [12.66] and [12.67], the invariant J_2 represents a somewhat averaged maximum shear stress at a point (see [12.3]), and p reflects the effect of hydrostatic pressure on the material. The terms Ap and Bp determine the rate of kinematic translation of the loading surface along the line $\sigma_1 = \sigma_2 = \sigma_3$ and the term τ^2 determines the rate of isotropic hardening of the subsequent loading surfaces.

Figure 12.13(c) presents the vertical cross-section of the loading surfaces on the plane $\sigma_3 = 0$ in the principal stress, $\sigma_1, \sigma_2, \sigma_3$-space as given by [12.66] and [12.67]. One such loading surface coincides with the limit or fracture surface of maximum strength and the other one coincides with the initial discontinuous surface. The initial discontinuous surface and the fracture surface can be determined from the uniaxial or biaxial test. The elastic domain is determined by using a prescribed value of offset strain as a measure of initial yield or discontinuity. Once the loading surfaces are determined, incremental plastic or discontinuous stress–strain relations can be obtained from the concept of normality condition.

12.4.3. Incremental stress–strain relations

Once the subsequent loading function is defined, incremental plastic or discontinuous stress–strain increment relations can be deduced directly from the normality condition. Denoting the general function by $f(\sigma_{ij})$, the normality of the plastic strain increment $d\epsilon_{ij}^p$ to the loading surface $f(\sigma_{ij})$ requires that in indicial notation:

$$d\epsilon_{ij}^p = d\lambda \frac{\partial f}{\partial \sigma_{ij}}$$

[12.70]

where $d\lambda$ is a positive incremental quantity depending upon the current state of stress, σ_{ij}, stress increment, $d\sigma_{ij}$, and the stress history. The value of $d\lambda$ is zero unless plastic or discontinuous deformation is taking place. In the p, J_2-form of stress–strain relation for a work-hardening material, as p and J_2 are increased beyond any previously established maximum value $f(p, J_2) = \tau^2$, $d\lambda$ is a function of p and J_2 multiplied by the increment of df:

$$d\lambda = G(p, J_2)\, df$$

[12.71]

where G similarly to f may be any positive scalar function of stress, strain and history of loading but is independent of $d\sigma_{ij}$.

Taking the inner product of [12.70] by itself and using [12.71], the scalar function G has the value:

$$G = \frac{\sqrt{d\epsilon_{ij}^p \, d\epsilon_{ij}^p}}{df \sqrt{\dfrac{\partial f}{\partial \sigma_{ij}} \dfrac{\partial f}{\partial \sigma_{ij}}}}$$

[12.72]

in which the term $(\partial f/\partial\sigma_{ij} \cdot \partial f/\partial\sigma_{ij})^{1/2}$ is a function of the current state of stress and can be evaluated directly from the current loading function. The quantity $\sqrt{d\epsilon^p_{ij}d\epsilon^p_{ij}}$ indicates the history of the straining of the material.

For simplicity, denote:

$$H = \frac{df}{\sqrt{d\epsilon^p_{rs}\,d\epsilon^p_{rs}}}$$

[12.73]

and use the relation:

$$df = \frac{\partial f}{\partial\sigma_{mn}}\,d\sigma_{mn}$$

[12.74]

The general stress–strain incremental relations resulting when the above equations are introduced into [12.70]:

$$d\epsilon^p_{ij} = \frac{\dfrac{\partial f}{\partial\sigma_{ij}}\dfrac{\partial f}{\partial\sigma_{mn}}}{H\,\sqrt{\dfrac{\partial f}{\partial\sigma_{rs}}\dfrac{\partial f}{\partial\sigma_{rs}}}}\,d\sigma_{mn}$$

[12.75]

Noting that the total strain increment, $d\epsilon_{ij}$, consists of two parts: elastic strain increment, $d\epsilon^e_{ij}$, and plastic strain increment, $d\epsilon^p_{ij}$:

$$d\epsilon_{ij} = d\epsilon^e_{ij} + d\epsilon^p_{ij}$$

[12.76]

Combining [12.76] with [12.75], the incremental elastic–plastic strain–stress relationship is obtained:

$$d\epsilon_{ij} = \left[H_{ijkl} + \frac{\dfrac{\partial f}{\partial\sigma_{ij}}\dfrac{\partial f}{\partial\sigma_{kl}}}{H\,\sqrt{\dfrac{\partial f}{\partial\sigma_{rs}}\dfrac{\partial f}{\partial\sigma_{rs}}}}\right]d\sigma_{kl}$$

[12.77]

where H_{ijkl} is the elastic compliance matrix and H is the quantity measuring the work hardening rate of the material. The value of H depends on the current state of stress, strain and its history.

12.4.4. Strain hardening rule

The form of the functions describing the strain hardening rate, H, can be ob-

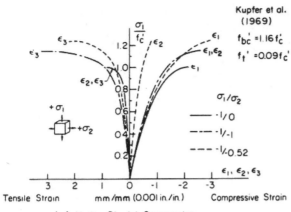

(a) Under Biaxial Compression

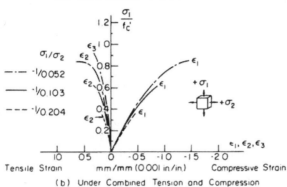

(b) Under Combined Tension and Compression

Fig. 12.17. Stress–strain relationships of concrete.

tained directly from [12.73], provided that the stress history and corresponding strain increments, $\sqrt{d\epsilon_{rs}^p \, d\epsilon_{rs}^p}$, are known from some simple tests. Herein, the experimental results reported by Kupfer et al. (1969) are used for the determination of the strain hardening rate function.

Fig. 12.17 shows the experimental stress–strain relations of concrete specimens 20 × 20 × 5 cm subjected to biaxial stress combinations in the compression and tension–compression regions. Defining the equivalent stress:

$$\sigma_e = \sqrt{f} \tag{12.78}$$

and the equivalent plastic or discontinuous strain:

$$\epsilon^p = \int d\epsilon^p = \int \sqrt{d\epsilon_{rs}^p \, d\epsilon_{rs}^p} \tag{12.79}$$

these experimental stress–strain relations are plotted in terms of σ_e and ϵ^p in Fig. 12.18. The strain hardening rate function, H, can be expressed in terms of the

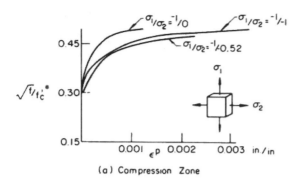

(a) Compression Zone

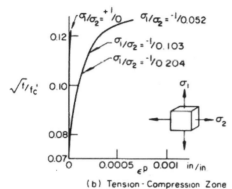

(b) Tension-Compression Zone

Fig. 12.18. Equivalent stress–strain relations.

slope of this equivalent stress–strain curve:

$$H = 2\sigma_e H' \qquad\qquad [12.80]$$

$$\text{where: } H' = \frac{d\sigma_e}{d\epsilon^p} \qquad\qquad [12.81]$$

Once this strain hardening rate function is determined, the stress–strain relations, [12.77], may be used for predicting concrete behavior under all loading paths. Using [12.77] and the strain hardening rate function from Fig. 12.18, the stress–strain relations corresponding to various uniaxial and biaxial tests are calculated in Figs. 12.19 and 12.20 and compared with experimental results reported by Kupfer et al. (1969). Good agreement is generally observed although there are some deviations between the theoretical and experimental stress–volumetric strain curves when the stress level is high.

It is seen that the $\sigma_e \sim \epsilon^p$ curves (Fig. 12.18) form a narrow band both in compression and tension–compression tests with the exception of the simple tension test. For simplicity, we may use two strain hardening functions, one for the

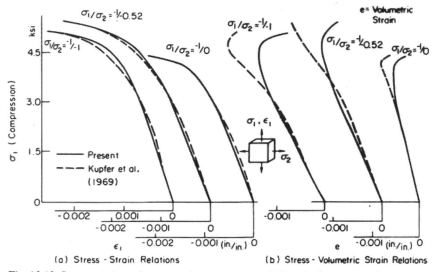

Fig. 12.19. Stress–strain and stress–volumetric strain relations in the compression zone.

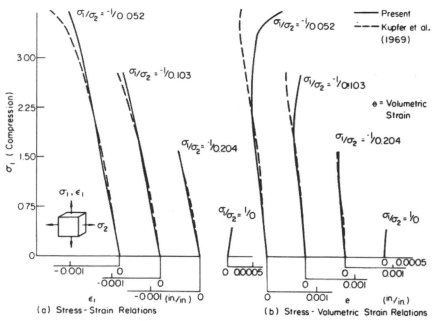

Fig. 12.20. Stress–strain and stress–volumetric strain relations in the tension–compression zone.

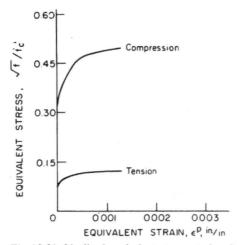

Fig. 12.21. Idealized equivalent stress strain relations.

compression region and the other for the tension–compression region (Fig. 12.21). Herein, we shall simplify it further by assuming that these two functions in both regions are identical in form to that of the uniaxial compression test (upper curve, Fig. 12.21). The theoretical stress–strain predictions based on this idealized strain hardening function are compared with experiments in Figs. 12.22 and 12.23 which seem capable of matching the principal characteristics of experimental results of stress-deformation behavior of concrete. In the calculations, the initial discontinuous surface is taken to be $0.6 f_c'$ where f_c' is the uniaxial compressive strength of concrete.

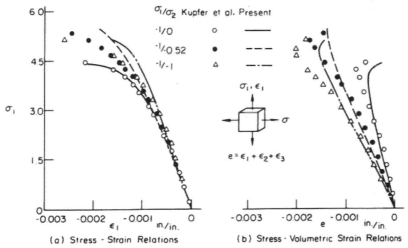

Fig. 12.22. Idealized stress–strain and stress–volumetric strain relations in the compression zone.

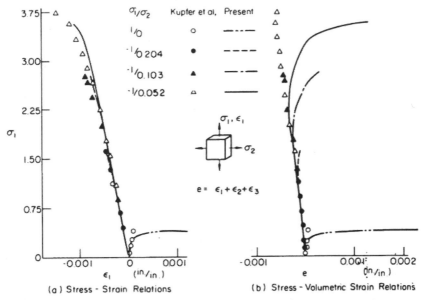

Fig. 12.23. Idealized stress–strain and stress–volumetric relations in the tension–compression zone.

12.4.5. Incremental stress–strain relations in matrix form

The description of the stress–strain increment relation given in the preceding sections is simple to visualize; however, in order to utilize this theory, it is necessary to express the stress–strain relations explicitly in terms of the proposed loading function. In the following, the stress–strain incremental relations associated with the proposed loading function are first formulated for a three-dimensional case and then reduced to the special cases of plane stress, plane strain and the axisymmetric case. These stress–strain relations are written in explicit matrix forms which can be used directly in the finite element method.

When the state of stress lies within the initial discontinuous surface, $f_0(\sigma_{ij}) = \tau_0^2$, the material behavior is assumed to be linear elastic. When the material is stressed beyond the initial discontinuous surface, a subsequent new discontinuous surface called the loading surface, $f(\sigma_{ij}) = \tau^2$, replaces the initial discontinuous surface. This new loading surface is specified by the new value $\tau^2 > \tau_0^2$, where τ continuously increases as discontinuous deformation proceeds. If the material is unloaded and then reloaded, no additional irrecoverable discontinuous deformation will occur until this new value τ^2 is reached. If straining is continued beyond this new surface, further discontinuity occurs and additional irrecoverable deformation results. In summary, at each stage of loading, there is an associated loading surface in stress space separating those states of stress which can be reached by purely elastic (reversible) changes from those outside which cannot be reached without producing some additional discontinuous irrecoverable deformation. Once the loading surface reaches the failure surface, the material cracks or crushes and the stresses drop abruptly to zero.

Accordingly, three types of stress–strain matrices are needed in the analysis, namely, linear elastic, elastic–plastic, and fractured ones. For elastic loading and unloading:

$$f(\sigma_{ij}) < \tau^2 \tag{12.82}$$

neutral loading:

$$f(\sigma_{ij}) = \tau^2 \text{ and } df = \frac{\partial f}{\partial \sigma_{ij}} d\sigma_{ij} = 0 \tag{12.83}$$

and plastic unloading:

$$f(\sigma_{ij}) = \tau^2 \text{ and } df = \frac{\partial f}{\partial \sigma_{ij}} d\sigma_{ij} < 0 \tag{12.84}$$

all stress–strain relations are linear elastic and can be expressed as:

$$\epsilon_{ij}^e = H_{ijkl}\,\sigma_{kl} \tag{12.85}$$

If the material is isotropic, the compliance matrix in terms of Young's modulus E and Poisson's ratio ν has the form:

$$\epsilon_{ij}^e = \frac{1+\nu}{E}\,\sigma_{ij} - \frac{\nu}{E}\,\sigma_{kk}\,\delta_{ij} \tag{12.86}$$

when the stress increment exceeds the current loading surface:

$$f(\sigma_{ij}) = \tau^2,\, df = \frac{\partial f}{\partial \sigma_{ij}}\,d\sigma_{ij} > 0 \tag{12.87}$$

the deformation consists of both elastic and plastic parts. The incremental stress–strain relations are given by [12.77]. When the stress state reaches the failure surface, $f_u(\sigma_{ij}) = \tau_u^2$, the stresses drop abruptly to zero and the stiffness of the material is zero.

Since the finite element displacement method is to be utilized here, we wish to develop an expression relating stress rates as a function of a strain rate.

The inversion of [12.85] offers no difficulty:

$$\sigma_{ij} = D_{ijkl}\,\epsilon_{kl}^e \tag{12.88}$$

the explicit form of the elastic rate relationship between the stress and strain is given by [12.8].

Noting that total strain increment, $d\epsilon_{ij}$ is the sum of elastic and plastic strain increments, [12.76], we can relate stress increment to total strain increment:

$$d\sigma_{kl} = D_{klij}\, d\epsilon^e_{ij} = D_{klij}(d\epsilon_{ij} - d\epsilon^p_{ij}) \tag{12.89}$$

or using [12.70] and [12.71], we have:

$$d\sigma_{kl} = D_{klij}\left(d\epsilon_{ij} - G\,\frac{\partial f}{\partial \sigma_{ij}}\, df\right) \tag{12.90}$$

In order to express the stress increment as a function of current stress and total strain increment, we use the identity:

$$df = \frac{\partial f}{\partial \sigma_{mn}}\, d\sigma_{mn} \tag{12.91}$$

The increment df resulting when [12.90] is introduced into [12.91]:

$$df = \frac{\dfrac{\partial f}{\partial \sigma_{mn}} D_{mnij}}{1 + G\,\dfrac{\partial f}{\partial \sigma_{vw}} D_{vwrs}\,\dfrac{\partial f}{\partial \sigma_{rs}}}\, d\epsilon_{ij} \tag{12.92}$$

Substituting [12.92] into [12.90], the incremental stress–strain relations result:

$$d\sigma_{kl} = \left[D_{klij} - \frac{D_{klmn}\dfrac{\partial f}{\partial \sigma_{mn}}\dfrac{\partial f}{\partial \sigma_{tu}} D_{tuij}}{\dfrac{1}{G} + \dfrac{\partial f}{\partial \sigma_{vw}} D_{vwrs}\dfrac{\partial f}{\partial \sigma_{rs}}} \right] d\epsilon_{ij} \tag{12.93}$$

Using [12.72] and [12.73], the desired incremental relationships between stress and strain are obtained:

$$d\sigma_{kl} = [D_{klij} - d\lambda\, \Phi_{klij}]\, d\epsilon_{ij} \tag{12.94}$$

$$\text{where:}\quad \frac{1}{d\lambda} = H\,\sqrt{\frac{\partial f}{\partial \sigma_{pq}}\frac{\partial f}{\partial \sigma_{pq}} + \frac{\partial f}{\partial \sigma_{vw}} D_{vwrs}\frac{\partial f}{\partial \sigma_{rs}}} \tag{12.95}$$

$$\text{and:}\quad \Phi_{klij} = D_{klmn}\frac{\partial f}{\partial \sigma_{mn}}\frac{\partial f}{\partial \sigma_{tu}} D_{tuij} \tag{12.96}$$

Equation [12.94] is the desired inversion of [12.77].

Introducing the loading functions [12.66] and [12.67], into [12.95] and [12.96], the incremental elastic–plastic stress–strain relations [12.94], are fully defined once $d\lambda$ and Φ_{klij} are known.

$$\frac{1}{d\lambda} = \frac{E}{(1+\nu)(1-2\nu)} \left[(1-2\nu)(2J_2+3\rho^2)+9\nu\rho^2\right]$$

$$+ H(1-\tfrac{1}{3}Bp)\left[\frac{(1+\nu)(1-2\nu)}{2}\right]^2 (2J_2+3\rho^2)^{1/2} \qquad [12.97]$$

$$\Phi_{klij} = \left[(1-2\nu)s_{ij}+(1+\nu)\delta_{ij}\rho\right]\left[(1-2\nu)s_{kl}+(1+\nu)\delta_{kl}\rho\right] \qquad [12.98]$$

in which $s_{ij} = \sigma_{ij} - p\delta_{ij}$ is the *stress deviator tensor* and the quantity ρ is defined as:

$$\rho = np + \tfrac{1}{3}(B+A\tau^2) \qquad [12.99]$$

where n is equal to 0 when stress state lying in the compression zone and is $-1/3$ when in the tension–compression zone.

Metals may be considered as a special case of concrete for which $A = B = n = 0$ and thus $\rho = 0$. When these values are substituted into [12.97] and [12.98], the resulting incremental relationships between stress and strain are identical to that of isotropic hardening of Von Mises material (Yamada et al., 1968).

Three-dimensional case
Referring to Fig. 12.24(a), the stress components in a Cartesian coordinates

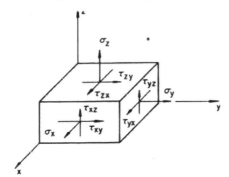

(a) Stress Components in the Cartesian Coordinates

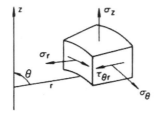

(b) Stress Components in the Polar Coordinates
Fig. 12.24. Stress components.

system are σ_x, σ_y, σ_z, τ_{xy}, τ_{yz}, and τ_{zx}, and the corresponding engineering strains are ϵ_x, ϵ_y, ϵ_z, γ_{xy}, γ_{yz}, and γ_{zx}. Equation [12.94] can be rewritten as:

$$
\begin{Bmatrix} d\sigma_x \\ d\sigma_y \\ d\sigma_z \\ d\tau_{xy} \\ d\tau_{yz} \\ d\tau_{zx} \end{Bmatrix} = \frac{E}{(1+\nu)(1-2\nu)}
\begin{bmatrix}
1-\nu-\omega\Phi_{11} & \nu-\omega\Phi_{12} & \nu-\omega\Phi_{13} \\
 & 1-\nu-\omega\Phi_{22} & \nu-\omega\Phi_{23} \\
 & & 1-\nu-\omega\Phi_{33} \\
\text{symmetric} & &
\end{bmatrix}
$$

$$
\begin{bmatrix}
-\omega\Phi_{14} & -\omega\Phi_{15} & -\omega\Phi_{16} \\
-\omega\Phi_{24} & -\omega\Phi_{25} & -\omega\Phi_{26} \\
-\omega\Phi_{34} & -\omega\Phi_{35} & -\omega\Phi_{36} \\
\frac{1-2\nu}{2}-\omega\Phi_{44} & -\omega\Phi_{45} & -\omega\Phi_{46} \\
 & \frac{1-2\nu}{2}-\omega\Phi_{55} & -\omega\Phi_{56} \\
 & & \frac{1-2\nu}{2}-\omega\Phi_{66}
\end{bmatrix}
\begin{Bmatrix} d\epsilon_x \\ d\epsilon_y \\ d\epsilon_z \\ d\gamma_{xy} \\ d\gamma_{yz} \\ d\gamma_{zx} \end{Bmatrix}
\quad [12.100]
$$

where we have denoted:

$$
\frac{1}{\omega} = [(1-2\nu)(2J_2+3\rho^2)+9\nu\rho^2] + \frac{H(1+\nu)(1-2\nu)}{E} [2J_2+3\rho^2]^{1/2} (1-\tfrac{1}{3}Bp)
$$

$$
\Phi_{11} = [(1-2\nu)(s_x+\rho)+3\nu\rho]^2
$$

$$
\Phi_{12} = [(1-2\nu)(s_x+\rho)+3\nu\rho] [(1-2\nu)(s_y+\rho)+3\nu\rho]
$$

$$
\Phi_{13} = [(1-2\nu)(s_x+\rho)+3\nu\rho] [(1-2\nu)(s_z+\rho)+3\nu\rho]
$$

$$
\Phi_{14} = [(1-2\nu)(s_x+\rho)+3\nu\rho] [(1-2\nu)\tau_{xy}]
$$

$$
\Phi_{15} = [(1-2\nu)(s_x+\rho)+3\nu\rho] [(1-2\nu)\tau_{yz}]
$$

$$
\Phi_{16} = [(1-2\nu)(s_x+\rho)+3\nu\rho] [(1-2\nu)\tau_{zx}]
$$

$$\Phi_{22} = [(1 - 2\nu)(s_y + \rho) + 3\nu\rho]^2$$

$$\Phi_{23} = [(1 - 2\nu)(s_y + \rho) + 3\nu\rho][(1 - 2\nu)(s_z + \rho) + 3\nu\rho]$$

$$\Phi_{24} = [(1 - 2\nu)(s_y + \rho) + 3\nu\rho][(1 - 2\nu)\tau_{xy}]$$

$$\Phi_{25} = [(1 - 2\nu)(s_y + \rho) + 3\nu\rho][(1 - 2\nu)\tau_{yz}]$$

$$\Phi_{26} = [(1 - 2\nu)(s_y + \rho) + 3\nu\rho][(1 - 2\nu)\tau_{zx}]$$

$$\Phi_{33} = [(1 - 2\nu)(s_z + \rho) + 3\nu\rho]^2$$

$$\Phi_{34} = [(1 - 2\nu)(s_z + \rho) + 3\nu\rho][(1 - 2\nu)\tau_{xy}]$$

$$\Phi_{35} = [(1 - 2\nu)(s_z + \rho) + 3\nu\rho][(1 - 2\nu)\tau_{yz}]$$

$$\Phi_{36} = [(1 - 2\nu)(s_z + \rho) + 3\nu\rho][(1 - 2\nu)\tau_{zx}]$$

$$\Phi_{44} = [(1 - 2\nu)\tau_{xy}]^2$$

$$\Phi_{45} = [(1 - 2\nu)\tau_{xy}][(1 - 2\nu)\tau_{yz}]$$

$$\Phi_{46} = [(1 - 2\nu)\tau_{xy}][(1 - 2\nu)\tau_{zx}]$$

$$\Phi_{55} = [(1 - 2\nu)\tau_{yz}]^2$$

$$\Phi_{56} = [(1 - 2\nu)\tau_{yz}][(1 - 2\nu)\tau_{zx}]$$

$$\Phi_{66} = [(1 - 2\nu)\tau_{zx}]^2 \qquad [12.101]$$

Axisymmetrical case

Referring to Fig. 12.24(b), the stress components in polar coordinate system are $\sigma_r, \sigma_\theta, \sigma_z$, and $\tau_{r\theta}$, and the corresponding engineering strains are $\epsilon_r, \epsilon_\theta, \epsilon_z$, and $\gamma_{r\theta}$. Equation [12.94] can be written as:

$$\begin{Bmatrix} d\sigma_r \\ d\sigma_\theta \\ d\sigma_z \\ d\tau_{r\theta} \end{Bmatrix} =$$

$$\frac{E}{(1+\nu)(1-2\nu)} \begin{bmatrix} 1-\nu-\omega\Phi_{11} & \nu-\omega\Phi_{12} & \nu-\omega\Phi_{13} & -\omega\Phi_{14} \\ & 1-\nu-\omega\Phi_{22} & \nu-\omega\Phi_{23} & -\omega\Phi_{24} \\ \text{symmetric} & & 1-\nu-\omega\Phi_{33} & -\omega\Phi_{34} \\ & & & \frac{1-2\nu}{2}-\omega\Phi_{44} \end{bmatrix} \begin{Bmatrix} d\epsilon_r \\ d\epsilon_\theta \\ d\epsilon_z \\ d\gamma_{r\theta} \end{Bmatrix}$$

[12.102]

where we have denoted:

$$\frac{1}{\omega} = [(1-2\nu)(2J_2+3\rho^2)+9\nu\rho^2]+\frac{H(1+\nu)(1-2\nu)}{E}[2J_2+3\rho^2]^{1/2}(1-\tfrac{1}{3}Bp)$$

$$\Phi_{11} = [(1-2\nu)(s_r+\rho)+3\nu\rho]^2$$

$$\Phi_{12} = [(1-2\nu)(s_r+\rho)+3\nu\rho][(1-2\nu)(s_\theta+\rho)+3\nu\rho]$$

$$\Phi_{13} = [(1-2\nu)(s_r+\rho)+3\nu\rho][(1-2\nu)(s_z+\rho)+3\nu\rho]$$

$$\Phi_{14} = [(1-2\nu)(s_r+\rho)+3\nu\rho][(1-2\nu)\tau_{r\theta}]$$

$$\Phi_{22} = [(1-2\nu)(s_\theta+\rho)+3\nu\rho]^2$$

$$\Phi_{23} = [(1-2\nu)(s_\theta+\rho)+3\nu\rho][(1-2\nu)(s_z+\rho)+3\nu\rho]$$

$$\Phi_{24} = [(1-2\nu)(s_\theta+\rho)+3\nu\rho][(1-2\nu)\tau_{r\theta}]$$

$$\Phi_{33} = [(1-2\nu)(s_z+\rho)+3\nu\rho]^2$$

$$\Phi_{34} = [(1-2\nu)(s_z+\rho)+3\nu\rho][(1-2\nu)\tau_{r\theta}]$$

$$\Phi_{44} = [(1-2\nu)\tau_{r\theta}]^2 \qquad\qquad\qquad [12.103]$$

Plane strain case

Suppose the z-axis is normal to the plane to which deformation is restricted, then the assumptions for plane strain are $\epsilon_z = \gamma_{xz} = \gamma_{yz} = 0$, [12.100] can be reduced to the form:

$$\begin{Bmatrix} d\sigma_x \\ d\sigma_y \\ d\tau_{xy} \end{Bmatrix} = \frac{E}{(1+\nu)(1-2\nu)} \begin{bmatrix} 1-\nu-\omega\Phi_{11} & \nu-\omega\Phi_{12} & -\omega\Phi_{13} \\ & 1-\nu-\omega\Phi_{22} & -\omega\Phi_{23} \\ \text{symmetric} & & \frac{1-2\nu}{2}-\omega\Phi_{33} \end{bmatrix} \begin{Bmatrix} d\epsilon_x \\ d\epsilon_y \\ d\gamma_{xy} \end{Bmatrix}.$$

[12.104]

where we have denoted:

$$\frac{1}{\omega} = [(1-2\nu)(2J_2 + 3\rho^2) + 9\nu\rho^2] + \frac{H(1+\nu)(1-2\nu)}{E}[2J_2 + 3\rho^2]^{1/2}(1 - \tfrac{1}{3}Bp).$$

$$\Phi_{11} = [(1-2\nu)(s_x + \rho) + 3\nu\rho]^2$$

$$\Phi_{12} = [(1-2\nu)(s_x + \rho) + 3\nu\rho][(1-2\nu)(s_y + \rho) + 3\nu\rho]$$

$$\Phi_{13} = [(1-2\nu)(s_x + \rho) + 3\nu\rho][(1-2\nu)\tau_{xy}]$$

$$\Phi_{22} = [(1-2\nu)(s_y + \rho) + 3\nu\rho]^2$$

$$\Phi_{23} = [(1-2\nu)(s_y + \rho) + 3\nu\rho][(1-2\nu)\tau_{xy}]$$

$$\Phi_{33} = [(1-2\nu)\tau_{xy}]^2 \qquad [12.105]$$

Plane stress case

Suppose the z-axis is normal to the plane to which stresses are zero, then the assumptions for plane stress are $\sigma_z = \tau_{xz} = \tau_{yz} = 0$, [12.100] can be reduced to the form:

$$\begin{Bmatrix} d\sigma_x \\ d\sigma_y \\ d\tau_{xy} \end{Bmatrix} = \frac{E}{1-\nu^2} \begin{bmatrix} 1-\omega\Phi_{11} & \nu-\omega\Phi_{12} & -\omega\Phi_{13} \\ & 1-\omega\Phi_{22} & -\omega\Phi_{23} \\ \text{symmetric} & & \frac{1-\nu}{2}-\omega\Phi_{33} \end{bmatrix} \begin{Bmatrix} d\epsilon_x \\ d\epsilon_y \\ d\gamma_{xy} \end{Bmatrix} \qquad [12.106]$$

where we have denoted:

$$\frac{1}{\omega} = [2(1-\nu)J_2 - (1-2\nu)s_z^2 - 2(1+\nu)\rho s_z + 2(1+\nu)\rho^2]$$

$$+ \frac{H(1-\nu^2)}{E}[2J_2 + 2\rho^2]^{1/2}(1 - \tfrac{1}{3}Bp)$$

$$\Phi_{11} = [(1-\nu)s_x - \nu s_z + (1+\nu)\rho]^2$$

$$\Phi_{12} = [(1-\nu)s_x - \nu s_z + (1+\nu)\rho][(1-\nu)s_y - \nu s_z + (1+\nu)\rho]$$

$$\Phi_{13} = [(1-\nu)s_x - \nu s_z + (1+\nu)\rho][(1-\nu)\tau_{xy}]$$

$$\Phi_{22} = [(1-\nu)s_y - \nu s_z + (1+\nu)\rho]^2$$

$$\Phi_{23} = [(1-\nu)s_y - \nu s_z + (1+\nu)\rho][(1-\nu)\tau_{xy}]$$

$$\Phi_{33} = [(1-\nu)\tau_{xy}]^2 \qquad [12.107]$$

12.5. FINITE ELEMENT FORMULATION

12.5.1. General

In the preceding sections we have written down stress–strain relations for soils as well as rock-like material such as concrete. To complete the theory, equations of equilibrium and equations of compatibility need be added. In principle, when appropriate boundary conditions are introduced, a solution can be determined. However, only in rare instances will it be feasible to obtain *exact* solutions. Almost always it will be necessary to approximate the answer on the basis of such methods as the *limit analysis* described in this book for load-carrying capacity determination or the now well-known method of *finite element* for a numerical solution of a progressive failure problem.

The *finite element method,* which is a computer-based solution technique, is extremely effective when applied with proper caution. Answers which are fully satisfactory for engineering analysis and design are obtained to problems for which the exact answer is impossibly difficult. The basic philosophy of finite element approach is to reduce the complex continuum from infinite degrees of freedom to a finite, if large, number of unknowns. Such a process of discretization was first successfully performed by the familiar method called *finite differences.* The finite element method provides an alternative to such a process. It appears to offer considerable advantages and its relatively simple logic makes it ideally suited for the computer.

This philosophy of discretizing a continuum into a finite number of elements is essentially the same as that of the *strength of materials* approach to a bar problem. In a bar problem, which is a well-established branch of mechanics of solids, an element of unit length and full cross-sectional area is taken as the basic building block. This element may be subjected to axial force, twisting moment, or bending moment separately or in combination. These loads are called the *generalized stresses* of the problem. The corresponding *generalized strains* are the elongation, or the angle of twist or the curvature. The very simple but powerful assumption is then made that the forms for generalized stress vs. generalized strain derived on the basis of certain *kinematic* assumptions of the element, may be applied to the infinitesimal length of the bar. Except for localized disturbances, all complexities of bar problems are reduced to elementary problems with the strength of materials assumption.

Similarily, the complexities of a continuous medium problem can be drastically simplified by separating the continuum into a number of finite elements with imaginary lines or surfaces such as the triangular elements shown in Figs. 5.1 and 5.5, or rectangular elements in Fig. 5.7., or quadrilateral elements shown in Figs. 5.9 and 5.10, or triangular ring elements in Fig. 5.36. These elements are assumed to be interconnected at a discrete number of nodal points situated on their boundaries. The displacements of these nodal points are the basic unknowns of the problem, just as in the strength of materials approach to a bar problem.

In short, the finite element method as applied to a continuum may be divided into three basic steps: (1) structural discretization; (2) evaluation of the element generalized stress–generalized strain relations; and (3) structural analysis of the element assemblage.

The structural discretization is the subdivision of the original continuous system into an assemblage of a finite number of discrete elements. Judgement is obviously required in making the subdivision because the analysis is now performed only on this substituted structure and the results are valid only to the extent that the behavior of the discretized structures simulates the actual situation.

12.5.2. Generalized stress–generalized strain relationships

Of interest to us here is the derivation of the generalized stress–generalized strain relations of the individual elements which, when combined with equilibrium equations of the individual elements and then superimposed, will provide the complete system of equations of the entire structure. The derivation of the generalized stress–generalized strain relations of the individual elements is based on virtual work equation.

Since the elements are interconnected only at a limited number of nodal points, the essential stress–strain characteristics of an element are represented by the generalized relationship between the forces $\{F\}$ at the nodal points and the corresponding nodal displacements $\{\delta\}$ resulting therefrom. Herein a derivation is made of the relationship of generalized stresses $\{F\}$ and generalized strains $\{\delta\}$ for a finite element having m nodal points. This force–displacement relationship may be expressed most conveniently by the stiffness matrix of the element and can be derived in the following manner.

Because plastic behavior is load path dependent and usually requires step-by-step calculations that follow the history of loading, it is only possible to establish the relationship between the infinitesimal generalized stress increments $\{\dot{F}\}$ or $\{dF\}$ and the infinitesimal generalized strain increments $\{\delta\}$ or $\{d\delta\}$, provided that the existing state of stress and strain is known. This relationship is introduced into the tangent stiffness matrix which will be derived herein. Since virtual work equation is utilized here, this relationship also satisfies the equilibrium equations, which, when superposed over all elements, will result in a stiffness relationship between the system of forces concentrated at the nodes and the system of displacements at the nodes. Once this stage has been reached the solution procedure can follow the standard structural routine described in many textbooks.

In order to satisfy compatibility conditions between the state of displacement within each "finite element" in terms of its nodal displacements, an admissible *displacement function* (or functions) must be assumed such that: (1) the function (or functions) now defines uniquely the state of strain within an element in terms of the nodal displacements; (2) the function (or functions) gives appropriate nodal displacements when the coordinates of the appropriate nodes are inserted; and (3) the function (or functions) satisfies the continuity of displacements with adjacent elements. Suitable linear functions of coordinates, for example, provide such a choice. We refer the reader to the book by Zienkiewicz (1971) for detailed discussions of this subject.

Consider now a typical finite element which is defined by nodes, i, j, m, etc., and straight line boundaries. Since displacement rates or increments, $\{\dot{u}(x,y,z)\}$, at any point within the element can be expressed as functions of the nodal displacement increments, $\{\dot{\delta}\}$:

$$\{\dot{u}\} = \mathbf{N} \; \{\dot{\delta}\} = [N_i N_j N_m \ldots] \begin{Bmatrix} \dot{\delta}_i \\ \dot{\delta}_j \\ \dot{\delta}_m \\ \vdots \end{Bmatrix} \qquad [12.108]$$

in which the components of matrix $\mathbf{N}$ are in general functions of the position and $\{\dot{\delta}\}$ represents a listing of nodal displacements for a particular element.

With the displacement increments known at all points within the element, the corresponding strain increment at any point in the element can be determined by the incremental strain–displacement relations given by [2.24]. Using this relation, we can express the strain increments in a finite element in terms of nodal displacement increments as follows:

$$\{\dot{\epsilon}\} = \mathbf{B} \; \{\dot{\delta}\} \qquad [12.109]$$

in which $\mathbf{B}$ is the strain–displacement relation matrix.

The incremental stress–strain relationship described in the preceding sections can be written in matrix form as:

$$\{\dot{\sigma}\} = \mathbf{D} \; \{\dot{\epsilon}\} \qquad [12.110]$$

where $\mathbf{D}$ is a matrix containing the appropriate constants of the material properties.

Let an arbitrary *virtual* displacement rate, $\{\dot{\delta}^*\}$, be applied at the nodes of the element. Denoting the transpose of a vector (matrix) by a superscript T, the external rate of virtual work done by the incremental nodal forces, $\{\dot{F}\}$, due to these virtual displacement increments is:

$$\{\dot{\delta}^*\}^T \{\dot{F}\} \qquad [12.111]$$

and the internal rate of virtual energy dissipation is:

$$\{\dot{\epsilon}*\}^T \{\dot{\sigma}\} \qquad\qquad [12.112]$$

where $\{\dot{\epsilon}*\}$ is related to $\{\dot{\delta}*\}$ by [12.109], and is independent of $\{\dot{\sigma}\}$, and $\{\dot{F}\}$ which equilibrates $\{\dot{\sigma}\}$.

Equating the external rate of virtual work to the internal rate of energy dissipation and using [12.109] and [12.110], we obtain:

$$\{\dot{\delta}*\}^T \{\dot{F}\} = \{\dot{\delta}*\}^T \left(\int_{vol} \mathbf{B}^T \mathbf{D} \mathbf{B} \, d(vol) \right) \{\dot{\delta}\} \qquad\qquad [12.113]$$

This is the virtual work equation [2.25] which is now written in matrix form. Since the virtual displacement rate, $\{\dot{\delta}*\}$ is assumed to be arbitrary, we have:

$$\{\dot{F}\} = \mathbf{k} \{\dot{\delta}\} \qquad\qquad [12.114]$$

in which the element stiffness matrix, $\mathbf{k}$, is defined by:

$$\mathbf{k} = \int_{vol} \mathbf{B}^T \mathbf{D} \mathbf{B} \, d(vol) \qquad\qquad [12.115]$$

The matrix $\mathbf{k}$ represents the tangent stiffness or the slope of the force—displacement curve of a finite element.

Once the element stiffness matrices are defined, the corresponding structural stiffness matrix can be assembled by the proper allocations of the element stiffness matrices:

$$\mathbf{K} = \Sigma \mathbf{k} \qquad\qquad [12.116]$$

The overall equilibrium relationship is then expressed as a set of simultaneous equations:

$$\{\dot{P}\} = \mathbf{K} \{\dot{\delta}\} \qquad\qquad [12.117]$$

where $\{\dot{P}\}$ is the total load increment vector and $\{\dot{\delta}\}$ is the total nodal displacement increment vector. These equations along with proper boundary conditions can be solved by any standard solution technique. Once the nodal displacement increments have been determined, the strain increments and thus the stress increments at any point of the element can be found from the relations given in [12.109] and [12.110], respectively.

12.6. INTEGRATION OF THE DISPLACEMENT RATE EQUILIBRIUM EQUATIONS

12.6.1. General

In the preceding section we have formulated the tangent stiffness of a generic finite element. A direct sum of element stiffnesses and load vectors yields a set of displacement rate equilibrium equations in the form of [12.117] for the discretized body. This section concerns the integration of these equations.

Since we employ here the so-called incremental plasticity theory in which the material response is load path dependent, we deal with displacement rates rather than displacements. If a nonlinear elastic material model had been considered, we would have had the option of formulating equations in terms of displacements and, perhaps, solving these equations iteratively for any applied load. Here we do not have that option and, although iterative techniques can be employed in the solution, we are essentially dealing with an integration procedure, not an iterative procedure.

Referring now to [12.117], matrix K is the current tangent stiffness of the discretized body and is a function of the current stress state and current configuration of the body. If plastic unloading is admitted, the tangent stiffness is also a function of the displacement rate vector. As before, the super dot denotes rate and in particular it implies differentiation with respect to a time like parameter denoted by t. An initial condition associated with [12.117] is:

$$\{\delta(t = 0)\} = \{0\} \tag{12.118}$$

Two numerical integration techniques are described here in which the applied load history (or applied displacement history) is divided into a finite number of increments. This method provides a numerically approximating step-by-step solution of [12.117].

12.6.2. Euler integration method (Fig. 12.25)

Perhaps the most obvious way of approximating the response in an increment is to use the *Euler integration method* (Ketter and Prawel, 1969) in which the tangent stiffness at the beginning of an increment is used to obtain a linear approximation for incremental response. For example, referring to Fig. 12.25(a), we suppose that the solution at point A is known and we wish to determine incremental displacements $\Delta\delta$ associated with the applied incremental load ΔP. We project along a tangent at point A to obtain an approximate solution denoted by point B. We can expect that after a number of increments the approximate solution will diverge from the true solution as indicated in Fig. 12.25(b).

In conjunction with a finite element approach, variations of the Euler integration method have been used by Pope (1965), Swedlow et al. (1965), Marcal and King (1967), and Yamada and Yoshimura (1968) to solve elastic–plastic problems. Pope (1965) used a modified Euler approach which accounts for unloading of previously plastified elements as well as yielding of previously elastic elements. Since this information can not be known a priori, an iterative scheme is used. Each

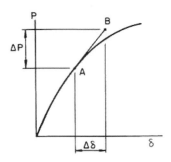
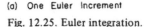

(a) One Euler Increment

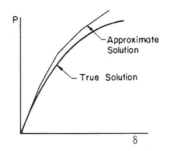

(b) Error Build-Up Associated With
     Euler Intergration

Fig. 12.25. Euler integration.

iteration involves the solution of a set of linear algebraic equations. Marcal and King (1967) used an approach suggested by Marcal (1965). Elements which yield for the first time during an increment are assigned a weighted average of elastic and elastic–plastic stiffness, and thus an iterative scheme is required. Yamada and Yoshimura (1968) used an Euler approach, however, only one element per increment was allowed to yield.

A comprehensive treatment of elastic–plastic-fracture behavior of various punch-indentation problems of concrete blocks is given in a dissertation by A.C.T. Chen (1973). The integration method used is the Euler method, allowing several elements to yield per each increment. The material model selected in this dissertation is the elastic–plastic strain hardening-fracture model for concrete (section 12.4). The finite element models selected are triangular plane and triangular ring elements representing plane and axisymmetric problems, respectively. The displacement function is linear over the element. We refer the reader to the dissertation for a detailed discussion of the integration procedure.

12.6.3. The mid-point integration method (Fig. 12.26)

Most of the numerical results presented in Chapter 5 are obtained by the so-called *mid-point integration method*. This technique has been used previously by Felippa (1966), Akyuz and Merwin (1968) and Fernandez and Christian (1971) to solve elastic–plastic, geometrically nonlinear problems. A non-linear one-dimensional load displacement curve is shown in Fig. 12.26(a). Presumably at point A the true solution is known and we wish to approximate incremental displacement $\Delta\delta$ associated with applied incremental load ΔP.

The mid-point integration rule is motivated by the idea that the secant stiffness, denoted by line AD, can probably be closely approximated by the tangent stiffness evaluated at mid-increment (half of the load increment). The mid-increment stiffness, of course, is not known but we can estimate it. Referring to Fig. 12.26(b), we

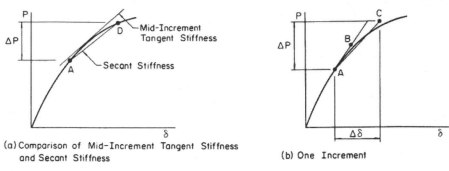

(a) Comparison of Mid-Increment Tangent Stiffness and Secant Stiffness

(b) One Increment

Fig. 12.26. Mid-point integration rule.

first apply half the incremental load and use the tangent stiffness evaluated at point A to approximate the mid-increment solution denoted by point B. We subsequently apply the complete incremental load and use the tangent stiffness evaluated at point B to obtain an approximate incremental solution denoted by point C. In each increment we thus solve two sets of linear, simultaneous, algebraic equations.

A comprehensive treatment of elastic–plastic large deformation response of clay to footing loads using this mid-point integration method is given in a dissertation by Davidson (1974). We refer the reader to the dissertation for a detailed discussion of the method. The material model selected in this dissertation is the extended Von Mises perfectly plastic model for soil (section 12.2). The finite element models selected are triangular and quadrilateral plane elements representing plane strain conditions. The displacement function is linear over the element.

12.7. EXAMPLE 1 – RIGID STRIP FOOTING ON A SOIL STRATUM

In this section we consider a number of variables associated with the finite element discretization and the mid-point integration technique. The problem used here is the same as that considered in Chapter 5 (sections 5.5, 5.6 and 5.7), that is, a 5 ft. wide rigid *strip* footing bearing on a 50 ft. deep soil stratum supported by a rigid rough base. The horizontal extent of the stratum was arbitrarily set at 50 ft. from the footing center and a smooth rigid boundary was prescribed. The material model selected is the extended Von Mises perfectly plastic model for soil (section 12.2). The following soil parameters were used:

$E = 500{,}000$ psf (24 MN/m^2) $c = 500$ psf (24 kN/m^2)

$\nu = 0.5$ $\varphi = 30°$

$\gamma = 0$

where γ is soil weight per unit volume. In sections 5.5–5.7, Chapter 5, elastic–

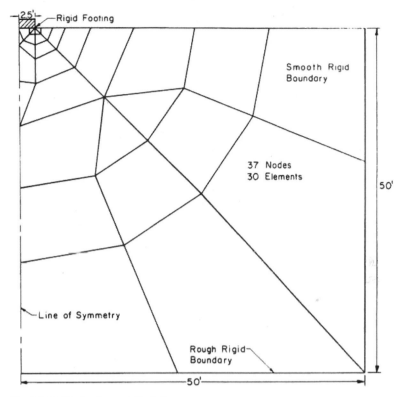

Fig. 12.27. Finite element Mesh 1.

plastic solutions of this problem for various soil parameters are presented for both small and large deformation analysis. Herein, we consider only small deformation analysis.

12.7.1. Finite element discretization

Three finite element meshes, shown in Figs. 12.27, 12.28 and 5.10, were utilized for the numerical experiments. Each mesh is composed of a number of triangular and quadrilateral regions. Three different arrangements of triangles were used to define stiffness for the quadrilateral regions. In the first arrangement, pictured in Fig. 12.29(a), the quadrilaterals are divided into two triangles with the dividing diagonal having the same orientation for all quadrilaterals. The second arrangement is shown in Fig. 12.29(b), and here the diagonals are staggered. In the third arrangement, shown in Fig. 12.29(c), the quadrilateral is subdivided into four triangles connected to a fifth node located at the quadrilateral centroid. The displacement function selected is a linear displacement expansion over the element (constant strain triangle).

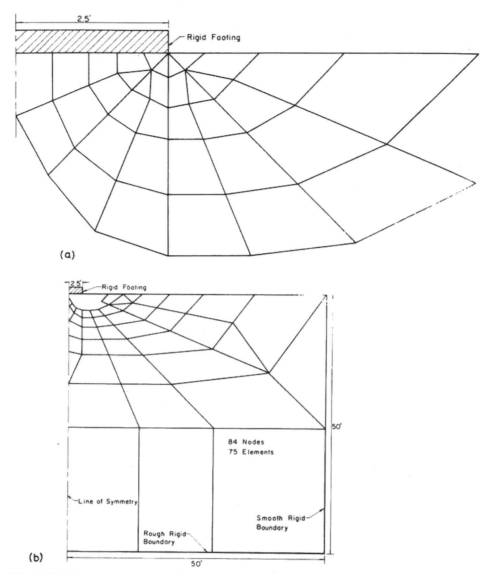

Fig. 12.28. Finite element Mesh 2. (a) Mesh detail near footing. (b) Mesh without details near footing.

All three meshes, as shown in Figs. 12.27, 12.28 and 5.10, are finest near the corner of the footing and get progressively coarser as the distance from the corner increases. It is not the aim here to capture the stress singularity at the corner since it is well known that it is difficult or perhaps impossible to do this with analytic finite element expansions. The aim here is simply to make the mesh fine where stress gradients are high.

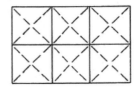

(a) Element Arrangement 1   (b) Element Arrangement 2   (c) Element Arrangement 3

Fig. 12.29. Three-element arrangements.

We discuss now the results of four numerical experiments in which the effects of mesh size, element arrangement, increment size, and stress scaling with associated equilibrium correction were investigated. Since the footing is assumed to be rigid, displacements rather than stresses are prescribed beneath the footing. For all of the numerical experiments the footing is assumed to be *smooth* and displacements are assumed to be *small*, that is, for the moment we consider only material nonlinearities. The theoretical bearing capacity obtained by Prandtl (1920) (see [3.81]) is q_0 = 15,040 psf (720 kN/m^2). We have shown in Chapter 6 (section 6.6.1) that the Prandtl solution is the true limit load for weightless soils with a Coulomb yield condition and the associated flow rule.

12.7.2. Effects of mesh size

Results for the three meshes are shown in Fig. 12.30 where we plot average stress beneath the footing, q, versus vertical displacement of the footing. Element arrangement 3 (Fig. 12.29) was used to compute quadrilateral stiffness for the quadrilateral regions shown in Figs. 12.27, 12.28 and 5.10, and an increment size of

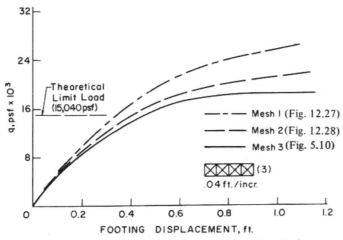

Fig. 12.30. Footing load displacement curves. Effect of mesh size.

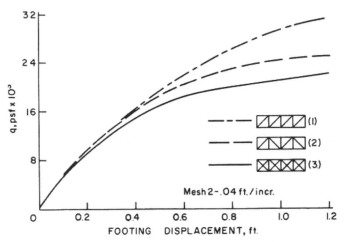

Fig. 12.31. Footing load displacement curves. Effect of element arrangement.

0.04 ft. of footing displacement per increment was used. Although the curves shown are smooth, there were some oscillations in the solutions, particularly at the higher loads. As might be expected, the finest mesh gave the softest response. Even with the finest mesh there is still a noticeable difference between the theoretical plastic limit load and that determined from the numerical finite element solution. If we take the limit load of the discretized body associated with mesh 3 (Fig. 5.10) to be 18,500 psf (886 kN/m^2), then this load is approximately 23% greater than the theoretical limit load.

12.7.3. Effects of element arrangement

Fig. 12.31 shows the results for the three element arrangements for the quadrilateral regions shown in Fig. 12.28 (mesh 2) and 0.04 ft. of footing displacement per increment. Most remarkable is the large difference in the solutions obtained from arrangements 1 and 2. Many writers have commented on the stress discontinuities between adjacent elements when constant strain triangles are used in finite element analysis (for a recent example see Owen et al., 1973). In regions where the stress gradient is high, element stresses tend to oscillate from element to element. This tendency is particularly noticeable in element arrangement 1 where stresses at the higher loads oscillate from tension to compression. The best solution was obtained with element arrangement 3, but we should note that computation time was approximately 40% greater than the time required for arrangements 1 and 2.

12.7.4. Effects of increment size

The effect of the size of the footing displacement increment is shown in

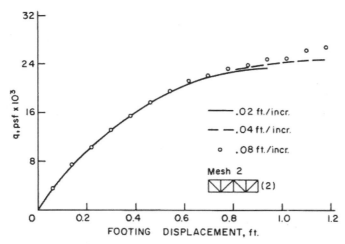

Fig. 12.32. Footing load displacement curves. Effect of increment size.

Fig. 12.32. Mesh 2 (Fig. 12.28) with element arrangement 2 for the quadrilateral regions was used for the three solutions. Three displacement increments were utilized, 0.02, 0.04 and 0.08 ft. per increment. Up to a footing displacement of about 0.7 ft., the solutions are essentially the same. After this the three curves diverge somewhat with the smallest increment size giving the softest response. We can see that the mid-point integration scheme (section 12.6.3) is not highly sensitive to increment size, at least at low and intermediate load levels.

12.7.5. Effect of stress scaling back to the yield curve

Consider for the moment an element which is found to be plastic at the beginning of an increment, and assume that the stress state lies exactly on the yield surface. A schematic diagram including a yield curve and the initial stress state, denoted by point A, is shown in Fig. 12.33(a). At the end of the increment the stress state, denoted by point B, probably lies somewhat outside of the yield surface. After a number of such load increments have been analyzed, the stress state may lie far enough from the yield curve to render the analysis of subsequent increments meaningless. A possible stress path produced by a number of increments is denoted by broken line ABC.

In order to correct this situation we scale stresses back to the yield surface at the end of each increment and at mid-increment. Since there is no unique way to scale, we arbitrarily require that the hydrostatic component and principal directions of the stress tensor remain unchanged. A schematic stress path associated with stress scaling is shown in Fig. 12.33(b). As can be seen in the figure, stresses are adjusted back to the yield surface at the end of an increment.

In general the scaled stresses can not be expected to satisfy the equilibrium

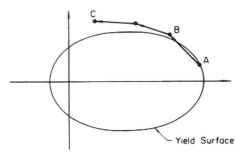

(a) Stress Path with no Scaling

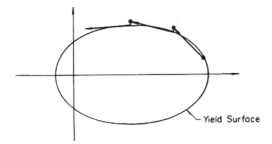

(b) Stress Path with Stress Scaling

Fig. 12.33. Stress scaling back to yield surface.

equations of the discretized body. We thus compute an equilibrium correction vector and apply this load vector, along with the prescribed loads, in the next increment. Strickland et al. (1971) have shown that correcting for nonequilibrating stresses significantly increases accuracy when used in conjunction with an incremental Euler integration approach. We show here that for elastic–plastic analysis, scaling stresses back to the yield surface and subsequently distributing the unbalanced nodal forces during the next increment is also computationally efficient.

In Fig. 12.34 two of the solutions were obtained without scaling stresses back to the yield surface after each increment. Thus, to within round-off and truncation errors, stresses determined at the end of each increment satisfied equilibrium. A third solution, shown by open circles in Fig. 12.34, was obtained using stress scaling and equilibrium correction. We note first of all that for the same increment size there is a noticeable difference in the solutions with and without scaling. We note secondly that the solution without scaling can be made to agree closely to the solution with scaling by using a very small increment size. We can thus conclude that for a small increase in computational effort, we get a significant increase in accuracy by using stress scaling with equilibrium correction.

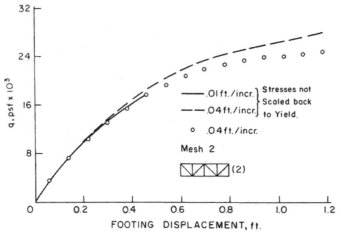

Fig. 12.34. Footing load displacement curves. Effect of stress scaling.

12.8. EXAMPLE 2 – PLANE STRAIN PUNCH-INDENTATION OF CONCRETE BLOCKS

12.8.1. Statement of the problem

To illustrate results deduced from the elastic–plastic strain hardening-fracture theory for concrete (section 12.4), we consider here the double-punch problem shown in Fig. 12.35(a). The only load applied to the block is by the two *rough* punches and the material is taken to be that of section 12.4. The plane strain

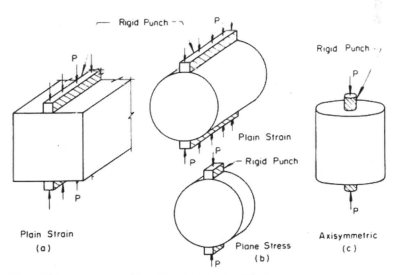

Fig. 12.35. Problems considered in Chapter 11 and here.

TABLE 12.1

Material constants (see [12.62] to [12.67])

	A_O (ksi)	A_u (ksi)	τ_0^2 (ksi^2)	τ_u^2 (ksi^2)	A (ksi) [12.68]	B (ksi^{-1}) [12.69]
Compression zone	0.699	1.165	1.754	4.873	0.437	0.149
Tension–compression zone	1.215	2.025	0.107	0.297	0.759	4.260

Young's modulus, E = 3791 ksi (26,138 MN/m^2).
Poisson's ratio, ν = 0.188.
Uniaxial compressive strength f'_c = 4.45 ksi (30.7 MN/m^2).
Uniaxial tensile strength, f'_t = 0.090 f'_c.
Equal biaxial compressive strength, f'_{bc} = 1.160 f'_c.

conditions are assumed in the analysis. The same punch-indentation problem of a Von Mises block has been considered in section 5.3, Chapter 5 (Fig. 5.3) which may be considered here as a special case of concrete for which $A = B = n = 0$ and $\rho = 0$ (see [12.68], [12.69] and [12.99]).

The finite element solutions for concrete cylinders under two punches representing plane strain (Fig. 12.35b) and axisymmetric conditions (Fig. 12.35c) are presented in Chapter 11. The material in those cases is taken to be identical to that of the present problem (Fig. 12.35a). The finite element models selected for all cases are triangular plane and triangular ring elements. The displacement field is linear over the element. The material constants defined in [12.62] to [12.67] are listed in Table 12.1. The initial discontinuous values for concrete under uniaxial compression, uniaxial tension and equal biaxial compression are taken to be 60% of that of the corresponding fracture values given in Table 12.1. The strain hardening modulus H' defined in [12.81] is taken to be that of the slope of the compression curve in Fig. 12.21 for both compression and tension–compression stress states. The finite element mesh used in the present problem is shown in Fig. 12.37. The dimensions used and the boundary conditions considered are the same as that of Fig. 5.3. The Euler integration method (section 12.6.2) is utilized for the numerical work. The numerical results are presented in Figs. 12.36–12.40.

12.8.2. Yield zones and fractured zones

Fig. 12.36 shows the load–displacement curve of a concrete block loaded by the double punch. Nonlinear behavior is seen to occur at a very early stage of loading. The plastic zone starts beneath the corner of the punch and is followed by yielding at the geometrical center of the block (this is probably due to the weak tensile property of concrete, Fig. 12.37a). In Fig. 12.37, the Arabic or Roman numbers

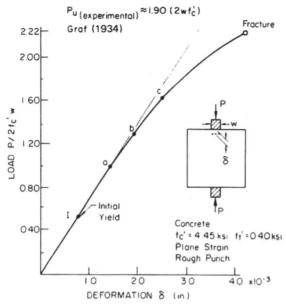

Fig. 12.36. The load–displacement curve of the punch-indentation of a concrete block under plane strain conditions.

marked in the elements indicate the order in which the elements become plastic or fracture during that specific loading interval. The dark and shaded areas indicate regions already yielded and fractured respectively previous to that loading interval. The stiffness of the load–displacement curve (Fig. 12.36) does not decrease significantly until point b is reached. Though most of the elements under the punch have become plastic at this stage of loading (Fig. 12.37b), their stiffnesses are still very close to the elastic stiffness except for the region directly beneath the corner of the punch where the stiffness is near zero. When the loading is increased slightly above point b, the element beneath the corner of the punch fractures, but the entire concrete block can still take further loading. It is interesting to note that some of the elements on the concrete free surface have become plastic (Fig. 12.37c) at this stage of loading because of the weak tensile strength of concrete.

The ultimate load obtained is $2.22 f_c' (2w)$ which is approximately 16% higher than that of the experimental value reported by Graf (1934). Near the point of collapse, the region beneath the corner of the punch is found to be completely fractured (Fig. 12.37d). Failure of the specimen is due to fracture rather than flow of concrete. Figure 12.38 shows the stiffness of the elements near the collapse load. The values given in the elements are the percentage of the stiffness of the elements near the collapse load to that of the elastic stiffness. Blank elements remain elastic and the minus sign indicates elements which have been plastically unloaded. As can be seen, elements beneath the punch, which have the stiffnesses less than 5% of that

$$\frac{P}{4f_c'w} = 0.27 \text{ to } 0.51$$

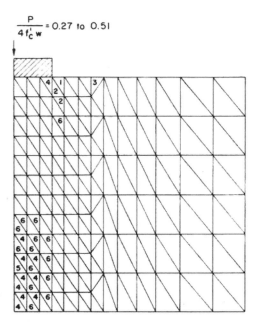

(a) From Point I to Point a

$$\frac{P}{4f_c'w} = 0.51 \text{ to } 0.66$$

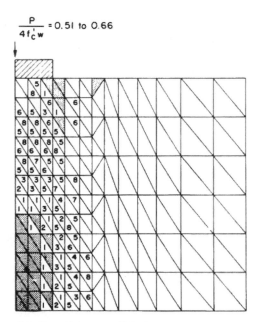

(b) From Point a to Point b

Fig. 12.37. The spreading of yield zones and fracture zones of the concrete block.

$$\frac{P}{4f_c'w} = 0.66 \text{ to } 0.833$$

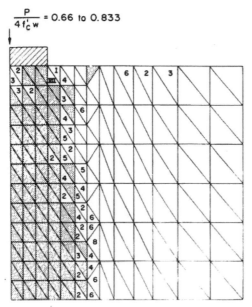

(c) Load Further Increased from Point b to Point c

$$\frac{P}{4f_c'w} = 0.833 \text{ to } 1.11$$

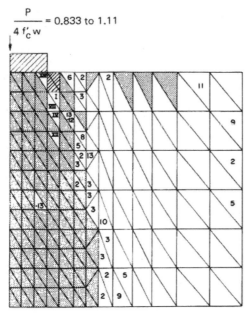

(d) Load Increased from Point c to the Point near Collapse

Fig. 12.37. Continued.

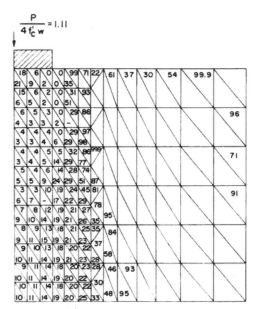

$$\frac{P}{4f_c' w} = 1.11$$

Fig. 12.38. The degree of softening of a concrete block near the collapse load.

of the elastic stiffness, form a wedge-shape zone. At this stage of loading, equations of equilibrium become singular and the solution to these equations cannot be obtained.

12.8.3. Contact stresses

The redistribution of the vertical contact stress along the punch–concrete interface AB and the vertical reaction stress along the horizontal symmetric plane of the block $A'C'$ are shown in Fig. 12.39. At initial yield load, the ratio of the vertical stress near the punch edge to that at the punch center is approximately

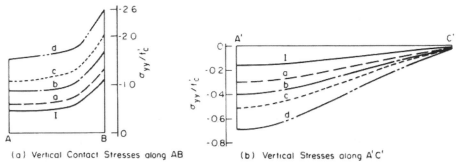

(a) Vertical Contact Stresses along AB   (b) Vertical Stresses along A'C'

Fig. 12.39. The redistribution of vertical contact stress along punch–concrete interface AB, and vertical reaction stress along horizontal symmetric plane $A'C'$ of the block.

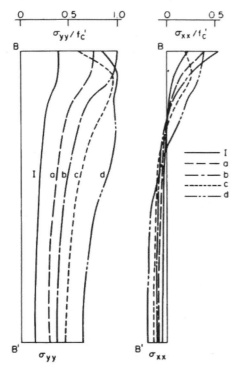

Fig. 12.40. The redistribution of the horizontal and vertical stress components along the vertical plane BB' passing through the edge of the punch.

2.57. As the loading increases, this ratio decreases and becomes 1.66 near the collapse load. Figure 12.40 shows the redistribution of the horizontal and vertical stress components σ_{xx} and σ_{yy} along the vertical plane BB' passing through the edge of the punch. The stresses everywhere along this plane are seen to increase as the punch load is increased except near the region where elements have fractured.

12.9. SUMMARY AND CONCLUSIONS

12.9.1. Summary

We have discussed here two elastic–plastic soil models; the perfectly plastic extended Von Mises model of Drucker-Prager and the isotropic strain hardening model of Roscoe and Burland. An explicit incremental constitutive matrix for plane strain conditions was presented for the Drucker-Prager model and an indicial expression for the Roscoe-Burland constitutive tensor (suitable for numerical analysis) was presented. We have also presented an elastic–plastic strain hardening-fracture model for rock-like material such as concrete. The incremental stress–strain relationships were derived for the three-dimensional case. Incremental constitutive matrices for plane stress, plane strain and axisymmetric conditions were then presented.

We have derived the generalized stress–generalized strain equilibrium relations of the individual finite elements. This equation, when superposed, provides the complete system of equilibrium equations of the entire structure. We have also reviewed two integration techniques suitable for elastic–plastic analysis and applied the mid-point integration method to a footing example and the Euler integration method to a concrete example.

In the first example, we consider a rigid strip footing on a soil stratum. The soil is modelled as an elastic–perfectly plastic extended Von Mises material. In the second example, we consider a plane strain concrete punch-indentation problem. The concrete is modelled as an elastic–plastic strain hardening-fracture material. Further numerical results of this two example problems were presented in Chapter 5 for soils and in Chapter 11 for concrete.

12.9.2. Conclusions

Much additional work is needed on each of the three material models discussed in this chapter and the many more which were omitted. Progressive failure of embankments, foundations, retaining wall, and similar configuration or concrete structures should be studied more extensively using these models. Although some analyses using Cambridge-type hardening models (Schofield and Wroth, 1968) have been briefly reported in the literature (Tang and Höeg, 1968; Zienkiewicz and Naylor, 1971), much more remains to be done.

Since strain softening is characteristic of many natural soils, analyses using strain softening stress–strain models would be extremely enlightening. In particular the effect of strain softening on the maximum load should be investigated. This kind of work is particularly necessary for sensitive clays which exhibit dramatic softening. Höeg (1972) and Zienkiewicz and Nayak (1972) have done some preliminary work in this area.

Since time-dependence is also an essential characteristic of many natural soils containing water, the time-independent idealization utilized in many material models may have little physical significance except in special cases. Time-independent description of soils or concrete is therefore a drastic idealization which should not be pushed too far. However, time-independent approximation will provide useful information over a wide range of conditions. In certain circumstances, such as in stability problems of soil mechanics over ordinary time scale, there appears to be reasonable justification for the adoption of a limit analysis approach based upon Coulomb's condition of yield in soils. This may prove to be as useful for soils as it has been for metals.

Much more of value will be uncovered as engineers who have need for particular results apply the methods of limit analysis described in this book to their own special problems. Work done at academic institutions will and should, in the main, deal with typical basic configuration and use various crude and refined material models so that physical insight is obtained and useful theorems and techniques for practical calculations may then be developed.

REFERENCES

Abdul-Baki, A. and Beik, L.A., 1970. Bearing capacity of foundations on sand. *J. Soil Mech. Found. Div., ASCE,* 96 (SM2); 545–559.

Absi, E., 1962. Soil mass in limit equilibrium. *Ann. Inst. Tech. Bat. Travaux, Publ.,* 177: 50 pp. (in French).

Akyuz, F.A., 1966. *Indentation of an Elasto-Plastic Half Space by a Rigid Punch.* Dissertation, Civil Engineering Department, Rice University, Houston, Texas.

Akyuz, F.A. and Merwin, J.E., 1968. Solution of nonlinear problems of elastoplasticity by finite-element method. *AIAA Journal,* 6 (10): 1825–1831.

Alexander, J.M., 1955. The effect of Coulomb friction in the plane-strain compression of a plastic–rigid material. *J. Mech. Phys. Solids,* 3: 233.

Allen, D.N. de G. and Southwell, R.V., 1950. Relaxation methods applied to engineering problems – XIV, plastic straining in two-dimensional stress systems. *Philos. Trans. R. Soc. London, Ser. A,* 242: 379–414.

Ang, A.H.S. and Harper, G.N., 1964. Analysis of contained plastic flow in plane solids. *J. Eng. Mech. Div., ASCE,* 90 (EM5): 397–418.

Argyris, J.H., 1965. Elastoplastic matrix displacement analysis of three-dimensional continua. *J. R. Aeronaut. Soc.,* 69: 633–636.

Arthur, J.R.F., James, R.G. and Roscoe, K.H., 1964. The determination of stress fields during plane strain of sand mass. *Géotechnique,* 14 (4): 283.

ASCE-WRC, 1971. *Plastic Design in Steel. ASCE Manual 41.* The Welding Research Council and the American Society of Civil Engineers, 2nd ed.

Bach, C. and Baumann, R., 1924. *Elastizität und Festigkeit.* Springer, Berlin, p. 217.

Baker, A.L.L., 1970. A criterion of concrete failure. *Inst. Civ. Eng., Proc.,* 45: 269–278.

Balla, A., 1960. The soil pressure on a sustaining wall. *Acta Tech.,* 29 (1/2): 99–115 (in French).

Barden, L., Ismail, H. and Tong, P., 1969. Plane-strain deformation of granular material at low and high pressures. *Géotechnique,* 19 (4): 441–452.

Belaenko, F.A., 1950. The stresses around a circular shaft in an elastic–plastic soil. *Izv. Akad. Nauk S.S.S.R.,* Otd. Tedhnicheskikh Nauk, Mekhanika i Mashinostroenie, Leningrad, pp. 914–925 (in Russian).

Berezantsev, V.G., 1948. Limit equilibrium of a medium having internal friction and cohesion in a stressed state, symmetrical to the axis *Appl. Math. Mech., (Prikl. Mat. Mekhanika),* 12: 99–100 (in Russian).

Berezantsev, V.G., 1952. *Axially Symmetric Problems of the Theory of Limiting Equilibrium of Granular Material.* Government Publishing House for Technical-Theoretical Literature, Moscow.

Berezantsev, V.G., 1955. Limit loads in indentation of cohesive soils by spherical and conical indenters *Izv. Akad. Nauk S.S.S.R., Otd. Tekh. Nauk,* 7: 70–74 (in Russian).

Bhattacharya, S.K., 1960. The geometry of a slip-line field in soils. *Géotechnique,* 1 (2): 75–81.

Biarez, J., Burel, M. and Wack, B., 1961. Contributions à l'étude de la force portante des foundations. *Proc. 5th Int. Conf. Soil Mech. Found. Eng., Paris.* Dunod, Paris, 1: 603–609.

Biot, M.A., 1965. *Mechanics of Incremental Deformations.* Wiley, New York.

Bishop, A.W., 1957. Discussion of soil properties and their measurement. *Proc. 4th Int. Conf. Soil Mech. Found. Eng.* Butterworths, London, III: 100.

Bishop, A.W., 1966. The strength of soils as engineering materials. *Géotechnique*, 16 (2): 91–128.

Bishop, A.W., 1971. The influence of progressive failure on the choice of method of stability analysis. *Géotechnique*, 21(2): 168–172.

Bishop, A.W. and Henkel, D.J., 1957. *The Measurement of Soil Properties in the Triaxial Test*. Arnold, London.

Bishop, J.F.W., 1953. On the complete solution to problems of deformation of a plastic–rigid material. *J. Mech. Phys. Solids*, 2: 43–53.

Bishop, J.F.W., Green, A.P. and Hill, R., 1956. A note on the deformable region in a rigid–plastic body. *J. Mech. Phys. Solids*, 4: 256–258.

Bjerrum, L., 1967. Progressive failure in slopes in overconsolidated plastic clay and clay shales. *J. Soil Mech. Found. Div., ASCE*, 93 (SM5), *Proc. Pap.*, 5456: 3–49.

Bland, D.R. and Naghdi, P.M., 1958. A compressible elastic, perfectly plastic wedge. *Trans. Am. Soc. Mech. Eng., Ser. E*, 80: 239–242.

Bleich, H.H. and Heer, E., 1963. Moving step-load on half-space of granular material. *J. Eng. Mech. Div., ASCE*, 89 (EM3): 97–130.

Bloem, D.L. and Stanton, W., 1952. Discussion of thermal expansion of aggregates and concrete durability, by J. Callan. *J. Am. Concrete Inst.*, 24 (4): 504-1–505-4; Proc., Vol. 48; Discussion, p. 48–33.

Boehler, J.P. and Sawczuk, A., 1970. Limit equilibrium of anisotropic soils. *J. Méch.*, 9 (1): 5–33 (in French).

Böker, R., 1915. Die Mechanik der bleibenden Formänderung in kristallinisch aufgebauten Körpern. *Mitteilungen über Forschungsarbeiten*, 175: 1–51.

Bonneau, M., 1947. Equilibrium limit and failure of continuous media (Equilibre limite rupture des milieux continus). *Ann. Ponts et Chaussées*, 117: 609–653; 769–801.

Booker, J.R., 1970. *Applications of Theories of Plasticity to Cohesive Frictional Soils*. Thesis, Univ. of Sydney.

Booker, J.R., 1972. A method of integration for the equations of plasticity of a weightless cohesive frictional material. *Q. J. Mech. Appl. Math.*, 25(1): 63–82.

Booker, J.R. and Davis, E.H., 1972. A note on a plasticity solution to the stability of slopes in inhomogeneous clays. *Géotechnique*, 22(3): 509–513.

Boussinesq, J., 1892. *C.R. Paris*, Vol. 114. The three-dimensional solution of this problem was also given by J. Boussinesq: see *Application des potentiels à l'étude de l'équilibre et du mouvement des solides élastiques*. Gauthier-Villars, Paris, 1885.

Bowles, J.E., 1970. *Engineering Properties of Soils and their Measurement*. McGraw-Hill, New York, p. 187.

Brace, W.F., 1969. The mechanical effects of pore pressure on fracturing of rocks. In: A.J. Baer and D.K. Norris (Editors), *Proc. Conf. Res. Tectonics, Geol. Surv. Canada Pap.*, 68-52: 113–123.

Brady, W.G. and Drucker, D.C., 1953. An experimental investigation and limit analysis of net area in tension. *Proc. ASME*, 296: 79.

Brady, W.G. and Drucker, D.C., 1955. Investigation and limit analysis of net area in tension. *Trans. ASCE* 120: 1133–1154.

Breen, J.J. and Stephens, J.E., 1966. Split cylinder test applied to bituminous mixtures at low temperatures. *J. Mater., ASTM*, 1, (1): 66–76.

Brinch Hansen, J., 1952. A general plasticity theory for clay. *Géotechnique*, 3(4): 154–164.

Brinch Hansen, J., 1953. *Earth Pressure Calculation*. Danish Technical Press, Copenhagen.

Brinch Hansen, J., 1955. Simpel beregning af fundamenters baereevne. *Ingenioren*, 64 (4): 95–100.

Brinch Hansen, J., 1958. Line ruptures regarded as narrow rupture zones, basic equations based on kinematic considerations. *Proc. Conf. Earth Pressure Problems, Brussels, 1958*, 1: 39–48.

Brinch Hansen, J., 1961. A general formula for bearing capacity. *Bull. Dan. Geotech. Inst.*, 11: 38–46.

Brinch Hansen J., 1965. An approximate method for the calculation of rupture figures in clay. *Proc. 6th Int. Conf. Soil Mechanics and Foundation Engineering, Montreal*, 2: 70–73.

Brinch Hansen, J., 1970. A revised and extended formula for bearing capacity. *Bull. Dan. Geotech. Inst.*, 28: 5–11.

Brooks, A.E. and Newman, K. (Editors), 1965. The structure of concrete and its behavior under load. *Proc. Int. Conf., London*. Cement and Concrete Association, 1968.

Brown, E.H., 1966. A theory for the mechanical behavior of sand. In: H. Gorther (Editor), *Proc. 11th Int. Congr. Appl. Mech., Munich, 1964*. Springer, Berlin, pp. 183–191.

Burland, J.B., 1965. The yielding and dilation of clay; correspondence. *Géotechnique*, 15 (2): 211–219.

Burland, J.B., 1969. Deformation of soft clay beneath loaded areas. *Proc. 7th Int. Conf. Soil Mech. Found. Eng., Mexico*, 1: 55–63.

Butovskaya, V.A., 1958. The applicability to clay soils of the hypothesis of the invariability of the volume in plastic deformation. *Tr. N.-i. Osovanii i Podzemn Sooruzh., Akad. Str-va i Arbitekt. S.S.S.R.*, 33: 70–76 (in Russian).

Buts'ko, Z.N., 1958. Determination of the bearing capacity of granular media. *Inzh. Sb. Akad. Nauk S.S.S.R.*, 26: 216–227 (in Russian).

Button, S.J., 1953. The bearing capacity of footings on a two-layer cohesive sub-soil. *Proc. 3rd Int. Conf. Soil Mech. Found. Eng., Zurich*, 1: 332–335.

Buyukozturk, O., Nilson, A.H. and Slate, F.O., 1971. Stress–strain response and fracture of a model of concrete in biaxial loading. *J. Am. Concrete Inst., Proc.*, 68 (8): 590–599.

Calladine, C.R., 1963. The yielding of clay. Correspondence *Géotechnique*, 13: 250–255.

Calladine, C.R., 1971. A microstructural view of the mechanical properties of saturated clay. *Géotechnique*, 21 (4): 391–415.

Campbell, A.D., 1957. Discussion of bearing capacity of concrete, by William Shelson. *J. Am. Concrete Inst.*, 29 (5): 405–414. *Proc.* 54, Discussion 12: 1185–1187.

Capurso, M., 1969. A general method for the incremental solution of elastic–plastic problems. *Meccanica*, 4 (4): 267–280.

Caquot, A., 1934. *Equilibre des massifs à frottement interne*. Gauthier-Villars, Paris, pp. 1–91.

Caquot, A. and Kerisel, J., 1948. *Tables for the Calculation of Passive Pressure, Active Pressure, and Bearing Capacity of Foundations*. Gauthier-Villars, Paris.

Caquot, A. and Kerisel, J., 1953. Sur le terme de surface dans le calcul des foundations en milieu pulvérulent. *Proc. 3rd Int. Conf. Soil Mech. Found. Eng.*, 1: 336–337.

Caquot, A. and Kerisel, J., 1956. *Traité de mécanique de sols*. Gauthier-Villars, Paris, 3rd edition.

Carneiro, F.L.L.B. and Aguinaldo, B., 1952. Tensile strength of concrete. *Union Testing Res. Lab. Mater. Struct., Bull.*, 13: 97–127.

Casagrande, A. and Carrillo, N., 1941–53, 1954. Shear failure of anisotropic materials. *J. Boston Soc. Civ. Eng., Contrib. Soil Mech.*,1941–1953.

Chadwick, P., 1959. The quasi-static expansion of a spherical cavity in metals and ideal soils. *Q. J. Mech. Appl. Math.*, 12 (1): 52–71.

Chaplin, T.K., 1969. Inner and outer plastic yield surfaces in clays. *Proc. 7th Int. Conf. Soil Mech. Found. Eng.*, 1: 73–80.

Cheatham, J.B. Jr., Paslay, P.R. and Fulcher, C.W.G., 1968. Analysis of the plastic flow of rock under a lubricated punch. *J. Appl. Mech.*, 35 (1): 87–94.

Chen, A.C.T., 1973. *Constitutive Relations of Concrete and Punch-indentation Problems*. Dissertation, Department of Civil Engineering, Lehigh University, Bethlehem, Pennsylvania.

Chen, A.C.T. and Chen, W.F., 1973. Constitutive relations of concrete and punch-indentation problems. *Fritz Eng. Lab. Rep.*, 370.11. Lehigh University, Bethlehem, Pennsylvania.

Chen, W.F., 1968a. On the rate of dissipation of energy in soils. *Soils Found.*, VIII (4): 48–51.

Chen, W.F., 1968b. Discussion on applications of limit plasticity in soil mechanics, by W.D. Liam Finn. *J. Soil Mech. Found. Div., ASCE*, 94 (SM2): 608–613.

Chen, W.F., 1969. Soil mechanics and theorems of limit analysis. *J. Soil Mech. Found. Div.*,

ASCE, 95 (SM2), *Proc. Pap.*, 6450: 493–518. – Closure of discussion, 1970, *ASCE*, 96: 1800.

Chen, W.F., 1970a. Plastic indentation of metal blocks by flat punch. *J. Eng. Mech. Div., ASCE*, 96 (EM3), *Proc. Pap.*, 7370: 353–363.

Chen, W.F., 1970b. Extensibility of concrete and theorems of limit analysis. *J. Eng. Mech. Div., ASCE*, 96 (EM3), *Proc. Pap.*, 7369: 341–352.

Chen, W.F., 1970c. Double punch test for tensile strength of concrete. *J. Am. Concrete Inst., Proc.*, 67 (2): 993–995.

Chen, W.F., 1970d. Discussion on circular and logarithmic spiral slip surfaces. *J. Soil Mech. Found. Div., ASCE*, 96 (SM1): 324–326.

Chen, W.F., 1973. Bearing strength of concrete blocks. *J. Eng. Mech. Div., ASCE*, 99 (EM6), *Proc. Pap.*, 10187: 1314–1321.

Chen, W.F. and Carson, J.L., 1971. Stress–strain properties of random wire reinforced concrete. *J. Am. Concrete Inst., Proc.*, 68, (11): 933–936.

Chen, W.F. and Carson, J.L., 1974. Bearing capacity of fiber reinforced concrete. *Int. Symp. Fiber Reinforced Concrete, ACI Spec. Publ.*, SP-44-12: 209–220.

Chen, W.F. and Colgrove, T.A., 1974. Double-punch test for tensile strength of concrete. *Transp. Res. Rec.*, 504: 43–50.

Chen, W.F. and Covarrubias, S., 1971. Bearing capacity of concrete blocks. *J. Eng. Mech. Div., ASCE, Proc.* EM5, *Pap.* 8421: 1413–1430.

Chen, W.F. and Dahl-Jorgensen, E., 1973. Stress–strain properties of polymer modified concrete. *Polymers in Concrete, ACI Spec. Publ.*, SP-40-17: 347–358.

Chen, W.F. and Dahl-Jorgensen, E., 1974. Polymer-impregnated concrete as a structural material. *Mag. Concrete Res.*, 26(86): 16–20.

Chen, W.F. and Davidson, H.L., 1973. Bearing capacity determination by limit analysis. *J. Soil Mech. Found. Div., ASCE*, 99 (SM6) *Proc. Pap.*, 9816: 433–449.

Chen, W.F. and Drucker, D.C., 1969. Bearing capacity of concrete blocks or rock. *J. Eng. Mech. Div., ASCE*, 95 (EM4), *Proc. Pap.*, 6742: 955–978.

Chen, W.F. and Giger, M.W., 1971. Limit analysis of stability of slopes. *J. Soil Mech. Found. Div., ASCE*, 97 (SM1) *Proc. Pap.*, 7828: 19–26.

Chen, W.F. and Rosenfarb, J.L., 1973. Limit analysis solutions of earth pressure problems. *Soils Found.*, 13(4): 45–60.

Chen, W.F. and Scawthorn, C.R., 1970. Limit analysis and limit equilibrium solutions in soil mechanics. *Soils Found.*, 10 (3): 13–49.

Chen, W.F. and Trumbauer, B.E., 1972. Double punch test and tensile strength of concrete. *J. Mater., ASTM*, 7(2): 148–154.

Chen, W.F., Giger, M.W. and Fang, H.Y., 1969. On the limit analysis of stability of slopes. *Soils Found.*, IX, 4: 23–32.

Chen, W.F., Snitbhan, N. and Fang, H.Y., 1975. Stability of slope in anisotropic nonhomogeneous soils. *Can. Geotech. J.*, 12, (1).

Christian, J.T., 1966. Two-dimensional analysis of stress and strain in soils, report 3: plane-strain deformation analysis of soil. *U.S. Army Eng. Waterw. Exp. Stn. Contr. Rep.*, 3-129.

Christian, J.T., 1968. Undrained stress distribution by numerical methods. *J. Soil Mech. Found. Div., ASCE*, 94(SM6), *Proc. Pap.*, 6423: 1333–1345.

Clough, R.W. and Woodward, R.J., III, 1967. Analysis of embankment stresses and deformations. *J. Soil Mech. Found. Div., ASCE*, 93(SM4), *Proc. Pap.*, 5329: 529–549.

Coenen, P.A., 1948. Fundamental equations in the theory of limit equilibrium. *Proc. 2nd Int. Conf. Soil Mech. Found. Eng., Rotterdam*, VII: 15–20.

Collins, I.F., 1969. The upper-bound theorem for rigid/plastic solids generalized to include Coulomb friction. *J. Mech. Phys. Solids*, 17: 323–338.

Collins, I.F., 1973. A note on the interpretation of Coulomb's analysis of the thrust on a rough retaining wall in terms of the limit theorems of plasticity theory. *Géotechnique*, 23(3): 442–447. Discussion by J.L. Justo, *Géotechnique*, 24(1): 106–108.

Cornelius, D.F., Franklin, R.E. and King, T.M.J., 1969. The effect of test method on the indirect tensile strength of concrete. *PRL Report LR 260*, Road Research Laboratory, Ministry of Transportation, Crowthorne, Berkshire.

Cornforth, D.H., 1964. Some experiments on the influence of strain conditions on the strength of sand. *Géotechnique*, 14: 143–167.

Coulomb, C.A., 1773. Sur une application des règles de maximis et minimis à quelques problèmes de statique relatifs à l'architecture. *Acad. R. Sci. Mém. Math. Phys. par divers savants*, 7: 343–382.

Coulomb, C.A., 1776. Essai sur une application des règles de maximis et minimis à quelques problèmes de statique relatifs à l'architecture. *Mém. Acad., R. Pres. Sav. Etr.*, Vol. 7: 343–382.

Cox, A.D., 1962. Axially symmetric plastic deformation in soils – II – indentation of ponderable soils. *Int. J. Mech. Sci.*, 4: 371–380.

Cox, A.D., 1963. The use of non-associated flow rules in soil plasticity. *R.A.R.D.E. Rep.*, B 2/63.

Cox, A.D., Eason, G. and Hopkins, H.G., 1961. Axially symmetric plastic deformation in soils. *Trans. R. Soc. London, Ser. A*, 254: 1.

Cunny, R.W. and Sloan, R.C., 1961. Dynamic loading machine and results of preliminary small-scale footing tests. *Symp. Soil Dyn., Spec. Tech. Publ.*, 305. American Society for Testing and Materials, pp. 65–77.

Dais, J.L., 1970a. An isotropic frictional theory for a granular medium with and without cohesion. *Int. J. Solids Struct.*, 6: 1185–1191.

Dais, J.L., 1970b. Nonuniqueness of collapse load for a frictional material. *Int. J. Solids Struct.*, 6: 1315–1319.

Dais, J.L., 1971. Soil indentation by translating flanged plate. *J. Eng. Mech. Div., ASCE*, 97 (EM4): 1057–1070.

Dantu, P., 1958. Study of the distribution of stresses in a two-component heterogeneous medium (Etude de la répartition des contraintes dans un milieu hétérogène à deux corps). *Symp. Non-Homogeneity in Elasticity and Plasticity, Warsaw*. Pergamon Press, London, pp. 443–451.

D'Appolonia, D.J., Poulos, H.G. and Ladd, C.C., 1971. Initial settlement of structures on clay. *J. Soil Mech. Found. Div., ASCE*, 97, (SM10): 1359–1377.

Davidson, H.L., 1974. *Elastic–Plastic, Large Deformation Response of Soil to Footing Load*. Dissertation, Department of Civil Engineering, Lehigh University, Bethlehem, Pennsylvania.

Davies, J.D., 1968. A modified splitting test for concrete specimen. *Mag. Concrete Res.*, 20 (64): 183–186.

Davies, J.D. and Bose, D.K., 1968. Stress distribution in splitting tests. *Proc. J. Am. Concrete Inst.*, 65 (8): 662–669.

Davis, E.H., 1967. A discussion of theories of plasticity and limit analysis in relation to the failure of soil masses. *Proc. 5th Aust.––N.Z. Conf. Soil Mech. Found. Eng.* Univ. of Auckland, Sigma, Part 1: 175–182.

Davis, E.H., 1968a. Theories of plasticity and the failure of soil masses. In: I.K. Lee (Editor), *Soil Mechanics, Selected Topics*. American Elsevier, New York, N.Y., pp. 341–354.

Davis, E.H., 1968b. *Soil mechanics – selected topics*. London, Butterworths, Chapter 6.

Davis, E.H. and Booker, J.R., 1971. The bearing capacity of strip footings from the standpoint of plasticity theory. *Proc. 1st Aust.–N.Z. Geomech. Conf. on Geomechanics, Melbourne*, pp. 275–282.

Davis, E.H. and Booker, J.R., 1973. The effect of increasing strength with depth on the bearing capacity of clays. *Géotechnique*, 23 (4): 551–563.

Davis, E.H. and Christian, J.T., 1971. Bearing capacity of anisotropic cohesive soil. *J. Soil Mech. Found. Div., ASCE*, 97 (SM5), *Proc. Pap.* 8146: 753–769.

Davis, E.H. and Woodward, R.S., 1949. Some laboratory studies of factors pertaining to the bearing capacity of soils. *Reprint No. 6*, Institute of Transportation and Traffic Engineering, University of California, Stanford, Calif.

De Beer, E.E., 1949. *Grondmechanica. II. Funderingen.* Standaard Boekhandel, Antwerp, pp. 41–51.

De Beer, E.E., 1965a. The scale effect on the phenomenon of progressive rupture in cohesionless soils. *Proc. 6th Int. Conf. Soil Mech. Found. Eng., Montreal,* II: 13–17.

De Beer, E.E., 1965b. Bearing capacity and settlement of shallow foundations on sand. *Symp. Bearing Capacity and Settlement of Foundations, Durham, N.C.,* pp. 15–33.

De Beer, E.E. and Ladanyi, B., 1961. Etude expérimentale de la capacité portante du sable sous des fondations circulaires établies en surface. *Proc. 5th Int. Conf. Soil Mech. Found. Eng., Paris,* Dunod, Paris, 1: 577–585.

De Beer, E.E. and Vesic, A.B., 1958. Etude expérimentale de la capacité portante du sable sous des fondations directes établies en surface. *Ann Trav. Publ. Belg.,* 59: 5–58.

De Jong, D.J.G., 1957. Graphical method for the determination of slip-line fields in soil mechanics. *Ingenieur,* 69 (29): 61–65 (in Dutch).

De Jong, D.J.G., 1958. Indefinitness in kinematics for friction materials. *Proc. Conf. on Earth Pressure Problems, Brussels,* 1: 55–70.

De Jong, D.J.G., 1959. *Statics and Kinematics in the Failure Zone of a Granular Material.* Thesis, Univ. of Delft, Delft.

De Jong, D.J.G., 1964. Lower-bound collapse theorem and lack of normality of strain-rate to yield surface for soils. In: J. Kravtchenko and P.M. Sirirys (Editors), *Rheology and Soil Mechanics. IUTAM Symp., Grenoble.* Springer, Berlin, 1966, pp. 69–75.

De Jong, D.J.G. and Geertsma, J., 1953. Stress distribution around vertically drilled holes in sandy terrain, internally supported by a heavy liquid. *Ingenieur,* 65 (9) (in Dutch).

Dembicki, E., 1963. A method of nonlinear approximation for the solution of limiting equilibrium problems in cohesive media. *Arch. Hydrotech.,* 10 (3): 367–472 (in French).

Dembicki, E., 1965. Determining the stress distribution along a retaining wall structure by method of characteristics: 4, Tables of passive earth pressure coefficient values in cohesive soils with triangular load distribution. *Archiwum Hydrotechniki,* 12 (2): 181–199 (in Polish).

Dembicki, E. and Negre, R., 1966. Distribution of stress of longitudinally supporting walls in a spatial, axisymmetric system. (in French), *Bull. Acad. Polon. Sci., Ser. Sci. Tech.,* 14 (8): 447–452.

Dembicki, E., Dravtchenko, J. and Sibille, E., 1964. Sur les solutions analytiques approchées des problèmes d'équilibre limite plan pour milieux cohérents et pèsants. *J. Mec.,* 3: 277–312.

Dembicki, E., Negre, R. and Stutz, P., 1968. Ecoulement dans un silo conique de révolution et déformation d'un échantillon dans l'essai triaxial pour un materiau à dilatation non standardisée. *Third Budapest Conf. Soil Mech.,* p. 35–53.

De Mello, V.F.B., 1969. Foundations of buildings in clay. State of the art volume. *Proc. 7th Int. Conf. Soil Mech. Found. Eng., Mexico City,* pp. 49–136.

Desai, C.S., 1971. Nonlinear analysis using spline functions. *J. Soil Mech. Found. Div., ASCE,* 97 (SM10): 1461–1480.

Desai, C.S. (Editor), 1972. Applications of the finite element method in geotechnical engineering. *Proc. Symp. Vicksburg, Miss., U.S. Army Corps of Engin., Waterway Exper. Stn.*

Desai, C.S. and Reese, L.C., 1970. Analysis of circular footings on layered soils. *J. Soil Mech. Found. Div., ASCE,* 96 (SM4): 1289–1310.

De Saint Venant, B., 1870. Mémoire sur l'établissement des équations differentielles des mouvements intérieurs opérés dans les corps solides ductiles au delà des limites où l'élasticité pourrait les ramener à leur premier état. *C.R. Acad. Sci., Paris,* 70: 473–480.

Deutsch, G.P. and Clyde, D.H., 1967. Flow and pressure of granular materials in silos. *J. Eng. Mech. Div., ASCE,* 93 (EM6): 103–125.

Dietrich, L. and Szczepinski, W., 1967. Plastic yielding of axially symmetric bars with non-symmetric v-notch. *Acta Mech.,* 4: 230–240.

Dietrich, L. and Szczepinski, W., 1969. A note on complete solutions for the plastic bending of notched bars. *J. Mech. Phys. Solids*, 17: 171–176.

Dietrich, L. and Turski, K., 1968. Carrying capacity of axially symmetric multi-grooved bars in tension. *Mech. Teor. Stosow.*, 6: 437–448 (in Polish).

DiMaggio, F.L. and Sandler, I.S., 1971. Material model for granular soils. *J. Eng. Mech. Div., ASCE*, 97 (EM3): 935–950.

Dismuke, T.D., Chen, W.F. and Fang, H.Y., 1972. *Tensile Strength of Rock by the Double-Punch Method. Rock Mechanics 4.* Springer, Berlin, pp. 79–87.

Doherty, W.P., Wilson, E. and Taylor, R.L., 1969. Stress analysis of axisymmetric solids utilizing higher-order quadrilateral finite elements. *Struct. Eng. Lab. Univ. Calif., Berkeley, Calif., Rep. 69-3.*

Donath, F.A., 1961. Experimental study of shear failure in anisotropic rocks. *Geol. Soc. Am. Bull.*, 72: 985–90.

Donath, F.A., 1970. Some information squeezed out of rock. *Am. Sci.*, 58 (1): 54–72.

Drescher, A. and Bojanowski, W., 1968. On the influence of stress path upon the mechanical properties of granular material. *Archiwum Inzynierii Ladowej*, 14 (3): 351–365 (in English).

Drescher, A. and Bujak, A., 1966. A study of kinematics of a granular body by indentation with a plane punch. *Rozprawy Inzynierskie*, 14 (2): 313–325 (in Polish).

Drescher, A., Kwaszczynska, K. and Mroz, Z., 1967. Statics and kinematics of a granular medium in the case of wedge indentation. *Archiwum Mechaniki Stosowanej*, 19 (1): 99–113.

Drucker, D.C., 1950. Stress–strain relations in the plastic range of metals – experiments and basic concepts. In: *Rheology – Theory and Applications, I.* Academic Press, New York, 1956, pp. 97–119.

Drucker, D.C., 1951. A more fundamental approach to stress–strain relations. *Proc. 1st U.S. Natl. Congr. Appl. Mech.* American Society of Mechanical Engineers, pp. 487–491.

Drucker, D.C., 1953. Limit analysis of two- and three-dimensional soil mechanics problems. *J. Mech. Phys. Solids*, 1: 217–226.

Drucker, D.C., 1954a. Coulomb friction, plasticity, and limit loads. *ASME, Trans.*, 76: 71–74.

Drucker, D.C., 1954b. Limit analysis and design. *Appl. Mech., Rev.*, 7: 421.

Drucker, D.C., 1954c. On obtaining plane strain and plane stress conditions in plasticity. *Proc. 2nd U.S. Natl. Congr. Appl. Mech.*, pp. 485–488.

Drucker, D.C., 1958a. Plastic design methods – advantages and limitations. *Trans., Soc. Nav. Arch. Mar. Eng.*, 65: pp. 172–190.

Drucker, D.C., 1958b. Variational principles in the mathematical theory of plasticity. *Proc. 1956 Symp. Appl. Math.* McGraw-Hill, New York, 8: 7–22.

Drucker, D.C., 1959. A definition of stable inelastic material. *ASME, Trans.*, 81: 101–106.

Drucker, D.C., 1961a. On stress–strain relations for soils and load carrying capacity. *Proc. 1st Int. Conf. Mech. Soil-Vehicle Systems, Turin.* Edizioni Minerva Tecnica, pp. 15–23.

Drucker, D.C., 1961b. On structural concrete and the theorems of limit analysis. *IABSE Publ., Engineering*, 21: 49–59.

Drucker, D.C., 1962a. Basic concepts in plasticity. In: W. Flügge (Editor), *Handbook of Engineering Mechanics.*, McGraw-Hill, New York, Chapter 46.

Drucker, D.C., 1962b. On the role of experiment in the development of theory. *Proc. 4th U.S. Natl. Congr. Appl. Mech. ASME*, pp. 15–33.

Drucker, D.C., 1966a. Concepts of path independence and material stability for soils. In: J. Kravtchenko and P.M. Sirieys (Editors), *Rhéol. Mécan. Soils Proc. IUTAM Symp. Grenoble.* Springer, Berlin, pp. 23–43.

Drucker, D.C., 1966b. The continuum theory of plasticity on the macroscale and the microscale. *Marburg Lecture, J. Mater., ASTM*, 1: 873–910.

Drucker, D.C. and Chen, W.F., 1968. On the use of simple discontinuous fields to bound limit loads. In: J. Heyman and F.A. Leckie (Editors), *Engineering Plasticity*. Cambridge University Press, pp. 129–145.

Drucker, D.C. and Prager, W., 1952. Soil mechanics and plastic analysis or limit design. *Q. Appl. Math.*, 10: 157–165.

Drucker, D.C., Greenberg, H.J., Lee, E.H. and Prager, W., 1951a. On plastic–rigid solutions and limit design theorems for elastic–plastic bodies. *Proc. 1st U.S. Natl. Congr. Appl. Mech. Chicago*, pp. 533–538.

Drucker, D.C., Greenberg, H.J. and Prager, W., 1951b. The safety factor of an elastic–plastic body in plane strain. *J. Appl. Mech.*, 73: 371–378.

Drucker, D.C., Greenberg, H.J. and Prager, W., 1952. Extended limit design theorems for continuous media. *Q. Appl. Math.*, 9: 381–389.

Drucker, D.C., Gibson, R.E. and Henkel, D.J., 1957. Soil mechanics and work-hardening theories of plasticity. *Trans. ASCE*, 122: 338–346.

Duncan, J.M. and Chang, C.-Y., 1970. Nonlinear analysis of stress and strain in soils. *J. Soil Mech. Found. Div., ASCE*, 96 (SM5): 1629–1653.

Eason, G. and Shield, R.T., 1960. The plastic indentation of a semi-infinite solid by a perfectly rough circular punch. *J. Appl. Math. Phys. (ZAMP)*, 11: 33–43.

Eden, W.J. and Bozozuk, M., 1962. Foundation failure of a silo on carved clay. *Eng. J.*, 45 (9): 54–57.

Eggleston, H.G., 1958. *Convexity*. Cambridge Univ. Press, London.

Elsamny, M.K. and Ghobarah, A.A., 1972. Stress field under slipping rigid wheel. *J. Soil Mech. Found. Div., ASCE*, 98 (SM1): 13–25.

Fang, H.Y. and Chen, W.F., 1971. New method for determination of tensile strength of soils. *Highw. Res. Rec.*, 354: 62–68.

Fang, H.Y. and Chen, W.F., 1972. Further studies of double-punch test for tensile strength of soil. *Proc. 3rd Southeast Asian Conf. Soil Eng., Hong Kong*. Univ. of Hong Kong, p. 211–215.

Fang, H.Y. and Hirst, T.J., 1970. Application of plasticity theory to slope stability problems. *Highway Res. Rec.*, 323: 26–38.

Fang, H.Y., Chen, W.F., Davidson, H.L. and Rosenfarb, J.L., 1974. Bibliography on Soil Plasticity. Envo Publ. Comp., Lehigh Valley, Pa.

Farmer, I.W., 1968. *Engineering Properties of Rocks*. Spon, London.

Feda, J., 1961. Research on the bearing capacity of loose soil. *Proc. 5th Int. Conf. Soil Mech. Found. Eng.* Dunod, Paris, 1: 635–642.

Federov, I.V., 1958. Certain problems in the elasto-plastic distribution of stresses in soil, associated with the analyses of foundations. *Inzhenernii Sb., Acad. Nauk S.S.S.R.*, Vol. 26: 205–215 (in Russian).

Felippa, C.L., 1966. *Refined Finite-element Analysis of Linear and Non-linear Two-dimensional Structures*. Thesis, University of California, Berkeley, Calif.

Fellenius, W., 1926. Mechanics of soils. *Statika Gruntov*, Gosstrollzdat, 1933.

Fellenius, W., 1927. *Erdstatische Berechnungen*. Ernst, Berlin.

Fellenius, W., 1936. Calculation of the stability of earth dams. *Trans. 2nd Congr. Large Dams, Washington*, 4: 9.2.

Fernandez, R.M. and Christian, J.T., 1971. Finite-element analysis of large strains in soils. *NASA Res. Rep.*, R71-37.

Finn, W.D.L., 1963. Boundary value problems in soil mechanics. *J. Soil Mech. Found. Div., ASCE*, 89 (SM5): 39–72.

Finn, W.D.L., 1967. Applications of limit plasticity in soil mechanics. *J. Soil Mech. Found. Div., ASCE*, 93 (SM5): 101–119.

Ford, H. and Lianis, G., 1957. Plastic yielding of notched strips under conditions of plane stress. *Z. Angew. Math. Phys.*, 8: 360–382.

Gallagher, R.H., 1972. Geometrically nonlinear finite-element analysis. *Proc. Spec. Conf. Finite-Element Methods in Civil Engineering*. McGill University, Montreal, pp. 3–33.

Gaponov, V.V., 1959. On displacements in friable soils in limiting equilibrium. *Prikladnaya Medhanika*, 5 (1): 65–74 (in Russian).

Geuze, E.C.W.A., 1963. The uniqueness of the Mohr-Coulomb concept in shear failure. *Laboratory Shear Testing of Soils, Spec. Tech. Publ.* No. 361, *ASTM*, p. 52–64.

Ghahramani, A. and Sabzevari, A., 1973. Load-displacement analysis of footings in dry sand. *Proc. 8th Int. Conf. Soil Mech. Found. Eng., Moscow*, Vol. 1 (3): 89–93.

Gibson, R.E. and Morgenstern, N., 1962. A note on the stability of cuttings in normally consolidated clays. *Géotechnique*, 12, (3): 212–216.

Girijavallabhan, C.V. and Reese, L.C., 1968. Finite-element method for problems in soil mechanics. *J. Soil Mech. Found. Div., ASCE*, 94 (SM2): 473–496.

Glushko, V.I., 1960. On the determination of stresses around a horizontal working. *Isv. Dnepropetr, Gorn. In-ta.*, 38: 5–10, 1959, *Referativnii Zhurnal Mekhanika Moscow*, No. 8, Revision 10770 (in Russian).

Goguel, J., 1947. Distribution of stresses around a cylindrical tunnel (Repartition des contraintes antons d'un tunnel cylindrique). *Ann. Ponts Chaussées*, 117: 157–188.

Goldshtein, L.M., 1969. Approximate solution of the problem of three-dimensional limiting equilibrium of soils. *Soil Mech. Found. Eng.*, 5: 323–329.

Goodier, J.N., 1932. Compression of rectangular blocks and the bending of beams by non-linear distributions of bending forces. *J. Appl. Mech.*, 54(18): 173–183.

Gorbunov-Possadov, M.I., 1965. Calculations for the stability of a sand bed by a solution combining the theories of elasticity and plasticity. *Proc. 6th Int. Conf. Soil Mech. Found. Eng., Montreal*, 2: 51–55.

Graf, O., 1934. Versuch mit Betonquadern zu Brückengelenken und Auflagern. *Ver. Dtsch. Ing. Mitt. Forschungsarb.n*, 232: 68 (1921). Über einige Aufgaben der Eisenbetonforschung aus alterer und neuerer Zeit. *Beton Eisen*, 33 (11): 165–173.

Graham, J., 1968. Plane plastic failure in cohesionless soil. *Géotechnique*, 18(3): 301–316.

Graham, J., 1974. Plasticity solutions to stability problems in sand. *Can. Geotech. J.*, 11: 238–247.

Graham, J. and Stuart, J.G., 1971. Scale and boundary effects in foundation analysis. *J. Soil Mech. Found. Div., ASCE*, 97 (SM11): 1533–1548.

Grasshoff, H., 1955. Settlement calculations for rigid foundations using the characteristic point. *Bauingenieur*, 30 (2): 53–54 (in German).

Green, A.P., 1951. A theoretical investigation of the compression of a ductile material between smooth flat dies. *Philos. Mag.*, 42: 900–918.

Green, A.P., 1953a. Bending of a wide bar with symmetrical deep wedge-shaped notches on both sides. *Q. J. Mech. Appl. Math.*, 6: 223–239.

Green, A.P., 1953b. The plastic yielding of notched bars due to bending. *Q. J. Mech. Appl. Math.*, 6: 223.

Green, A.P., 1954. On the use of hodographs in problems of plane plastic strain. *J. Mech. Phys. Solids*, 2: 73–80.

Green, A.P., 1962. Two-dimensional problems. In: W. Flügge (Editor), *Handbook of Engineering Mechanics*, McGraw-Hill, New York, Chapter 50.

Gudehus, G., 1972. Elastic–plastic constitutive equations for dry sand. *Arch. Mech. Stosow.*, 24(3): 395–402.

Gvozdev, A.A., 1960. The determination of the value of the collapse load for statically indeterminate systems undergoing plastic deformation. *Proc. Conf. Plastic Deformation, 1936.* Akademiia Nauk S.S.S.R., Moscow-Leningrad, 1938, p. 19. Translated from the Russian by R.M. Haythornthwaite, *Int. J. Mech. Sci.*, 1: 322–335.

Haar, A. and Von Karman, T., 1909. Zur Theorie der Spannungszustände in plastischen und sandartigen Medien. *Nachr. Ges. Wiss. Gött., Math.-Phys. Kl.*, 1909: 204–218.

Haisler, W.E., Strickland, J.A. and Stebbins, F.J., 1972. Development and evaluation of solution procedures for geometrically nonlinear structural analysis. *AIAA Journal*, 10 (3): 264–272.

Hambly, E.C. and Roscoe, K.H., 1969. Observations and predictions of stresses and strains during plane strain of "wet" clays. *Proc. 7th Int. Conf. Soil Mech. Found. Eng., Mexico*, 1: 173–181.

Hannant, D.J., 1972. The tensile strength of concrete: a review paper. *Struct. Eng.*, 50 (7): 253–257.

Hannant, D.J. and Fredrick, C.O., 1968. Failure criteria for concrete in compression. *Mag. Concrete Res.*, 20 (64): 137–144.

Hansen, B., 1965. *A Theory of Plasticity for Ideal Frictionless Materials.* Teknisk Forlag, Copenhagen, 471 pp.

Hansen, B., 1969. Bearing capacity of shallow strip footings in clay. *Proc. 7th Int. Conf. Soil Mech. Found. Eng., Mexico City*, 2: 107–113.

Hansen, B. and Christensen, N.H., 1969. Discussion of theoretical bearing capacity of very shallow footings, by L.A. Larkin. *J. Soil Mech. Found. Div. ASCE*, 95 (SM6): 1568–1572.

Hartmann, F., 1968. On a new earth pressure theory, *Bautechnik*, 45 (9): 307–313 (in German).

Hartz, B.J. and Nathan, N.D., 1971. Finite-element formulation of geometrically nonlinear problems of elasticity. In: R.H. Gallagher, Y. Yamada and J.T. Oden (Editors), *Recent Advances in Methods of Structural Analysis and Design. Proc. Japan–U.S. Seminar, Tokyo.* Univ. of Alabama Press, Alabama.

Hashiguchi, K., 1969. Theories of a velocity field for plastic soils. *Trans. JSCE*, 1 (2): 405–421.

Hata, S., Ohta, H. and Yoshitani, S., 1969. On the state surface of soils. *Trans. JSCE*, 1 (2): 479–499.

Hawkins, N.M., 1968. The bearing strength of concrete loaded through rigid plates. *Mag. Concrete Res.*, 20 (62): 31–40.

Haythornthwaite, R.M., 1960. Mechanics of the triaxial test for soils. *J. Soil Mech. Found. Div. ASCE*, 86 (SM5): 35–62.

Haythornthwaite, R.M., 1961a. Methods of plasticity in land locomotion studies. *Proc. 1st Int. Conf. Mech. Soil-Vehicle Systems, Turin.* Edizion Minerva Tecnica, pp. 3–19.

Haythornthwaite, R.M., 1961b. Range of yield condition in ideal plasticity. *J. Engin. Mech. Div., ASCE*, 87 (EM6): 117–133.

Hencky, H., 1923. Über einige statisch bestimmte Fälle des Gleichgewichts in plastichen Körpern. *Z. Angew. Math. Mech.*, 3: 241–251.

Henkel, D.J., 1959. The relationship between the strength, pore water pressure and volume change characteristics of saturated clays. *Géotechnique*, 9: 119–135.

Henkel, D.J., 1960. The relationship between the effective stresses and water content of saturated clays. *Géotechnique*, 10: 41–54.

Herrman, L.R., 1965. Elasticity equations for incompressible and nearly incompressible materials by a variational theorem. *Am. Inst. Aeronaut. Astronaut. J.*, 3: 1896–1900.

Heyman, J., 1966. The stone skeleton. *Int. J. Solids Struct.*, 2: 249–279.

Heyman, J., 1972. *Coulomb's Memoir on Statics – An Essay in the History of Civil Engineering.* Cambridge University Press, London, 212 pp.

Hibbitt, H.D., Marcal, P.V. and Rice, J.R., 1970. A finite-element formulation for problems of large strain and large displacement. *Int. J. Solids Struct.*, 6 (8): 1069–1086.

Hill, R., 1949. The plastic yielding of notched bars under tension. *Q. J. Mech. Appl. Math.*, 2: 40–52.

Hill, R., 1950. *The Mathematical Theory of Plasticity.* Clarendon Press, Oxford.

Hill, R., 1951. On the state of stress in a plastic–rigid body at the yield point. *Philos. Mag.*, 42: 868–875.

Hill, R., 1952. A note on estimating yield point loads in a plastic–rigid body. *Philos. Mag.*, 43: 353–355.

Hill, R., 1956. On the problem of uniqueness in the theory of a rigid–plastic solid. *J. Mech. Phys. Solids*, 4: 247; 5: 1, 153 and 302.

Hill, R., Lee, E.H. and Tupper, S.J., 1947. The theory of wedge indentation of ductile materials. *Proc. R. Soc. London, Ser. A*, 188: 273–289.

Hodge, P.G. Jr., 1950. Approximate solutions of problems of plane plastic flow. *J. Appl. Mech., ASME*, 17: 257–264.

Hodge, P.G. Jr. and Prager, W., 1951. Limit design of reinforcements of cut-outs in slabs. *Brown Univ. Rep.* B11-2 to Office of Naval Research.

Höeg, K., 1972. Finite-element analysis of strain-softening clay. *J. Soil Mech. Found. Div., ASCE,* 98 (SM1): 43–58.

Höeg, K. and Balakrishna, H.A., 1970. Dynamic strip load on elastic–plastic soil. *J. Soil Mech. Found. Div., ASCE,* 96 (SM2): 429–438.

Höeg, K., Christian, J.T. and Whitman, R.V., 1968. Settlement of strip load on elastic–plastic soil. *J. Soil Mech. Found. Div., ASCE,* 94(SM2): 431–445.

Hofmeister, L.D., Greenbaum, G.A. and Evensen, D.A., 1971. Large strain, elastoplastic finite-element analysis. *AIAA Journal,* 9 (7): 1248–1254.

Hoshino, K., 1948. A fundamental theory of plastic deformation and breakage of soil. *Proc. 2nd Int. Conf. Soil Mech. Found. Eng. Rotterdam,* 1: 93–103.

Hoshino, K., 1957. A general theory of mechanics of soils. *Proc. 4th Int. Conf. Soil Mech. Found. Eng., London,* Vol. 1, Butterworths, London, 1958.

Hunter, J.H. and Schuster, R.L., 1968. Stability of simple cuttings in normally consolidated clays. *Géotechnique,* 18(3): 372–378.

Huntington, W.C., 1961. *Earth Pressures and Retaining Walls.* Wiley, New York.

Hvorslev, M.J., 1936. Conditions of failure for remolded cohesive soils. *Proc. 1st Int. Conf. Soil Mech. Found. Eng., Cambridge, Mass.,* 3: 51–53. Reprinted by Harvard Press, 1965.

Hyland, M.W. and Chen, W.F., 1970. Bearing capacity of concrete blocks. *J. Am. Concrete Inst.,* 67: 228–236.

Il'Yushin, A.A., 1955. Modern Problems in the theory of plasticity. *Vestnik Moscov. Univ.,* No. 4-5: 101–113 (in Russian).

Il'Yushin, A.A., 1961. O postulate plastichnosti (On the postulate of plasticity). *Prikl. Mat. Mekh.,* 25: 503–507.

Inoue, N., 1952. Application of gas dynamical method to soil mechanics and theory of plasticity. *J. Phys. Soc. Japan,* I, II, Vol. 7 (6): 604–618.

Inoue, N., 1953. Discontinuous solutions in soil mechanics. *Proc. 2nd Japan Natl. Congr. Appl. Mech., 1952;* National Committee for Theoretical Applied Mechanics, p. 23–27.

Ivcovic, M. and Radenkovic, D., 1957. On application of the limit analysis in soil mechanics. *9th Int. Congr. Appl. Mech., Brussels,* VIII: 204–205.

Ivlev, D.D., 1958. On the general equations of the theory of ideal plasticity and of statics of granular media. *Appl. Math. Mech.,* 22 (1): 119–128.

Ivlev, D.D., 1959. On relations defining plastic flow under Tresca's condition of plasticity and its generalizations. *Sov. Phys-Doklady,* 4 (1): 217–220.

Ivlev, D.D. and Martynova, T.N., 1961. Fundamental relations in the theory of loose anisotropic media. *Zh. Prikl., Mekh. Tokh. Fiz. (PMTF),* 2: 116–121 (in Russian).

Izbicki, R.J., 1972a. On the axially symmetric plastic deformation of Coulomb medium. *Bull. Acad. Pol. Sci., Ser. Sci. Tech.,* 20(7-8): 547–555.

Izbicki, R.J., 1972b. General yield condition – plane deformation. *Bull. Acad. Pol. Sci., Ser. Sci. Tech.,* 20(7-8): 557–564.

Jacobs, J.A., 1950. Relaxation methods applied to problems of plastic flow. *Philos. Mag., Ser.* 7, 41: 347–361 and 458–467.

Jaky, J., 1947. Stability of earthworks in the plastic state, I (Sur la stabilité des masses de terre complètement plastiques, I). *Publ. Hung. Tech. Univ. (Muegyeteme Kozlemenyek), Budapest,* 2: 129–151.

Jaky, J., 1948a. Stability of earthworks in the plastic state, II. (Sur la stabilité des masses de terre complètement plastiques, II). *Publ. Hung. Tech. Univ. (Muegyeteme Kozlemenyek), Budapest,* 1: 34–56.

Jaky, J., 1948b. Stability of earthworks in the plastic state. *Publ. Hung. Tech. Univ. (Muegyeteme Kozlemenyek), Budapest,* 3: 158–172 (in French).

Jaky, J., 1953. Network of slip lines in soil stability. *Acta Tech., Acad. Sci. Hung., Budapest,* 6 (1/2): 25–38.

James, R.G. and Bransby, D.L., 1970. Experimental and theoretical investigation of a passive earth pressure problem. *Géotechnique*, 20 (1): 17–37.

James, R.G. and Bransby, D.L., 1971. A velocity field for some passive pressure problems. *Géotechnique*, 21 (1): 61–83.

Jenike, A.W. and Shield, R.T., 1959. On the plastic flow of Coulomb solids beyond original failure. *J. Appl. Mech.*, 26: 599–602.

Jeske, T., 1968. On the kinematics of a granular medium in the case of equilibrium state under plane strain conditions. *Archiwum Mechaniki Stosowanej*, 20 (2): 211–224.

Jumikis, A.R., 1961. The shape of rupture surface in dry sand. *Proc. 5th Int. Conf. Soil Mech. Found. Eng.*, *Paris*, Dunod, 1: 693–698.

Kachanov, L.M., 1971. *Foundations of the Theory of Plasticity*. American Elsevier, New York, p. 197.

Kamenov, S., 1973. Some problems of non-linear stress–strain analysis of soils. *Stavebnicky cas.*, 21 (6/8): 589–598.

Kameswara Rao, N.S.V. and Krishnamurthy, S., 1972. Bearing capacity factors for inclined loads. *J. Soil Mech. Found. Div.*, *Proc., ASCE*, 98 (SM11).

Kaplan, M.F., 1963. Strains and stresses of concrete at initiation of cracking and near failure. *J. Am. Concrete Inst.*, 60: 853–879.

Karafiath, L.L., 1957. An analysis of new techniques for the estimation of footing sinkage in soils. U.S. Army Ordinance Corps, Land Locomotion Research Branch, Research and Development Division, *ATC Report* 18, p. 32.

Karafiath, L.L. and Nowatzki, E.A., 1970. Stability of slopes loaded over a finite area. *Highway Res. Rec.*, 323: 14–25.

Ketter, R.L. and Prawel, S.P., 1969. *Modern Methods of Engineering Computation*. McGraw-Hill, New York.

Kezdi, A., 1956. Results of theoretical research. *Proc. 2nd Conf. Building of the Hungarian Academy of Sciences*. (Foundation and Soil Mech. Sect., Magyar Tudomanyos Akademia Muszaki Tudomanyok, Osztalyanak Kozlemenyei), 19 (1/3): 71–84.

Kirkpatrick, W.M., 1957. The condition of failure for sands. *Proc. 4th Int. Conf. Soil Mech. Found. Eng.*, *London*. Butterworth, 1: 172–178.

Kjellman, W., 1936. Report on an apparatus for consumate investigation of the mechanical properties of soils. *Proc. 1st Int. Conf. Soil Mech. Found. Eng.*, *Cambridge, Mass.*, 2: 16–20. Reprinted by Harvard Press, 1965.

Ko, H.-Y. and Davidson, L.W., 1973. Bearing capacity of footings in plane strain. *J. Soil Mech. Found. Div. ASCE, Proc.*, 99 (SM1): 1–23.

Ko, H.-Y. and Scott, R.F., 1967a. Deformation of sand in hydrostatic compression. *J. Soil Mech. Found. Div.*, *ASCE*, 93 (SM3), *Proc. Pap.* 5245: 137–156.

Ko, H.-Y. and Scott, R.F., 1967b. Deformation of sand in shear. *J. Soil Mech. Found. Div.*, *ASCE*, 93 (SM5), *Proc. Pap.* 5470: 283–310.

Ko, H.-Y. and Scott, R.F., 1968. Deformation of sand at failure. *J. Soil Mech. Found. Div.*, *ASCE*, 94 (SM4), *Proc. Pap.* 6028: 883–898.

Ko, H.-Y. and Scott, R.F., 1973. Bearing capacities by plasticity theory. *J. Soil Mech. Found. Div.*, *ASCE*, 99 (SM1), *Proc. Pap.* 9497: 25–43.

Kogan, B.I. and Lupashko, A.A., 1970. Stability analysis of slopes. *Soil Mech. Found. Eng.*, 3: 153–157.

Kooharian, A., 1952. Limit analysis of voussior (segmental) and concrete arches. *J. Am. Concrete Inst.*, *Proc.*, 24: 317–328.

Kötter, F., 1888. Über das Problem der Erddruckbestimmung. *Verh. Phys. Ges. Berlin*, 7: 1–8.

Kötter, F., 1903. Die Bestimmung des Druckes an gekrümmten Gleitflächen, eine Aufgabe aus der Lehre vom Erddruck. *Monatsber. Akad. Wiss. Berlin*, 1903: 229–233.

Kupfer, H., Hilsdorf, H.K. and Rusch, H., 1969. Behavior of concrete under biaxial stresses. *J. Am. Concrete Inst.*, *Proc.*, 66(8): 656–666.

Kuznetsov, S.V., 1961. Interaction of tectonic pressures and gas pressure in a coal seam. *Zhurnal Prikladnoi Mekhaniki i Tekhnicheskoi Fiziki*, (PMFT), No. 4: 57–77 (in Russian).

Kwaszczynska, K., Mroz, Z. and Drescher, A., 1969. Analysis of compression of short cylinders of Coulomb material. *Int. J. Mech. Sci.*, 11: 145–158.

Ladanyi, B. and Roy, A., 1971. Some aspects of bearing capacity of rock mass *7th Can. Symp. Rock Mech., Edmonton*, p. 25–27.

Lade, P.V. and Duncan, J.M., 1973. Cubical triaxial tests on cohesionless soil. *J. Soil Mech. Found. Div., Proc., ASCE*, 99 (SM10): 793–812.

Larkin, L.A., 1968. Theoretical bearing capacity of very shallow footings. *J. Soil Mech. Found. Div., ASCE*, 94 (SM6), *Proc. Pap.* 6258: 1347–1357.

Larkin, L.A., 1970. On plastic analysis and the bearing capacity of circular foundations on granular soils. In: D. Frederick (Editor), *Developments in Theoretical and Applied Mechanics*, 4. Pergamon Press, London, pp. 503–517.

Larkin, L.A., 1972. Upper-bound analysis of rough footing indentation. *J. Eng. Mech. Div., ASCE*, 98 (EM2): 493–496.

Lee, C.H. and Kobayashi, S., 1970. Elastoplastic analysis of plane strain and axisymmetric flat punch indentation by the finite-element method. *Int. J. Mech. Sci.* 12: 349–370.

Lee, E.H., 1952. On the significance of the limit theorems for a plastic-rigid body. *Philos. Mag.*, VII (43): 549–560.

Lee, I.K. and Herington, J.R., 1971a. The effect of wall movement on active and passive pressures. *Uniciv Rep.* R-71, Univ. of New South Wales.

Lee, I.K. and Herington, J.R., 1971b. Stresses beneath granular embankments. *Proc., 1st Australia–N.Z. Conf. Geomech., Melbourne*, 1: 291–297.

Lee, I.K. and Herington, J.R., 1972. A theoretical study of the pressures acting on a rigid wall by a sloping earth or rock fill. *Geotechnique*, 22 (1): 1–26.

Lee, I.K. and Ingles, O.G., 1968. Strength and Deformation of Soils and Rocks. In: I.K. Lee (Editor), *Soil Mechanics Selected Topics*. American Elsevier, New York, p. 195–294.

Lee, K.L., 1970. Comparison of plane strain and triaxial tests on sand. *J. Soil Mech. Found. Div., ASCE*, 96 (SM3), *Proc. Pap.* 7276: 901–921.

Lenoe, E.M., 1966. Deformation and failure of granular media under three-dimensional stresses. *Exp. Mech.*, 6 (2): 99–104.

Lewin, P.I. and Burland, J.B., 1970. Stress probe experiments on saturated normally consolidated clay. *Geotechnique*, 20 (1): 38–56.

Liu, T.C.Y., Nilson, A.H. and Slate, F.O., 1971. Stress–strain response and fracture of concrete in biaxial compression. *Report*, Department of Structural Engineering, School of Civil Engineering, Cornell University, No. 339.

Liu, T.C.Y., Nilson, A.H. and Slate, F.O., 1972. Biaxial stress–strain relations for concrete. *J. Struct. Div., ASCE*, 98 (ST5), *Proc. Paper* 8905: 1025–1034.

Lippmann, H., 1971. Plasticity in rock mechanics. *Int. J. Mech. Sci.*, 13 (4): 291–297.

Livneh, M., 1965. The theoretical bearing capacity of soils on a rock foundation. *Proc. 6th Int. Conf. Soil Mech. Found. Eng., Montreal.* Univ. of Toronto Press, 2: 122–126.

Livneh, M. and Greenstein, J., 1973. The bearing capacity of footings on nonhomogeneous clays. *Proc. 8th Int. Conf. Soil Mech. Found. Eng., Moscow*, 1 (3): 151–153.

Livneh, M. and Shklarsky, E., 1962a. The bearing capacity of asphaltic concrete carpets surfacing. *Proc. Int. Conf. Structural Design of Asphalt Pavements, Ann Arbor*, pp. 345–353.

Livneh, M. and Shklarsky, E., 1962b. The splitting test for determination of bituminous concrete strength. *Proc. AAPT*, 31: 457–476.

Livneh, M. and Shklarsky, E., 1965. Equations of failure stresses in materials with anisotropic strength parameters. *Highway Res. Rec.*, 74: 44–45.

Lo, K.Y., 1965. Stability of slopes in anisotropic soils. *J. Soil Mech. Found. Div. ASCE*, 91 (SM4), *Proc. Pap.* 4405: 85–106.

Lo, K.Y. and Lee, C.F., 1973. Stress analysis and slope stability in strain-softening materials. *Geotechnique*, 23 (1): 1–11.

Lomize, G.M. and Kryzhanovskii, A.L., 1966. Fundamental relations of the stress state and the strength of sandy ground. *Soil Mech. Found. Eng.*, 3: 165–169.

Lomize, G.M. and Kryzhanovskii, A.L., 1967. On the strength of sand. *Proc. Geotech. Conf.*, *Oslo*, 1: 215–219.

Lundgren, H. and Mortensen, K., 1953. Determination by the theory of plasticity of the bearing capacity of continuous footings on sand. *Proc. 3rd. Int. Conf. Soil Mech. Found. Eng., Zurich*, 1: 409–412.

Lysmer, J., 1970. Limit analysis of plane problems in soil mechanics. *J. Soil Mech. Found. Div., ASCE*, 96 (SM4), *Proc. Pap.* 7416: 1311–1334.

Malysev, M., 1971. Slip lines and particle displacement trajectories in cohesionless media. *Osnov. Fund. Mech. Grunt.*, 13 (6): 1.

Malyshev, M.V., 1969. Application of the Hubert-Mises-Botkin strength criterion to unconsolidated sands. *Soil Mech. Found. Eng.*, 5: 302–307.

Malyshev, M.V., Zaretsky, Y.K., Shirokov, V.N. and Cheremnikh, V.A., 1973. Interaction of rigid foundations with a base that deforms nonlinearly. *Proc. 8th Int. Conf. Soil Mech. Found. Eng., Moscow*, 1 (3): 155–159.

Mandel, J., 1966. Equation of flow in ideal soils in plain deformation and the concept of double slip. *J. Mech. Phys. Solids*, 14 (6): 303–308 (in French).

Mandel, J. and Luque, R.F., 1970. Fully developed plastic shear flow of granular materials. *Geotechnique*, 20 (3): 277–307.

Marais, G., 1969. Stresses in wedges of cohesionless materials formed by free discharge at the apex. *J. Eng. Ind., Trans., ASME, Ser. B*, 91 (2): 345–352.

Marcal, P.V., 1965. A stiffness method for elastic–plastic problems. *Int. J. Mech. Sci.*, 7: 229–238.

Marcal, P.V., 1969. Finite-element analysis of combined problems of non-linear material and geometric behavior. In: E. Sevin (Editor), *Computational Approaches in Applied Mechanics*. ASME, New York, pp. 133–149.

Marcal, P.V. and King, I.P., 1967. Elastic–plastic analysis of two-dimensional stress systems by the finite-element method. *Int. J. Mech. Sci.*, 9: 143–155.

Martin, H.C., 1965. On the derivation of stiffness matrices for the analysis of large deflection and stability problems. *Proc. Conf. Matrix Methods in Structural Mechanics, Wright-Patterson Air Force Base*, AFFDL-TR-66-80, pp. 697–716.

Massau, J., 1899. *Mémoire sur l'intégration des équations aux dérivées partielles*. Reprinted as Edition du Centenaire, Delporte, Mons, 1952.

McCutcheon, J.O., Mirza, M.S. and Mufti, A.A., 1972. *Proc. Spec. Conf. Finite-Element Methods in Civil Engineering*. McGill University, Montreal.

McHenry, D. and Karni, J., 1958. Strength of concrete under combined tensile and compressive stress. *J. ACI*, 54 (10): 829–840.

Mendelson, A., 1968. *Plasticity: Theory and Application*. MacMillan, New York, pp. 164–212.

Mendelson, A. and Manson, S.S., 1959. Practical solution of plastic deformation problems in elastic–plastic range. *NACA Tech. Note*, 4088.

Meyerhof, G.G., 1948. An investigation of the bearing capacity of shallow foundations on dry sand. *Proc. 2nd Int. Conf. Soil Mech. Found. Eng., Rotterdam*, 1: 237–243.

Meyerhof, G.G., 1951. The ultimate bearing capacity of foundations. *Geotechnique*, 2 (4): 301–332.

Meyerhof, G.G., 1953. The bearing capacity of concrete and rock. *Mag. Concrete Res.*, 4 (12): 107–116.

Meyerhof, G.G., 1955. Influence of roughness of base and ground-water conditions on the ultimate bearing capacity of foundations. *Geotechnique*, 5 (3): 227–242.

Meyerhof, G.G., 1961. The ultimate bearing capacity of wedge-shaped foundations. *Proc. 5th Int. Conf. Soil Mech. Found. Eng., Paris*. Dunod, Paris, 2: 105–109.

Meyerhof, G.G., 1963. Some recent research on the bearing capacity of foundations. *Can. Geotech. J.*, 1 (1): 16–26.

Meyerhof, G.G. and Chaplin, T.K., 1953. The compression and bearing capacity of cohesive layers. *Proc., Br. J. Appl. Phys.*, 4(1): 20–26.

Miastkowski, J., 1969. Plastic straining of v-notched elements in tension. *Mech. Teor. Stosow.*, 7: 81–98 (in Polish).

Miastkowski, J. and Szczepinski, W., 1969. Limit analysis of strips weakened by single row of holes. *Mech. Teoret. Stosow.*, 7: 335–352 (in Polish).

Mitchell, J.K., 1960. Fundamental aspects of thixotropy in soils. *J. Soil Mech. Found. Div.*, *ASCE*, 86 (SM3): 19–52.

Mitchell, R.J., 1970. On the yielding and mechanical strength of Leda clays. *Can. Geotech. J.*, 7 (3): 297–312.

Mizuno, T., 1948. On the bearing power of soil in a two-dimensional problem. *Proc. 2nd Int. Conf. Soil Mech. Found. Eng.*, *Rotterdam*, 3: 44–46.

Mizuno, T., 1953. On the bearing power of soil under a uniformly distributed circular load. *Proc. 3rd Int. Conf. Soil Mech. Found. Eng.*, *Zurich*, 1: 446–449.

Mohr, O., 1882. Über die Darstellung des Spannungszustandes und des Deformationszustandes eines Körperclements. *Zivilingenieur*, 28: 113–156.

Moseley, C.H., 1833. New principle in statics, called the principle of least pressure. *Philos. Mag.*, 3 (16): 285–288.

Mroz, Z., 1963. Non-associated flow laws in plasticity. *J. Mécanique*, 2: 21–42.

Mroz, Z., 1964. On non-linear flow laws in the theory of plasticity. *Bull., Acad. Polit. Sci., Ser. Sci. Tech.*, 12: 531–539.

Mroz, Z., 1967. Graphical solutions of axially symmetric problems of plastic flow. Z. *Angew. Math. Phys.*, 18: 219–236.

Mroz, Z., 1972. Mathematical models of inelastic concrete behavior. In: M.Z. Cohn (Editor), *Inelasticity and Non-Linearity in Structural Concrete*. Symposium, Univ. of Waterloo Press, pp. 47–72.

Mroz, Z. and Drescher, A., 1969. Limit plasticity approach to some cases of flow of bulk solids. *J. Eng. Ind.*, 51: 357–364.

Mroz, Z. and Kwaszcynska, K., 1968. Axially symmetric plastic flow of soils treated by the graphical method. *Archiwum Inzynierii Ladowej*, 14 (1): 27–37 (in English).

Muhs, E., 1965. On the phenomenon of progressive rupture in connection with the failure behavior of footings on sand. *Proc. 6th Int. Conf. Soil Mech. Found. Eng.*, *Montreal*, Univ. of Toronto Press, 3: 436–440.

Muskhelishvili, N.I., 1963. *Some Basic Problems of the Mathematical Theory of Elasticity*. Noordhoff, Groningen, pp. 475–477.

Nadai, A., 1950. *Theory of Flow and Fracture of Solids, 1*. McGraw-Hill, New York, p. 207.

Nadai, A., 1963. *Theory of Flow and Fracture of Solids, 2*. McGraw-Hill, New York, p. 470.

Narain, J., Saran, S. and Nandakumaran, P., 1969. Model study of passive pressure in sand. *J. Soil Mech. Found. Div. ASCE*, 95 (SM4): 969–983.

Negre, M.R. and Stutz, P., 1970. Contribution à l'étude des fondations de révolution dans l'hypothèse de la plasticité parfaite. *Int. J. Solids Struct.*, 6, (1): 69–90.

Nelissen, L.J.M., 1972. Biaxial testing of normal concrete. *Rep. Stevin Lab., Technol. Univ. Delft*, 18 (1).

Newman, K., 1968. Criterion for the behavior of plain concrete under complex states of stress. In: A.E. Brooks and K. Newman (Editors), *The Structure of Concrete. Proc. Int. Conf.*, *London, 1965*. Cement and Concrete Association, London, pp. 255–274.

Nikolaevskii, V.N., 1968. On the formulation of the defining equations for a plane flow of a continuous medium with dry friction. *J. Appl. Math. Mech.*, 32 (5): 959–962.

Nilsson, S., 1961. The tensile strength of concrete determined by splitting tests on cubes. *RILEM Bull., New Ser.* 11: 63–67.

Novotortsev, V.I., 1938. Application of the theory of plasticity to problems of determining the bearing capacity of building foundations. *Izv.*, VNIG 22.

Oden, J.T., 1969. Finite-element applications in nonlinear structural analysis. *Proc. Symp. Applications of Finite Element Methods in Civil Engineering*, Vanderbuilt University, Nashville, Tennessee, pp. 419–456.

Oden, J.T., 1972. *Finite Elements of Nonlinear Continua.* McGraw-Hill, New York, pp. 254–261.

Odenstad, S., 1963. Correspondence, *Géotechnique,* 13(2): 166–170.

Ohde, J., 1938. Zur Theorie des Erddruckes unter besonderer Berücksichtigung der Erddruckverteilung (The theory of earth pressure with special reference to earth pressure distribution). *Bautechnik,* 16: 150–761.

Ohta, H. and Hata, S., 1971. A theoretical study of the stress–strain relations for clays. *Soils Found.,* Vol. II (3): 65–90.

Ohta, H. and Hata, S., 1973. Immediate and consolidation deformations of clay. *Proc. 8th Int. Conf. Soil Mech. Found. Eng., Moscow,* 1 (3): 193–196.

Oladapo, I.O., 1964. Extensibility and modulus of rupture of concrete. Structural Research Laboratory, Technical University of Denmark, *Bulletin,* 18.

Owen, D.R.J., Nayak, G.L., Kfouri, A.P. and Griffiths, J.R., 1973. Stresses in a partially yielded notched bar – an assessment of three alternative programs. *Int. J. Numer. Methods Eng.,* 6: 63–73.

Oz, A.C., 1969. Limit analysis of forces exerted on soil shoving equipment. *Istanbul Teknik Universitesi bulteni,* 22 (1): 1–6.

Palmer, A.C., 1966. A limit theorem for materials with non-associated flow laws. *J. Mécanique,* 5 (2): 217–222.

Palmer, A.C., 1967. Stress–strain relation for clay: an energy theory. *Géotechnique,* 17 (4): 348–358.

Palmer, A.C. (Editor), 1973. *Proceedings of the Symposium on the Role of Plasticity in Soil Mechanics, Cambridge, U.K.* Cambridge Univ. Engineering Dept.

Palmer, A.C., Maier, G. and Drucker, D.C., 1967. Normality relations and convexity of yield surfaces for unstable materials or structural elements. *ASME Trans., Ser. E,* 89: 464–470.

Pariseau, W.G., 1966. A new view of the ideal plasticity of soils and unconsolidated rock materials. *Int. J. Rock Mech. Mining Sci.,* 3 (4): 307–317.

Pariseau, W.G., 1969. Gravity flows of ideally plastic materials through slots. *J. Eng. Ind. Trans. ASME, Ser. B,* 91 (2): 414–422.

Parry, R.H.G., 1960. Triaxial compression and extension tests in remoulded saturated clay. *Géotechnique,* 10: 166–180.

Parry, R.H.G. (Editor), 1971. Stress–strain behavior of soils. *Proc. Roscoe Mem. Symp., Cambr. Univ.* Foulis, Henley-on-Thames.

Parsons, D.H., 1956. Plastic flow with axial symmetry using the Von Mise flow criterion. *Proc. London Math. Soc.,* 6 (24): 610–625.

Paslay, P.R. and Weidler, J.B., 1969. Analysis of triaxial test for granular soils. *J. Eng. Mech. Div., ASCE,* 95 (EM3): 587–609.

Paslay, P.R., Cheatham, J.B. and Fulcher, C.W.G., 1968. Plastic flow of rock under a pointed punch in plane strain. *ASME Trans., Ser. E,* 35: 95–101.

Paul, B., 1962. Macroscopic criteria for plastic flow and brittle fracture. In: H. Liebowitz (Editor), *Fracture, an Advanced Treatise,* 2. Academic Press, London, pp. 313–496.

Pohle, W., 1951. Lastübertragung auf Stahlpfähle, *Bauingenieur,* 26 (9): 257–259.

Poorooshasb, H.B. and Roscoe, K.H., 1961. The correlation of the results of shear tests with varying degrees of dilation. *Proc. 5th Int. Conf. Soil Mech. Found. Eng., Paris,* Dunod, Paris, 1: 297–302.

Poorooshasb, H.B. and Roscoe, K.H., 1963. A graphical approach to the problem of the stress–strain relationship of normally consolidated clays. *ASTM Spec. Tech. Publ.* 361, Laboratory Shear Testing of Soils, p. 258–259.

Poorooshasb, H.B., Holubec, I. and Sherbourne, A.N., 1966. On quasi-static yielding of a cohesionless granular medium. *Proc. 5th U.S. Natl. Congr. Appl. Mech.,* p. 582.

Poorooshasb, H.B., Holubec, I. and Sherbourne, A.N., 1967. Yielding and flow of sand in triaxial compression. Part I, *Can. Geotech. J.,* 3 (4): 176–190; Parts II and III, *Can. Geotech. J.,* 4 (4): 376–397.

Pope, G.G., 1965. The application of the matrix displacement method in plane elastoplastic problems. *Proc. Conf. Matrix Methods in Structural Mechanics, Wright-Patterson Air Force Base*, AFFDL-TR-66-80, pp. 635–654.

Popovics, S., 1970. Review of stress–strain relationships for concrete. *J. Am. Concrete Inst.*, 67 (3): 243–248.

Poulos, H.G. and Davis, E.H., 1974. *Elastic Solutions for Soil and Rock Mechanics.* Wiley, New York.

Powell, M.J.D., 1964. An efficient method for finding the minimum of a function on several variables without calculating derivatives. *Comput. J.*, 7: 155–164.

Prager, W., 1948. Discontinuous solutions in the theory of plasticity. *Courant Anniversary Volume*, Interscience, New York, pp. 289–299.

Prager, W., 1950. On the boundary value problems of the mathematical theory of plasticity. *Proc. Int. Congr. Mathematicians, Cambridge, Mass.*, 2: 297–303.

Prager, W., 1952a. The general theory of limit design. *Proc. 8th Int. Congr. Appl. Mech., Istanbul.* Faculty of Science, Univ. Istanbul, II: 65–72.

Prager, W., 1952b. On the kinematics of soils. *Colloques Junius Massau, Comite National de Mécanique Théorique et Appliquée, Brussels*, pp. 3–8.

Prager, W., 1953. A geometrical discussion of the slipline field in plane plastic flow. *K. Tek. Högskol. Hand.*, 65.

Prager, W., 1955a. The theory of plasticity: a survey of recent achievements. James Clayton Lecture. *Proc. Inst. Mech. Eng.*, 169: 41.

Prager, W., 1955b. Discontinuous fields of plastic stress and flow. *Proc. 2nd U.S. Natl. Congr. Appl. Mech., ASME*, pp. 21–32.

Prager, W. and Hodge, P.G., 1968. Plane strain: specific problems. *Theory of Perfectly Plastic Solids.* Dover, New York, pp. 169–173.

Pramborg, B.O., 1961. Plastic equilibrium in soil, *Proc. 5th Int. Conf. Soil Mech. Found. Eng., Paris*, Dunod, Paris, 2: 459–463.

Prandtl, L., 1920. Über die Härte plastischer Körper. *Nachr. K. Ges. Wiss. Gött., Math.-Phys. Kl.*, 1920: 74–85.

Prandtl, L., 1921. Über die Eindringungsfestigkeit (Härte) plastischer Baustoffe und die Festigkeit von Schneiden. *Z. Angew. Math. Mech.*, 1 (1): 15–20.

Purushothamaraj, P., Ramiah, B.K. and Venkatakrishna Rao, K.N., 1974. Bearing capacity of strip footings in two layered cohesive-friction soils. *Can. Geotech. J.*, 11 (32): 32–45.

Radenkovic, D., 1961. Théorie des charges limitées, extension à la mécanique des sols. Seminaires de Plasticité. *Ecole Polytechnique, Publ. Sci. Tech.*, 116.

Radhakrishnan, N. and Reese, L.C., 1970. A review of applications of the finite-element method of analysis to problems in soil and rock mechanics. *Soils Found.*, X (3): 95–112.

Rankine, W.J.M., 1857. On the stability of loose earth. *Philos. Trans. R. Soc. London*, 147: 9.

Rechea, M. and Krizek, R.J., 1971. On certain plasticity solutions in soil mechanics. *J. Franklin Inst.*, 292 (3): 153–167.

Reddy, A.S. and Srinivasan, R.J., 1967. Bearing capacity of footings on layered clays. *J. Soil Mech. Found. Div., ASCE*, 93 (SM2): 83–99.

Reissner, H., 1924. Zum Erddruckproblem. In: C.B. Biezend and J.M. Burgers (Editors), *Proc. 1st Int. Congr. Appl. Mech., Delft*, pp. 295–311.

Rendulic, L., 1936. Relation between void ratio and effective principal stresses for a remolded silty clay. *Proc. 1st Int. Conf. Soil Mech. Found. Eng., Cambridge*, Mass., 3: 48–51. Reprinted by Harvard Press, 1965.

Rhines, W.J., 1969. Elastic–plastic foundation model for punch-shear failure. *J. Soil Mech. Found. Div., ASCE*, 95 (SM3): 819–828.

Richart, F.E., Brandtzaeg, A. and Brown, R.L., 1928. A study of the failure of concrete under combined compressive stresses. *Eng. Exp. Stn., Univ. Illinois, Bull.*, 185: 104.

Robinson, G.S., 1967. Behavior of concrete in biaxial compression. *J. Struct. Div., ASCE*, 93 (ST1), *Proc. Pap.* 5090: 71–86.

Rochette, P.A., 1961. Earth pressures on structures and mobilized shear resistance. *Proc. 15th Can. Soil Mech. Conf., Tech. Mem.*, 73: 3–59.

Romualdi, J.P. and Mandel, J.A., 1964. Tensile strength of concrete affected by uniformly distributed and closely spaced short lengths of wire reinforcement. *J. Am. Concrete Inst., Proc.*, 61 (6): 657–670.

Roscoe, K.H., 1966. Discussion of "Lower bound collapse theorem and lack of normality of strain rate to yield surface for soils" by G. de Josselin de Jong. In: J. Kravtchenko and P.M. Sirirys (Editors), *Rheology and Soil Mechanics. Proc. IUTAM Symp., Grenoble.* Springer, Berlin, p. 75–78.

Roscoe, K.H., 1967. Shear strength of soft clay. *Proc. Geotech. Conf., Oslo*, 2: 120–122.

Roscoe, K.H., 1970. The influence of strains in soil mechanics. Tenth Rankine lecture. *Geotechnique*, 20 (2): 129–170.

Roscoe, K.H. and Burland, J.B., 1968. On the generalized stress–strain behavior of wet clay. In: J. Heyman and F.A. Leckie (Editors), *Engineering Plasticity.* Cambridge University Press, London, pp. 535–609.

Roscoe, K.H. and Poorooshasb, H.B., 1963. A theoretical and experimental study of strains in triaxial compression tests on normally consolidated clays. *Geotechnique*, 13(1): 12–38.

Roscoe, K.H. and Schofield, A.N., 1963. Mechanical behavior of an idealized "wet-clay". *Proc. Europ. Conf. Soil Mech., Wiesbaden*, 1: 47–54.

Roscoe, K.H. and Thurairajah, A., 1966. On the uniqueness of yield surfaces for wet clays. In: J. Kravtchenko and P.M. Sirirys (Editors), *Rheology and Soil Mechanics. Proc. IUTAM Symp., Grenoble.* Springer, Berlin, p. 364–384.

Roscoe, K.H., Schofield, A.N. and Wroth, C.P., 1958. On the yielding of soils. *Géotechnique*, 8(1): 22–53.

Roscoe, K.H., Schofield, A.N. and Thurairajah, A., 1963a. Yielding of soils in states wetter than critical. *Géotechnique*, 13(3): 211–240.

Roscoe, K.H., Schofield, A.N. and Thurairajah, A., 1963b. An evaluation of test data for selecting a yield criterion for soils. *Spec. Tech. Publ. ASTM*, 361: 111–128.

Roscoe, K.H., Bassett, R.H. and Cole, E.R.L., 1967. Principal axes observed during shear of a sand. *Proc. Geotech. Conf., Oslo*, 1: 231–237.

Rowe, P.W., 1969. Progressive failure and strength of a sand mass. *Proc. 7th Int. Conf. Soil Mech., Mexico.* Sociedad Mexicana de Mecánico de Suelos, 1: 341–349.

Rowe, P.W. and Peaker, K., 1965. Passive earth pressure measurements. *Géotechnique* 15 (1): 57.

Sacchi, G. and Save, M.A., 1968. A note on the limit loads of non-standard materials. *Meccanica*, 3: 43–45.

Save, M.A. and Massonnet, C.E., 1972. *Plastic Analysis and Design of Plates, Shells and Disks.* North Holland, Amsterdam.

Sawczuk, A., 1967. On yield criteria and incipient plastic motion of soils. *Acta Mech.*, 4 (3): 308–314.

Sawczuk, A. and Stutz, P., 1968. On formulation of stress–strain relations for soils at failure. *Z. Angew. Math. Phys.*, 19 (5): 770–778.

Schlechtweg, H., 1958. Two-dimensional problem of a plastic material obeying Coulomb law of yield. *Zamm*, 38 (3/4): 139–148 (in German).

Schofield, A.N., 1961. The development of lateral force of sand against the vertical face of a rotating model foundation. *Proc., 5th Int. Conf. Soil Mech., Paris*, Dunod, Paris, 2: 479–484.

Schofield, A.N. and Wroth, C.P., 1968. *Critical State Soil Mechanics.* McGraw-Hill, New York.

Schultze, E., 1948. Composition and resolution of slip lines. *Abh. Bodenmech. Grundbau*, 1948: 34–45 (in German).

Schultze, E., 1952. Resistance of soil foundations to oblique base pressure. *Bautechnik*, 29 (12): 336–342 (in German).

Schultze, E., 1961. Distribution of stress beneath a rigid foundation. *Proc. 5th Int. Conf. Soil Mech. Found. Eng., Paris.* Dunod, Paris, 1: 807.

Scordelis, A.C., 1972. Finite-element analysis of reinforced concrete systems. *Proc., Spec. Conf.: Finite-Element Methods in Civil Engineering.* McGill University, Montreal.

Scott, R.F., 1963. *Principles of Soil Mechanics.* Addison-Wesley, Reading, Mass., Chapter 9, p. 422.

Scott, R.F. and Ko, H.Y., 1969. Stress-deformation and strength characteristics — State-of-the-art report. *7th Int. Conf. Soil Mech. Found. Eng., State of the Art Volume,* p. 1–47.

Selig, E.T. and McKee, K.E., 1961. Static and dynamic behavior of small footings. *J. Soil Mech. Found. Div., ASCE,* 87 (SM6) *Proc. Pap.* 3020: 29–47.

Shibata, T. and Karube, D., 1965. Influence of the variation of the intermediate principal stress on the mechanical properties of normally consolidated clays. *Proc. 6th Int. Conf. Soil Mech. Found. Eng., Monteal.* Univ. of Toronto Press, 1: 359–363.

Shield, R.T., 1953. Mixed boundary value problems in soil mechanics. *Q. Appl. Math.,* 11 (1): 61–75.

Shield, R.T., 1954a. Plastic potential theory and the Prandtl bearing capacity solution. *J. Appl. Mech.,* 21 (2): 193–194.

Shield, R.T., 1954b. Stress and velocity fields in soil mechanics. *J. Math. Phys.,* 33 (2): 144–156.

Shield, R.T., 1955a. On Coulomb's law of failure in soils. *J. Mech. Phys. Solids,* 4(1): 10–16.

Shield, R.T., 1955b. On the plastic flow of metals under conditions of axial symmetry. *Proc. R. Soc. London, Ser. A,* 233: 267.

Shield, R.T., 1955c. The plastic indentation of a layer by a flat punch. *Q. Appl. Math.,* XIII (1): 27–46.

Shield, R.T., 1960. The plastic indention of a semi-infinite solid by a perfectly rough punch. *Z. Angew. Math. Phys.,* XI: 33–42.

Shield, R.T. and Drucker, D.C., 1953. The application of limit analysis to punch-indentation problems. *J. Appl. Mech., ASME,* 75: 453–460.

Shield, R.T. and Ziegler, H., 1958. On Prager's hardening rule. *Z. Angew. Math. Phys.,* IXa: 260–276.

Sih, G.C. (Editor), 1973. Methods of analysis and solutions of crack problems. *Application of the Finite-Element Method to the Calculation of Stress Intensity Factors,* Noordhoff, Groningen, pp. 426–483.

Skempton, A.W., 1942. An investigation of the bearing capacity of a soft clay soil. *J. Inst. Civ. Eng.,* 18: 307–321.

Skempton, A.W., 1951. The bearing capacity of clays. *Proc. Building Res. Congr., London,* Institute of Civil Engineers, pp. 180–189.

Skempton, A.W., 1960. Effective stress in soils, concrete and rocks. *Proc. Conf. Pore Pressure and Suction in Soils, London.* Butterworths, London, pp. 4–16.

Smith, I.M., 1970. Incremental numerical solution of a simple deformation problem in soil mechanics. *Géotechnique,* 20 (4): 357–372.

Smith, I.M. and Kay, S., 1971. Stress analysis of contractive or dilative soils. *J. Soil Mech. Found. Div., ASCE,* 97, (SM7): 981–997.

Snitbhan, N., 1975. *Plasticity Solutions of Slopes in Anisotropic, Inhomogeneous soil.* Thesis Lehigh University, Bethlehem, Pa. University Microfilms, Ann Arbor, Michigan.

Snitbhan, N., Chen, W.F. and Fang, H.Y., 1975. Slope stability analysis of layered soils, *Proc. 4th Southeast Asian Conf. Soil Eng., Kuala Lumpur, Malaysia.*

Sobotka, Z., 1958. Non-homogeneity in elasticity and plasticity. In: W. Olszak (Editor), *Proc. IUTAM Symp., Warsaw.* Pergamon Press, London, p. 227.

Sobotka, Z., 1960. Some axially symmetrical and three-dimensional problems of the limit states of non-homogeneous continuous matters. In: F. Rolla and W.T. Koiter (Editors), *Proc. 10th Int. Congr. Appl. Mech., Stresa.* Elsevier, Amsterdam, pp. 277–280.

Sobotka, Z., 1961a. Axially symmetrical and three-dimensional limiting states of non-homogeneous soils and other continuous media. *Archiwum Mechaniki Stosowanej, Warsaw,* 13 (2): 151–175.

Sobotka, Z., 1961b. On a new approach to the analysis of limit states in soils and in other continuous media. *Bull. Acad. Polon. Sci.*, 9 (2): 85–93.

Sobotka, Z., 1961c. The slip lines and surfaces in the theory of plasticity and soil mechanics. *Appl. Mech. Rev.*, 14 (10): 753–759.

Sobotka, Z., 1964. Problèmes de l'équilibre limite des sols no homogènes. In: *Rheology and Soil Mechanics. Proc. IUTAM Symp., Grenoble, 1964.* Springer, Berlin, p. 160–163.

Sokolovskii, V.V., 1948. Limiting equilibrium of rocks in conditions of plane stress. *Bull. Acad. Sci. U.S.S.R., Ser. Tech. Sci.*, 9: 1361–1370 (in Russian).

Sokolovskii, V.V., 1949. Equations of plastic equilibrium for plane stress. *Prikl. Mat. Mekh.*, 13: 219–221 (in Russian).

Sokolovskii, V.V., 1951. On the limit equilibrium of granular media. *Prikl. Mat. Mekh.*, 15 (6): 689–708 (in Russian).

Sokolovskii, V.V., 1952a. On an approximate method in statics of granular media. *Prikl. Mat. Mekh.*, 16 (2): 246–248 (in Russian).

Sokolovskii, V.V., 1952b. Theory of limit equilibrium of soils and its application to the analysis of hydrotechnical structures. *Izv. Akad. Nauk S.S.S.R. Otd. Tekh. Nauk*, 6: 809–823 (in Russian).

Sokolovskii, V.V., 1955. On the stability of foundation beds of laminated cohesionless material. *Inzh., Sb., Akad. Nauk S.S.S.R.*, 22: 74–82 (in Russian).

Sokolovskii, V.V., 1957. Some problems of soil pressure. *Proc. 4th Int. Conf. Soil Mech. Found. Eng., London*, Butterworths, London, 2: 239–242.

Sokolovskii, V.V., 1961. Limit equilibrium of loose media for small angles of internal friction. *Inzh. Sb. akad. Nauk S.S.S.R.*, 31: 119–122 (in Russian).

Sokolovskii, V.V., 1965. *Statics of Granular Media.* Pergamon Press, New York.

Spencer, A.J.M., 1962. Perturbation methods in plasticity. III, Plane strain of ideal soils and plastic solids with body forces. *J. Mech. Phys. Solids*, 10: 165–177.

Spencer, A.J.M., 1964. A theory of the kinematics of ideal soils under plane-strain conditions. *J. Mech. Phys. Solids*, 12: 337–351.

Spencer, E., 1964. The movement of soil beneath model foundations. *Civ. Eng. Public Works Rev.*, June, pp. 728–731; July, pp. 878–881.

Spencer, E., 1969. Circular and logarithmic spiral slip surfaces. *J. Soil Mech. Found. Div., ASCE*, 95 (SM1): 227–234.

Steirnamm, I.Y., 1956. The pressure distribution under a foundation in the presence of a plastic region. *Sb. Tr. Mosk. Inzh.-Stroit In-ta*, 14: 32–56 (in Russian).

Strickland, J.A., Haisler, W.E. and Von Riesemann, W.A., 1971. Geometrically nonlinear structural analysis by direct stiffness method. *J. Struct. Div., ASCE*, 97 (ST9): 2229–2314.

Stroganov, A.S., 1965. Analysis of plane plastic deformation of soils. *Sov. Eng. J. S*, 4: 573–577.

Stuart, J.G., 1962. Interference between foundations with special reference to surface footings in sand. *Géotechnique*, 12 (1): 15–22.

Subtahmanyam, G., 1967. The effect of roughness of footings on bearing capacity. *J. Indian Natl. Soc. Soil Mech. Found. Eng.*, 6 (1): 33–45.

Sundara, R.I.K.T. and Yogananda, C.V., 1966. A three-dimensional stress distribution problem in the anchorage zone of a post-tensioned concrete beam. *Mag. Concrete Res.*, 18 (55): 75–84.

Sundara, R.I.K.T., Chandrashekhara, K. and Krishnaswamy, K.T., 1965. Strength of concrete under biaxial compression. *J. Am. Concrete Inst., Proc.*, 62 (2): 239–249.

Swanson, S.R. and Brown, W.S., 1972. The influence of state of stress on the stress–strain behaviour of rocks. *J. Basic Eng. ASME, Ser. D*, 94 (1): 238–241.

Swedlow, J.L., Williams, M.L. and Yang, W.H., 1965. Elastoplastic stresses and strains in cracked plates. *Proc. 1st Int. Conf. Fract., Sendai.* Jap. Soc. Strength and Fracture of Materials, 1: 259–282.

Sylwestrowicz, W., 1953. Experimental investigation of the behavior of soil under a punch or footing. *J. Mech. Phys. Solids*, 1: 258–264.

Symonds, P.S., 1962. Limit analysis. In: W. Flügge (Editor), *Handbook of Engineering Mechanics.* McGraw-Hill, New York, Chapter 49.

Szczepinski, W. and Miastkowski, J., 1968. Plastic straining of notched bars with intermediate thickness and small shoulder ratio. *Int. J. Non-Linear Mech.,* 3: 83–97.

Szczepinski, W., Dietrich, L., Drescher, E. and Miastkowski, J., 1966. Plastic flow of axially symmetric notched bars pulled in tension. *Int. J. Solid Struct.,* 2: 543–554.

Tabor, D., 1951. *The Hardness of Metals.* Clarendon Press, Oxford.

Takagi, S., 1962. Plane plastic deformation of soils. *J. Eng. Mech. Div., ASCE,* 88 (EM3): 107–151.

Tan, E.K., 1948. Stability of soil slopes. *Trans. ASCE,* 113: 139.

Tang, W.H. and Höeg, K., 1968. Two-dimensional analysis of stress and strain in soils, report 5: plane-strain loading of a strain-hardening soil. *U.S. Army Eng. Waterw. Exp. Stn., Contr. Rep.* 3–129.

Taylor, D.W., 1948. *Fundamentals of Soil Mechanics.* Wiley, New York.

Taylor, M.A. 1971. General behavior theory for cement pastes, mortars, and concrete. *J. Am. Concrete Inst., Proc.,* 68: (10): 756–762.

Terzaghi, K., 1943. *Theoretical Soil Mechanics.* Wiley, New York.

Terzaghi, K. and Peck, R.B., 1967. Plastic equilibrium in soils. In: *Soil Mechanics in Engineering Pratice.* Wiley, New York, 2nd edition, pp. 219–223.

Thompson, M.R., 1965. The split-tensile strength of lime-stabilized soils. *Highw. Res. Rec.,* 92: 11–23.

Thomsen, E.G., 1957. A new method for the construction of Hencky-Prandtl nets. *J. Appl. Mech.,* 24; *Trans. ASME, Ser. E.,* 79: 81–84.

Thurairajah, A. and Roscoe, K.H., 1965. The correlation of triaxial compression test data on cohesionless granular material. *Proc. 6th Int. Conf. Soil Mech. Found. Eng., Montreal.* Univ. of Toronto Press, 1: 377–381.

Timoshenko, S., 1934. *Theory of Elasticity.* McGraw-Hill, New York, pp. 104–108.

Torre, C., 1948. State of stresses in a heavy soil mass. *Proc. 2nd Int. Conf. Soil Mech. Found. Eng., Rotterdam,* 3: 57–61.

Tresca, H., 1868. Mémoire sur l'écoulement des corps solides. *Mémoires présentés par divers savants à l'Académie des Sciences,* 18: 733–799.

Tschebotarioff, G.P., 1951. *Soil Mechanics, Foundations and Earth Structures.* McGraw-Hill, New York, pp. 222–226.

Tschebotarioff, G.P. and Johnson, E.G., 1953. *The Effects of Restraining Boundaries on the Passive Resistance of Sand.* Princeton University, New Jersey.

Tschebotarioff, G.P., Ward, E.R. and DePhilippe, A.A., 1953. The tensile strength of disturbed and recompacted soils. *Proc. 3rd Int. Conf. Soil Mech. Found. Eng.,* 3: 207–210.

Urbano, G., 1973. On the bearing capacity of inclined foundations under inclined loads. *Riv. Ital. Geotecn.,* 7 (1): 7–21.

Valliappan, S., 1969. Discussion. *J. Soil Mech. Found. Div., ASCE,* 95 (SM2): 676–678.

Valliappan, S. and Doolan, T.F., 1972. Nonlinear stress analysis of reinforced concrete. *J. Struct. Div., ASCE, Proc. Pap.* 8845: 885–898.

Van Mierlo, W.C., 1965. Earth and rock pressures. *Proc. 6th Int. Conf. Soil Mech. Found. Eng., Montreal.* Univ. of Toronto Press, 3: 513–539.

Verghese Chummar, A., 1972. Bearing capacity theory from experimental results. *J. Soil Mech. Found. Div., ASCE,* 98(SM12): 1311–1324.

Vesic, A.S., 1963. Bearing capacity of deep foundations in sand. *Highw. Res. Rec.,* 39: 112–153.

Vesic, A.S., 1973. Analysis of ultimate loads of shallow foundations. *Proc., ASCE,* 99 (SM1): 45–73.

Vesic, A.S., Banks, D.C. and Woodard, J.M., 1965. An experimental study of dynamic bearing capacity of footings on sand. *Proc. 6th Int. Conf. Soil Mech. Found. Eng., Montreal.* Univ. Toronto Press, II: 209–213.

Vile, G.W.D., 1968. The strength of concrete under short-term static biaxial stress. In: A.E. Brooks and K. Newman (Editors), *The Structure of Concrete and its Behavior under Load. Proc. Int. Conf., London, 1965.* Cement and Concrete association, London, pp. 275–288.

Von Karman, Th., 1911. Festigkeitsversuche unter allseitigem Druck. *Z. Ver. Dtsch. Ing.*, 42.

Von Mises, R., 1928. Mechanik der plastischen Formänderung von Kristallen. *Z. Angew. Math. Mech.*, 8: 161–185.

Vyalov, S.S., 1951. Limit equilibrium of weak soils on a rigid base. *Izv. Akad. Nauk S.S.S.R., Otd. Tech. Nauk, Mekh. Mashinostr.*, 5: 813–828 (in Russian).

Wästlund, G., 1937. New evidence regarding the basic strength properties of concrete. *Betong*, 22 (3): 189.

Wei, R., 1964. The plastic potential of normally consolidated clay. *J. Hydraul. Eng.*, 6: 9–20.

Weidler, J.B. and Paslay, P.R., 1969. Analytical description of behavior of granular media. *J. Eng. Mech. Div., ASCE*, 95 (EM2): 379–395.

Weidler, J.B. and Paslay, P.R., 1970. Constitutive relations for inelastic granular medium. *J. Eng. Mech. Div., ASCE*, 96 (EM4): 395–406.

Weigler, H. and Becker, G., 1963. Investigation into strength and deformation properties of concrete subjected to biaxial stresses. *Dtsch. Ausschuss Stahlbeton, Bull.*, 157: 66.

West, J.M. and Stuart, J.G., 1965. Oblique loading resulting from interference between surface footings on sand. *Proc. 6th Int. Conf. Soil Mech. Found. Eng., Montreal.* Univ. Toronto Press, II: 214–217.

Westergaard, H.M., 1940. Plastic state of stress around a deep well. *Contrib. Soil Mech.*, 1935–1940, reported from Boston Society of Civil Engineers, Vol. 27–28: 1.

Westwood, D. and Wallace, J.F., 1960. Upper-bound values for the loads on a rigid-plastic body in plane strain. *J. Mech. Eng. Sci.*, 2: 178.

Whitman, R.V. and Healy, K.A., 1962. Shear strength of sands during rapid loadings. *J. Soil Mech. Found. Div., ASCE*, 88, (SM2) *Proc. Pap.* 3102: 99–132.

Winter, G., 1960. Properties of steel and concrete and the behavior of structures. *J. Struct. Div., ASCE*, 86 (ST2), *Proc. Pap.* 2384: 33–62.

Winzer, A. and Carrier, G.F., 1948. The interaction of discontinuity surfaces in plastic fields of stress. *J. ASME Trans.*, 48: 261–264.

Winzer, A. and Carrier, G.F., 1949. Discontinuities of stress in plane plastic flow. *ASME Trans.*, 71: 346–348.

Wright, P.J.F., 1955. Comments on an indirect tensile test on concrete cylinders. *Mag. Concrete Res.*, 7 (20): 87–96.

Wroth, C.P., 1958. Soil behavior during shear. *Engineering*, 186 (4829): 409–413.

Wroth, C.P., 1965. The prediction of shear strains in triaxial tests on normally consolidated clays. *Proc. 6th Int. Conf. Soil Mech. Found. Eng., Montreal.* Univ. of Toronto Press, 1: 417–420.

Wroth, C.P. and Bassett, R.H., 1965. A stress–strain relationship for the shearing behavior of a sand. *Géotechnique*, 15 (1): 32–56.

Wroth, C.P. and Loudon, P.A., 1967. The correlation of strains within a family of triaxial tests on overconsolidated samples of kaolin. *Proc. Geotech. Conf., Oslo*, 1: 159–163.

Wu, T.H., Loh, A.K. and Malvern, L.E., 1963. Study of failure envelopes of soils. *J. Soil Mech. Found. Div., ASCE*, 89 (SM1): 145–181.

Yamada, Y., Yoshimura, N. and Sakurai, T., 1968. Plastic stress–strain matrix and its application for the solution of elastic–plastic problem by the finite-element method. *Int. J. Mech. Sci.*, 10: 343–354.

Yamaguchi, H., 1959a. A theory on the velocity field in the plastic flow of granular materials. *Trans. Japan Soc. Civ. Eng.*, July, p. 8–16 (in Japanese).

Yamaguchi, H., 1959b. Application of Kötter equation to theoretical soil mechanics. *Trans. Japan Soc. Civ. Eng.*, 65 (1): 1–9 (in Japanese).

Yamaguchi, H., 1967. Discontinuity of stress and velocity in the rigid plastic field in soil mechanics. *Soils Found.*, 7 (3): 54–64.

Yokawa, Y., Yamagata, K. and Nagaoka, H., 1968. Bearing capacity of a continuous footing set in two-layered ground. *Soils Found.*, 8 (3): 1–31.

Yong, R.N. and Mckyes, E., 1971. Yield and failure of clay under triaxial stresses. *J. Soil Mech. Found. Div., ASCE*, 97 (SM1): 159–176.

Ziegler, H., 1969. On the plastic potential in soil mechanics. *Z. Angew. Math. Phys.*, 20 (5): 659–675 (in German).

Zienkiewicz, O.C. and Cheung, Y.K., 1967. *The Finite-Element Method in Structural and Continuum Mechanics.* McGraw-Hill, New York.

Zienkiewicz, O.C., 1970. The finite-element method: from intuition to generality. *Appl. Mech. Rev.*, 23 (3): 249–256.

Zienkiewicz, O.C., 1971. *The Finite-Element Method in Engineering Science.* McGraw-Hill, New York.

Zienkiewicz, O.C. and Nayak, G.C., 1971. A general approach to problems of large deformation and plasticity using isoparametric elements. *Proc. 3rd Conf. Matrix Methods in Structural Mechanics, Wright-Patterson Air Force Base.*

Zienkiewicz, O.C. and Nayak, G.C., 1972. Elastoplastic stress analysis. A generalization for various constitutive relations including strain softening. *Int. J. Numer. Methods Eng.*, 5: 113–135.

Zienkiewicz, O.C. and Naylor, D.J., 1971a. The adaption of critical state soil mechanics theory for use in finite elements. In: R.H.G. Parry (Editor), *Stress–strain Behavior of Soils.* Foulis, Henley-on-Thames, pp. 537–547.

Zienkiewicz, O.C. and Naylor, D.J., 1971b. Finite-element studies of soils and porous media. In: J.T. Oden and E.R.A. Oliveira (Editors), *Finite Element Methods in Continuum Mechanics.* Lectures, delivered at Advanced Study Institute held at Lisbon, pp. 459–493.

Zienkiewicz, O.C., Valliappan, S. and King, I.P., 1969. Elastoplastic solutions of engineering problems: initial-stress finite-element approach. *Int. J. Numer. Methods Eng.*, 1: 75–100.

Zisman, W.A., 1933. Compressibility and anisotropy of rocks at and near the earth's surface. *Proc. Natl. Acad. Sci.*, 19: 666–679.

AUTHOR INDEX